Machine Learning and Big Data-enabled Biotechnology

Further Volumes of the "Advanced Biotechnology" Series: Published:

Villadsen, J. (ed.)

Fundamental Bioengineering

2016
Print ISBN: 978-3-527-33674-6

Love, J. Ch. (ed.)

Micro- and Nanosystems for Biotechnology

2016
Print ISBN: 978-3-527-33281-6

Wittmann, Ch., Liao, J.C. (eds.)

Industrial Biotechnology

Microorganisms (2 Volumes)

2017
Print ISBN: 978-3-527-34179-5

Wittmann, Ch., Liao, J.C. (eds.)

Industrial Biotechnology

Products and Processes

2017
Print ISBN: 978-3-527-34181-8

Nielsen, J., Hohmann, S. (eds.)

Systems Biology

2017
Print ISBN: 978-3-527-33558-9

Smolke, C.

Synthetic Biology

Parts, Devices, and Applications

2018
Print ISBN: 978-3-527-33075-1

Chang, H. N.

Emerging Areas in Bioengineering

2018
Print ISBN: 978-3-527-34088-0

Lee, M. L., Kildgaard, H. F.

Cell Culture Engineering

Recombinant Protein Production

2019
Print ISBN: 978-3-527-81140-3

Rehm. B. H. A., Moradali, M. F.

Biopolymers for Biomedical and Biotechnological Applications

2021
Print ISBN: 978-3-527-34530-4

Hudson, P.

Cyanobacteria Biotechnology

2021
Print ISBN: 978-3-527-34714-8

Lee, S. Y., Nielsen, J., Stephanopoulos, G. (Eds.)

Metabolic Engineering

2021
Print ISBN: 978-3-527-34662-2

Chang, M. W. (ed.)

Principles in Microbiome Engineering

2022
Print ISBN: 978-3-527-34725

Machine Learning and Big Data-enabled Biotechnology

Edited by Hal S. Alper

WILEY-VCH

Volume Editor

Prof. Hal S. Alper
The University of Texas at Austin
McKetta Department of Chemical Engineering
200 E. Dean Keeton St.
Austin
TX 78712-1589
USA

Series Editors

Prof. Dr. Sang Y. Lee
KAIST
373-1, Guseong-Dong
291 Daehak-ro, Yuseong-gu
34141 Daejeon
South Korea

Prof. Dr. Jens Nielsen
Chalmers University
Department of Chemical and Biological Engineering
Kemivägen 10
412 96 Göteborg
Sweden

Prof. Dr. Gregory Stephanopoulos
Massachusetts Institute of Technology
Department of Chemical Engineering
Massachusetts Ave 77
Cambridge, MA 02139
United States

Cover Design: Wiley
Cover Image: © issaro prakalung/Shutterstock

Typesetting: Straive, Chennai, India

Library of Congress Card No.: applied for

British Library Cataloguing-in-Publication Data A catalogue record for this book is available from the British Library.

Bibliographic information published by the Deutsche Nationalbibliothek
The Deutsche Nationalbibliothek lists this publication in the Deutsche Nationalbibliografie; detailed bibliographic data are available on the Internet at <http://dnb.d-nb.de>.

Print ISBN: 978-3-527-35474-0
ePDF ISBN: 978-3-527-85052-5
ePub ISBN: 978-3-527-85051-8
oBook ISBN: 978-3-527-85053-2

Printed and bound by
CPI Group (UK) Ltd, Croydon, CR0 4YY

C9783527354740_130126

Contents

Preface

Biotechnology has always revolved around the collection of large datasets. Advances in the 'omics era have certainly increased the quantity and quality of this data. Now, even just a single experiment can generate an overwhelming amount of data. At the same time, emerging approaches in machine learning are helping to contextualize these datasets and provide a path forward toward making biological engineering a predictive, data-driven science. This book highlights major advances in the application of machine learning to data-driven biotechnology. This book is roughly organized according to the following topics: (1) understanding function directly from the DNA-sequence level, (2) understanding protein function, structure, and arrangement into metabolic pathways, (3) understanding whole-cell performance and 'omics data, and (4) understanding biofoundries and bioreactors. These chapters represent the current state-of-the-art for combining machine learning and big data.

1

From Genome to Actionable Insights in Biotechnology

James Morrissey, Benjamin Strain, and Cleo Kontoravdi

Imperial College London, Department of Chemical Engineering, London SW7 2AZ, UK

1.1 Introduction

Systems biology facilitates the understanding of biological processes [1, 2]. It provides a framework for contextualizing and understanding high-throughput data, allowing structured and meaningful insights to be gained. With the increasing generation of high-throughput omics data and usage of machine learning (ML) data analysis tools, it becomes ever more important to develop robust systems biology tools.

One of the key tools in a systems biologist's repertoire is a network. Networks in biology represent the flow and interactions between components such as genes, proteins, reactions, and metabolites, where network nodes correspond to these biological entities and edges define their interactions, such as biochemical conversion, regulation, and physical binding [3]. Networks improve interpretation and understanding of complex biological phenomena, but, most importantly, in the context of big data, they provide structure. This structure can help turn highly complex data from a "black box" to multiscale models with predictive and interpretable capabilities. Networks allow underlying biological mechanisms to be understood during big data approaches. This structural framework is particularly valuable in biotechnology, where datasets like genomics, transcriptomics, proteomics, and metabolomics must be carefully integrated and analyzed. Even if they are "smaller" in scale, both in network size and connectivity, than those in fields like image recognition or natural language processing, their biological complexity requires equal, if not greater, interpretative attention [4, 5].

By leveraging the structure provided by biological networks, researchers can generate predictions that help elucidate how cellular systems function under various conditions. These predictions, such as flux distributions, regulatory responses, or growth outcomes, offer mechanistic insight into the underlying biology [6]. With this understanding, targeted interventions can be designed, such as modifying gene expression, adjusting nutrient feeds, or engineering metabolic

Machine Learning and Big Data-enabled Biotechnology, First Edition. Edited by Hal S. Alper.

pathways. Crucially, these interventions often lead to positive outcomes, including improved productivity, robustness, or efficiency in biotechnological applications. Thus, networks serve not only as interpretative tools but also as platforms for driving practical, data-informed decisions.

The most widely utilized type of biological network is the metabolic network [7]. These networks describe the flow (or flux) of metabolites through biochemical reactions within a cellular system. Depending on the application, metabolic networks can range from small, pathway-specific subsystems to genome-scale reconstructions that aim to represent the entirety of an organism's metabolic capabilities [8]. These large-scale reconstructions are referred to as genome-scale metabolic models (GEMs). GEMs provide a computational framework for probing metabolism by predicting intracellular fluxes under defined conditions. By integrating experimental data into a metabolic network, GEMs enable simulation of cellular behavior, guiding hypothesis generation, strain engineering, and process optimization in biotechnology [9, 10]. GEMs also serve as a central scaffold for incorporating other biological networks, enabling the integration of multiomics data and supporting comprehensive, data-driven approaches [11, 12].

In this chapter, we explore how biological networks, particularly metabolic networks, can be constructed from genomic data, validated and refined, transformed into predictive models, and ultimately used to generate actionable insights in biotechnology. We highlight how high-throughput omics and ML tools can be used to enhance interpretability, constrain solution spaces, and improve predictive power to enable biotechnology applications.

1.2 From Genome to Network

Genomic data provides the foundational blueprint for all cellular processes. However, to extract actionable insight, this static information must be translated into dynamic representations of biological function. Networks offer a powerful way to achieve this, linking genes to their roles in metabolism, regulation, signaling, and molecular interactions [7].

From a genome sequence, various biological networks can be reconstructed, such as metabolic networks describing biochemical reactions, gene regulatory networks (GRNs) that capture transcriptional control, signaling networks that map cellular communication, and protein–protein interaction (PPI) networks that reveal the physical interactions within the proteome [3]. Each of these networks offers a different layer of understanding, and together they form a comprehensive systems-level view of the cell.

This section outlines how these networks are constructed from genomic data, with a focus on metabolic networks, either using bottom-up approaches that start from gene annotations and known biological functions or using top-down approaches that integrate omics data to refine or contextualize existing network structures. These reconstructions are the foundation for building models that can interpret data, generate predictions, and guide interventions in biotechnology.

Chapter 2 provides detailed information on how network models can be constructed from genomic information using computational algorithms and manual approaches.

1.2.1 Metabolic Networks

Among the various biological networks that can be reconstructed from genomic data, metabolic networks are the most widely applied. Metabolic networks represent the biochemical reactions linking metabolites, genes, enzymes, and reactions in the cell system [13]. This can either be a subset of biochemical reactions (e.g. just focusing on core functions) or can be genome-scale. In recent years, GEMs have been created for a multitude of organisms [14] due to the availability of high-throughput omics data and increased applications of these models in systems biology.

1.2.1.1 Bottom-Up Approaches for Network Reconstruction

The bottom-up approach is a lengthy and manual approach to GEM creation, but it is recommended to create high-quality GEMs from scratch. The reader is pointed to a protocol [8], which contains an in-depth protocol for GEM reconstruction. In this section, we summarize the key steps as illustrated in Figure 1.1. A GEM reconstruction begins with a genome annotation for the organism of interest. To create a draft metabolic network, metabolic reactions can be extracted from the genome annotation using gene ontology (GO) [16], enzyme commission (EC) numbers [17], and biochemical reaction databases such as the Kyoto Encyclopedia of Genes and Genomes (KEGG) [18] and BRaunschweig ENzyme DAtabase (BRENDA) [19]. Chapters 4 and 5 discuss how protein function can be extracted using these methods as well as other ML frameworks.

This draft annotation is then subject to manual curation to scrutinize every gene and reaction entry. The entries must be relevant to the organism of interest, for example, ensuring correct cofactor and substrate specificity for each enzyme, as well as the correct gene localization. It is preferable to have literature or experimental data to support the presence and function of genes and reactions. When data is lacking, phylogenetically close organisms can be used. During manual curation, a confidence score is useful for assessing the amount of information available for each entry. Gene-protein-reaction (GPR) associations indicate which genes are required for reaction to occur, which states the presence of isozymes and enzyme complexes. The GPRs must be manually refined using databases and literature searches.

The next step is to ensure correct reaction stoichiometry. Metabolites in databases (and hence the draft reconstruction) are represented with uncharged formulas, but their protonation state varies depending on the pH of the cellular environment. The charged formula is determined based on the pKa values of functional groups, which can be predicted using computational tools or literature for the correct pH in the subcellular location. Once the charged formulas are assigned, the correct reaction stoichiometry is established by ensuring mass and charge balance across reactions, incorporating protons and water where necessary. Correct balancing is crucial to avoid artificial energy generation. Correct reaction

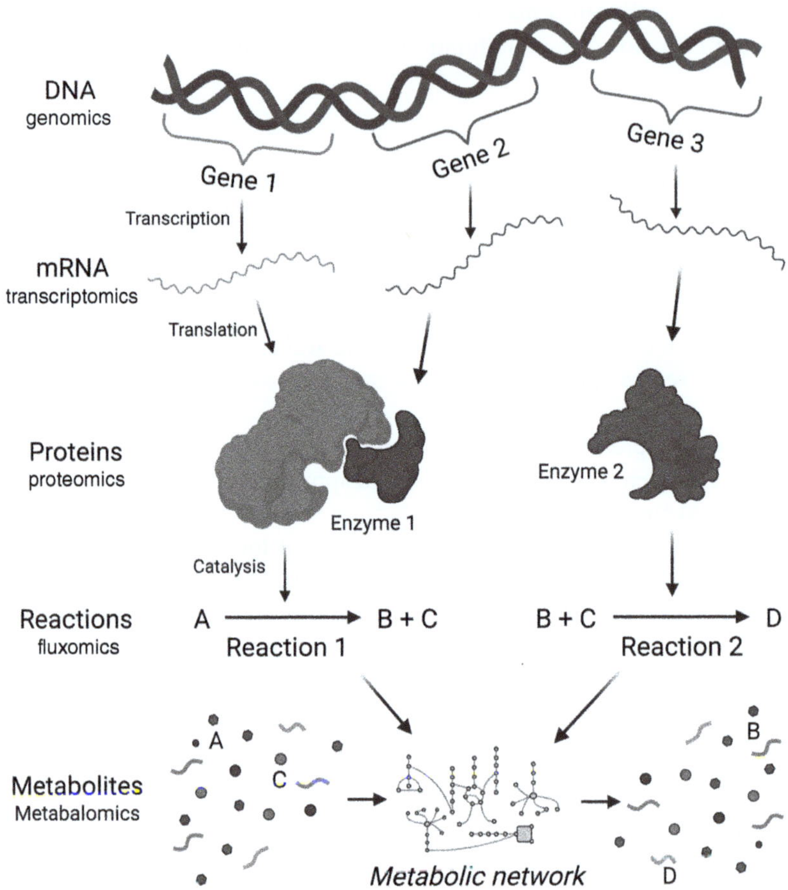

Figure 1.1 Key omics data and steps in the bottom-up construction of a metabolic network from an organism's genome. *Source*: Strain et al. [15] / Elsevier / CC BY 4.0.

directionality is also essential for preventing irreversible reactions from running backward, which would lead to incorrect predictions and thermodynamically infeasible loops (futile cycles), which are discussed in Section 1.3.2. Correct directionality is determined using existing biochemical data, but when this is unavailable, Gibbs free energy change can be obtained from databases or calculated using group contribution methods [20]. To complete and validate the network, further reactions must be added, which are discussed in Section 1.3.1.

1.2.1.2 Top-Down Approaches for Network Reconstruction

In contrast to building draft networks from scratch, top-down approaches to GEM reconstruction rely on pre-existing networks, knowledge, and omics data to infer metabolic networks. The top-down approach to GEM reconstruction is typically faster and more automated than the bottom-up method and is well-suited for generating draft models, particularly when high-throughput data is available. These approaches begin with existing genome annotations and rely on automated pipelines that map annotated genes to known reactions in curated metabolic databases such as KEGG, MetaCyc [21], or ModelSEED [22]. Automated draft reconstructions can be rapidly generated using tools such as CarveMe [23] or Pathway Tools [24].

Top-down approaches can also be used to build context-specific models by integrating omics data into existing generic GEMs [25, 26]. Algorithms like GIMME [27], iMAT [28], and CORDA [29] filter and adjust reaction content based on expression thresholds or activity likelihood, producing models tailored to particular tissues, conditions, or phenotypes.

1.2.2 Networks Beyond Metabolism

GEM reconstructions focus on metabolic functions, but systems biology encompasses a broader range of network reconstructions that capture other aspects of cellular function. These include gene regulation, signaling, protein interaction, and transcription–translation networks, each offering complementary insights into cellular behavior [3].

These network types can be integrated into multilayer models, as discussed in Section 1.5.4, to improve predictive accuracy and capture cellular complexity. While GEMs are well-established, the ongoing development of GRNs, signaling, and transcription-translation models is crucial for building more comprehensive whole-cell models.

GRNs describe the interactions between transcription factors (TFs), regulatory elements, and their target genes [30, 31]. They are often inferred from gene expression data, using statistical methods (e.g. mutual information, Bayesian inference) or curated from literature. The matrix formalism has been used to describe the functional states of such systems, providing a structured way to analyze regulatory dynamics. A key example is the genome-scale reconstruction of the *Escherichia coli* transcriptional regulatory system, which enables simulation of transcriptional responses to environmental changes [32].

Signaling networks map how extracellular signals lead to cellular responses through cascades of molecular interactions. These can be constructed from curated pathway databases or derived from proteomics data but are less commonplace than other networks [33]. PPI networks represent the physical interactions between proteins within a cell [34]. Unlike metabolic or regulatory networks, these are typically undirected and do not reflect stoichiometry or directionality but instead capture the structural and functional connectivity of the proteome. PPIs are essential for nearly all cellular processes, including enzyme complexes, signaling cascades, structural assemblies, and transport mechanisms.

To model the impact of gene expression beyond regulation, genome-scale reconstructions of transcriptional and translational machinery have been developed. For instance, multiscale models of metabolism and expression (ME models) integrate metabolism with macromolecular expression systems, accounting for RNA polymerase, ribosomes, and enzyme synthesis [35, 36]. These ME models provide a platform to simulate how cellular resources are allocated between growth and gene expression and are discussed further in Section 1.5.4.

1.3 From Draft to Functional Network

In order to turn the draft network into a functional model, several refinements have to occur. These include adding additional reactions to the network, which are not represented by particular gene(s), such as sink reactions, spontaneous reactions, and the biomass reaction, but are required to create a metabolic model. These reactions are key; they provide a termination point of reaction pathways to prevent infinite accumulation of reaction flux, as well as provide routes for experimental data to be integrated into GEMs.

In addition, some basic network validation functionality checks must be implemented to verify the integrity of the network. These are elaborated in the following section.

1.3.1 Additional Reactions

1.3.1.1 Exchange Reactions

Exchange reactions represent the transfer of metabolites between the cell and its extracellular environment. The exchange reactions include sugars, amino acids, water, ammonia, sulfur, and many other potential compounds known to be taken up or secreted by the cell. The exchange reaction of metabolite A is modeled in the form:

$$EX_A : A_e \leftrightarrow$$

The format of the exchange reactions is such that a negative flux represents uptake from the extracellular environment and a positive flux represents secretion. Metabolites that are secretion-only are modeled as irreversible reactions to prevent a negative flux.

1.3.1.2 Demand Reactions

Demand reactions are additional sinks added to model terminal metabolites without secretion. This can represent metabolites required by the cell for normal function, desired products, e.g. antibodies or generic demand reactions to replace specific biosynthetic pathways, which may not be understood. The demand reaction of metabolite A in the cytosol is modeled in the form:

$$DM_A : A_c \rightarrow$$

1.3.1.3 Transport Reactions

GEMs also contain location information for metabolites, allowing the inclusion of transport reactions between compartments. For example, A_c refers to metabolite A in the cytosol (c) and A_e refers to metabolite A in the extracellular space (e). Other locations modeled include [m] for mitochondria, [x] for peroxisome, [r] for endoplasmic reticulum, and [g] for the Golgi apparatus. Thus, transport reactions are modeled in the form:

$$A_e \leftrightarrow A_c$$

1.3.1.4 Spontaneous Reactions

Spontaneous reactions are those that occur without enzymatic catalysis and therefore are not associated with specific genes; hence, these are not added to the network during the initial reconstruction phase. They ensure network completeness by allowing certain transformations to proceed even when no known enzyme is responsible. Spontaneous reactions are typically irreversible and should be added when they have at least one metabolite connecting them to the rest of the reconstruction, to avoid too many dead-end metabolites caused by spontaneous reactions. Examples include the spontaneous degradation of dihydroxyacetone phosphate (DHAP) and glyceraldehyde-3-phosphate (G3P) to form methylglyoxal [37] as well as metal-catalyzed reactions such as Fenton chemistry (H_2O_2 and Fe^{2+} reacting to generate hydroxyl radicals) [38].

1.3.1.5 Nongrowth Associated ATP Maintenance

Nongrowth associated ATP maintenance (NGAM) reactions account for the energy required by the cell to sustain basic functions unrelated to biomass production, such as ion transport, turnover of macromolecules, and maintenance of membrane potential. A fixed lower bound is set on this reaction's flux to reflect the minimum ATP demand necessary for cell viability, based on experimental estimates, for example, 7.6 and 5.9 mmol g^{-1} hr^{-1} for *E. coli* and Chinese hamster ovary (CHO) cells, respectively [39, 40]. This reaction is typically modeled as follows:

$$H_2O + ATP \rightarrow ADP + H^+ + P$$

Note that growth-associated maintenance (GAM), which accounts for the energy necessary to replicate a cell, e.g. for macromolecular synthesis (e.g. proteins, DNA, and RNA), is included in the biomass reaction outlined in the next section.

1.3.1.6 Biomass Reaction

A fundamental component of GEMs is the addition of an *ad hoc* biomass reaction [41]. The overall cell composition (amino acids, lipids, carbohydrates) is measured, and its constituent components are used as an input to this biomass reaction in the form. The flux through this reaction is used as an approximation of cell growth rate; as such, the molar composition of the biomass inputs sums up to one gram of dry cell weight (gDCW). The resulting units of this reaction are then h, allowing comparison with experimentally measured cell growth rates.

$$\sum_{i \in I^{\text{biomass}}} S_{i,\text{biomass}} X_i + ATP + H_2O \rightarrow \text{biomass} + ADP + H^+ + P$$

where X_i is a biomass precursor (e.g. amino acid, lipid, carbohydrate) and $S_{i,\text{biomass}}$ is the stoichiometric coefficient for precursor *i*. The biomass reaction also requires an amount of ATP and produces ADP, hydrogen ions, and phosphate.

1.3.2 Network Validation

Validation of the model structure is an essential step to ensure that the GEM accurately represents the biological system it is intended to capture. Structural validation aims to detect errors such as missing reactions, incorrect associations, or stoichiometric imbalances, which could lead to inaccurate predictions.

1.3.2.1 Manual Screening

Manual screening involves meticulously curating each reaction in the model to verify its accuracy, ensuring that reaction stoichiometry and the GPR associations are correct and reflect the known biology of the organism. This step is especially important because computational algorithms used to construct GEMs may introduce errors or make assumptions that lead to inaccurate or incomplete GPR associations. For example, the manually curated genome-scale reconstruction of the metabolic network of *Bacillus megaterium*, a microorganism widely used in industrial biotechnology, is a prime case where manual curation applied after automated reconstruction ensured accurate reflection of the organism's metabolism and high predictive accuracy [42]. Correcting errors at this stage prevents propagation of inaccuracies in subsequent analyses, ultimately improving the predictive power and utility of the model.

1.3.2.2 Screening for Dead-End Reactions and Blocked Metabolites

A critical step in GEM validation is the identification and resolution of dead-end reactions and blocked metabolites. Dead-end reactions are those that either lack substrates or fail to produce products that can be utilized by subsequent reactions. Blocked metabolites are compounds that cannot be consumed or produced by any reaction in the network. These issues typically arise due to incomplete or inaccurate model construction and can significantly distort model predictions by preventing key metabolic pathways from functioning [43].

The set of blocked reactions in a metabolic network is identified by determining whether each reaction can carry any flux under given conditions. This is done by maximizing the flux for each reaction while maintaining the network's stoichiometry. If the maximum flux for a reaction is zero, it is considered unusable or blocked. Although they may appear functional in a static view of the network, these reactions are inactive in practice due to incomplete or inconsistent pathway integration [44].

Flux variability analysis (FVA) is a powerful technique for uncovering such issues. An extension of flux balance analysis (FBA), FVA examines the range of flux values that reactions can carry while optimizing for a specific cellular objective [45]. FVA is particularly useful for identifying reactions that, despite being part of the network, are blocked or inactive in certain conditions due to gaps in pathway connectivity or stoichiometric errors.

Another commonly used technique is FastGapFill [46], which identifies reactions that must be added to restore the functionality of blocked pathways in a way that minimizes alterations to the existing network. FastGapFill operates by proposing the inclusion of missing reactions based on data from external reaction databases (e.g. KEGG) or from closely related organisms. The algorithm works under the assumption that the number of changes should be minimal to preserve the integrity of the model, only adding reactions that are most likely to occur naturally in the organism.

1.3.2.3 Infinite Loops

Infinite loops refer to metabolic cycles that can carry flux indefinitely without the net consumption or production of metabolites (Figure 1.2). These loops often arise from structural errors, such as incorrect reaction stoichiometries, missing thermodynamic constraints, or improperly connected reactions. These cycles can lead to implausible outcomes. For example, the model might predict a perpetual flux through ATP-generating pathways without any corresponding substrate consumption, violating basic thermodynamic laws [47].

Infinite loops typically occur in linear programming-based methods such as FBA, where reactions are optimized for specific objectives (e.g. biomass production or ATP generation). In the absence of appropriate constraints, the model may inadvertently favor these loops.

Detecting and eliminating infinite loops is essential for ensuring the model behaves realistically. Several techniques are used to identify and address infinite loops in genome-scale models. Much like dead-end reactions, infinite loops can also be detected by analyzing flux variability across different environmental conditions. Using FVA, reactions that carry flux under unlikely or unrealistic scenarios, pointing to the presence of infinite loops, can be identified. Reactions with unusually wide ranges of flux under varying conditions are often indicative of these erroneous cycles.

Adding thermodynamic constraints is one of the most effective ways to prevent infinite loops. Thermodynamically constrained FBA (tFBA) [48] imposes directionality constraints on irreversible reactions and ensures that reactions with unfavorable Gibbs free energy do not carry flux. Loopless-COBRA (COnstraint-Based Reconstruction and Analysis) is an extension of traditional FBA that specifically aims to eliminate flux through futile cycles or loops [49]. It introduces additional constraints that prevent reactions from forming internal cycles that can carry flux without contributing to the objective function (e.g. biomass production). Finally, manual identification of highly active reactions and pathways is another approach to detect potential infinite loops that fail to consume or produce net metabolites. Additionally, it can help identify dead-end reactions, which are often linked to infinite loops.

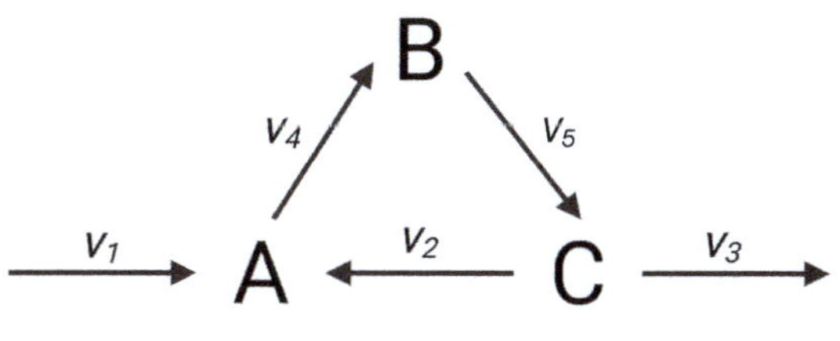

Figure 1.2 A depiction of an infinite cycle. If uptake of metabolite *A*, v_1, is constrained to an experimental value, the demand flux of metabolite *C*, v_3, is limited to this uptake i.e. $v_1 = v_3$. *However,* fluxes v_2, v_4 and v_5 can take any value and still produce a feasible solution. This becomes an issue when these predicted fluxes are important as part of metabolic analysis, for example if the user wants to find a value for conversion of *A* to *B*, v_4.

1.3.2.4 Leaks and Siphons

Leaks and siphons can lead to biologically inaccurate predictions, particularly in the context of energy and metabolite balances [50]. Leaks occur when a model predicts the loss of metabolites or energy carriers, such as ATP or nicotinamide adenine dinucleotide hydride (NADH), without accounting for their regeneration or the presence of corresponding inputs. This leads to the unrealistic depletion of key molecules, disrupting the energy balance and resulting in misleading simulations, such as predicting cell death or metabolic failure when, biologically, the organism could still thrive.

Siphons are pathways or sets of reactions in the model where certain metabolites accumulate or are depleted without affecting the overall balance of the system. These siphons represent "traps" where flux is directed through reactions, but the corresponding consumption and production of metabolites are not balanced, leading to unbounded fluxes or unrealistic metabolite accumulation. Siphons are often problematic in energy metabolism, where they can allow energy carriers (e.g. ATP, NADH) to be produced or consumed without appropriate regulation, distorting the model's predictions of cellular growth or metabolic output.

One approach to identifying leaks is performing mass and charge balance checks across all reactions. Metabolites in a properly balanced model should not disappear or be generated without corresponding inputs and outputs. This method helps detect reactions that may lack stoichiometrically correct inputs (substrates) or outputs (products), which can lead to leaks where energy or metabolites are lost or accumulated.

A simple but effective approach to testing for leaks involves the addition of synthetic exchange reactions to the model. These synthetic reactions are designed to "leak" specific metabolites or energy carriers from the system, mimicking the effect of a metabolic leak. By adding and constraining these reactions, researchers can test if energy or key metabolites are leaving the system inappropriately. If the model shows flux through these synthetic reactions, it indicates the presence of a leak, requiring further investigation into the reactions involved.

The fast leak test function within the COBRA toolbox [51] systematically checks if a metabolic model can produce metabolites from nothing when all exchange reactions are closed, meaning no uptake is allowed. It identifies any leaking metabolites by maximizing the flux through each exchange reaction. If secretion flux is detected, it signals a leak. Optionally, the function can also test for leaks through demand reactions. It returns a list of leaking metabolites, the closed model, and the flux vector for exchange reactions, helping pinpoint and resolve problematic reactions.

1.4 From Functional Network to Model

Once a functional network has been constructed and validated, it can be transformed into a computational model capable of simulating cellular behavior. This requires formulating the network as a mathematical problem, typically using *constraint-based* methods such as FBA. These approaches enable the prediction of metabolic flux distributions under defined conditions and objectives, providing a powerful tool for integrating high-throughput datasets to understand cell behavior.

1.4.1 Flux Balance Analysis

FBA is the most widely used method to analyze metabolic networks [52, 53]. FBA calculates the flow of metabolites through the system in a linear optimization program, thereby making it possible to make metabolic flux predictions, including growth rate, recombinant protein production, accumulation of toxic metabolites, media uptake/secretion, effects of genetic engineering, influence of enzymes on phenotype, and much more. The principle of FBA is a material balance on all metabolites within the GEM. The (molar) material balance on metabolite A would be:

$$\frac{\mathrm{d}[A]}{\mathrm{d}t} = \sum_{j \in J} S_{A,j} v_j$$

where $[A]$ is the concentration of metabolite A in units of mmol per gram dry cell weight (gDCW), t is the time (hr), $S_{A,j}$ is the stoichiometric coefficient of A in reaction j, from the stoichiometric matrix, and v_j is the flux through reaction j in units of mmol $\mathrm{gDCW}^{-1}\ \mathrm{hr}^{-1}$.

In kinetic models of metabolism, reaction flux v_j is replaced with kinetic expressions, leading to a system of nonlinear ordinary differential equations (ODEs), with metabolite concentrations as the free variables. Instead, FBA makes a pseudo-steady-state assumption, meaning the concentration of metabolites is constant, hence a steady state with no accumulation. This assumption is appropriate in many conditions, as metabolic flux far exceeds accumulation [54], permitting cell culture to be divided into periods of pseudo-steady state. This leads to the elimination of the left-hand side and v_j becomes the free variable, leading to the general equation:

$$\frac{\mathrm{d}[C_i]}{\mathrm{d}t} = \sum_{j \in J} S_{ij} v_j = 0 \text{ for } i \in I$$

where C_i is the intracellular concentration of metabolite i and S_{ij} is the stoichiometric coefficient of metabolite i in reaction j. With this formulation, the number of equations is equal to the number of metabolites, i, and the number of variables is equal to the number of reactions, j. Since most metabolites participate in many reactions, the number of variables far exceeds the number of equations. This means there are multiple degrees of freedom and the system is underdetermined. In general, an underdetermined system of linear equations has an infinite number of solutions. However, in optimization problems that are subject to linear equality constraints, only one of these solutions is relevant, namely the one giving the highest or lowest value of an objective function.

In its most general form, FBA is a constrained linear optimization problem, with an objective function to maximize (or minimize) flux through a particular reaction. The most frequently used reaction in the objective function is the biomass reaction. The assumption behind a biomass objective function is that the cell is allocating all resources toward growth; hence, the cell will prioritize placing fluxes in reactions that will lead to the highest biomass production.

The reaction fluxes are also subject to upper and lower bounds, v_j^{LB} and v_j^{UB}. The selection of these bounds is a key factor in making accurate predictions using FBA. This generalized form of FBA is shown below: e

$$\text{maximise } v_{\text{objective}}$$

$$\sum_{j \in J} S_{i,j} v_j = 0 \text{ for } i \in I$$

$$v_j^{\text{LB}} \leq v_j \leq v_j^{\text{UB}} \text{ for } j \in J$$

1.4.2 Flux Variability Analysis

Solutions to the (FVA) problem are not unique since multiple flux distributions can yield the unique optimal objective value. As a variation on FBA, FVA captures the feasible range of reaction fluxes that satisfy an optimal objective value [45, 55]. The method calculates the minimum and maximum allowable fluxes through each reaction using the linear program:

$$\text{maximise (and minimise)} \, v_j$$

$$\sum_{j \in J} S_{i,j} v_j = 0 \text{ for } i \in I$$

$$v_j^{\text{LB}} \leq v_j \leq v_j^{\text{UB}} \text{ for } j \in J$$

$$v_{\text{objective}} = v_{\text{objective}}^{\text{optimal}}$$

For a network with N reactions, $2N$ optimization problems would be solved. The feasible region of all fluxes is described by an N-dimensional polytope. FVA is a convenient way to inscribe the smallest box around this feasible region. FVA cannot find the N-dimensional polytope itself since changing the flux of a reaction for the same objective function typically requires that the remaining fluxes in the network change as well. Figure 1.3 shows the feasible region and FVA solution space.

FVA can also be used to find blocked reactions in the metabolic network for a given environmental condition, as discussed in Section 1.3.2. These are reactions for which the minimum and maximum values identified by FVA are both zero. Blocked reactions can be caused by reaction constraints and also due to gaps in the metabolic network, for example, metabolites for which there is no production or consumption path in the network (dead-end metabolites). Identifying these blocked reactions is useful for spotting dead-end metabolites and also reducing the FBA search space by pre-setting the value of these reactions to zero.

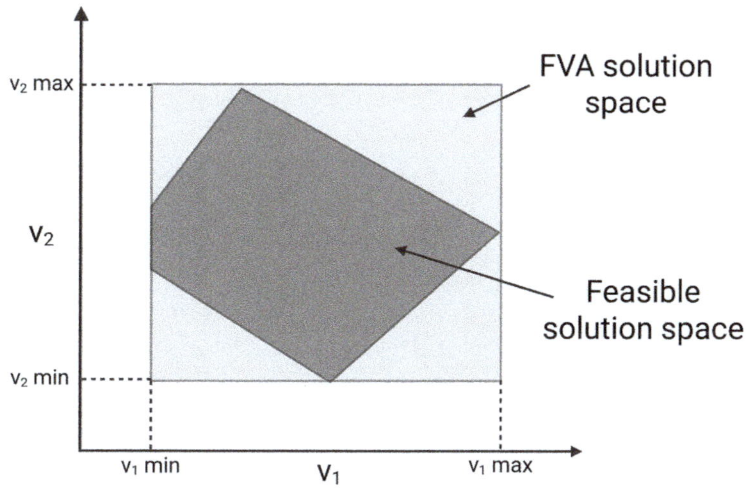

Figure 1.3 FVA solution space.

1.4.3 Flux Sampling

While FBA provides a flux distribution that maximizes (or minimizes) a defined objective, and FVA gives the range of feasible fluxes for each reaction, both are limited in that they do not characterize the full solution space. In reality, the space of possible flux distributions satisfying the steady state mass balance and other biological constraints is vast and underdetermined. Flux sampling methods aim to explore this space by generating many feasible solutions, providing a more comprehensive and probabilistic view of the model's behavior [56–58].

Flux sampling algorithms such as Hit-and-Run [59], Artificial Centering Hit-and-Run [60], and Coordinate Hit-and-Run with Rounding [61] operate by iteratively drawing flux vectors that satisfy the network constraints to fall within the defined flux bounds. This produces a representative sample of the solution space, rather than a single point estimate, enabling statistical analysis of fluxes across different pathways and conditions.

The key advantage of flux sampling is its ability to reveal the range and distribution of possible flux values without being tied to a specific optimization objective [56] as shown in Figure 1.4. This is particularly useful in situations where the biological objective is unclear or where the system does not necessarily prioritize growth, such as in stress conditions, nondividing cells, or engineered strains with altered metabolism. Sampling can highlight which pathways are consistently active, identify flux correlations or trade-offs between competing pathways, and uncover alternative metabolic strategies that achieve similar phenotypes.

Flux sampling also allows integration with ML approaches by generating high-dimensional datasets of flux states that can be clustered, classified, or

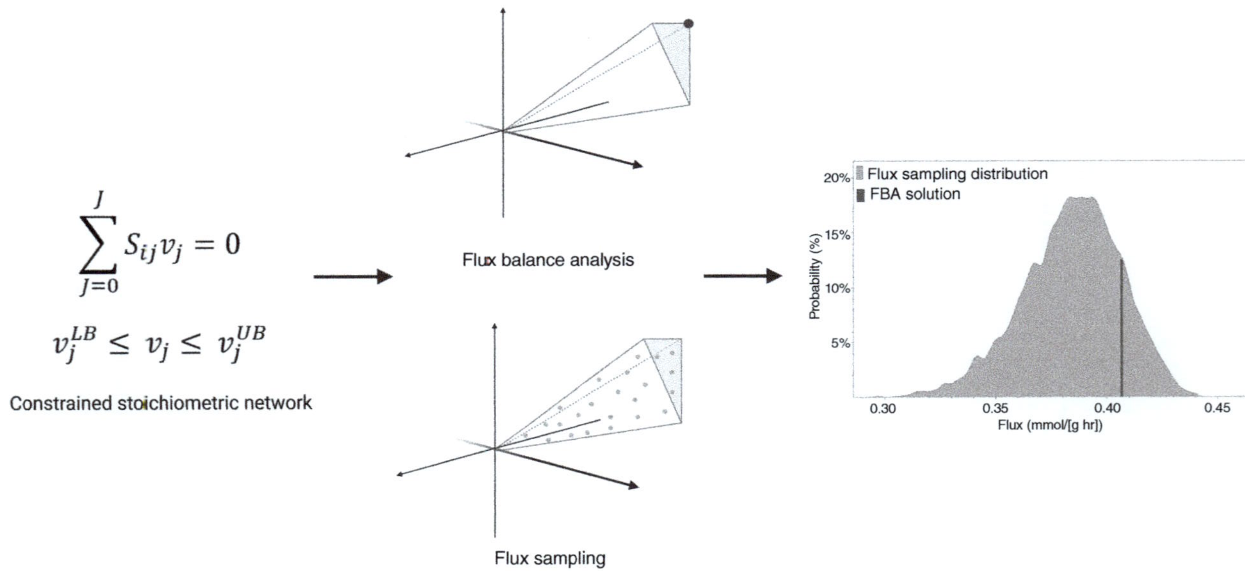

Figure 1.4 Flux sampling vs FBA approaches to finding a flux solution. FBA optimizes for a particular metabolic objective, yielding a single flux distribution, while flux sampling samples the feasible solution space without the assumption of an objective. As a result, flux sampling generates a probability distribution and provides deeper insights into metabolism. This example flux distribution is for GAPDH (glyceraldehyde 3-phosphate dehydrogenase) with $N = 5{,}000$.

regressed against phenotypic traits [62–64]. This can be particularly valuable in systems with complex regulation or when trying to predict emergent properties from large design spaces.

However, flux sampling is computationally intensive, especially for large-scale GEMs with thousands of reactions. Efficient sampling requires advanced linear algebra methods and careful preprocessing to ensure numerical stability and uniform coverage of the solution space. Furthermore, interpreting the resulting high-dimensional data can be challenging and often requires dimensionality reduction or statistical summarization. Unlike FBA, which offers a clear optimization criterion, flux sampling does not indicate which distribution is "best" or "most likely" in a biological sense; it just shows what could happen, given the model constraints.

Unlike FBA, which identifies a single flux distribution by optimizing a defined objective function (e.g. maximizing growth), flux sampling generates a large ensemble of feasible flux states without prioritizing any particular solution. As such, the results are inherently nondeterministic, reflecting the full range of possible metabolic behaviors permitted by the model constraints. Further analysis is required to extract biological insight from these distributions, such as identifying frequently active reactions, flux correlations, or dominant metabolic pathways that may represent physiologically relevant states [65].

1.5 From Model to *In Silico* Predictions

Once a functional and validated GEM is constructed, it can be used as a powerful *in silico* tool to simulate and analyze cellular behavior. This step marks the transition from a basic biochemical network model to a predictive platform capable of generating biological insights. *In silico* predictions can guide metabolic engineering, strain design, feed optimization, and hypothesis generation across various biotechnological applications discussed in Section 1.6. However, the quality and reliability of these predictions are highly dependent on the quality of constraints applied to the model, the selected optimization strategy, and the integration of relevant biological data. This section discusses tools and techniques to integrate omics data, systems knowledge, ML, and other biological networks to generate reliable predictions.

1.5.1 Constraints

A metabolic model represents all the possible metabolic capabilities of an organism under any condition. In order to make accurate, condition-specific predictions, biological constraints must be applied to reduce the feasible solution space. These constraints are applied to the same models defined in Section 1.4 (FBA, flux sampling) but provide more context and refinement to predictions by defining which reactions are likely to be active under the given experimental or physiological context. The more biologically realistic the constraints, the closer the predicted fluxes will be to true cellular behavior.

Various omics datasets can be used to impose these constraints, as summarized in Table 1.1. Metabolomics data, when converted to uptake and secretion rates, directly constrain exchange reactions and are commonly used in all FBA approaches. Transcriptomics allows inference of reaction activity via gene expression levels, with methods like MADE [67], MOOMIN [68], and SPOT [81] converting expression data into condition-specific models through GPR associations. Thermodynamic constraints, using Gibbs free energy change, assign directionality to reactions and prevent infeasible loops; approaches like TFA [48], CycleFreeFlux [78], and loopless-COBRA [49] incorporate these principles.

Enzyme kinetic and physical capacity constraints account for cellular limits on abundance and catalytic capacity of enzymes; methods such as FBAwMC [72], MOMENT [73], and GECKO [76] reflect proteome constraints, crowding, and energetic budgets, thereby moving beyond stoichiometry to better capture cellular priorities.

1.5.2 Objective Function

In optimization-based models (of which FBA is the most popular), the objective function is required to achieve a predict flux distribution, yet the choice of objective function is not always clear. Objective functions are a useful component of metabolic models, as they can guide flux toward realistic and appropriate solutions in an underdetermined solution space. However, a poor choice can lead to flux predictions moving away from *in vivo* reaction rates and reducing model validity.

The most commonly applied objective in metabolic models is biomass maximization, based on the assumption that cells naturally evolve to prioritize growth. However, in some conditions (especially under stress, nutrient limitation, or engineered phenotypes) and will more complex cell types, this assumption may not hold. Omics data can be integrated into metabolic models to help guide objective functions to generate more realistic solutions. Similar to the constraining methods discussed above, these can use transcriptomics, proteomics, to guide fluxes that best fit with the omics data provided.

Furthermore, objectives can be inferred directly from the data rather than assumed or guided by the data. Methods such as ObjFind [82], BOSS [83], invFBA [84], and SCOOTI [85] use experimental fluxes, omics data, or single-cell profiles to mathematically deduce plausible objective functions that are consistent with observed phenotypes. These approaches help overcome the limitations of assuming a fixed objective like biomass, particularly in nonproliferative cells or systems with poorly defined cellular priorities. Multiobjective formulations and pareto based methods also allow simultaneous consideration of competing biological goals, such as balancing growth with energy efficiency or survival, providing a more nuanced and flexible modeling framework.

Examples of these methods are summarized in Table 1.2.

1.5.3 Validating In Silico Predictions

The utility of a GEM lies in its ability to make accurate predictions of biological behavior under defined conditions. Predictive accuracy can be assessed through

Table 1.1 Examples of methods using omics data and biological knowledge to constrain metabolic models.

Method/name	Description	Original application	Reference
Standard FBA	Uptake reactions are bound by experimental measurements.	Growth prediction in *E. coli*.	Varma and Palsson [53]
E-Flux (Expression-Flux)	Directly binds reactions by gene expression.	Prediction of drug impacts on mycolic acid production in *Mycobacterium tuberculosis*.	Colijn et al. [66]
MADE (Metabolic Adjustment by Differential Expression)	MADE uses the statistical significance of change in expression between conditions to determine if the genes are switched on or off.	Model the metabolic adjustments seen in the transition from fermentative to glycerol-based respiration in *S. cerevisiae*.	Jensen and Papin [67]
MOOMIN (Mathematical explOration of Omics data on a MetabolIc Network)	MOOMIN uses a Bayesian posterior probability of differential gene expression to assess if genes are switched on or off.	Model diauxic shift from glucose to glycerol in *S. cerevisiae*.	Pusa et al. [68]
LBFBA (Linear Bound FBA)	Finds a linear relationship between gene expression and reaction flux. Uses this relationship to bind reactions in new conditions.	Predicting intracellular fluxes in *E. coli* and *S. cerevisiae*.	Tian and Reed [69]
mCADRE (Metabolic Context-specificity Assessed by Deterministic Reaction Evaluation)	Creates context-specific GEMs by ranking and pruning reactions from the network.	Creating context-specific GEMs for 126 human tissue and cell types.	Wang et al. [70]
ccFBA (carbon-constrained FBA)	Constrains each reaction by a mass balance on the total available carbon.	Genome-scale metabolic model of CHO cells.	Lularevic et al. [71]
TFA (Thermodynamic-based Flux Analysis)	Reaction directionality is determined by Gibbs free energy change.	Genome-scale metabolic model of *E. coli* cells.	Henry et al. [48]

(Continued)

Table 1.1 (Continued)

Method/name	Description	Original application	Reference
FBAwMC (FBA with Molecular Crowding)	Reactions are constrained by maximum volume of enzymes inside the cell or compartments.	Prediction of growth rate of wild-type and mutant *E. coli* cells.	Beg et al. [72]
MOMENT (MetabOlic Modeling with ENzyme kineTics)	Reactions are constrained by maximum concentration of enzymes.	Predict growth rates and intracellular reactions of *E. coli*.	Adadi et al. [73]
ecFBA (enzyme capacity FBA)	Reactions are constrained by protein abundance, using enzyme turnover numbers and molecular weights.	Prediction of central carbon metabolism fluxes in CHO cells and lactate metabolism.	Yeo et al. [74]
CAFBA (Constrained Allocation FBA)	Biosynthetic costs associated with growth are accounted for through a single additional genome-wide constraint.	Predictions on the rate of acetate excretion and growth in *E. coli*.	Mori et al. [75]
GECKO (GEM with Enzymatic Constraints using Kinetics and Omics data)	Including enzymes as part of GEM, thus limiting fluxes by protein abundance.	Describing *S. cerevisiae* phenotypes under high enzymatic pressure conditions, such as alternative carbon sources, stress, or overexpressing a specific pathway.	Sánchez et al. [76]
RBA (Resource Balance Analysis)	Accounts for the distribution of proteins across translation apparatus, enzymatic reactions, and production of microcontents as a limitation on cell growth.	Built for purely bacterial applications, no results were presented in the original paper.	Goelzer and Fromion [77]
Multiscale Models of Metabolism and Macromolecular Expression (ME)	Integration of macromolecular machinery into metabolic network.	Prediction of growth in *E. coli*.	Thiele et al. [35]

Loopless-COBRA	Applies mixed-integer linear programming to prevent cyclic fluxes without thermodynamic data.	Impact on the solution space of an *H. pylori* network.	Schellenberger et al. [49]
CycleFreeFlux	Post-process flux distributions to remove cyclic fluxes.	Impact on the solution space of an *E. coli* network in glucose-limited aerobic medium.	Desouki et al. [78]
Regulatory FBA	Incorporates gene regulatory logic (e.g. Boolean rules) to simulate regulatory control of metabolic fluxes.	Gene expression and metabolic adaptation in *E. coli* and *M. tuberculosis*.	Covert et al. [79]
NEXT-FBA (Neural-net EXtracellular Trained Flux Balance Analysis)	Correlates exometabolomic data with ^{13}C-labeled intracellular fluxomic data to compute upper and lower bounds for intracellular reaction fluxes for constraining GEMs.	Constraining CHO cell GEM for improved accuracy of intracellular flux distribution.	Morrissey et al. [80]

Table 1.2 Examples of metabolic objective methodologies, including those guided or inferred from omics data.

Method	Description	Original application	Reference
pFBA (parsimonious FBA)	Minimizes the sum of all fluxes in the network.	Applied to *E. coli* to better understand genotype-phenotype reactionship.	Lewis et al. [86]
IMAT (Integrative Metabolic Analysis Toolbox)	Categorizes reactions as low/moderate/high expression. MILP maximizes the number of reactions whose activity is consistent with their expression state.	Method applied to a toy model as an example.	Zur et al. [28]
SPOT (Simplified Pearson cOrrelation with Transcriptomic data)	Uses a Pearson product-moment correlation to maximize the link between gene expression and flux.	Predicting intracellular flux distribution in *E. coli* and *S. cerevisiae*.	Kim et al. [81]
GIMME (Gene Inactivity Moderated by Metabolism and Expression)	GIMME minimizes usage of low-expression reactions while keeping a metabolic objective (e.g. biomass and productivity) above a certain threshold.	Produce context-specific metabolic networks for *E. coli* for several different conditions. Produce genome-scale models for particular human cells.	Becker and Palsson [27]
MOMA (Minimization Of Metabolic Adjustment)	Predicts post-perturbation fluxes by minimizing the distance from the wild-type state (typically using Euclidean distance in flux space).	Predicting flux distributions in *E. coli* mutants with disrupted pyruvate kinase.	Segrè et al. [87]
PROM (Probabilistic Regulation Of Metabolism)	Integrates gene regulatory networks and transcriptomics into GEMs by estimating the probability that a gene is active and adjusting flux bounds accordingly.	Context-specific modeling in *E. coli* and *M. tuberculosis* to predict condition-specific metabolic states.	Chandrasekaran and Price [88]

ObjFind	Finds the objective function that best matches experimental flux data by assigning metabolite weights.	*E. coli* in aerobic and anaerobic conditions to compare metabolic objectives.	Burgard and Maranas [82]
invFBA (inverse FBA)	Two-step optimization problems that minimize the error between measured and predicted objective functions in linear, quadratic.	Applied fluxes measured in the central carbon metabolism of ancestral and evolved *E. coli* strains.	Zhao et al. [84]
BOSS (Biological Objective Solution Search)	Searches over randomly generated objectives to find the one best matching experimental fluxes.	*S. cerevisiae* central metabolic network to evaluate which objective reaction it infers.	Gianchandani et al. [83]
SCOOTI (Single-Cell Optimization Objective and Trade-off Inference)	Infers cell-specific metabolic objectives and trade-offs by integrating GEMs with bulk or single-cell omics data.	Studying embryogenesis and cell cycle with single-cell transcriptomics and flux models.	Lin et al. [85]
CellFIE (Cellular Functions InferencE)	Summarize flux solutions of transcriptomics-based context-specific models with metabolic subsystems to provide scores for metabolic tasks.	Determining ranking of cellular tasks for various human tissues.	Richelle et al. [89]
ParTi (Pareto Task Inference)	Identifies transcriptomic “archetypes” and links them to metabolic tasks using Pareto fronts.	Modeling cancer transcriptomes to infer key metabolic tasks and drug sensitivity.	Hart et al. [90]

comparisons with experimental data across multiple axes, depending on the context of the model and the application. In the following sections, we outline several key benchmarks for evaluating model performance.

1.5.3.1 Growth Rate Predictions

A key initial benchmark for model validation is to assess predictive accuracy and to compare the predicted growth rate (through biomass reaction maximization) to experimentally measured growth rates under matched culture conditions. This provides a direct readout of how well the model reflects the overall metabolic capacity of the cell and the appropriateness of the biomass reaction. Deviations between predicted and observed growth can signal poor model formulation, missing constraints, and inaccurate biomass reaction construction.

1.5.3.2 Amino Acid Auxotrophies

Amino acid auxotrophies (where an organism is unable to synthesize an amino acid that it needs for growth) can be assessed by limiting the exchange reactions for specific amino acids and observing whether the model becomes infeasible when running an FBA biomass maximization. This simulates conditions where the organism is unable to produce a particular amino acid, reflecting its dependency on external sources for growth. Accurate models should correctly predict known auxotrophies, where the absence of an essential nutrient renders the organism nonviable.

For example, in CHO cells, auxotrophies for amino acids like proline, cystine, and arginine have been reported. In one study, the authors evaluated several CHO models by restricting the exchange reactions for these amino acids [91]. The models were tested to see if growth would be inhibited when proline, cystine, or arginine was unavailable, reflecting known auxotrophies. Accurate prediction of these dependencies demonstrates that the model successfully captures the metabolic requirements of CHO cells, making it a useful tool for optimizing culture conditions or improving strain robustness in industrial applications.

1.5.3.3 Gene Essentialities

Gene essentiality analysis is a critical method for validating genome-scale models by determining whether the inactivation of a gene leads to cell death, as predicted by the model. This is achieved by deleting reactions associated with a particular gene and testing the model's viability. The process simulates the knockout of genes that encode essential enzymes or metabolic functions, allowing researchers to assess whether the model accurately predicts the organism's survival under such conditions.

To validate these predictions, *in silico* gene knockouts are often compared to experimental results, such as those generated through CRISPR knockout screens. CRISPR technology enables precise gene deletions *in vivo*, providing experimental data on which genes are essential for cell survival [92]. A robust model will correctly predict these essential genes, while discrepancies may indicate gaps in the model, such as missing pathways or incomplete regulatory information. For example, the iML1515 model of *E. coli* showed 93.4% accuracy in predicting gene essentiality when tested under minimal media containing 16 different carbon sources [93]. This high level of agreement between computational predictions and experimental data emphasizes the quality of this model and the utility of gene deletion screens in validating genome-scale models and identifying areas where models can be improved.

1.5.3.4 Known Host Traits

Models can also be benchmarked by their ability to reproduce known metabolic or physiological characteristics of the host organism. For example, the Crabtree effect in yeast or the Warburg effect in cancer cells is a well-documented trait that should be captured by accurate GEMs. Additionally, uptake and secretion profiles of amino acids, glucose, or waste products can be compared directly to extracellular metabolomic data. The more traits the model can replicate without overfitting, the more confidence one can have in its biological fidelity.

1.5.3.5 Intracellular Predictive Accuracy

A crucial aspect of genome-scale models is their ability to predict intracellular fluxes, providing insights into the metabolic state of the cell at a genome-wide level. These predictions enable researchers to understand cellular metabolism in detail, facilitating the design of industrial processes and the engineering of optimized strains. To ensure that these predictions are reliable, it is essential to validate the model's ability to accurately predict intracellular fluxes.

One of the most effective methods for this validation is through comparisons with experimental ^{13}C-metabolic flux analysis (MFA) data. By incorporating ^{13}C-labeled substrates and tracking their incorporation into metabolic products, researchers can derive actual flux distributions within the cell, providing a robust dataset for comparison with GEM predictions [94, 95].

Due to the inherent issues with multiple optimal solutions using FBA to simulate metabolism in GEMs, flux sampling is recommended as a method for generating intracellular flux predictions (see Section 1.4.3). Flux sampling generates a range of possible solutions, allowing researchers to assess the variability of flux distributions rather than relying on a single optimal solution. These sampled flux distributions can then be compared with ^{13}C-MFA results to assess the accuracy of the model's predictions [96].

This approach has been successfully applied in the literature for industrially relevant cells. In one body of work, researchers compared the flux predictions from CHO cell GEMs with ^{13}C-MFA data, providing insights into the model's reliability in predicting intracellular metabolic activity [91]. These comparisons are critical for ensuring the model's utility in guiding cell line engineering and optimizing biotechnological processes and must be considered a part of any model evaluation workflow.

1.5.4 Toward Multilayer, Multiscale Metabolic Networks

While GEMs have provided a powerful framework for modeling metabolism, they only capture one layer of cellular function. In reality, biological systems operate through an interplay of multiple processes, including transcription, translation,

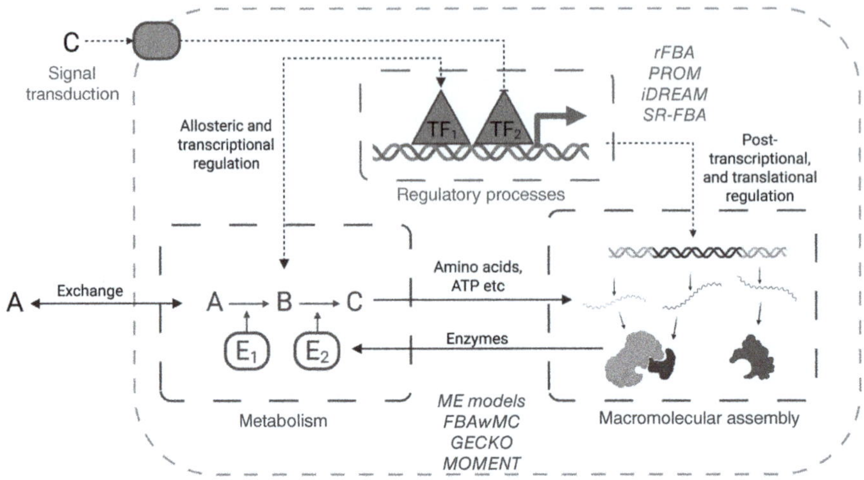

Figure 1.5 Overview of multi-layer modelling of Cellular Function. Core metabolic networks can be constrained by regulatory processes through TFs. ME models consider the interaction between transcription and translation machinery, the resources needed to synthesize them, and how they interact with metabolic function. Simpler resource allocation methods like GECKO consider enzyme kinetics and abundance, but not the full macromolecular machinery. Signal transduction influences network function. In this example, the accumulation of metabolite *C* in the extracellular environment may trigger a cascade to activate TF2 and prevent the expression of the gene required to catalyze the *B*→*C* reaction to prevent overproduction of metabolite *C*. Diagram created in BioRender.

signaling, and regulation. To move toward a more complete understanding of cell behavior, efforts are focused on developing multilayer and multiscale models that integrate these biological processes with metabolic networks, as illustrated in Figure 1.5.

1.5.4.1 Integrating Gene Regulatory Networks

While metabolic networks offer a great method to integrate omics data to simulate cell behavior and assess genetic perturbations, they do not capture any regulatory interactions. For example, *E. coli* has hundreds of TFs that can turn genes on or off in a combinatorial fashion. In more complex organisms, these regulatory processes extend to a myriad of posttranslational and posttranscriptional interactions that govern cellular behavior.

The first approach to incorporate transcriptional regulation into metabolic models, regulatory FBA (rFBA) [79] combines Boolean regulatory networks with FBA. Regulatory constraints (e.g. presence or absence of TFs) are used to turn reactions on or off, enabling simulation of regulatory effects on metabolism under dynamic environmental conditions. Though simplistic in its logic layer, rFBA provides a tractable method to include gene regulation and has been used in *E. coli*, *M. tuberculosis*, and *S. cerevisiae*.

PROM [88] improves upon rFBA by relaxing the Boolean logic and estimating the probability that a regulatory gene/TF is active and then adjusting the flux bounds of associated reactions accordingly. This probabilistic framework has been applied to various organisms, including *M. tuberculosis*, offering condition-specific metabolic predictions based on transcriptional profiles. Other methods include IDREAM [97], SR-FBA [98], iFBA [99], and TRFBA [100].

1.5.4.2 Integrating Transcription and Translation

Beyond transcriptional regulation, transcription and translation interact with metabolism through their dependence on metabolic precursors (e.g. RNA, amino acids, ATP) for macromolecule synthesis, while the resulting proteins catalyze metabolic reactions. Capturing this dependency can impose additional constraints and offer deeper insight into metabolic networks.

ME models address this by integrating GEMs with the transcriptional and translational machinery required to produce metabolic enzymes [35, 36]. These models impose global, growth-related regulatory constraints on metabolism and have outperformed standalone GEMs in predicting phenotypes such as growth, metabolic fluxes, and gene expression levels [36]. ME models do not explicitly capture specific regulatory mechanisms, which is an important area for development of multiscale ME models.

1.5.4.3 Integrating Signaling Networks

Cells respond to changing environments by adjusting gene expression and metabolic activity, a process tightly coordinated by signal transduction pathways. These pathways relay external cues to intracellular targets, influencing regulatory and metabolic networks. Integrating signaling into metabolic and regulatory models can provide context and impose additional constraints, improving predictive accuracy. However, reconstructing intracellular signaling networks remains a major challenge, with large-scale maps available for only a limited number of organisms [99, 101]. Since independent characterization of these pathways is both time-consuming and resource-intensive, a more scalable approach is to leverage high-throughput datasets to infer how environmental changes influence cellular coordination.

1.5.4.4 Multicellular and Multitissue Models

Biological function often emerges not from isolated cells but from coordinated interactions across tissues and cell types. As such, there is growing interest in extending GEMs beyond the single-cell level to simulate multicellular and multitissue systems. One approach involves connecting multiple tissue-specific GEMs via shared extracellular compartments, enabling simulation of metabolite exchange across organs [102]. This strategy has been applied in whole-body metabolic models of humans, where liver, muscle, adipose, and gut models interact to study systemic metabolism, such as in diabetes or fasting responses [103].

In the context of biotechnology, coculture microbial cell cultures have been studied [104], as well as combining mammalian and microbe GEMs to study host–microbiome interactions [105]. Cancer research has leveraged tumor–stromal metabolic models to capture nutrient competition and microenvironmental effects [106]. Frameworks such as the human metabolic atlas and resources like the virtual metabolic human database have facilitated the construction of such models by providing tissue-specific expression data and curated reaction sets [107, 108].

1.5.4.5 Multiscale Bioreactor Models

In the above sections, "multiscale" has referred to layering biological processes to gain deeper insight into each layer individually. In the context of biotechnology, it may be prudent to add an additional scale, the bioreactor. To link molecular models with population or bioreactor-scale behavior, multiscale models integrate intracellular networks with higher-level process models describing growth, nutrient consumption, and bioreactor conditions [109–113]. To make these simulations computationally tractable, reduced-order models or simplified GEMs are often used in place of full-scale metabolic reconstructions. Dynamic FBA (dFBA) approaches [114] integrate time-dependent environmental changes with FBA simulations over discrete intervals. Model reduction techniques, such as elementary mode analysis or using core metabolic subnetworks, allow faster simulation while retaining essential metabolic features [115]. As cells exhibit heterogeneity in metabolic activity in cell culture due to asynchronous growth, cell cycle states, and location, population balance models can be used [116].

1.6 From Predictions to Actionable Insights in Biotechnology

Metabolic networks and constraint-based modeling offer powerful tools for simulating metabolism, but their true value lies in enabling real-world applications across industrial biotechnology. This chapter section focuses on translating *in silico* predictions into practical strategies for metabolic engineering, process design, and bioprocess control. To further enhance these strategies, the integration of mechanistic and data-driven approaches is increasingly adopted.

Hybridization of network-based and data-driven tools allows the biological context and structure of network approaches to be combined with the pattern recognition strengths of data-driven methods, offering enhanced predictive power and interpretability. Broadly, hybrid approaches are used in two ways: either to inform and constrain network models using data-driven predictions or to use outputs from network models as features for data-driven algorithms (summarized in Figure 1.6). These two strategies provide complementary routes to deepen biological insight and improve model utility in biotechnology. In the following section, we highlight several key examples that demonstrate their application.

1.6.1 Metabolic Engineering

GEMs formalize biological knowledge and enable the design and quantitative comparison of metabolic engineering strategies *in silico*. They are a smart test bed for screening such strategies efficiently before selecting a subset for experimental investigation as they capture the complexity of cellular systems and the intricate network

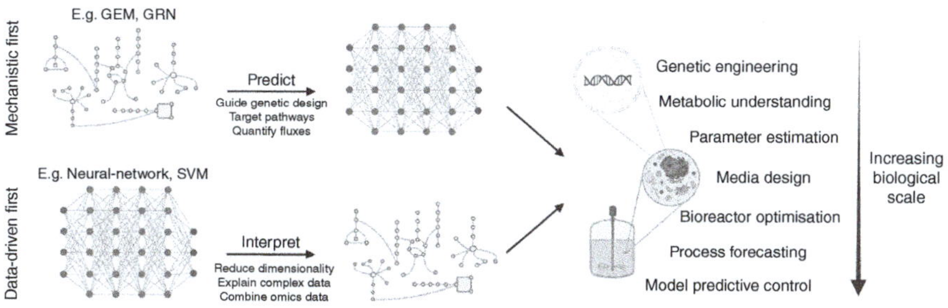

Figure 1.6 Schematic summarizing how data-driven and mechanistic approaches can be hybridized to generate insights in biotechnology. Hybridization can be mechanistic first, meaning mechanistic predictions (such as flux from a GEM) can be inputted to data-driven tools. Alternatively, data-driven tools can be used to contextualize and interpret complex data to inform mechanistic models. Both approaches have yielded actionable insights in biotechnology. Diagram created in BioRender.

of biochemical reactions that are candidates for manipulation. By integrating large datasets and creating custom GEMs, researchers can then apply a variety of novel optimization algorithms for target identification as detailed in Table 1.3, reducing the trial-and-error of experimental methods and accelerating the development of improved strains.

These algorithms primarily use mixed-integer linear programming (MILP) to identify candidate reactions (and therefore genes) for manipulation, e.g. knockout, knockdown, or overexpression. For example, OptKnock [117] uses a bilevel optimization framework to suggest gene knockouts that link the overproduction of a desired product (metabolic or recombinant protein) to cellular growth. A number of other tools have expanded on this concept [118], with subsequent algorithms incorporating knock-ins of non-native functions from extensive reaction databases [126]. Most of these efforts were originally developed for and applied to microbial cell systems, which are described by smaller stoichiometric networks. Moving to plant or mammalian cell systems, where the networks expanded from 100s to 1,000s of equations, makes the application of these algorithms more computationally demanding [127].

1.6.2 Cell Line Development and Metabolic Profiling

In parallel to algorithmic approaches, researchers have also used GEMs to guide cell line development in a data-driven fashion. For example, tailored CHO cell GEMs solved using pFBA have been used to identify key metabolic reactions that distinguish low- and high-producing cell lines [128]. The authors identified glucose-6-phosphate dehydrogenase (G6PD) and PDH as potential biomarkers, as well as a bottleneck at malate dehydrogenase (MDH) in low producers, in line with previously published work [129]. Similarly, Huang and Yoon [130] studied transcriptomic and metabolomic data using a CHO GEM to identify the metabolic pathways that determine recombinant protein productivity in two clones grown in batch cultures. Their analysis confirmed the importance of G6PD and PDH in high

Table 1.3 Summary of key optimization algorithms for strain design.

Algorithm	Description	Reference
OptKnock	Suggests gene knockouts based on a bilevel optimization framework that couples the desired overproduction target (i.e. the biopharmaceutical) to growth.	Burgard et al. [117]
OptReg	Extension of OptKnock that suggests reactions for regulation (up, down, etc.) based on a bilevel optimization framework that couples the fluxes of two different reactions.	Pharkya and Maranas [118]
OptForce	Extension of OptReg that contrasts the metabolic flux patterns observed in an initial strain and a strain overproducing the chemical at the target yield.	Ranganathan et al., [119]
GDLS	Uses logical search to look for multiple paths to improve the production of metabolites.	Lun et al. [120]
OptORF	Identifies optimal number of metabolic and regulatory gene knockouts/overexpressions to couple production with growth.	Kim and Reed [121]
k-OptForce	Identifies regulators in transcription factors and metabolic genes by accounting for gene expression in its objective function.	Chowdhury et al. [122]
OptRam	Extension of OptForce that uses kinetic equations to better calculate steady-state fluxes of metabolic networks.	Shen et al. [123]
gcOpt	Predicts coupling strength of two metabolic reactions.	Alter and Ebert [124]
OptCouple	Analyzes both genetic engineering strategies with process engineering solutions for the overproduction of a metabolite.	Jensen et al. [125]
OptStrain	Assesses knock-ins of non-native functionalities from a comprehensive database of reactions.	Pharkya et al. [126]

Source: Antonakoudis et al., [11] / with permission of Elsevier.

producers and revealed a tricarboxylic acid (TCA) cycle bottleneck at succinyl-CoA synthetase. Yusufi et al. [131] validated GEM predictions with metabolomic data, identifying metabolites associated with energy metabolism and oxidative phosphorylation, the concentration of which is elevated in protein-producing CHO cells. In a study across different bioreactor scales ranging from 10to 1,000 L, Vodopivec et al. [132] employed a GEM to analyze metabolomics data and drive our understanding of the differences in metabolic states that occur during upstream process scale-up. Kol et al. [133] used a CHO cell GEM that includes the secretory pathway to quantify the secretion cost of abundant host cell proteins (HCPs). They used the results to rank HCP as knockout targets and subsequently implemented sequential knockouts of burdensome HCPs achieving a 40–70% reduction in HCP content in the supernatant. In addition to improving cell growth and productivity, these modifications generated clean feedstock for downstream purification steps.

1.6.3 Media and Feed Design

Multiomics datasets and derived GEMs have also found application in the design of cell culture media and supplements. GEMs can be used to elucidate the impact of changes in the culture environment on intracellular flux distribution and therefore identify bottlenecks for cell growth or the production of a metabolite/recombinant product. This understanding can then be used to propose improved process strategies, including media and feed formulations. An example of such an application is the work of Huang et al., in which a GEM together with transcriptomic data were used to identify metabolic pathways that are upregulated in high-producing CHO cell clones [134]. Their analysis highlighted branched-chain amino acids as a key source of TCA cycle intermediates, which led to the design of a new cell culture feed rich in valine and leucine. The feed was tested experimentally and was demonstrated to result in a significant increase in recombinant protein titer. A similar study used a CHO-DG44 GEM to reveal a link between folate supplementation in media and lactate secretion. This mechanism balances increased lactate production by suppressing TCA cycle flux with reduced lactate production through decreased glutamine/asparagine anaplerosis [135]. GEMs have also been used to understand the metabolism of asparagine and aspartate for feed ratio optimization [136], to design potential feed supplements for recombinant protein titer optimization [137], and to reduce the accumulation of toxic metabolic byproducts like ammonia by improving feed formulation [138]. Additionally, GEMs have been used to identify succinate dehydrogenase (SDH) as a bottleneck in early cell culture, leading to media reformulation to include coenzyme q10 (ubiquinone) for enhanced FAD+ regeneration [139].

The analysis of large fluxomic datasets resulting from metabolic models can be facilitated by statistical or ML methods toward new knowledge generation. To this end, Ramos et al. analyzed the results of a CHO-K1 GEM with principal component analysis (PCA) to design an efficient growth medium that minimized by-product accumulation [140]. Previously, isotope labeling experimental datasets analyzed using small-scale metabolic models to infer the flux of key metabolic pathways have been further interrogated using a variety of ML techniques. For example, Wu et al. developed a platform that applied support vector machines (SVMs), k-nearest neighbors, and decision trees to literature data from nearly 100 ^{13}C-FMA studies on heterotrophic bacteria [141]. The platform predicts fluxomes as a function of bacterial species, substrate type, growth rate, oxygen conditions, and cultivation methods. Nandi et al. [142] extended the methodology to include gene sequencing and expression, network topology, and flux-based features. By also accounting for environmental factors, the model was able to capture the minimal set of genes that are essential in any given environment.

NEXT-FBA [80] is another recent example of using ML methods to understand complex biological data. NEXT-FBA uses neural networks trained on metabolomic and fluxomic data to understand the relationship between a cell's internal and external dynamics. This understanding is used to compute constraints on GEMs that create more refined and accurate predictions. NEXT-FBA was used to identify the causes of ammonia accumulation in a CHO cell process, determining feed and metabolic engineering process improvements.

1.6.4 Gene Essentiality

Understanding how genetic perturbations lead to phenotypic changes is a key application of metabolic networks. Combined data-driven and mechanistic approaches have been used to identify essential genes. For example, Plaimas et al. used SVMs on an *E. coli* FBA network with genomic and transcriptomic data to identify essential enzymes [143]. Acencio and Lemke extended this by applying decision trees to topological, compartmental, and functional features, improving gene essentiality prediction and revealing phenotype determinants [144].

Graph-based learning is another emerging data-driven-first approach, where metabolic reaction networks – constructed from empirical data or draft reconstructions – are treated as graphs and analyzed using graph neural networks (GNNs). Nodes may represent metabolites or reactions, and edges capture biochemical or topological relationships. By training GNNs on these graphs with omics data as node or edge features, researchers can predict reaction essentiality, flux control points, or context-specific pathway activation [145]. This approach retains biological structure without requiring a fully curated GEM, making it particularly useful for less-characterized organisms or exploratory studies where network topology and data integration are the primary drivers of insight.

1.6.5 Kinetic Parameter Estimation

Another application of data-driven-first applications is parameter estimation. A major barrier to developing large-scale kinetic or enzyme-constrained network models is the lack of high-quality kinetic parameters. As an early solution, data-driven methods were used to predict kinetic values to parameterize network models. For instance, a statistical regression method was used to estimate kinetic parameters within appropriate experimental ranges [146]. More recently, ML algorithms were trained on a dataset incorporating enzyme biochemistry, protein structure, and network context to predict enzyme turnover numbers, used in proteome-constrained models [147, 148].

1.6.6 Process Monitoring and Forecasting

A key focus of the biopharmaceutical industry has been a move from quality by testing to quality by design (QbD) in process development and manufacturing (ICH Q8 guideline). Mathematical model-based process monitoring, forecasting, and control is an important part of the QbD approach. Models designed for cell culture control must be complex enough to understand the link between critical process parameters (CPPs) and both the key performance indicators (KPIs) and the critical quality attributes (CQAs) in the process, but as the same time simple enough to allow fast action by the controller. Mathematical models for mammalian cell culture process control can be data-driven, based on simple mechanistic relations, or more complex metabolic models [11, 149].

GEMs offer a unique perspective on process control as they link CPPs and KPIs/CQAs by modeling cell metabolism, rather than treating cells as a black box for the conversion of substrates to products using a set of macroscopic reactions without considering intracellular behavior. GEMs can offer insights that macroscopic models cannot and therefore improving forecasting and control. CHO cell GEMs have been used to predict amino acid concentrations by applying minimization of uptake rate objective combined with forecasted viable cell densities [150]. In addition, Schinn et al. combined CHO cell GEM flux sampling with statistical models to describe time-course amino acid consumption [151], while Antonakoudis et al. used GEM-computed fluxes of nucleotide sugars as inputs to an ML model that predicted the glycosylation profile of the protein product [152].

However, several limitations hinder the practical use of network models in process control. Their complexity often exceeds what is necessary for actionable control, increasing computational demand without proportional benefit. The underdetermined nature of these models leads to broad solution spaces, reducing predictive precision unless constrained appropriately. Furthermore, the reliance on steady-state assumptions limits forecasting accuracy over longer culture durations, necessitating dynamic modeling approaches. Lastly, GEMs alone cannot capture the impact of key CPPs such as temperature, pH, and toxic metabolite accumulation on metabolism. To overcome these issues, simplification strategies, dynamic extensions, and integration with empirical or data-driven methods (such as reinforcement learning or hybrid MPC frameworks) are essential for effective implementation.

In this direction, metabolic predictions from a small-scale CHO cell metabolic model have been integrated into dynamic models of production bioreactors [109, 112]. Finally, Gopalakrishnan et al. (2024) [110] proposed the use of a data-driven classifier for categorizing cells into "growth" or "production" states, each with distinct metabolic objectives. Using this term, a GEM was embedded within a system of ODEs to predict metabolite uptake and secretion throughout a fed-batch CHO cell culture, successfully capturing the coexistence of two cellular states.

References

1 Ideker, T., Galitski, T., and Hood, L. (2001). A new approach to decoding life: systems biology. *Annu. Rev. Genomics Hum. Genet.* 2: 342–372.

2 Kitano, H. (2002). Systems biology: a brief overview. *Science* 295: 1662–1664.

3 Barabási, A.L. and Oltvai, Z.N. (2004). Network biology: understanding the cell's functional organization. *Nat. Rev. Genet.* 5: 101–113.

4 Berger, B., Peng, J., and Singh, M. (2013). Computational solutions for omics data. *Nat. Rev. Genet.* 14: 333–346.

5 Pal, S., Mondal, S., Das, G. et al. (2020). Big data in biology: the hope and present-day challenges in it. *Gene Rep.* 21: 100869.

6 Lewis, N.E., Nagarajan, H., and Palsson, B.O. (2012). Constraining the metabolic genotype-phenotype relationship using a phylogeny of in silico methods. *Nat. Rev. Microbiol.* 10: 291–305.

7 Palsson, B. (2006). *Systems Biology: Properties of Reconstructed Networks.* Cambridge University Press.

8 Thiele, I. and Palsson, B. (2010). A protocol for generating a high-quality genome-scale metabolic reconstruction. *Nat. Protoc.* 5: 93–121.

9 Fang, X., Lloyd, C.J., and Palsson, B.O. (2020). Reconstructing organisms in silico: genome-scale models and their emerging applications. *Nat. Rev. Microbiol.* 18: 731–743.

10 Zielinski, D.C., Patel, A., and Palsson, B.O. (2020). The expanding computational toolbox for engineering microbial phenotypes at the genome scale. *Microorganisms* 8: 2050.

11 Antonakoudis, A., Barbosa, R., Kotidis, P., and Kontoravdi, C. (2020). The era of big data: genome-scale modelling meets machine learning. *Comput. Struct. Biotechnol. J.* 18: 3287, 3300.

12 Passi, A., Tibocha-Bonilla, J.D., Kumar, M. et al. (2022). Genome-scale metabolic modeling enables in-depth understanding of big data. *Metabolites* 12: 14.

13 Francke, C., Siezen, R.J., and Teusink, B. (2005). Reconstructing the metabolic network of a bacterium from its genome. *Trends Microbiol.* 13: 550–558.

14 Gu, C., Kim, G.B., Kim, W.J. et al. (2019). Current status and applications of genome-scale metabolic models. *Genome Biol.* 20: 121.

15 Strain, B., Morrissey, J., Antonakoudis, A., and Kontoravdi, C. (2023). Genome-scale models as a vehicle for knowledge transfer from microbial to mammalian cell systems. *Comput. Struct. Biotechnol. J.* 21: 1543–1549.

16 Ashburner, M., Ball, C.A., Blake, J.A. et al. (2000). Gene ontology: tool for the unification of biology. *Nat. Genet.* 25: 25–29.

17 Barrett, A.J. (1995). Enzyme Nomenclature. Recommendations 1992. *Eur. J. Biochem.* 232: 1.

18 Kanehisa, M. and Goto, S. (2000). KEGG: Kyoto Encyclopedia of genes and genomes. *Nucleic Acids Res.* 28: 27–30.

19 Placzek, S., Schomburg, I., Chang, A. et al. (2017). BRENDA in 2017: new perspectives and new tools in BRENDA. *Nucleic Acids Res.* 45: D380–D388.

20 Jankowski, M.D., Henry, C.S., Broadbelt, L.J., and Hatzimanikatis, V. (2008). Group contribution method for thermodynamic analysis of complex metabolic networks. *Biophys. J.* 95: 1487–1499.

21 Caspi, R., Billington, R., Fulcher, C.A. et al. (2018). The MetaCyc database of metabolic pathways and enzymes. *Nucleic Acids Res.* 46: D633–D639.

22 Henry, C.S., Dejongh, M., Best, A.A. et al. (2010). High-throughput generation, optimization and analysis of genome-scale metabolic models. *Nat. Biotechnol.* 28: 977–982.

23 Machado, D., Andrejev, S., Tramontano, M., and Patil, K.R. (2018). Fast automated reconstruction of genome-scale metabolic models for microbial species and communities. *Nucleic Acids Res.* 46: 7542–7553.

24 Karp, P.D., Paley, S.M., Krummenacker, M. et al. (2009). Pathway tools version 13.0: integrated software for pathway/genome informatics and systems biology. *Briefings Bioinf.* 11: 40–79.

25 Gopalakrishnan, S., Joshi, C.J., Valderrama-Gómez, M. et al. (2023). Guidelines for extracting biologically relevant context-specific metabolic models using gene expression data. *Metab. Eng.* 75: 181–191.

26 Richelle, A., Chiang, A.W.T., Kuo, C.C., and Lewis, N.E. (2019). Increasing consensus of context-specific metabolic models by integrating data-inferred cell functions. *PLoS Comput. Biol.* 15: e1006867.

27 Becker, S.A. and Palsson, B.O. (2008). Context-specific metabolic networks are consistent with experiments. *PLoS Comput. Biol.* 4: e1000082.

28 Zur, H., Ruppin, E., and Shlomi, T. (2010). iMAT: an integrative metabolic analysis tool. *Bioinformatics* 26: 3140–3142.

29 Schultz, A. and Qutub, A.A. (2016). Reconstruction of tissue-specific metabolic networks using CORDA. *PLoS Comput. Biol.* 12: e1004808.

30 Babu, M.M., Luscombe, N.M., Aravind, L. et al. (2004). Structure and evolution of transcriptional regulatory networks. *Curr. Opin. Struct. Biol.* 14: 283–291.

31 Blais, A. and Dynlacht, B.D. (2005). Constructing transcriptional regulatory networks. *Genes Dev.* 19: 1499–1511.

32 Covert, M.W., Knight, E.M., Reed, J.L. et al. (2004). Integrating high-throughput and computational data elucidates bacterial networks. *Nature* 429: 92–96.

33 Türei, D., Korcsmáros, T., and Saez-Rodriguez, J. (2016). OmniPath: guidelines and gateway for literature-curated signaling pathway resources. *Nat. Methods* 13: 966–967.

34 Safari-Alighiarloo, N., Taghizadeh, M., Rezaei-Tavirani, M. et al. (2014). Protein-protein interaction networks (PPI) and complex diseases. *Gastroenterol Hepatol Bed Bench.* 7: 17.

35 Thiele, I., Fleming, R.M.T., Que, R. et al. (2012). Multiscale Modeling of metabolism and macromolecular synthesis in *E. Coli* and its application to the evolution of codon usage. *PLoS One* 7: e45635.

36 O'Brien, E.J., Lerman, J.A., Chang, R.L. et al. (2013). Genome-scale models of metabolism and gene expression extend and refine growth phenotype prediction. *Mol. Syst. Biol.* 9: 693.

37 Allaman, I., Bélanger, M., and Magistretti, P.J. (2015). Methylglyoxal, the dark side of glycolysis. *Front. Neurosci.* 9: 23.

38 Held, K.D., Sylvester, K.C., Hopcia, K.L., and Biaglow, J.E. (1996). Role of Fenton chemistry in thiol-induced toxicity and apoptosis. *Radiat. Res.* 145: 542–553.

39 Selvarasu, S., Ow, D.S.W., Lee, S.Y. et al. (2009). Characterizing *Escherichia coli DH5α* growth and metabolism in a complex medium using genome-scale flux analysis. *Biotechnol. Bioeng.* 102: 923–934.

40 Széliová, D., Štor, J., Thiel, I. et al. (2020). Inclusion of maintenance energy improves the intracellular flux predictions of CHO. *PLoS Comput. Biol.* 17: e1009022.

41 Feist, A.M. and Palsson, B.O. (2010). The biomass objective function. *Curr. Opin. Microbiol.* 13: 344–349.

42 Aminian-Dehkordi, J., Mousavi, S.M., Jafari, A. et al. (2019). Manually curated genome-scale reconstruction of the metabolic network of *bacillus megaterium* DSM319. *Sci. Rep.* 9: 18762.

43 Price, N.D., Reed, J.L., and Palsson, B. (2004). Genome-scale models of microbial cells: evaluating the consequences of constraints. *Nat. Rev. Microbiol.* 2: 886–897.

44 Reed, J.L., Vo, T.D., Schilling, C.H., and Palsson, B.O. (2003). An expanded genome-scale model of Escherichia coli *K-12* (iJR904 GSM/GPR). *Genome Biol.* 4: R54.

45 Mahadevan, R. and Schilling, C.H. (2003). The effects of alternate optimal solutions in constraint-based genome-scale metabolic models. *Metab. Eng.* 5: 264–276.

46 Thiele, I., Vlassis, N., and RMT, F. (2014). FASTGAPFILL: efficient gap filling in metabolic networks. *Bioinformatics* 30: 2529–2531.

47 Beard, D.A., Liang, S.D., and Qian, H. (2002). Energy balance for analysis of complex metabolic networks. *Biophys. J.* 83: 79–86.

48 Henry, C.S., Broadbelt, L.J., and Hatzimanikatis, V. (2007). Thermodynamics-based metabolic flux analysis. *Biophys. J.* 92: 1792–1805.

49 Schellenberger, J., Lewis, N.E., and Palsson, B. (2011). Elimination of thermodynamically infeasible loops in steady-state metabolic models. *Biophys. J.* 100: 544–553.

50 Fleming, R.M.T., Haraldsdottir, H.S., Minh, L.H. et al. (2023). Cardinality optimization in constraint-based modelling: application to human metabolism. *Bioinformatics* 39: btad450.

51 Heirendt, L., Arreckx, S., Pfau, T. et al. (2019). Creation and analysis of biochemical constraint-based models using the COBRA Toolbox v.3.0. *Nat. Protoc.* 14: 639–702.

52 Orth, J.D., Thiele, I., and Palsson, B.O. (2010). What is flux balance analysis? *Nat. Biotechnol.* 28: 245–248.

53 Varma, A. and Palsson, B.O. (1994). Stoichiometric flux balance models quantitatively predict growth and metabolic by-product secretion in wild-type *Escherichia coli* W3110. *Appl. Environ. Microbiol.* 60: 3724–3731.

54 Srivastava, S. and Chan, C. (2008). Application of metabolic flux analysis to identify the mechanisms of free fatty acid toxicity to human hepatoma cell line. *Biotechnol. Bioeng.* 99: 399–410.

55 Gudmundsson, S. and Thiele, I. (2010). Computationally efficient flux variability analysis. *BMC Bioinf.* 11: 489.

56 Schellenberger, J. and Palsson, B. (2009). Use of randomized sampling for analysis of metabolic networks. *J. Biol. Chem.* 5457–5461.

57 Herrmann, H.A., Dyson, B.C., Vass, L. et al. (2019). Flux sampling is a powerful tool to study metabolism under changing environmental conditions. *npj Syst. Biol. Appl.* 5: 32.

58 Wiback, S.J., Famili, I., Greenberg, H.J., and Palsson, B. (2004). Monte Carlo sampling can be used to determine the size and shape of the steady-state flux space. *J. Theor. Biol.* 228: 437–447.

59 Smith, R.L. (1984). Efficient Monte Carlo procedures for generating points uniformly distributed over bounded regions. *Oper. Res.* 32: 1296–1308.

60 Kaufman, D.E. and Smith, R.L. (1998). Direction choice for accelerated convergence in hit-and-run sampling. *Oper. Res.* 46: 84–95.

61 Haraldsdóttir, H.S., Cousins, B., Thiele, I. et al. (2017). CHRR: coordinate hit-and-run with rounding for uniform sampling of constraint-based models. *Bioinformatics* 33: 1741–1743.

62 Zampieri, G., Vijayakumar, S., Yaneske, E., and Angione, C. (2019). Machine and deep learning meet genome-scale metabolic modeling. *PLoS Comput. Biol.* e1007054.

63 King, Z.A., Lloyd, C.J., Feist, A.M., and Palsson, B.O. (2015). Next-generation genome-scale models for metabolic engineering. *Curr. Opin. Biotechnol.* 35: 23–29.

64 Kim, M., Rai, N., Zorraquino, V., and Tagkopoulos, I. (2016). Multi-omics integration accurately predicts cellular state in unexplored conditions for *Escherichia coli*. *Nat. Commun.* 7: 13090.

65 Bordel, S., Agren, R., and Nielsen, J. (2010). Sampling the solution space in genome-scale metabolic networks reveals transcriptional regulation in key enzymes. *PLoS Comput. Biol.* 6: e1000859.

66 Colijn, C., Brandes, A., Zucker, J. et al. (2009). Interpreting expression data with metabolic flux models: predicting *Mycobacterium tuberculosis* mycolic acid production. *PLoS Comput. Biol.* 5: e1000489.

67 Jensen, P.A. and Papin, J.A. (2011). Functional integration of a metabolic network model and expression data without arbitrary thresholding. *Bioinformatics.* 27: 541–547.

68 Pusa, T., Ferrarini, M.G., Andrade, R. et al. (2020). MOOMIN – mathematical explOration of'omics data on a MetabolIc network. *Bioinformatics.* 36: 514–523.

69 Tian, M. and Reed, J.L. (2018). Integrating proteomic or transcriptomic data into metabolic models using linear bound flux balance analysis. *Bioinformatics.* 34: 3882–3888.

70 Wang, Y., Eddy, J.A., and Price, N.D. (2012). Reconstruction of genome-scale metabolic models for 126 human tissues using mCADRE. *BMC Syst. Biol.* 6: 153.

71 Lularevic, M., Racher, A.J., Jaques, C., and Kiparissides, A. (2019). Improving the accuracy of flux balance analysis through the implementation of carbon availability constraints for intracellular reactions. *Biotechnol. Bioeng.* 116: 2339–2352.

72 Beg, Q.K., Vazquez, A., Ernst, J. et al. (2007). Intracellular crowding defines the mode and sequence of substrate uptake by *Escherichia coli* and constrains its metabolic activity. *PNAS* 104: 12663–12668.

73 Adadi, R., Volkmer, B., Milo, R. et al. (2012). Prediction of microbial growth rate versus biomass yield by a metabolic network with kinetic parameters. *PLoS Comput. Biol.* 8: e1002575.

74 Yeo, H.C., Hong, J., Lakshmanan, M., and Lee, D.Y. (2020). Enzyme capacity-based genome scale modelling of CHO cells. *Metab. Eng.* 60: 138–147.

75 Mori, M., Hwa, T., Martin, O.C. et al. (2016). Constrained allocation flux balance analysis. *PLoS Comput. Biol.* 12: e1004913.

76 Sánchez, B.J., Zhang, C., Nilsson, A. et al. (2017). Improving the phenotype predictions of a yeast genome-scale metabolic model by incorporating enzymatic constraints. *Mol. Syst. Biol.* 13: 935.

77 Goelzer, A. and Fromion, V. (2011). Bacterial growth rate reflects a bottleneck in resource allocation. *Biochim. Biophys. Acta, Gen. Subj.* 1810: 978–988.

78 Desouki, A.A., Jarre, F., Gelius-Dietrich, G., and Lercher, M.J. (2015). CycleFreeFlux: efficient removal of thermodynamically infeasible loops from flux distributions. *Bioinformatics* 31: 2159–2165.

79 Covert, M.W., Schilling, C.H., and Palsson, B. (2001). Regulation of gene expression in flux balance models of metabolism. *J. Theor. Biol.* 213: 73–88.

80 Morrissey, J., Barberi, G., Strain, B. et al. (2025). NEXT-FBA: a hybrid stoichiometric/data-driven approach to improve intracellular flux predictions. *Metab. Eng.* 91: 130–144. https://www.sciencedirect.com/science/article/pii/S1096717625000461.

81 Kim, M.K., Lane, A., Kelley, J.J., and Lun, D.S. (2016). E-Flux2 and sPOT: validated methods for inferring intracellular metabolic flux distributions from transcriptomic data. *PLoS One* 11: e0157101.

82 Burgard, A.P. and Maranas, C.D. (2003). Optimization-based framework for inferring and testing hypothesized metabolic objective functions. *Biotechnol. Bioeng.* 82.

83 Gianchandani, E.P., Oberhardt, M.A., Burgard, A.P. et al. (2008). Predicting biological system objectives de novo from internal state measurements. *BMC Bioinf.* 9: 670–677.

84 Zhao, Q., Stettner, A.I., Reznik, E. et al. (2016). Mapping the landscape of metabolic goals of a cell. *Genome Biol.* 17: 109.

85 Lin, D.-W., Zhang, L., Zhang, J., and Chandrasekaran, S. (2025). Inferring metabolic objectives and trade-offs in single cells during embryogenesis. *Cell Syst.* 16: 101164. https://www.sciencedirect.com/science/article/pii/S2405471224003703.

86 Lewis, N.E., Hixson, K.K., Conrad, T.M. et al. (2010). Omic data from evolved *E. coli* are consistent with computed optimal growth from genome-scale models. *Mol. Syst. Biol.* 6: 390.

87 Segrè, D., Vitkup, D., and Church, G.M. (2002). Analysis of optimality in natural and perturbed metabolic networks. *PNAS* 99: 15112–15117.

88 Chandrasekaran, S. and Price, N.D. (2010). Probabilistic integrative modeling of genome-scale metabolic and regulatory networks in *Escherichia coli* and *Mycobacterium tuberculosis*. *PNAS* 107: 17845–17850.

89 Richelle, A., Kellman, B.P., Wenzel, A.T. et al. (2021). Model-based assessment of mammalian cell metabolic functionalities using omics data. *Cell Reports Methods*. 1: 100040.

90 Hart, Y., Sheftel, H., Hausser, J. et al. (2015). Inferring biological tasks using Pareto analysis of high-dimensional data. *Nat. Methods* 12: 233–235.

91 Strain, B., Morrissey, J., Antonakoudis, A., and Kontoravdi, C. (2023). How reliable are Chinese hamster ovary (CHO) cell genome-scale metabolic models? *Biotechnol. Bioeng.* 120: 2460–2478.

92 Wang, T., Birsoy, K., Hughes, N.W. et al. (2015). Identification and characterization of essential genes in the human genome. *Science* 350: 1096–1101.

93 Monk, J.M., Lloyd, C.J., Brunk, E. et al. (2017). iML1515, a knowledgebase that computes *Escherichia coli* traits. *Nat. Biotechnol.* 904–908.

94 Young, J.D., Walther, J.L., Antoniewicz, M.R. et al. (2008). An elementary metabolite unit (EMU) based method of isotopically nonstationary flux analysis. *Biotechnol. Bioeng.* 99: 686–699.

95 Antoniewicz, M.R. (2015). Methods and advances in metabolic flux analysis: a mini-review. *J. Ind. Microbiol. Biotechnol.* 42: 317–325.

96 Quek, L.E., Wittmann, C., Nielsen, L.K., and Krömer, J.O. (2009). Open FLUX: efficient modelling software for 13 C-based metabolic flux analysis. *Microb. Cell Fact.* 8: 25.

97 Wang, Z., Danziger, S.A., Heavner, B.D. et al. (2017). Combining inferred regulatory and reconstructed metabolic networks enhances phenotype prediction in yeast. *PLoS Comput. Biol.* 13: e1005189.

98 Shlomi, T., Eisenberg, Y., Sharan, R., and Ruppin, E. (2007). A genome-scale computational study of the interplay between transcriptional regulation and metabolism. *Mol. Syst. Biol.* 3: 101.

99 Covert, M.W., Xiao, N., Chen, T.J., and Karr, J.R. (2008). Integrating metabolic, transcriptional regulatory and signal transduction models in *Escherichia coli*. *Bioinformatics* 24: 2044–2050.

100 Motamedian, E., Mohammadi, M., Shojaosadati, S.A., and Heydari, M. (2017). TRFBA: an algorithm to integrate genome-scale metabolic and transcriptional regulatory networks with incorporation of expression data. *Bioinformatics* 33: 1057–1063.

101 Carrera, J., Estrela, R., Luo, J. et al. (2014). An integrative, multi-scale, genome-wide model reveals the phenotypic landscape of Escherichia coli. *Mol. Syst. Biol.* 10: 735.

102 Bordbar, A., Feist, A.M., Usaite-Black, R. et al. (2011). A multi-tissue type genome-scale metabolic network for analysis of whole-body systems physiology. *BMC Syst. Biol.* 5: 180.

103 Thiele, I., Sahoo, S., Heinken, A. et al. (2020). Personalized whole-body models integrate metabolism, physiology, and the gut microbiome. *Mol. Syst. Biol.* 16: e8982.

104 Zomorrodi, A.R. and Maranas, C.D. (2012). OptCom: a multi-level optimization framework for the metabolic modeling and analysis of microbial communities. *PLoS Comput. Biol.* e1002363.

105 Heinken, A., Sahoo, S., Fleming, R.M.T., and Thiele, I. (2013). Systems-level characterization of a host-microbe metabolic symbiosis in the mammalian gut. *Gut Microbes* 4: 28–40.

106 Zhang, C. and Hua, Q. (2016). Applications of genome-scale metabolic models in biotechnology and systems medicine. *Front. Physiol.* 6: 413.

107 Noronha, A., Modamio, J., Jarosz, Y. et al. (2019). The virtual metabolic human database: integrating human and gut microbiome metabolism with nutrition and disease. *Nucleic Acids Res.* 47: D614–D624.

108 Robinson, J.L., Kocabaş, P., Wang, H. et al. (2020). An atlas of human metabolism. *Sci. Signaling* 13: eaaz1482.

109 Monteiro, M., Fadda, S., and Kontoravdi, C. (2023). Towards advanced bioprocess optimization: a multiscale modelling approach. *Comput. Struct. Biotechnol. J.* 21: 3639–3655.

110 Gopalakrishnan, S., Johnson, W., Valderrama-Gomez, M.A. et al. (2024). COSMIC-dFBA: a novel multi-scale hybrid framework for bioprocess modeling. *Metab. Eng.* 82: 183–192.

111 Mahnert, C., Oyarzún, D.A., and Berrios, J. (2024). Multiscale modelling of bioprocess dynamics and cellular growth. *Microb. Cell Fact.* 23: 315. https://doi.org/10.1186/s12934-024-02581-0.

112 Lázaro, J., Jansen, G., Yang, Y. et al. (2022). Combination of genome-scale models and bioreactor dynamics to optimize the production of commodity chemicals. *Front. Mol. Biosci.* 9: 855735.

113 Yatipanthalawa, B.S., Wallace Fitzsimons, S.E., Horning, T. et al. (2024). Development and validation of a hybrid model for prediction of viable cell density, titer and cumulative glucose consumption in a mammalian cell culture system. *Comput. Chem. Eng.* 184: 108648.

114 Mahadevan, R., Edwards, J.S., and Doyle, F.J. (2002). Dynamic flux balance analysis of diauxic growth in *Escherichia coli*. *Biophys. J.* 83: 1331–1340.

115 von Kamp, A. and Klamt, S. (2014). Enumeration of smallest intervention strategies in genome-scale metabolic networks. *PLoS Comput. Biol.* 10: e1003378.

116 Karra, S., Sager, B., and Karim, M.N. (2010). Multi-scale modeling of heterogeneities in mammalian cell culture processes. *Ind. Eng. Chem. Res.* 49: 7990–8006.

117 Burgard, A.P., Pharkya, P., and Maranas, C.D. (2003). OptKnock: a Bilevel programming framework for identifying gene knockout strategies for microbial Strain optimization. *Biotechnol. Bioeng.* 84: 647–657.

118 Pharkya, P. and Maranas, C.D. (2006). An optimization framework for identifying reaction activation/inhibition or elimination candidates for overproduction in microbial systems. *Metab. Eng.* 8: 1–13.

119 Ranganathan, S., Suthers, P.F., and Maranas, C.D. (2010). OptForce: an optimization procedure for identifying all genetic manipulations leading to targeted overproductions. *PLoS Comput. Biol.* 6: e1000744.

120 Lun, D.S., Rockwell, G., Guido, N.J. et al. (2009). Large-scale identification of genetic design strategies using local search. *Mol. Syst. Biol.* 5: 296.

121 Kim, J. and Reed, J.L. (2010). OptORF: optimal metabolic and regulatory perturbations for metabolic engineering of microbial strains. *BMC Syst. Biol.* 4: 53.

122 Chowdhury, A., Zomorrodi, A.R., and Maranas, C.D. (2014). K-OptForce: integrating kinetics with flux balance analysis for Strain design. *PLoS Comput. Biol.* 10: e1003487.

123 Shen, F., Sun, R., Yao, J. et al. (2019). Optram: in-silico strain design via integrative regulatory-metabolic network modeling. *PLoS Comput. Biol.* 15: e1006835.

124 Alter, T.B. and Ebert, B.E. (2019). Determination of growth-coupling strategies and their underlying principles. *BMC Bioinf.* 20: 447.

125 Jensen, K., Broeken, V., Hansen, A.S.L. et al. (2019). OptCouple: joint simulation of gene knockouts, insertions and medium modifications for prediction of growth-coupled strain designs. *Metab. Eng. Commun.* 8: e00087.

126 Pharkya, P., Burgard, A.P., and Maranas, C.D. (2004). OptStrain: a computational framework for redesign of microbial production systems. *Genome Res.* 14: 2367–2376.

127 Long, M.R., Ong, W.K., and Reed, J.L. (2015). Computational methods in metabolic engineering for strain design. *Curr. Opin. Biotechnol.* 34: 135–141.

128 Calmels, C., Arnoult, S., Ben Yahia, B. et al. (2019). Application of a genome-scale model in tandem with enzyme assays for identification of metabolic signatures of high and low CHO cell producers. *Metab. Eng. Commun.* 9: e00097.

129 Chong, W.P.K., Reddy, S.G., Yusufi, F.N.K. et al. (2010). Metabolomics-driven approach for the improvement of Chinese hamster ovary cell growth: overexpression of malate dehydrogenase II. *J. Biotechnol.* 147: 116–121.

130 Huang, Z. and Yoon, S. (2020). Identifying metabolic features and engineering targets for productivity improvement in CHO cells by integrated transcriptomics and genome-scale metabolic model. *Biochem. Eng. J.* 159: 107624.

131 Yusufi, F.N.K., Lakshmanan, M., Ho, Y.S. et al. (2017). Mammalian systems biotechnology reveals global cellular adaptations in a recombinant CHO cell line. *Cell Syst.* 4: 530–542.

132 Vodopivec, M., Lah, L., Narat, M., and Curk, T. (2019). Metabolomic profiling of CHO fed-batch growth phases at 10, 100, and 1,000 L. *Biotechnol. Bioeng.* 116: 2720–2729.

133 Kol, S., Ley, D., Wulff, T. et al. (2020). Multiplex secretome engineering enhances recombinant protein production and purity. *Nat. Commun.* 11: 1908.

134 Huang, Z., Xu, J., Yongky, A. et al. (2020). CHO cell productivity improvement by genome-scale modeling and pathway analysis: application to feed supplements. *Biochem. Eng. J.* 160: 107638.

135 Yeo, H.C., Park, S.Y., Tan, T. et al. (2022). Combined multivariate statistical and flux balance analyses uncover media bottlenecks to the growth and productivity of Chinese hamster ovary cell cultures. *Biotechnol. Bioeng.* 119: 1740–1754.

136 Pang, K.T., Hong, Y.F., Shozui, F. et al. (2024). Genome-scale modeling of CHO cells unravel the critical role of asparagine in cell culture feed media. *Biotechnol. J.* http://dx.doi.org/10.22541/au.171105421.18572875/v1.

137 Fouladiha, H., Marashi, S.A., Torkashvand, F. et al. (2020). A metabolic network-based approach for developing feeding strategies for CHO cells to increase monoclonal antibody production. *Bioprocess. Biosyst. Eng.* 43: 1381–1389.

138 Park, S.Y., Choi, D.H., Song, J. et al. (2023). Debottlenecking and reformulating feed media for improved CHO cell growth and titer by data-driven and model-guided analyses. *Biotechnol. J.* 18: 2300126.

139 Hong, J.K., Choi, D.H., Park, S.Y. et al. (2022). Data-driven and model-guided systematic framework for media development in CHO cell culture. *Metab. Eng.* 73: 114–123.

140 Ramos, J.R.C., Oliveira, G.P., Dumas, P., and Oliveira, R. (2022). Genome-scale modeling of Chinese hamster ovary cells by hybrid semi-parametric flux balance analysis. *Bioprocess. Biosyst. Eng.* 45: 1889–1904.

141 Wu, S.G., Wang, Y., Jiang, W. et al. (2016). Rapid prediction of bacterial heterotrophic fluxomics using machine learning and constraint programming. *PLoS Comput. Biol.* 12: e1004838.

142 Nandi, S., Subramanian, A., and Sarkar, R.R. (2017). An integrative machine learning strategy for improved prediction of essential genes in *Escherichia coli* metabolism using flux-coupled features. *Mol. Biosyst.* 13: 1584–1596.

143 Plaimas, K., Mallm, J.P., Oswald, M. et al. (2008). Machine learning based analyses on metabolic networks supports high-throughput knockout screens. *BMC Syst. Biol.* 2: 67.

144 Acencio, M.L. and Lemke, N. (2009). Towards the prediction of essential genes by integration of network topology, cellular localization and biological process information. *BMC Bioinf.* 10: 290.

145 Hasibi, R., Michoel, T., and Oyarzún, D.A. (2024). Integration of graph neural networks and genome-scale metabolic models for predicting gene essentiality. *npj Syst. Biol. Appl.* 10: 24.

146 Borger, S., Liebermeister, W., and Klipp, E. (2006). Prediction of enzyme kinetic parameters based on statistical learning. *Genome Inform.* 17: 80–87.

147 Heckmann, D., Lloyd, C.J., Mih, N. et al. (2018). Machine learning applied to enzyme turnover numbers reveals protein structural correlates and improves metabolic models. *Nat. Commun.* 9: 5252.

148 Ferreira, M.A.d.M., da Silveira, W.B., and Nikoloski, Z. (2024). Protein constraints in genome-scale metabolic models: data integration, parameter estimation, and prediction of metabolic phenotypes. *Biotechnol. Bioeng.* 121: 915–930.

149 Tsopanoglou, A. and Jiménez del Val, I. (2021). Moving towards an era of hybrid modelling: advantages and challenges of coupling mechanistic and data-driven models for upstream pharmaceutical bioprocesses. *Curr. Opin. Chem. Eng.* 32: 100691.

150 Chen, Y., Liu, X., Anderson, J.Y.L. et al. (2022). A genome-scale nutrient minimization forecast algorithm for controlling essential amino acid levels in CHO cell cultures. *Biotechnol. Bioeng.* 119: 435–451.

151 Schinn, S.M., Morrison, C., Wei, W. et al. (2021). A genome-scale metabolic network model and machine learning predict amino acid concentrations in Chinese hamster ovary cell cultures. *Biotechnol. Bioeng.* 118: 2118–2123.

152 Antonakoudis, A., Strain, B., Barbosa, R. et al. (2021). Synergising stoichiometric modelling with artificial neural networks to predict antibody glycosylation patterns in Chinese hamster ovary cells. *Comput. Chem. Eng.* 154: 107471.

2

Automated Approaches for the Development of Genome-Scale Metabolic Network Models

Emma M. Glass[1], *Deborah A. Powers*[1], *and Jason A. Papin*[1,2]

[1] *University of Virginia, Department of Biomedical Engineering, 415 Lane Road, Charlottesville, VA 22904, USA*
[2] *University of Virginia, Department of Biochemistry & Molecular Genetics, 1415 Jefferson Park Avenue, Charlottesville, VA 22908, USA*

2.1 Introduction

Metabolism is a system of biochemical pathways that convert nutrients into energy, complex biomolecules, and waste; together these conversions enable essential cellular functions for growth and homeostasis. Consequently, improving our understanding of how metabolism affects phenotypes for cellular function and growth can have important implications for health and disease. Two distinct but complementary approaches are often taken when studying metabolism: deep characterization of individual biochemical pathways as well as broad characterization of a whole metabolic system. A systems approach can be particularly valuable for linking the mechanisms of individual reactions and pathways to whole metabolic systems, probing the cellular genotype–phenotype relationships and guiding disease treatments.

Genome-scale metabolic network models (GSMs), as introduced in Chapter 1, are valuable tools for analyzing metabolic states in a variety of organisms. GSMs are a mathematical framework that represents the metabolic network of an organism; they are constructed using a set of gene-protein-reaction (GPR) rules formulated from the whole-genome sequence of an organism and a stoichiometric, mass-balanced matrix of all the metabolic reactions of an organism. There is a generally accepted and standardized protocol for generating metabolic models from genome sequencing data, including four main steps as follows: (1) Draft reconstruction, (2) refinement of reconstruction, (3) conversion of reconstruction into a compatible format, and (4) network evaluation [1]. Following reconstruction and curation, high-throughput omics data, such as proteomics, transcriptomics, and metabolomics, can be integrated with the reconstructed networks to create contextualized models with enhanced biological plausibility and predictability (Figure 2.1).

The first GSM was created manually in 1999 for the opportunistic bacterial pathogen *Haemophilus influenzae* [2]. By 2019, merely two decades later, there

Machine Learning and Big Data-enabled Biotechnology, First Edition. Edited by Hal S. Alper.

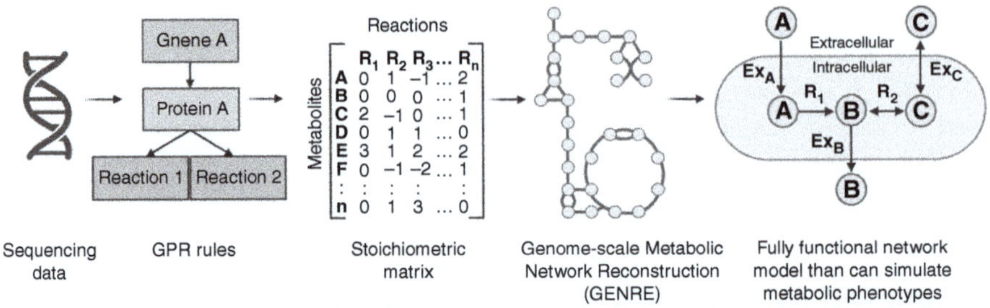

Figure 2.1 Schematic representation of the general pipeline for GSM development from sequencing data to a fully functional network model capable of simulating metabolic phenotypes. R = reaction, Ex = Exchange reaction.

were over 6,000 GSMs created through both manual methods and automated computational pipeline [3]. However, in the past five years, explosive development of automated GSM tools has led to the publication of several large collections of GSMs, including assembly of gut organisms through reconstruction and analysis (AGORA1) [4] (773 GSMs), AGORA2 [5] (7,302 GSMs), PATHGENN [6] (914 GSMs), EMBL GEMs [7] (5,587 GSMs), MetaGEM [8] (447 GSMs), and most recently APOLLO [9] (247,092 GSMs). Currently, including the APOLLO resource, there are over 250,000 GSMs of prokaryotic and eukaryotic cells, resulting in novel discoveries related to human health. For example, GSMs from the AGORA1 and AGORA2 collections have been used to discover links between the gut microbiome and colorectal cancer [10], elucidate how metabolism is modulated in microbial irritable bowel disease communities [11], predict personalized short-chain fatty acid production profiles in the human gut [12], growth dependencies of *Aspergillus fumigatus* strains on the lung microbiome [13], and many more clinically relevant applications. These examples highlight the potential impact GSMs can have on solving important biomedical problems.

In this chapter, we will first discuss the process for manually and automatically creating a GSM from a genome sequence. Then, we will provide an overview of tools in the field that have been developed for automated GSM construction, as well as tools that can be used to modify and visualize GSMs. Finally, we will discuss applications of large GSM collections that led to important biological discoveries, as well as current gaps in the field and future directions.

2.2 Manual GSM Creation

To generate a GSM from a genome sequence, it is necessary to consider necessary to consider GPR associations. Not all genes have a one-to-one relationship with their corresponding enzyme or metabolic reaction. For example, multiple genes could code for multiple protein subunits that come together to form one enzyme complex, or one protein could be used in multiple reactions. Therefore, it is necessary to precisely define the relationships between genes, proteins, and reactions.

After formulating the GPR rules, we must define the list of biochemical reactions in the network. Prior to the advent of genomics, these reactions were identified through experimental studies: an enzyme was isolated from an organism, and its function was demonstrated. This type of biochemical experiment provides the strongest evidence for the presence of a given metabolic reaction in an organism. However, using genomic tools, we can now functionally annotate whole-genome sequences based on DNA sequence homology to determine which reactions to include in the GSM. Since few organisms have extensive experimentally derived biochemical information available, GSM reconstruction today relies heavily on the annotated genome sequences to identify metabolic reactions present in a given organism.

After determining GPR-associations and identifying the metabolic reactions for inclusion in the GSM, we build a stoichiometric (S) matrix using the stoichiometric coefficients of the reactions that will comprise our network model. The columns of the S matrix represent individual reactions, the rows represent each compound, and the values in the S matrix are integers corresponding to the stoichiometric coefficients of each compound in each reaction. The architecture of the S matrix allows for the storage of both chemical information and network information, and explains how reactions within the network are interconnected. In general, the S matrix represents a reaction map that is topologically nonlinear: there are joint reactions between nodes in reversible reactions. Once completed, the S matrix represents the connected network of biochemical reactions in your organism.

After reconstructing the biological network in the form of the S matrix, we can compute possible functional network states directly related to cellular phenotypes. Cells are subjected to constraints arising from natural laws, environmental limitations, and physical space, limiting the range of possible behaviors. In our GSM, we can impose constraints on the network to obtain a set of solutions that are representative of a variety of functional states of the cell. We can capture different solution spaces and cellular phenotypes by altering what we are constraining and how stringently the constraints are imposed on the network.

2.3 Automated GSM Development

2.3.1 General Approach for Automated GSM Methods

In recent years, there has been significant effort in developing automated methods for GSM development based on the manual GSM construction framework described above (delineating GPR rules, selecting reactions, constructing the stoichiometric matrix, imposing constraints). Typically, automated network reconstruction pipelines follow several major steps for generating a high-quality draft model from a genome sequence: (1) BLAST search, (2) reference biochemical databases, (3) reaction inclusion, (4) gap-filling step (optional). The output of these automated tools is a GSM that is either simulation-ready or that can be further manually curated prior to simulation (Figure 2.2).

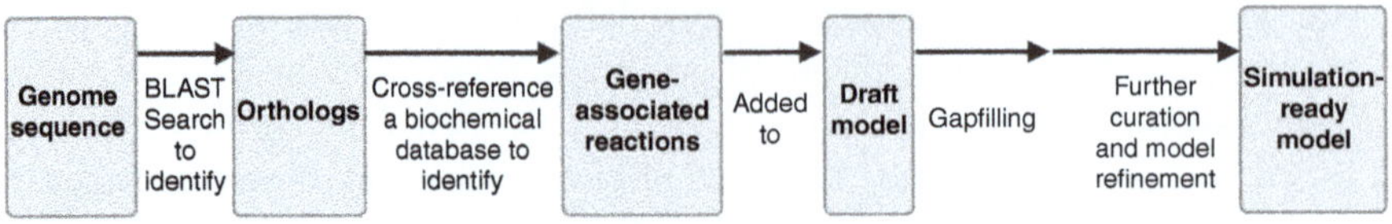

Figure 2.2 General workflow for automated GSM tools.

Automated GSM methods begin with either a raw or annotated genome sequence, which can be obtained from a variety of publicly available genome sequence databases such as GenBank [14], BV-BRC [15], and RefSeq [16]. The genome is then aligned to a biochemical database using a BLAST search algorithm to find orthologs. Subsequently, gene-associated reactions are identified using a chemical reaction database like KEGG [17], BiGG [18], or MetaCyc [19], which provides comprehensive mappings between genes, proteins, and metabolic reactions. These gene-associated reactions are then added to a draft GSM. Oftentimes, draft GSMs built only from gene-associated reactions will have network gaps, leading to impaired pathway connectivity and ultimately incorrect simulations and predictions. To correct these errors, some automated GSM construction methods include a gap-filling step to establish network connectivity by adding reactions to the GSM in either a probabilistic, machine-learning-based, optimization-based, or heuristic-based manner, depending on the method at hand. The final GSM is then typically saved in the Systems Biology Markup Language (SBML) format or JavaScript Object Notation (JSON) as these are community standards for cross-tool compatibility.

In this chapter, we will discuss methods for automated GSM development. We will discuss de-novo reconstruction tools including those that begin with (1) a raw genome, protein, or metagenome sequence, (2) annotated sequences, (3) biochemical databases, and (4) existing models. Additionally, we will discuss several tools that can be used to modify and visualize existing GSMs. A summary of all GSM methods discussed in this chapter is available in Table 2.1.

2.3.2 GSM Construction Tools

2.3.2.1 From Raw Sequences

ModelSEED ModelSEED [20] is a web-based platform for automated GSM reconstruction that is tailored to prokaryotes. The input to the ModelSEED tool is either an assembled genome sequence or a list of SEED subsystems. The genome sequence or subsystems are then annotated using the Rapid Annotations using Subsystems Technology (RAST) annotation server. The GPR associations for each reaction in the network are based on the functional role assigned to the genes during RAST annotation which are subsequently mapped to reactions in the ModelSEED biochemistry database. After GPR association, a preliminary reconstruction is generated by adding all reactions associated with one or more enzymes encoded in the organism's genome, as well as a set of spontaneous reactions that do not require enzymatic catalysis. After preliminary draft GSM construction, reactions are added back to the

Table 2.1 Table of tools for automated reconstruction and analysis.

	Input	Output	Implementation	Gap-filling method	Organism type	Reference	Year published
ModelSEED	Assembled genome sequence OR list of SEED subsystems	SBML, Excel	Web (with GUI)	"Auto-completion"; addition of a minimal reaction set	Prokaryote	[20]	2010
CoReCo	Genome sequence	SBML with enzyme probability annotations	Python (with command line interface)	Gapless by nature	Prokaryote, Fungi	[21]	2014
metaGEM	Raw metagenome sequence (FASTQ)	SBML GSM	Snakemake (with command line interface)	Gapless by nature, uses CarveMe () for model building	Prokaryote (suited for multispecies communities)	[8]	2021
Merlin	Raw amino acid or nucleotide FASTA	SBML GSM	Java (with GUI)	Identifies gaps but does not employ a gap-filling algorithm	Prokaryote, Eukaryote	[22]	2015
MS2/OMEGGA on KBASE	Genome sequence	SBML GSM	Web (with GUI)	Addition of a minimal reaction set for biomass production	Prokaryote, Plant	[23]	2018
CarveMe	Pre-annotated DNA or protein FASTA file	SBML GSM	Python	Gapless by nature	Prokaryote	[7]	2018
Reconstructor	Pre-annotated amino acid FASTA file	SBML	Python (with command line interface/ package importation)	Addition of a reaction set that minimizes total flux	Prokaryote	[24]	2023

(Continued)

Table 2.1 (Continued)

	Input	Output	Implementation	Gap-filling method	Organism type	Reference	Year published
pyFBA	Pre-annotated genome sequence	SBML	Python	Five optional methods for gap-filling	Prokaryote	[25]	2016
Raven	Existing GSM of organism closely related to organism being reconstructed OR list of KEGG or MetaCyc reactions	SBML, YAML, Excel	MATLAB	Optional gap-filling algorithm included	Prokaryote, Eukaryote, Fungi, Tissue	[26]	2018
AutoKEGGRec	KEGG organism ID	List of KEGG reactions, GPR rules, and annotation information stored as MATLAB variable	MATLAB	None	Prokaryote, Eukaryote	[27]	2018
rBioNet	List of reactions or existing GSM	Biochemical reconstruction in SBML, spreadsheet, or MATLAB variable format	MATLAB	None	Prokaryote, Eukaryote	[28]	2011
FAME	KEGG pathways or an existing model	SBML GSM, Excel	Web (With GUI)	None	Prokaryote	[29]	2012

MicrobesFlux	KEGG organism	SBML	Web (With GUI)	Optional gap-filling method available	Prokaryote	[30]	2012
MetaDraft	Protein FASTA sequence or GenBank file	SBML	Python (with GUI)	None	Prokaryote, Eukaryote	[31]	2020
GEMSiRV	Existing SBML GSM	SBML	Web (with GUI)	None	Prokaryote, Eukaryote	[32]	2012
SuBliMinaL	Existing SBML GSM	SBML	Java (with command line interface)	None	Prokaryote, Eukaryote	[33]	2011
PathwayTools	Existing SBML GSM	SBML	Java (with GUI)	None	Prokaryote, Eukaryote	[34]	2016
AuReMe	Genome GenBank files, or protein sequence FASTA, existing SBML GSM, pathway/gene databases, gene expression info, growth media composition, metabolic targets, biomass components, databases of reactions and metabolites in CSV format	SBML	Python (with command line interface)	Optional gap-filling method	Prokaryote, Eukaryote	[35]	2018
MEMOSys	Existing SBML GSM	SBML, Excel	Web (with GUI)	None	Prokaryote, Eukaryote, Fungi	[36]	2014
MetExplore	Existing SBML GSM	SBML	Web (with GUI)	None	Prokaryote, Eukaryote	[37]	2018

model in a gap-filling step called "Auto-completion." In this step, a minimal set of reactions is added through an optimization process so that the model can simulate growth *in silico*.

Following auto-completion, all models are simulation-ready; however, ModelSEED has additional built-in methods for model optimization. These methods include Biolog consistency analysis, gene essentiality consistency analysis, and GapFill/GapGen steps. These model optimization methods help ensure that the final GSM is as biologically plausible as possible.

CoReCo Comparative metabolic reconstruction (CoReCo) [21] is a python-based software that simultaneously generates GSMs of multiple related species and has been shown to be specifically useful for the generation of fungal GSMs. This comparative approach is key for generating GSMs of species that are phylogenetically distant from extensively studied model species, when the genome sequence has low coverage, or when only incomplete genome reference sequences are available. The first phase in CoReCo GSM creation is to quantify evidence of known enzymes in a species. Given the input of a genome sequence, homologous sequences are found by both BLAST and Global Trace Graph (GTG, for more distant homologs). The quantitative results of BLAST and GTG are integrated into a Bayesian network model, resulting in the probability of an enzyme in each present and hypothetical ancestral species. The topology of the Bayesian network is based on the phylogenetic tree, and conditional probability distributions describe the probability of an enzyme in each ancestral species. In the second phase, a metabolic model is constructed for the present and ancestral species. Using the enzyme probabilities computed in phase one, CoReCo creates a highly probable network of reactions in the KEGG namespace. Only highly probable reactions are included unless inclusion of a highly probable reaction would leave a gap in the model, in which case, a low-probability reaction would be included in its place. The result of CoReCo is a gapless GSM in SBML format containing annotations with enzyme probabilities.

metaGEM metaGEM [8] reconstructs GSMs directlyfrom raw metagenome sequences. The pipeline used in metaGEM enables modeling of communities of multiple species directly from metagenome sequences, with automation of all steps, from extraction of context-specific prokaryotic GSMs from metagenome-assembled genomes (MAGs) to community-level flux balance analysis (FBA) simulations. The metaGEM pipeline is designed and implemented in Snakemake (an open-source bioinformatics workflow engine) and can be run by end-users in the command-line interface.

The input to metaGEM is raw FASTQ files, and the first step in the metaGEM pipeline includes metagenome reconstruction. After assembly of the MAGs, FBA-ready GSMs are constructed using CarveME (see 3.2.2.1) and then quality checked using MEMOTE. Additionally, the Species METabolic ANAlysis (SMETANA) framework is integrated into the metaGEM pipeline, and can be used for sample-specific community interaction modeling. The pipeline can also perform automatic assignment of taxonomic classification to the reconstructed MAGs and GSMs, calculate relative/absolute abundance for generated MAGs, estimate growth rates for high-coverage MAGs, and perform pangenome analysis. The metaGEM pipeline is versatile, combining many existing high-quality metagenomic and GSM tools into one user-friendly pipeline.

Merlin Metabolic models reconstruction using genome-scale information (Merlin) [22] is a user-friendly Java-based GEM creation tool that can be used for any organism that has a sequenced genome. Merlin can annotate a prokaryotic or eukaryotic genome with both enzymatic and transport functions, and build a compartmentalized draft GSM with minimum user interaction.

The input to Merlin is a raw amino acid or nucleotide FASTA file. Enzymatic functions are then assigned to genes using the homology search tools BLAST and HMMER. An enzyme commission (EC) number and corresponding confidence level are then assigned to each gene based on a weighted score. The user can also define an EC inclusion threshold, allowing for more specific annotation curation. After enzymatic function assignment, transporter annotations and compartment predictions are determined. Following the genome annotation step, model building begins with loading an internal biochemical database, which includes compound, glycan, drug, reaction, module, pathway, and enzyme information from KEGG. GPR rules are defined for all genes in the genome by cross-referencing the biochemical database to identify gene-associated reactions. Finally, there is a module in Merlin that identifies gaps in the model but does not employ a gap-filling algorithm to fill those gaps. Ultimately, the final GSM created with Merlin includes only gene-associated reactions. The final SBML GSM with minimum information required in the annotation of models (MIRIAM) [38] annotations can be exported and used for simulation or modification with other tools.

MS2/OMEGGA on KBase KBase [23] is the United States Department of Energy Systems Biology Knowledgebase, which is an open-source, web-based, and user-friendly software and data platform. KBase has an internal database of over 90,000 microbial genomes from RefSeq, 50 plant genomes from Protozyme, >30,000 reactions and compounds from KEGG, BIGG, and MetaCyc. This data can be integrated and analyzed through over 160 applications in Kbase, which offer diverse scientific functionality for metagenome assembly, comparative genomics, and RNA-seq processing, among others. Most importantly, in the context of this chapter, KBase offers a user-friendly web-based app for generating GSMs from genome sequences called MS2/OMEGGA, which is based on the ModelSEED reconstruction pipeline (see 3.2.1.1).

In the MS2/OMEGGA pipeline, genomes are first annotated using the RAST annotation server, and then GPR rules are determined. The functionally annotated genes are then mapped to reactions in the ModelSEED database. Next, the draft model is created by adding all gene-associated reactions to a GSM. In the KBase reconstruction pipeline, there is an optional gap-filling step for identifying the minimal set of biochemical reactions to add to the draft model to enable the simulation of biomass synthesis on a specified media. After the model is created, it can be analyzed with other KBase tools or downloaded for analysis with external tools.

2.3.2.2 From Pre-annotated Sequences

CarveME CarveME [7] is a command-line, python-based tool for GSM generation using a top-down approach. In general, GSMs are created using a bottom-up approach where reactions are iteratively added to a draft model to gap-fill and enable the simulation of growth under different media conditions. The alternative top-down approach used by CarveMe begins with a universal metabolic model, which is "carved" to remove reactions and metabolites not predicted to be present in given organisms based on their genome sequence.

The CarveMe workflow begins with the creation of a universal draft model, which is a simulation-ready model consisting of all biochemical reactions in the BiGG database. The user input is a pre-annotated DNA or protein FASTA file; raw genome files are not supported. DIAMOND is used to align the input genome to the BiGG database to find all homologous genes. The homologs identified are mapped to reactions using GPR associations. Reactions are scored based on whether there is direct genetic evidence for the enzymes involved in the reaction. The universal model is then carved to create an organism-specific model by solving a mixed integer linear programming (MILP) optimization problem that maximizes the presence of high-score reactions (those with genetic evidence) and minimizes the presence of low-score reactions (those without genetic evidence). The result is a "carved" organism-specific model. After generating the organism-specific model, there are options for post-reconstruction modifications, including ensemble generation of models based on random weighting of reactions without genetic evidence, addition of experimental constraints, and media-based gap-filling. The output of CarveMe is a simulation-ready GSM in SBML format, the community standard.

Reconstructor Reconstructor [24] is a Python-based tool for GSM construction from an annotated genome sequence that creates GSMs that are directly compatible with COBRApy (does not require the use of additional compatibility modules). Reconstructor GSMs are built using the ModelSEED namespace, allowing them to be used in conjunction with models created from a variety of other tools.

Reconstructor begins by taking a pre-annotated protein FASTA file and aligning it to the KEGG database using DIAMOND BLASTp to identify all homologous genes which are then mapped to reactions in the ModelSEED namespace. These gene-associated reactions are then added to a draft GSM. Reconstructor then uses a parsimonious FBA gap-filling approach to identify the set of non-gene-associated reactions that minimizes flux through the network, resulting in the most biologically parsimonious network. After pFBA-based gap-filling, a subsequent media-based gap-filling step can be performed to add additional media-specific transport reactions to the GSM. Reconstructor can take three types of input: (1) an annotated amino acid FASTA file, (2) the output of a DIAMOND BLASTp search, and (3) an existing SBML GSM for further media-based gap-filling, allowing for reconstructor to be integrated into computational pipelines in a variety of capacities. The result of reconstructor is a simulation-ready SBML file.

pyFBA pyFBA [25] is a Python-based open-source software package where functional annotations can be used to build GSMs ready for simulation that are backed by the ModelSEED biochemistry database. No specific annotation type is required for pyFBA (RAST, PROKKA, BG7, Blast2GO, BASys annotations are all valid), but it does require annotation to connect genes to enzymes to reactions for defining GPR rules.

The pyFBA algorithm begins by converting functional roles provided by the genome annotation information into reactions with the use of a biochemistry database. If using RAST annotations, the SEED database provides a connection between function, enzyme complexes, reactions, and compounds in a consistent namespace. pyFBA identifies ModelSEED reactions and compounds that map to the functional roles provided by the annotations. This list of gene-associated reactions and compounds is converted to a stoichiometric matrix, then unconstrained export or secretion reactions are defined and added to the stoichiometric matrix. This process results in a draft model that is ready for FBA simulation. If the results of the FBA simulation do not produce a growing model, the model can be gapfilled using a variety of methods: (1) a list of 109 reactions known to be in every model are first tried, (2) media-based exchange reactions are added, (3) add reactions that are proposed to complete a subsystem in the model, (4) add reactions that connect orphan compounds, and (5) add reactions that are known to be present based on annotations of closely related organisms. The result of the pyFBA iterative gapfilling process is a growing, simulation-ready, SBML GSM.

2.3.2.3 From Reaction Database Information

Raven Raven [26] is a MATLAB toolbox for GSM reconstruction, curation, constraint-based modeling, and simulation. Using Raven, GSMs of a variety of organisms, including bacteria, archaea, the human gut microbiome, eukaryotic microalgae, parasites, fungi, and human tissues, can be created. Raven uses either the MetaCyc database (which contains only experimentally verified pathways) or the KEGG database for GSM reconstruction. Raven can create GSMs (1) based on protein homology to an existing template GSM, or (2) *de novo* using reaction databases.

The first approach requires an existing high-quality GSM of an organism that is phylogenetically closely related to the organism you are reconstructing. Protein homology is inferred using a bidirectional BLASTp, and the results from the BLASTp are used to build the new draft GSM. Alternatively, *de novo* reconstructions can be built using either the KEGG or MetaCyc databases. KEGG-based reconstruction uses either KEGG-supplied annotations (over 5,000 genomes are available) or queries protein sequences for similarity to hidden Markov models trained on the genes in KEGG. If building a MetaCyc-based GSM, protein sequences are queried with BLASTp for homology to enzymes curated in MetaCyc. Regardless of the method, the draft model that results is further curated with a gap identification and gap-filling method to result in a high-quality network reconstruction that is simulation-ready.

AutoKEGGRec Automated generation of genome-scale metabolic draft reconstructions based on KEGG (AutoKEGGRec) [27] is a MATLAB-based tool that creates draft GSMs from KEGG organism IDs and integrates seamlessly with COBRA Toolbox v3. The draft GSM created by AutoKEGGRec is not intended to be fully functional. Rather, it is a starting point for further manual curation and simulations with other tools in the COBRA Toolbox.

AutoKEGGRec begins by providing a KEGG organism ID as input, which then queries the link between EC numbers and the corresponding genes from KEGG for each organism. Then, the linkage between reactions and EC numbers is retrieved. Every reaction in KEGG that is associated with the queried KEGG organism ID is retrieved and related to the corresponding genes in the target organism. At this point, each reaction, along with its annotation and the corresponding compound annotations, is downloaded from KEGG to be used in draft model reconstruction. The output of AutoKEGGRec is a list of KEGG reactions, associated GPR rules, and annotation annotation information, which in a MATLAB structure variable. This model can then be used with the COBRA Toolbox for gap-filling and curation efforts. Importantly, this model is not intended to be fully functional, but a first draft network reconstruction based on the KEGG database, which is created through a set of simple procedures.

rBioNet rBioNet [28] is a MATLAB package that implements QA/QC measures to generate high-quality biochemical network reconstructions that can be readily converted to COBRA Toolbox-compatible GSMs. The rBioNet pipeline consists of three parts: (1) a metabolite creator and corresponding database (MetDB), (2) a reaction creator with corresponding reaction database (RxnDB), and (3) a reconstruction creator.

Metabolites can be created in three ways: (1) manual addition of metabolite information, (2) loading of a tab-delimited file containing all information, and (3) loading metabolite information from other GSM reconstructions. Secondly, reactions can be added in the same three ways. Importantly, reactions can only contain metabolites contained in MetDB. Then, the user can create a reconstruction based on the reaction information provided in the RxnDB, and import metadata to the reconstruction if necessary. After creating the biochemical network reconstruction, the user can assign each reaction a gene association through GPRs. To accomplish this, the user must provide a tab-delimited file consisting of the organism's gene annotation information. Following GPR associations, the reconstruction can be benchmarked to find dead-end metabolites, and finally exported in a variety of formats, including SBML, a spreadsheet, or a data structure compatible with COBRA Toolbox. Importantly, rBioNet generatesi8u a biochemical network reconstruction, not a model. The reconstruction can be constrained using the COBRA Toolbox to generate a condition-specific, simulation-ready model.

FAME Flux analysis and modeling environment (FAME) [29] is a web-based one-stop-shop stoichiometric modeling tool that incorporates creation, editing, running, analysis/visualization into a user-friendly interface. While FAME is no longer maintained as a functioning tool, certain aspects of this tool are worth discussing to inspire the next generation of modern/functioning tools with similar characteristics.

In the FAME workflow, the user can either generate their own model based on KEGG information or upload an existing model. When generating a new model, the user can select a subset of pathways from KEGG to avoid, including unnecessary reactions that may be present in existing GEMS. FAME also stores KEGG ID information to find more information from KEGG if the need arises during the generation process. After model loading or generation, FAME has incorporated tools to edit flux bounds, edit reactions, recognize orphan metabolites, and assign reactions to compartments. Once the model is curated, FAME also has built-in optimization and analysis tools, including FBA and flux variability analysis (FVA), which are supported by PySCeS-CBM [39]. Additionally, the gene association workbench allows the user to simulate knockout mutants by taking advantage of the gene association information encoded in the model metadata. Finally, FAME incorporates a pathway visualization module to more deeply explore the biochemistry involved in the model. One of the most attractive features of FAME is the web-based point-and-click interface, which makes GSM creation and analysis more accessible to a wider range of users with both computational and non-computational scientific backgrounds.

MicrobesFlux MicrobesFlux [30] is a web-based platform for semiautomatic generation of GSMs for annotated microorganisms in the KEGG database. MicrobesFlux can download the enzymatic reactions, metabolites, and metabolic networks associated with ~1,200 unique species from the KEGG database and convert the metabolic network information into a draft GSM. The draft model can be converted to a constraint-based flux model to predict carbon fluxes in microbial metabolism, and can ultimately be exported in SBML format. Importantly, MicrobesFlux is an installation-free platform that enables accessibility of metabolic network modeling to biologists without existing programming knowledge to develop models for organisms in the KEGG database.

The MicrobesFlux pipeline is dependent on the KGML (organism-specific metabolic networks) and KEGG LIGAND (enzymatic reactions and metabolites) databases for model generation. To generate a functional GSM, organism-specific metabolic networks are first downloaded from KEGG, which can then be formulated as an FBA model for simulation. MicrobesFlux also provides a suite of tools for model customization and gap-filling; however, it is important to note that the GSMs created with MicrobesFlux are still considered draft models. After GSM customization, constraint-based flux analysis is performed using optimization solvers, and the resulting flux distributions and model are recorded in SBML format, which is provided to the user. Overall, MicrobesFlux can perform three major tasks: (1) high-throughput metabolic model generation, (2) customized metabolic model drafting, and (3) constraint-based flux analysis in steady and dynamic states.

2.3.2.4 Based on Existing GSMs

MetaDraft MetaDraft [31] is a downloadable Python-based software with a graphical user interface (GUI) that creates a GSM based upon a previously manually curated model. MetaDraft has an extensive user interface that allows for informed model curation through visualization of gene BLAST scores, reactions, and metabolites, which can be edited and removed from the reconstruction.

MetaDraft takes a protein sequence in the form of a FASTA or GenBank file as input. MetaDraft works by comparing the protein input to a list of template models curated from the BiGG database. The user can select a set of template models, which will then form the basis of the metaproteome. Then, the user can blast their input protein sequence against the metaproteome created by the set of chosen template models. The BLAST results can then be viewed to select gene matches from the metaproteome that the user would like to include in their reconstruction based on score values. Each gene that is selected has an existing GPR association, and the user can view and edit the gene-associated reactions for further model curation. The reconstruction can then be converted and exported in SBML format for further analysis and simulation with other tools.

GEMSiRV GEMSiRV [32] is a freely available, downloadable software platform for genome-scale metabolic simulation, reconstruction, and visualization. GEMSiRV comes with a repository of existing metabolic models, reference databases, metabolic maps, and analysis toolboxes, which can be used as the starting point for network construction and simulation. GEMSiRV can run on local computers/servers without relying on a web-based interface. Importantly, GEMSiRV can also import existing models for further curation and gap-filling, and can incorporate experimental data for model contextualization.

GEMSiRV consists of three main modules: (1) network reconstruction, (2) simulation, and (3) visualization. A new GSM can be built using existing GSMs of closely related species as a reference. Given a user-defined list of orthologous genes, GEMSiRV will extract corresponding reactions that conform to the gene-protein reaction associations defined in the reference model. Building a GSM based on an existing model helps reduce the effort required for manual model construction. After construction of the draft GSM, GEMSiRV has built-in simulation modules which users can use for biological hypothesis generation, including dead-end metabolite identification, objective optimization, FVA [40], robustness analysis, gene/reaction essentiality analysis, and gene deletion analysis. Finally, the visualization module allows for editing and visualizing the metabolic networks of interest based on KEGG pathways.

SuBliMinaL The SuBliMinaL [33] toolbox helps facilitate the GSM reconstruction process by providing independent modules to support common reconstruction tasks like generating a draft reconstruction, mass/charge balancing reactions, adding transport reactions, a biomass function, and converting to the proper format for analysis with third-party tools. The modules in SuBliMinaL can be chained to create a complete reconstruction pipeline or just be used to modify an existing model in SBML format.

Initial pathways for a given organism are extracted from KEGG and MetaCyc using the KEGG-extract and MetaCyc-extract modules. These pathways are then converted into SBML draft models, which can be further modified and curated. Alternatively, an existing model can be loaded into the software and modified through various modules. Taking the organism-specific pathways, SuBLiMinaL performs an annotation step, which cross-references model components with databases such as GO [41, 42], ChEBI [43], and UniProt [44], updating the model for downstream processing. SuBLiMinaL also supports merging KEGG- and MetaCyc-based models to create one consolidated model. There is also a protonate module that predicts metabolite pKa and charge states, subsequently updating these in the model. The Balance module automates the process of mass and charge balancing of reactions through a mixed integer linear programming approach. Since SuBliMinaL generates non-compartmentalized reconstructions, the compartmentalized module predicts the compartment where metabolites reside based upon protein localization information from UniProt. The transport reaction module adds import reactions to the model to allow the uptake of metabolites from the growth medium. Finally, the biomass module assigns a biomass reaction to the reconstruction, and performs reformatting to the model that allows FBA simulations with the COBRA Toolbox. The result of the pipeline is a simulation-ready GSM that can be used for biological hypothesis generation.

2.3.2.5 GSM Modification and Visualization Tools

PathwayTools PathwayTools [34] is a software environment designed for managing, analyzing, simulating, and visualizing integrated genome, pathway, and regulatory data collections. PathwayTools supports the creation and curation of organism-specific databases, including pathway-genome databases (PGDBs), which include information on genes, proteins, as well as metabolic and regulatory networks. PathwayTools includes a MetaFlux module, which allows for the construction and execution of steady-state metabolic flux models. MetaFlux simulates gene and reaction knockouts, which allows users to explore metabolic dynamics and dependencies within an organism. PathwayTools is also equipped with an API tool for executing more complex queries and data mining, which allows users to execute more custom analyses. One of the most important aspects of the PathwayTools program is the ability for users to visualize entire networks through cellular overview diagrams, and it even allows for mapping gene expression data or other experimental data onto the network. The ability of PathwayTools to integrate genomic, chemical, and regulatory information makes it a powerful tool for enriched metabolic modeling and visualization.

AuReMe AuReMe (Automatic Reconstruction of Metabolic Models) [35] is a pipeline tool and workflow controller that supports flexible reconstruction and customization of GSMs by integrating a suite of existing academic-grade tools. While AuReMe does not generate models, it enables users to compile tools into customizable pipelines tailored to different data sources and project requirements. Importantly, AuReMe maintains a complete record of methods used and metadata for pipeline reproducibility. This tool supports key GSM creation practices like annotation, gap-filling, manual curation, and various model analysis and simulation methods, making AuReMe adaptable for automated and manual modeling approaches. AuReMe works with prominent metabolic databases like BiGG, ModelSEED, and MetaCyc, and can also access KEGG if the user has a license available. Additionally, AuReMe offers broad input flexibility, accepting genome GenBank files, protein sequence FASTA files, metabolic models in SBML format, pathway/genome databases, gene expression data, growth media composition, metabolic targets, biomass components, and databases of reactions and metabolites in CSV format. AuReMe's comprehensive integration and export functionality, which includes exporting models in SBML format for compatibility with the COBRA toolbox, allows users to construct detailed, transparent, and easily shareable metabolic models tailored to diverse research needs.

MEMOSys MEMOSys (metabolic model research and development system) [36] is a comprehensive, web-based platform that facilitates the management, storage, and development of GSMs. MEMOSys is designed to enhance model modification workflows rather than generating models from scratch, making it ideal for users who work with evolving or derivative metabolic models. It provides a robust database for GEMs, enabling users to create hierarchical model structures where changes can propagate through related models, allowing for efficient updating and streamlined version control across model generationsOne major feature of MEMOSys is its capability to annotate model components, including reactions, metabolites, genes, and cellular compartments, using the minimum information required in the annotation of biochemical models (MIRIAM) standard [17]. This integration includes links to external databases, enhancing model utility by allowing researchers to leverage external biochemical, genetic, and regulatory information. The database also supports metadata extraction from SBML files, making it easier to incorporate and store detailed annotations.

MEMOSys currently hosts 20 publicly available GSMs, which can be exported in standardized formats for further analysis with external tools. The platform's recent updates have expanded compatibility, allowing installation on Linux, Mac, and Windows operating systems, and it is now available as a fully configured virtual machine image, reducing setup complexity for users. This virtual image configuration not only simplifies deployment but also makes MEMOSys more user-friendly, offering researchers a ready-to-use environment for constructing, modifying, and managing metabolic models based on existing GSMs.

MetExplore MetExplore [37] is a collaborative, web-based application that supports the curation, annotation, visualization, and sharing of draft metabolic models. Designed to enhance previously reconstructed models, MetExplore provides users with advanced tools for refining and annotating metabolic networks, improving their accuracy and utility for downstream analysis. Researchers can perform key analytical functions directly within the platformincluding FBA, FVA, and assessments of gene and reaction essentiality, which facilitate deeper exploration of metabolic dependencies and bottlenecks. MetExplore also offers a comprehensive visualization system, enabling users to create graphical representations of selected subnetworks and to visualize GPR relationships within the network. This visual feature is particularly useful for mapping out specific areas of interest and understanding the functional roles of genes and reactions in the model. By integrating curation, annotation, analysis, and visualization capabilities in a single platform, MetExplore supports collaborative metabolic model refinement, making it an invaluable tool for researchers aiming to share, improve, and explore draft metabolic models collaboratively.

Table of 20 computational tools that can be used for automated genome-scale metabolic network reconstruction and further analysis, curation, or pathway visualization. Strengths and weaknesses of each tool are outlined in the main text. Tools are listed in the table in the order that they appear in the text.

2.4 Applications of Automatically-Generated GSM Collections

Automated GSM tools enable easy, accessible reconstruction of metabolic networks by drastically reducing the time required for reconstruction and eliminating the need for extensive manual curation. Beyond the ease-of-use, automated GSMs enable scientists to ask high-level questions and generate biological hypotheses about large groups of organisms (e.g. all pathogens, all members of the gut microbiome), which would be difficult to achieve with manually curated GSMs of individual organisms. For example, automated GSM methods can enable large-scale community modeling, allowing for predictions about interactions between community members. Recently, several large collections of automatically generated GSMs have been created and have since been used to address a variety of biological questions and pose biological hypotheses that could not have otherwise been addressed with individual manually curated GSMs.

2.4.1 AGORA1 – 773 GSMs

The AGORA collection [4] consists of 773 automatically generated GSMs of human gut bacteria, representing 205 genera and 605 species. This collection of GSMs was created with the purpose of inferring the metabolic repertoire of the whole human gut microbiome. This collection of automatically generated GSMs was one of the first, created in 2016. Since then, the AGORA collection has been a keystone resource in the metabolic network modeling community.

The AGORA resource has been used in many studies that address biological questions related to the microbiome and gut health. One interesting application used GSMs from the AGORA collection to identify metabolic functions of gut microbes associated with efficacy of tumor necrosis factor agonists in patients with irritable bowel disease. The 773 GSMs in the AGORA database were used to predict metabolic biochemical interactions between species in the human gut microbiome of inflammatory bowel disease (IBD) and rheumatic diseases compared to healthy controls. Using the GSMs, they determined that antagonistic interactions in the gut microbiome were reduced at the beginning of IBD treatment and restored toward the end of treatment. Furthermore, increased competition among species was observed in inflammatory conditions, suggesting that the gut microbiome community is less stable in an inflammatory state. The results of this study were partially validated with clinical data, demonstrating the use of the AGORA resource for making biologically meaningful computational predictions [45].

Another study used the AGORA collection to identify metabolic functions in the gut microbiome associated with Parkinson's disease. 243 GSMs from the AGORA database were selected based upon the species presence in stool samples of Parkinson's patients and healthy controls. Using the GSMs and community modeling techniques, they identified altered metabolite secretion potentials in Parkinson's disease patients. Specifically, they observed an age-dependent association with Parkinson's disease patients and Cysteine-Glycine production. Although the correlation was not particularly strong, this study represents a step toward a systems-biology approach to identifying the interplay between the gut microbiome and neurological disorders [46].

2.4.2 EMBL GEMs – 5,587 GSMs

The EMBL GEMs (genome-scale models) collectionconsists of 5,587 models of bacterial metabolism created with the CarveMe tool [7]. The GSMs in this collection represent the metabolic diversity across bacterial life at the strain level. At the time of its creation (2017), this was the largest collection of GSMs. Since 2017, the EMBL collection has been used to explore a variety of biological topics, two of which we discuss below.

The CarveMe-generated EMBL GEMs have been previously used in community metabolic modeling applications. A study published in 2024 [47] used 331 EMBL GEMs of gut bacteria to identify competitive, neutral, and mutualistic interactions within the gut microbiome community and between the microbiome and intestinal epithelial cells. Using a multi-objective optimization approach, they demonstrated that two organisms in nutritional co-limitation can attain better fitness together than alone by taking advantage of mutualistic interactions. Additionally, they demonstrated that diet plays an important role in modulating this mutualistic interaction, emphasizing the implication of diet composition in the function of the gut microbiome.

The EMBL GEM collection has also been used in another study to gain a deeper understanding of tryptophan metabolism across members of the gut microbiome. In this study, all 5,587 GSMs in the EMBL collection were used to develop a comprehensive database and analytics platform that is dedicated to tryptophan metabolism within a host context by using metabolism information from the EMBL GSMs. The database of tryptophan metabolism information, TrpNet, contains data on tryptophan metabolism involving 130 reactions, 108 metabolites, and 91 enzymes across 1,246 human gut bacterial species and 88 mouse gut bacterial species [48].

2.4.3 MetaGEM – 447 GSMs

During the validation of the MetaGEM tool (3.2.1.3), the authors generated a collection of 447 metagenome-based GSMs that are available on a public Zenodo repository. These 477 GSMs were created based on metagenome shotgun samples from five separate previously published metagenomic studies. While the MetaGEM tool has been used by many in the field, the GSM collection is relatively underutilized. Nonetheless, the creators of MetaGEM used their collection of GSMs to model human gut communities to validate their pipeline.

In this validation study, the collection of GSMs was used to investigate potential microbial metabolic interactions in the gut microbiome of healthy individuals and individuals with type 2 diabetes. They identified 22 compound exchanges that were significantly different between disease groups. Furthermore, they identified 27 donor microbes and 27 receiver microbes that had had significantly different cross-feeding interactions than the rest of the community members in diabetes vs healthy control patients (such as *Bifidobacterium*, *Faecalibacterium*, *Roseburia*, and *Ruminococcus* species). These results are consistent with some previous experimental results, suggesting that this collection of microbes was able to discern metabolic differences in diabetes and healthy control microbiomes. Moreover, by directly modeling community-level microbiomes, they identified potential interaction differences in type 2 diabetes microbiomes independent of abundance estimates, emphasizing the need for models that simulate functions of the microbiome[8].

2.4.4 AGORA2–7,302 GSMs

The AGORA2 collectionis an expanded version of the AGORA resource, containing 7,302 GSMs of human gut microorganisms, including 1,738 species and 25 phyla. This AGORA2 collection significantly expands upon the number of models present in the original AGORA collection. This collection of bacterial GSMs has been used in a variety of studies to address biological questions related to the gut microbiome, the lung microbiome, the Atlantic salmon gut microbiome, obesity, colorectal cancer, and food microbiology, among others.

As an example, one study published in 2023 used the *A. fumigatus* strains present in the AGORA2 database to gain a better understanding of their growth dependencies in the lung microbiome. More specifically, they used the AGORA2 GSMs to identify core and unique reactions across the *A. fumigatus* genome, identify reactions that have differential metabolic activity between environmental and clinical strain-specific GSMs, and corroborate experimental validation data. Using the GSMs, the authors ultimately determined that aromatic amino acids provided by other lung bacteria might have beneficial growth implications for *A. fumigatus* in the context of cystic fibrosis. Additionally, they determined that there are 53 reactions that have significant flux changes in cystic fibrosis conditions, and several of these reactions are related to the shikimate pathway. Together, these results suggest that aromatic amino acids are possibly important for fungal colonization and have implications in the severity of cystic fibrosis lung disease [13].

2.4.5 PATHGENN – 914 GSMs

The PATHGENN collectioncontains 914 GSMs of all known human bacterial pathogens with high-quality genome sequences, including 345 species and 9 phyla. This collection was created using the Reconstructor tool and is currently the only publicly available collection of pathogenic GSMs. The PATHGENN collection was used to identify uniquely targetable genes in bacterial species sharing a physiological niche. Beyond this application, the PATHGENN collection has the potential to be a valuable resource in the future for antimicrobial identification, computational simulation of pathogens that are difficult to grow *in vitro*, and pathogen pangenome analysis [7].

2.5 Future Directions for the Field of Automated GSM Development

Automated GSM tools have been valuable for fast generation of GSMs of thousands of bacterial species, enabling rapid biological hypothesis generation and metabolic simulations. While GSM construction has come a long way since the manual creation of the *H. influenzae* GSM in 1999, there is still room for improvement. Current automated GSM methods enable rapid creation of GSMs but do not always produce GSMs that are high-quality, simulation-ready, compatible with COBRA-based analysis tools, or are user-friendly. Because of these shortcomings, there is an increasing interest in making these automated tools generate higher-quality GSMs by integrating machine learning techniques into model generation, incorporating an iterative curation process, and improving maintenance of tools, standardization of GSM collections, and quality assurance methods.

Integration of machine learning techniques into automated GSM methods could allow for higher-quality automated GSM reconstruction. Most often, machine learning has been integrated into GSM analysis tools for metabolic flux phenotype predictions after the generation of the models. However, few studies have worked to integrate machine learning methods into the GSM creation step. One study has used unsupervised machine learning, like PCA, on metabolomics data to identify key metabolites that can be used as constraints for GSM construction [49]. In the future, there could be an opportunity to use deep learning methods to integrate multi-omic data into GSMs in an automated manner to increase the biological relevance and quality of generated GSMs. However, this idea has not been incorporated into any existing automated GSM development tools.

Another way to improve the quality of automatically generated GSMs is by using iterative model curation processes through automated experimental validation.

Carbon-source utilization experiments using Biolog plates have been one way to corroborate computational model predictions with *in vitro* growth data. While this approach provides valuable information for model curation, it is not an automated process. A recently developed approach uses AI systems for performing autonomous experiments to more comprehensively map microbial metabolism. This method enables high-throughput experimental data collection through AI-driven experiments [50]. This method could be used to develop an experimentally driven approach for automatically generating high-quality, experimentally validated, physiologically relevant GSMs. Integrating model curation through experimentation into automated GSM tools could be the key to generating the next generation of high-quality models of metabolism.

While many tools for automated GSM development have been created, some of these tools are poorly maintained and currently nonfunctional. Furthermore, not all automated GSM tools create models with the same output file format, often requiring additional compatibility modules to use analysis tools like COBRA Toolbox or COBRApy. These points underscore the need for community standards [51]. The community standards for quality assurance including MEMOTE and FROG have been invaluable for model benchmarking allowing researchers to assess the quality of the models they are generating with automated GSM methods. Developing additional standards within or beyond MEMOTE and FROG for tool maintenance and standard GSM output file formats would allow researchers to have access to a wider variety of functional tools and be able to ensure that the models they are generating are simulation-ready models.

2.6 Conclusion

The first GSM (*H. influenzae*) was created manually in 1999. Since then, thousands of bacterial and eukaryotic GSMs have been created both manually and with automated GSM tools. Automated GSM creation tools have proven to be invaluable for generating large, high-quality collections of GSMs rapidly, which can then be used for biological hypothesis generation or further in silico analysis. The integration of machine learning methods into automated GSM methods, iterative model curation processes like AI-driven automated experimentation, and community standards for tool maintenance and model format will allow these methods to become more accurate and more powerful.

References

1 Thiele, I. and Palsson, B. (2010). A protocol for generating a high-quality genome-scale metabolic reconstruction. *Nat. Protoc.* 5 (1): 93–121.

2 Edwards, J.S. and Palsson, B.O. (1999). Systems properties of the *Haemophilus influenzae* Rd metabolic genotype*. *J. Biol. Chem.* http://www.jbc.org.

3 Gu, C., Kim, G.B., Kim, W.J. et al. (2019). Current status and applications of genome-scale metabolic models. *Genome Biol.* 20: 121.

4 Magnúsdóttir, S., Heinken, A., Kutt, L. et al. (2017). Generation of genome-scale metabolic reconstructions for 773 members of the human gut microbiota. *Nat. Biotechnol.* 35 (1): 81–89.

5 Heinken, A., Hertel, J., Acharya, G. et al. (2023). Genome-scale metabolic reconstruction of 7,302 human microorganisms for personalized medicine. *Nat. Biotechnol.* 41 (9): 1320–1331.

6 Glass, E.M., Dillard, L.R., Kolling, G.L. et al. (2024). Niche-specific metabolic phenotypes can be used to identify antimicrobial targets in pathogens. *PLoS Biol.* 22 (11): e3002907.

7 Machado, D., Andrejev, S., Tramontano, M., and Patil, K.R. (2018). Fast automated reconstruction of genome-scale metabolic models for microbial species and communities. *Nucleic Acids Res.* 46 (15): 7542–7553.

8 Zorrilla, F., Buric, F., Patil, K.R., and Zelezniak, A. (2021). MetaGEM: reconstruction of genome scale metabolic models directly from metagenomes. *Nucleic Acids Res.* 49 (21): e126.

9 Heineken, A., Hulshof, T.O., Nap, B. et al. (2025). APOLLO: a genome-scale metabolic reconstruction resource 247,092 diverse human microbes spanning multiple continents, age groups, and body sites. *Cell Syst.* 16 (2): 101196.

10 Ternes, D., Tsenkova, M., Pozdeev, V.I. et al. (2022). The gut microbial metabolite formate exacerbates colorectal cancer progression. *Nat. Metab.* 4 (4): 458–475.

11 Yilmaz, B., Juillerat, P., Øyås, O. et al. (2019). Microbial network disturbances in relapsing refractory Crohn's disease. *Nat. Med.* 25 (2): 323–336.

12 Quinn-Bohmann, N., Wilmanski, T., Sarmiento, K.R. et al. (2024). Microbial community-scale metabolic modelling predicts personalized short-chain fatty acid production profiles in the human gut. *Nat. Microbiol.* 9 (7): 1700–1712.

13 Mirhakkak, M.H., Chen, X., Ni, Y. et al. (2023). Genome-scale metabolic modeling of *Aspergillus fumigatus* strains reveals growth dependencies on the lung microbiome. *Nat. Commun.* 14 (1): 4369.

14 Benson, D.A., Cavanaugh, M., Clark, K. et al. (2012). GenBank. *Nucleic Acids Res.* 41 (Database issue): D36–D42.

15 Olson, R.D., Assaf, R., Brettin, T. et al. (2023 Jan 6). Introducing the Bacterial and Viral Bioinformatics Resource Center (BV-BRC): a resource combining PATRIC, IRD, and ViPR. *Nucleic Acids Res.* 51 (D1): D678–D689.

16 Pruitt, K.D., Tatusova, T., and Maglott, D.R. (2006). NCBI reference sequences (RefSeq): a curated non-redundant sequence database of genomes, transcripts and proteins. *Nucleic Acids Res.* 35 (1): D61–D65.

17 Kanehisa, M. and Goto, S. (2000). KEGG: kyoto encyclopedia of genes and genomes. *Nucleic Acids Res.* 28 (1): 27–30.

18 Schellenberger, J., Park, J.O., Conrad, T.M., and Palsson, B.O. (2010). BiGG: a Biochemical Genetic and Genomic knowledgebase of large scale metabolic reconstructions. *BMC Bioinf.* 11: 213.

19 Caspi, R., Billington, R., Keseler, I.M. et al. (2019). The MetaCyc database of metabolic pathways and enzymes – a 2019 update. *Nucleic Acids Res.* 48 (D1): D445–D453.

20 Henry, C.S., Dejongh, M., Best, A.A. et al. (2010). High-throughput generation, optimization and analysis of genome-scale metabolic models. *Nat. Biotechnol.* 28 (9): 977–982.

21 Pitkänen, E., Jouhten, P., Hou, J. et al. (2014). Comparative genome-scale reconstruction of gapless metabolic networks for present and ancestral species. *PLoS Comput. Biol.* 10 (2): e1003465.

22 Dias, O., Rocha, M., Ferreira, E.C., and Rocha, I. (2015). Reconstructing genome-scale metabolic models with merlin. *Nucleic Acids Res.* 43 (8): 3899–3910.

23 Arkin, A.P., Cottingham, R.W., Henry, C.S. et al. (2018). KBase: the United States department of energy systems biology knowledgebase. *Nat. Biotechnol.* 36: 566–569.

24 Jenior, M.L., Glass, E.M., and Papin, J.A. (2023). Reconstructor: a COBRApy compatible tool for automated genome-scale metabolic network reconstruction with parsimonious flux-based gap-filling. *Bioinformatics* 39 (6): btad367.

25 Cuevas, D.A., Edirisinghe, J., Henry, C.S. et al. (2016). From DNA to FBA: how to build your own genome-scale metabolic model. *Front. Microbiol.* 7: 907.

26 Wang, H., Marcišauskas, S., Sánchez, B.J. et al. (2018). RAVEN 2.0: a versatile toolbox for metabolic network reconstruction and a case study on Streptomyces coelicolor. *PLoS Comput. Biol.* 14 (10): e1006541.

27 Karlsen, E., Schulz, C., and Almaas, E. (2018). Automated generation of genome-scale metabolic draft reconstructions based on KEGG. *BMC Bioinf.* 19 (1): 467.

28 Thorleifsson, S.G. and Thiele, I. (2011). rBioNet: A COBRA toolbox extension for reconstructing high-quality biochemical networks. *Bioinformatics* 27 (14): 2009–2010.

29 Boele, J., Olivier, B.G., and Teusink, B. (2012). FAME, the flux analysis and modeling environment. *BMC Syst. Biol.* 6: 8.

30 Feng, X., Xu, Y., Chen, Y., and Tang, Y.J. (2012). MicrobesFlux: a web platform for drafting metabolic models from the KEGG database. *BMC Syst. Biol.* 6: 94.

31 Olivier BG. MetaDraft [Github Pages] Amsterdam (NL) [updated 2020 May 11; cited 2025 Mar 4]. Available from: https://systemsbioinformatics.github.io/cbmpy-metadraft/

32 Liao, Y.C., Tsai, M.H., Chen, F.C., and Hsiung, C.A. (2012 Jul). GEMSiRV: a software platform for GEnome-scale metabolic model simulation, reconstruction and visualization. *Bioinformatics* 28 (13): 1752–1758.

33 Swainston, N., Smallbone, K., Mendes, P. et al. (2011). The SuBliMinaL toolbox: automating steps in the reconstruction of metabolic networks. *J. Integr. Bioinform.* 8 (2): 186. http://journal.imbio.de.

34 Karp, P.D., Latendresse, M., Paley, S.M. et al. (2016). Pathway tools version 19.0 update: software for pathway/genome informatics and systems biology. *Briefings Bioinf.* 17 (5): 877–890.

35 Aite, M., Chevallier, M., Frioux, C. et al. (2018). Traceability, reproducibility and wiki-exploration for "à-la-carte" reconstructions of genome-scale metabolic models. *PLoS Comput. Biol.* 14 (5): e1006146.

36 Pabinger, S., Snajder, R., Hardiman, T. et al. (2014). MEMOSys 2.0: an update of the bioinformatics database for genome-scale models and genomic data. *Database* 2014: bau004.

37 Cottret, L., Frainay, C., Chazalviel, M. et al. (2018). MetExplore: collaborative edition and exploration of metabolic networks. *Nucleic Acids Res.* 46 (W1): W495–W502.

38 Novere, N.L., Finney, A., Hucka, M. et al. (2005). Minimum information requested in the annotation of biochemical models (MIRIAM). *Nat. Biotechnol.* 23: 1509–1515.

39 Olivier, B.G., Rohwer, J.M., and Hofmeyr, J.S. (2005). Modeling cellular systems with PySCeS. *Bioinformatics* 21 (4): 560–561.

40 Gudmundsson, S. and Thiele, S. (2010). Computationally efficient flux variability analysis. *BMC Bioinf.* 11: 489.

41 Ashburner, M., Ball, C.A., Blake, J.A. et al. (2000). Gene ontology: tool for the unification of biology. *Nat. Genet.* 25: 25–29.

42 The Gene Ontology Consortium, Aleksander, S.A., Balhoff, J. et al. (2023). The gene ontology knowledgebase in 2023. *Genetics* 224 (1): iyad031.

43 Degtyarenko, K., Matos, P., Ennis, M. et al. (2008). ChEBI: a database and ontology for chemical entities of biological interest. *Nucleic Acids Res.* 36 (1): D344–D350.

44 The UniProt Consortium (2017). UniProt: the universal protein knowledgebase. *Nucleic Acids Res.* 45 (D1): D158–D169.

45 Aden, K., Rheman, A., Waschina, S. et al. (2019). Metabolic functions of gut microbes associate with efficacy of tumor necrosis factor antagonists in patients with inflammatory bowel diseases. *Gastroenterology* 157 (5): 1279–1292.

46 Baldini, F., Hertel, J., Sandt, E. et al. (2020). Parkinson's disease-associated alterations of the gut microbiome predict disease-relevant changes in metabolic functions. *BMC Biol.* 18: 62.

47 Lambert, A., Budinich, M., Mahé, M. et al. (2024). Community metabolic modeling of hots-microbiota interactions through multi-objective optimization. *iScience* 27 (6): 110092.

48 Lu, Y., Chong, J., Shen, S. et al. (2022). TrpNet: understanding tryptophan metabolism across gut microbiome. *Metabolites* 12 (1): 10.

49 Zampieri, G., Vijayakumar, S., Yaneske, E., and Angione, C. (2019). Machine and deep learning meet genome-scale metabolic modeling. *PLoS Comput. Biol.* 15 (7): e1007084.

50 Dama, A.C., Kim, K.S., Leyva, D.M. et al. (2023). BacterAI maps microbial metabolism without prior knowledge. *Nat. Microbiol.* 8 (6): 1018–1025.

51 Carey, M.A., Drager, A., Beber, M.E. et al. (2020). Community standards to facilitate development and address challenges in metabolic modeling. *Mol. Syst. Biol.* 16 (8): e9235.

3

Machine-Guided Approaches for Synthetic Biology Part Design

Marc Amil[1], *Leandro N. Ventimiglia*[1], *and Aleksej Zelezniak*[1,2,3]

[1]*King's College London, Randall Centre for Cell & Molecular Biophysics, New Hunt's House, Guy's Campus, SE1 1UL, London, UK*
[2]*Chalmers University of Technology, Department of Biology and Biological Engineering, Kemivägen 10, SE-412 96, Gothenburg, Sweden*
[3]*Vilnius University, Institute of Biotechnology, Life Sciences Centre, Sauletekio al. 7, LT10257, Vilnius, Lithuania*

3.1 Introduction

Synthetic biology aims to engineer biological organisms by designing and combining genetic parts to create complex systems that are used for diverse biotechnological applications [1], including the production of biofuels [2, 3], pharmaceuticals [4], and industrial chemicals [5]. Central to synthetic biology is the concept of standardized genetic parts – DNA sequences that encode defined biological functions. While Chapters 1 and 2 have demonstrated ways to extract genome-scale metabolic functions from these DNA sequences, this chapter seeks to look at the functional parts of the DNA itself. These parts include promoters that initiate transcription, ribosome binding sites (RBSs) that facilitate translation initiation, coding sequences that determine transcription rate, protein structure and function, terminators that signal transcription termination, and enhancers that regulate RNA transcription. For instance, combining a promoter, a coding sequence, and a terminator yields a gene expression cassette. These modular components can be further assembled into higher-order genetic biological circuits to perform complex tasks within engineered organisms like oscillators or toggle switches [6, 7]. A well-known example is the genetic toggle switch, which uses two repressible promoters (such as *Ptrc-2* and *PLtetO-1*) and their matching repressor genes (*lacI* and *tetR*), each blocking the other's activity. This setup allows cells to stably switch between two states ON (expressing GFP) and OFF (not expressing GFP) – and has been used to create a memory device that records exposure to a chemical such as isopropyl-beta-d-1-thiogalactopyranoside (IPTG), which temporarily inactivates *lacI*, flipping the switch from one state to the other [7]. Resources such as the iGEM Registry of Standard Biological Parts [8] and SynBioHub [9] have facilitated this process by curating extensive libraries of characterized genetic parts, promoting standardization and reuse.

Machine Learning and Big Data-enabled Biotechnology, First Edition. Edited by Hal S. Alper.

Access to high-quality, diverse libraries of functional genetic elements – such as promoters, enhancers, and terminators – is essential for building robust and tunable genetic systems. One common strategy to generate such diversity involves mutagenesis-based methods, including error-prone PCR and saturation mutagenesis [10–13]. These approaches introduce mutations that are subsequently screened using selection methods, forming the basis of the iterative process known as directed evolution [14, 15]. Directed evolution mimics natural selection in the laboratory by applying repeated cycles of diversification, expression, and selection to enrich for improved variants, without requiring prior knowledge of beneficial mutations [16, 17]. While highly effective, conventional directed evolution approaches remain time-intensive, requiring multiple rounds of screening [18]. Because mutations are introduced at random, a vast majority of variants are either nonfunctional or fail to meet design specifications, needing extensive screening efforts to isolate rare beneficial mutations. For example, when mutating proteins, up to 70% of random single amino-acid substitutions typically result in a decline of protein activity, out of which 50% are completely deleterious to protein function [19–21]. In *E. coli,* replacing a native promoter with random ∼100 bp sequences showed that only ∼10% of random sequences could drive measurable transcription, whereas ∼90% produced no expression [22]. Similarly, random DNA can lead to active gene expression [23], but the sheer size of the sequence space presents a major challenge, requiring screening hundreds of millions of DNA constructs. For example, a 100 bp promoter has $\sim 1.6 \times 10^{60}$ possible variants, which is comparable to the number of atoms in the observable universe [24], making exhaustive screening impractical. When the goal is to improve promoter-driven expression, modeling the random mutation process shows that fewer than 0.5% of variants lead to the desired outcome [25]. Additionally, because each evolutionary step builds upon the last, the process is inherently path-dependent, meaning that each mutation selected in one round influences which future mutations are possible or likely to succeed. If an early mutation provides a small benefit but blocks access to a much better solution (due to needing to go through intermediate, less-fit variants), the process may miss out on reaching the best possible outcome. While experimental strategies for directed evolution are rapidly advancing [26] and remain a go-to method for biological sequence engineering, they have been extensively reviewed elsewhere [27]. This chapter instead focuses on emerging data-driven approaches employing deep neural networks for DNA sequence design.

To reduce experimental burden and guide design efforts more effectively, computational tools have been developed to simulate genetic part performance *in silico*. These include tools such as the RBS Calculator [28, 29], RBS Designer [30], UTR Designer [31], and EMOPEC [32]. By modeling molecular interactions such as ribosome–mRNA binding, mRNA secondary structure, and thermodynamic properties, these algorithms estimate translation initiation rates and thus aim to predict gene expression levels. For instance, the RBS Calculator [28] uses a thermodynamic model of bacterial translation initiation to predict the Gibbs free energy change associated with ribosome binding, thereby estimating the relative translation initiation rate of a given mRNA sequence. It supports both reverse engineering – predicting initiation rates from existing sequences – and forward engineering – designing synthetic RBSs to achieve user-specified expression levels, with predictions spanning over five orders of magnitude. However, despite its sophistication, the RBS Calculator is primarily tailored for *Escherichia coli* and assumes equilibrium thermodynamics, which may not hold for highly structured RNAs or in non-model organisms. Additionally, it does not account for other regulatory elements such as RNA-binding proteins. Consequently, genetic parts designed with these tools may exhibit unpredictable behavior across diverse hosts or environmental conditions, limiting their portability and predictive power.

Despite progress in computational tools for rational sequence design, conventional statistical and mechanistic models remain limited due to incomplete understanding of molecular mechanisms, host-specific assumptions, and poor generalizability across diverse biological contexts [33]. As a result, previous generation computational methods failed to replace traditional experimental approaches involving large-scale screening, which remain the standard in biological parts engineering. On the other hand, over the past decade, our ability to read biology has dramatically improved. Advances in single-cell sequencing and high-resolution profiling of gene expression and chromatin accessibility across species, tissues, and cell types have generated vast datasets [34–37]. This wealth of information, combined with fast-paced developments in machine learning and progress in artificial intelligence (AI) has enabled the modeling of nonlinear complex sequence-to-expression relationships, to design enhancers [38], tune promoters [39–41], design entire genes [25, 42], control tissue specific expression [43, 44], and even the design of synthetic regulatory DNA tailored to specific human cell types [45, 46].

Mechanism-agnostic, AI-driven approaches are transforming sequence design – from trial-and-error screening to a predictive, programmable engineering discipline. Leveraging large-scale biological datasets, deep neural networks can learn subtle sequence–function relationships, enabling the design of regulatory DNA elements for gene expression control without solely relying on exhaustive experimental screening. These advances have given rise to model-guided sequence design – a framework in which trained predictors assess the functional fitness of candidate sequences (e.g. expression strength or cell-type specificity) and help iteratively guide the design of sequences with greater functionality. In the following section, we introduce neural network-based frameworks for predictive and generative sequence design. We discuss how commonly used architectures, such as convolutional neural networks (CNNs) [47, 48], are employed to interpret DNA functionality and extract regulatory features like sequence motifs [49]. We also highlight the integration of generative deep learning (DL) models, such as generative adversarial networks (GANs) [50, 51], diffusion models [52, 53], and transformer-based language models [54], which can generate novel, functional regulatory sequences by learning directly from existing data. These data-driven strategies reduce reliance on large-scale experimental approaches like random mutagenesis and directed evolution, marking a significant shift toward programmable and predictive design of biological parts.

3.2 Model-Guided Sequence Design Using Deep Learning

In the model-guided sequence design framework, candidate DNA sequences are generated and evaluated to identify optimal designs for specific biological functions, such as maximizing protein expression. This framework typically integrates two neural network-based components: generative models, which produce novel sequences, and predictive models, which assess the functional potential of each design (Figure 3.1d).

The predictive model acts as an *oracle*, guiding the generative model by scoring sequences based on desired biological outcomes. Predictive supervised models operate in two modes: (i) classification, which assigns discrete labels (e.g. transcription factor binding: yes/no), and (ii) regression, which predicts continuous variables such as gene expression levels [42, 55, 56] or translational efficiency metrics like mean ribosome load (MRL) [57]. CNNs are commonly used for regression tasks due to their effective ability to learn local sequence features relevant to biological function.

On the generative side, recent advances have demonstrated the use of DL architectures, including GANs, diffusion models, and transformer-based language models (e.g. Generalized Pre-trained Transformers, GPTs), to sample biologically plausible and diverse DNA sequences. Unlike random mutagenesis approaches that often yield nonfunctional or biologically implausible variants, these models learn from functional sequence distributions and improve design efficiency, thereby reducing experimental screening burdens (Figure 3.1c). While there are many other generative models used in genomics research, such as variational autoencoders [58, 59], here, we explicitly focus on models that have been experimentally applied to DNA part design.

3.2.1 Predictive Models for DNA Function: CNNs in Regulatory Sequence Analysis

Decoding the *cis*-regulatory code – the sequence rules that govern gene expression – remains one of the central challenges in genomics. With the rise of large-scale functional genomics datasets, machine learning models, particularly DL architectures, have emerged as powerful tools to decode these regulatory elements directly from DNA sequence. Among these, CNNs have become foundational for predicting key regulatory phenomena such as transcription factor binding [48, 60], chromatin accessibility [61], gene expression [55, 56], and even 3D genome architecture [62, 63]. CNN-based models not only match or exceed prior methods in predictive power but also offer interpretable insights into the sequence motifs and interactions underlying gene regulation [49]. Notably, CNNs have been extended to model outputs from massively parallel reporter assays (MPRA), providing a direct link between sequence and regulatory activity, allowing the design of novel synthetic regulatory elements [64, 65]. In this chapter, we will focus specifically on the application of CNNs to model gene expression to predict quantitative metrics such as transcripts per million (TPM) from regulatory DNA sequences.

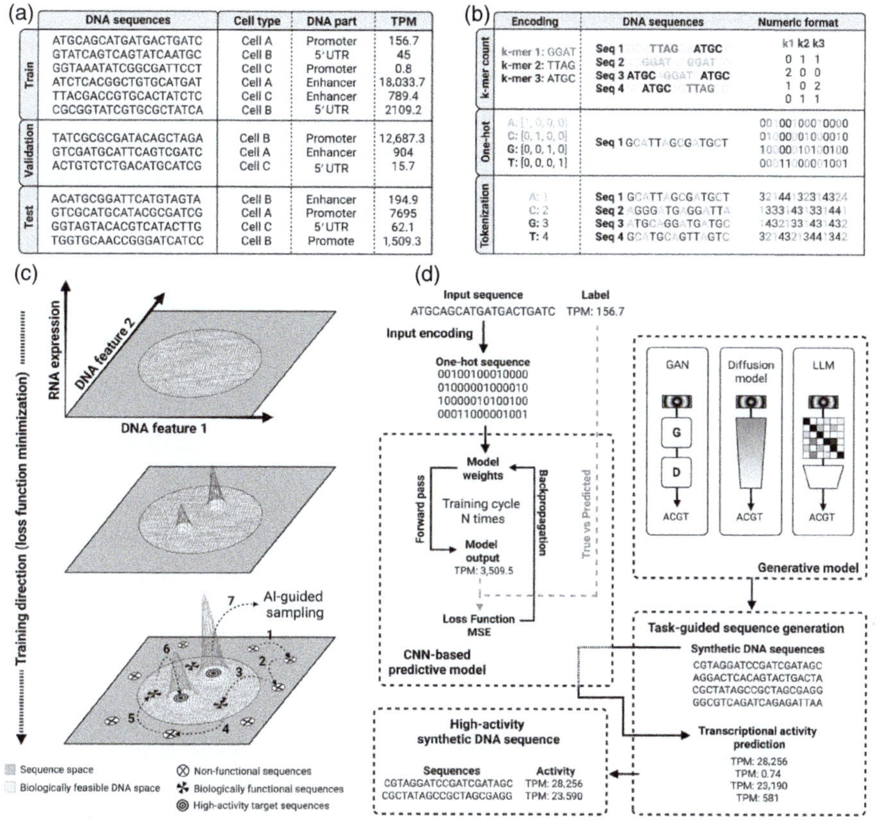

Figure 3.1 Overview of framework for predictive and generative modeling of regulatory DNA sequences. (a) Data splits in a representative curated dataset of regulatory DNA sequences. The sequences are annotated with categorical labels (e.g. cell type, element class) for classification tasks and continuous measurements (transcriptional activity as TPM) for regression. The full dataset is stratified into training, validation, and held-out test subsets to support model fitting, hyperparameter optimization, and rigorous performance evaluation. (b) Data encoding strategies. Three common sequence-to-numeric conversions used as training inputs: k-mer counting (frequency vectors of short subsequences), one-hot encoding (binary channel representation of A, C, G, and T) at each position, and tokenization (integer mapping of nucleotides or k-mers for transformer-style models). (c) RNA expression landscape during synthetic DNA generation. Conceptual illustration of the transcriptional activity landscape over DNA-sequence feature space of regulatory sequences. Each point on the X-Y surface represents a DNA sequence in feature space, with height corresponding to a measured RNA expression (TPM). Steps 1–6: directed-evolution workflow. The process consists in the random mutations of candidate DNA sequences. The selection of each mutation is not guided by a known transcriptional landscape, so directed evolution requires several mutation-screening iterations to find a sequence with high activity. Step 7: a deep generative model, guided by a trained "oracle," learns to approximate and sample directly from peaks of the transcriptional activity landscape, yielding novel high-activity sequences in fewer iterations. (d) Generative models overview. Encoded DNA sequences (such as one-hot vectors) are propagated through a CNN-based model to predict transcriptional activity. This predictor then serves as an "oracle" to guide the training of generative models (such as GANs, diffusion models, or LLM) toward synthesizing regulatory sequencespredicted to reside in the highest-activity regions of the learned landscape. *Source*: BioRender.

3.2.1.1 Data Considerations for Supervised Learning on Genomic Sequences

DL is inherently both data-centric and algorithm-centric – successful outcomes rely not only on model architecture but also on careful dataset preparation. All supervised learning models share a common framework: a dataset is assembled from input variables, such as genomic DNA sequences or sequences derived from MPRAs, and paired with target gene expression values, often measured using RNA-seq data (e.g. TPM values).

When using genomic data, model inputs typically consist of regulatory DNA sequences associated with each gene – such as enhancers, promoters, and untranslated regions (UTRs). The dataset is usually split into training (80–90%), validation (5–10%), and test (5–10%) sets (Figure 3.1a,b). The training set is used to learn model parameters, while the validation set monitors performance during training (e.g. for hyperparameter tuning). The test set is strictly reserved for final evaluation and must not be exposed to the model during training or validation to ensure an unbiased assessment of generalization.

While we do not cover general statistical and machine learning principles or optimization strategies here (see recommended literature [66–68] as well as other chapters as other models relate to different sectors of understanding biological function), we emphasize several data-specific considerations that are crucial when working with biological sequence data (especially for the characterization of parts) to avoid misleading results:

1) **Bias in the Target Variable**
   Expression measurements like TPM are influenced by gene length. A naive model may appear to perform well by simply learning gene length rather than true regulatory determinants. To avoid this, it is common to regress out gene length from TPM values using linear models prior to training.
2) **Bias in Input Sequences**
   RNA-seq and other sequencing technologies may introduce GC-content bias due to amplification steps [69]. These biases must be corrected using standard normalization protocols to avoid training the model on technical artifacts [70].
3) **Distribution Consistency**
   The training, validation, and test sets should follow similar distributions. Otherwise, performance on the test set may not reflect real-world generalization.
4) **Sequence Redundancy and Data Leakage**
   Data leakage happens when the data used to train a machine learning algorithm happens to have the information you are trying to predict. A frequent pitfall in genomics is sequence redundancy: similar or identical sequences appearing in multiple data splits. This leads to data leakage, where the model "cheats" by memorizing rather than learning generalizable features. To mitigate this, sequences should be clustered by identity (e.g. 80–90%) and split in a way that prevents similar sequences from crossing between training, validation, and test sets.
5) **Data Transformation Leakage**
   Preprocessing steps such as scaling, normalization, and imputation must be fit only on the training data and applied separately to validation and test sets.

Applying transformations globally (i.e. using information from the full dataset) introduces data leakage and inflates performance metrics [71].

By adhering to these practices, one ensures that neural networks trained on DNA sequence data generalize robustly and are evaluated fairly, avoiding artifacts that could confound biological interpretation [72].

3.2.1.2 Primer on Convolutional Neural Networks for Supervised Genomic Sequence Modeling

Once the dataset has been properly curated, the next step is to train the supervised predictive model. The process of training a CNN, also summarized in Figure 3.2, begins with a forward pass, where the information flows from the

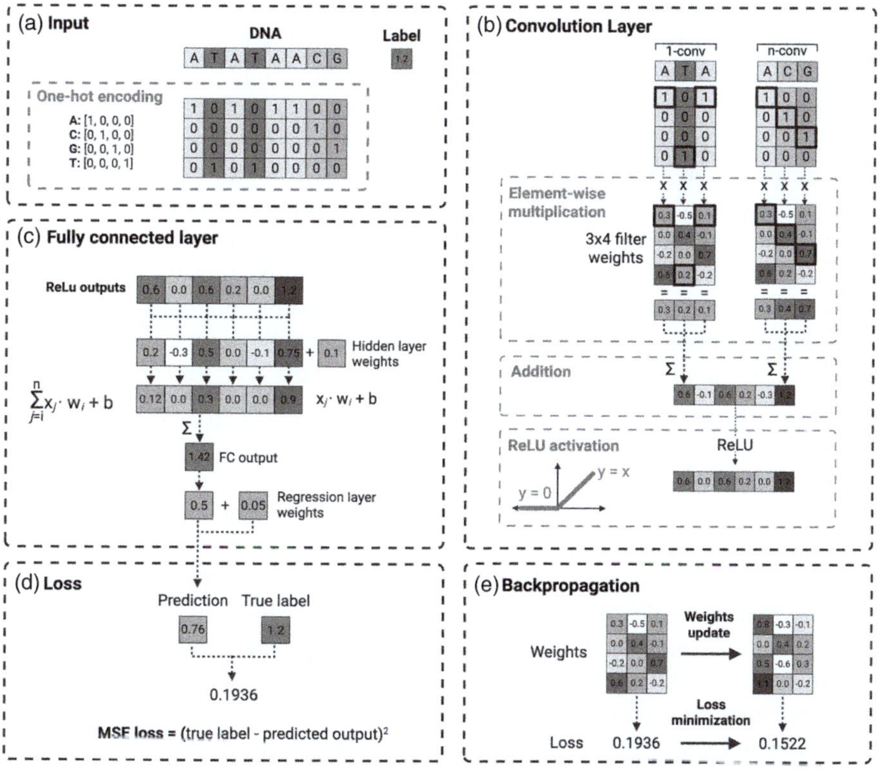

Figure 3.2 CNN-based transcriptional activity prediction. (a) The raw DNA sequence ATATAACG is used as input. Each nucleotide is one-hot encoded into a 4-element binary vector, resulting in an 8×4 matrix (b) A 3×4 convolutional kernel slides across the matrix, performing dot products at each position to generate local activations. The ReLU activation function is applied to the convolutional output, setting all negative values to zero. (c) The resulting activation map is flattened and passed through a fully connected layer with predefined weights and a bias term to produce a scalar output of 1.42. This output pass through a single neuron regression layer to output the final prediction value of 0.76. (d) The loss function quantifies the difference between the model prediction and the true label value of 1.2. (e) During backpropagation the model weights are optimized to minimize the loss value in the next training iteration. *Source*: BioRender.

input DNA sequence through all the model layers to output a prediction value. Prior to training, each nucleotide in a DNA sequence is converted into a numeric, machine-readable one-hot vector format. For instance, A is encoded as [1, 0, 0, 0], C as [0, 1, 0, 0], G as [0, 0, 1, 0], and T as [0, 0, 0, 1]. As a result, a sequence such as ATATAACG is represented as an 8-by-4 binary matrix, where each row corresponds to a nucleotide and each column to a channel in the input tensor (Figure 3.2a). This matrix is then passed through a convolutional layer composed of learnable filters or kernels (Figure 3.2b). These filters act as pattern detectors, scanning across k-mer subsequences to compute dot products between their internal weights and the corresponding one-hot encoded window. In the example in Figure 3.2, the CNN applies a single 3×4 kernel across the DNA sequence. Each kernel position produces a scalar output. These outputs reflect early representations of biologically significant features, where larger values represent more significant features – such as transcription factor binding sites (TFBS) or splicing enhancers. To introduce nonlinearity, a Rectified Linear Unit (ReLU)) [73] activation function is applied. ReLU retains all positive values and replaces negative activations with zero, effectively filtering out weak or irrelevant signals and allowing the network to focus on salient motifs. The result is a refined activation map, which is then flattened into a one-dimensional vector (the activation vector) to prepare it for downstream processing. Next, this vector is passed into a fully connected (dense) layer (Figure 3.2c), which integrates localized sequence features into a global representation. Specifically, each neuron in the dense layer multiplies each value from the activation vector by the corresponding weight, sums these products, and adds a bias term. For example, using the values shown in Figure 3.2c, the calculation is:

$$(0.6 \times 0.2) + (0.0 \times -0.3) + (0.6 \times 0.5) + (0.2 \times 0.0) + (0.0 \times -0.1)$$
$$+ (1.2 \times 0.75) + 0.1 = 1.42$$

By combining information from all positions in this way, each neuron learns how different local patterns – such as motifs detected at various positions – collectively influence the final prediction. This approach allows the CNN to capture sophisticated syntactic logics and represent them as a scalar output, which often corresponds to a biologically relevant score such as a predicted gene expression level or the probability that a sequence has regulatory activity. This output is compared to the true, known target value using a loss function – in this case, mean squared error (MSE). For example, if the true label is 1.2 and the model predicts 0.76, the squared error is 0.1936 (Figure 3.2d).

In the second stage of CNN training, or backpropagation, the model adjusts its internal parameters to minimize the loss (Figure 3.2e). Backpropagation starts from the computed loss and works backward through the network, calculating how each weight in the convolutional and fully connected layers influences the final prediction. Specifically, the network computes gradients – mathematical values indicating the direction and magnitude by which each neuron's weight should change to minimize the loss function. Using gradient descent, the CNN updates these weights incrementally to improve its predictions. Over multiple iterations, the network fine-tunes itself, becoming progressively better at recognizing biologically meaningful patterns, thereby enhancing its predictive accuracy and interpretability.

CNNs have been successfully applied to predict the levels of mRNA expression from DNA sequence alone [55]. Using a dataset of over 20,000 RNA-seq experiments across seven species, Zrimec *et al.* [42] trained a model that achieved a predictive score of $R^2 = 0.82$ in *S. cerevisiae*. This paper demonstrates that gene regulatory structure alone encodes a substantial portion of expression variability. Crucially, this model identified over 2,000 regulatory motifs, including both known TFBS and novel sequence elements distributed across promoters, UTRs, and terminators. The model also uncovered rules of co-occurring motifs – specific combinations of motifs modulating expression changes spanning over three orders of magnitude. These findings highlight the potential of CNNs to extract biologically meaningful patterns and motif interactions from sequence alone, thanks to their ability to integrate local motif signals to infer global expression outcomes.

3.2.2 Generative Sequence Modeling

After discussing the predictive component of the model-guided sequence design framework using CNNs, we now turn to the generative component, which typically involves designing sequences using Generative AI (GenAI). GenAI methods have become powerful tools for designing synthetic biological parts, with growing applications across genomics. The rise of GenAI has been fueled by tools such as ChatGPT and DALL·E, which can generate human-like text and realistic images based on simple prompts tailored to the user's needs. These same underlying technologies are proving powerful in biology, as DNA and proteins can be represented as text – DNA with bases "A," "C," "T," and "G," and proteins being represented as amino acid letters. This representation allows GenAI models to generate functional biological sequences in much the same way as they are able to generate language or art. Because of their ability to produce novel sequences with desired properties, GenAI models are increasingly being used to generate DNA, RNA, and protein sequences [74–77] with specific functional properties, from the design of optimized gene regulatory elements to functional promoters. For instance, PromoGen [78], a GPT-2-based model, enables the *de novo* design of prokaryotic promoters. Pretrained on over 120,000 promoter sequences from 27 organisms, PromoGen can be fine-tuned for species-specific tasks, capturing critical regulatory motifs such as the conserved −10 and −35 elements. It generates promoter sequences and has demonstrated strong performance in *Bacillus subtilis*, where over 80% of its predicted promoters were experimentally validated as active – some surpassing the strength of native promoters. Beyond regulatory elements, generative AI is also advancing codon optimization, traditionally performed using simple frequency tables that overlooked sequence context and regulatory features. Transformer-based models like CodonTransformer [79] provide a context-aware, host-specific approach, translating amino acid tokens (e.g. [M_UNK], [A_UNK]) into optimized codon tokens (e.g. [M_ATG], [A_GCC]) while maintaining GC balance and avoiding inhibitory motifs. Trained on over one million genes from 164 species, CodonTransformer allows the generation of a complete coding DNA sequence optimized for a chosen species. These models allow a more context-aware codon selection, enabling precise, high-yield expression for synthetic biology applications across diverse species.

Importantly, GenAI application extends well beyond individual DNA parts to the generation of entire molecular systems and even whole genomes. Evo [80], a multimodal genomic foundation model, exemplifies this broader potential by generating not only discrete elements like promoters or coding sequences but also complex functional assemblies such as CRISPR-Cas protein-RNA complexes and IS200/IS605 transposon systems. Evo is even capable of producing plausible megabase-scale genome sequences with coherent gene architecture, appropriate coding density, and operon-like structures at single-nucleotide resolution. This represents a transformative shift from designing isolated components to modeling and engineering genome-wide systems, positioning GenAI as a key enabler for constructing biological functions across synthetic biology systems. In the following sections, we explore how GenAI models can be applied to DNA to design novel biological parts.

3.2.2.1 Data Preparation for Unsupervised Learning

Successful generative modeling of biological sequences – using approaches such as GANs, diffusion models, or transformer-based language models – relies heavily on rigorous dataset preparation. Unlike supervised learning, where models are trained on labeled datasets to map input sequences (e.g. promoters) to outputs (e.g. gene expression levels), unsupervised generative models learn statistical patterns directly from unlabeled sequence data, typically directly using the raw genome FASTA file. Although labels are not required, data preparation remains essential – much like in CNNs – and involves removing redundancy, data leakage, and cleaning the sequence data. First, raw genomic DNA sequences must undergo quality control to eliminate sequencing errors and ambiguous nucleotides (e.g. "N" characters), which can impair learning. Another challenge in generative modeling is gene length variability, especially when models require fixed-length inputs. This is addressed by standardizing sequence lengths through trimming or padding. Padding typically involves appending neutral tokens (e.g. "N" characters) to shorter sequences to ensure uniform input lengths. For instance, if a model requires 512-base inputs, a 430-base sequence would be padded with 82 bases. Finally, to avoid training bias, datasets may be balanced by upsampling underrepresented sequence classes or downsampling overrepresented ones. In summary, careful data preparation is essential for unsupervised learning, enabling generative models to produce biologically plausible, diverse, and functional sequences.

3.2.2.2 GANs for the Design of Biological Sequences

GANs [50] are generative DL approaches for designing biologically meaningful DNA sequences by modeling the distribution of real genomic elements. A GAN is composed of two neural networks: the Generator (G), which is trained to generate realistic synthetic DNA sequences, and the Discriminator (D), which evaluates whether a given sequence is real or artificially generated. As shown in Figure 3.3a, the process of training a GAN begins with sampling a random latent vector $z \sim \mathcal{N}(0,1)$. This vector is then passed into the Generator, where it undergoes a linear transformation ($logits = wz + b$) to produce raw scores or logits that are reshaped into a position-by-nucleotide matrix representing a synthetic DNA sequence. A softmax activation function is then applied across each row of this matrix – where each row represents a position-specific distribution over A, C, G, and T – to convert the logits into probabilities over the four DNA bases, and the resulting probability matrix is passed directly to the Discriminator. The Generator and Discriminator are trained in an adversarial loop: the Discriminator updates its weights to better learn to distinguish real from generated sequences, while the Generator adjusts its parameters to produce outputs that are increasingly difficult for the Discriminator to detect. This dynamic is visualized in Figure 3.3a, right panel, where the light grey distribution representing synthetic sequences gradually shifts to align with the dark gray distribution of real sequences. After applying a gradient descent algorithm over many training iterations, the Generator progressively learns to synthesize sequences indistinguishable from natural biological sequences. During inference, an already trained GAN is used to synthesize a discrete DNA sequence for inspection or downstream use. At this stage, the position-by-nucleotide matrix returned by the Generator is first used to generate a nucleotide sequence by calculating the nucleotide with the highest probability at each position of the matrix. Then, this synthetic sequence is passed into the Discriminator, which assigns a score indicating the likelihood that the sequence is real. High scores suggest the sequence is from the true data distribution, while low scores indicate it is synthetic.

The iterative adversarial training of a GAN involves alternating optimization steps for the Generator and the Discriminator. First, the Discriminator is optimized to accurately distinguish real sequences from generated "fake" ones. Mathematically, this optimization step is performed by calculating the gradients of the Discriminator loss function (L_D), defined as:

$$L_D = -[\log D(x_{\text{real}}) + \log(1 - D(G(z)))]$$

$D(x_{real})$ represents the predicted probability that an input sequence x_{real} drawn from the training dataset is real. $D(G(z))$ denotes the probability that a synthetic sequence – generated from a random latent vector z via the Generator – is classified as real. The first term $logD(x_{real})$ rewards the Discriminator for correctly identifying real sequences, while the second term $\log(1 - D(G(z))$ rewards it for correctly identifying generated sequences as synthetic. Therefore, the Discriminator loss is

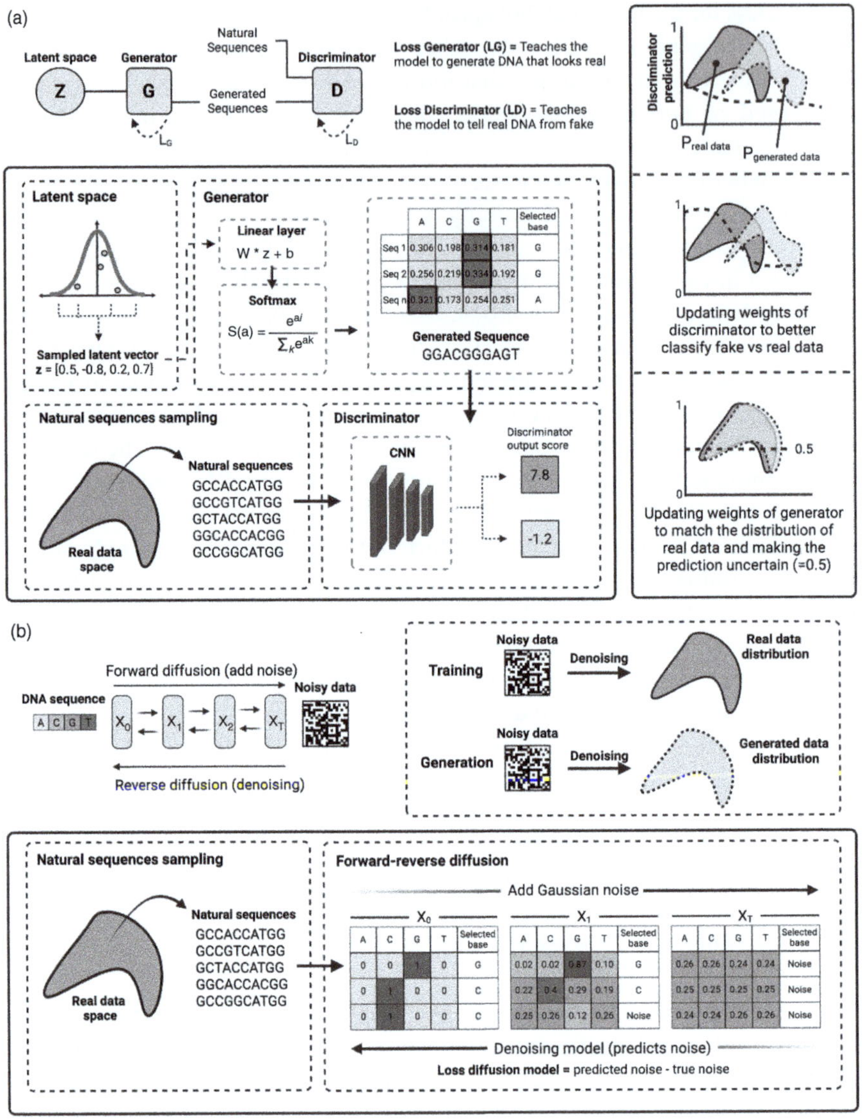
(a)
Latent space
Generator
Natural Sequences
Generated Sequences
Discriminator
Z
G
D
Loss Generator (LG) = Teaches the model to generate DNA that looks real
Loss Discriminator (LD) = Teaches the model to tell real DNA from fake
Latent space
Sampled latent vector
z = [0.5, -0.8, 0.2, 0.7]
Generator
Linear layer
W * z + b
Softmax
S(a) = e^{aj} / Σ_k e^{ak}
Selected base
Seq 1 0.306 0.198 0.314 0.181 G
Seq 2 0.256 0.219 0.334 0.192 G
Seq n 0.321 0.173 0.254 0.251 A
Generated Sequence
GGACGGGAGT
Natural sequences sampling
Natural sequences
GCCACCATGG
GCCGTCATGG
GCTACCATGG
GGCACCACGG
GCCGGCATGG
Real data space
Discriminator
CNN
Discriminator output score
7.8
-1.2
Discriminator prediction
P real data
P generated data
Updating weights of discriminator to better classify fake vs real data
0.5
Updating weights of generator to match the distribution of real data and making the prediction uncertain (=0.5)
(b)
Forward diffusion (add noise)
DNA sequence
Noisy data
Reverse diffusion (denoising)
Training
Noisy data
Denoising
Real data distribution
Generation
Generated data distribution
Natural sequences sampling
Natural sequences
GCCACCATGG
GCCGTCATGG
GCTACCATGG
GGCACCACGG
GCCGGCATGG
Real data space
Forward-reverse diffusion
Add Gaussian noise
Denoising model (predicts noise)
Loss diffusion model = predicted noise - true noise

Figure 3.3 GAN and diffusion models. (a) GAN workflow for generating synthetic DNA sequences. Top Left panel, GAN overview. A latent vector z is passed through the Generator (G), which outputs a synthetic sequence. The Discriminator (D) evaluates whether a given sequence comes from real data distribution or if it was generated. Right panel, Generator learns real data distribution. Initially, the generated sequence distribution (light grey) differs significantly from the real sequence distribution (dark grey). After iterative training, the generator learns the real data distribution and generates synthetic sequences that resemble real data. Bottom panel, GAN training. The Generator takes a latent vector z as input to compute logits (logits=W*z+b). Logits are passed through a softmax layer to calculate the nucleotide probabilities at each position. The generated sequences are one-hot encoded and passed through the Discriminator (a CNN itself), which outputs the probability of a sequence being drawn from the real data distribution. These scores guide the adversarial training loop, allowing the Generator to iteratively refine its outputs toward biologically realistic DNA sequences. (b) Diffusion model workflow for DNA sequence generation. Top panel, diffusion process overview. A real DNA sequence is one-hot encoded and subjected to a forward diffusion process where Gaussian noise is incrementally added over time. The denoising network learns to reverse this process, distilling the noisy representation back into a structured sequence. Top right panel, training vs generation. During training the model learns to map noise to a real DNA sequences. During generation the model denoises random generated noise into a realistic synthetic DNA sequence. Bottom panel. The model begins with uniform base probabilities (~0.25 each). At each timestep (X0, X1, ⋯, XT) the probabilities are adjusted toward a biologically plausible distribution. Ultimately, the most probable nucleotide is selected at each position to form the final sequence. This denoising process enables generation of novel, diverse DNA sequences that capture key biological motifs and patterns without replicating training examples. *Source*: BioRender.

minimized by giving a low probability to generated sequences that look random (e.g. "GGGGGGGGG"); if it mistakenly gives a high probability, the loss increases. Gradient descent updates the Discriminator's weights and aims to minimize L_D, thus enhancing its accuracy in classifying "natural" sequences. Subsequently, the Generator is optimized using the gradients computed from its own loss function (L_G), aimed at minimizing the ability of the Discriminator to recognize generated sequences as synthetic, or in other words, the Generator loss L_G measures how well the Generator can fool the Discriminator:

$$L_G = -\log D(G(z))$$

Here, $D(G(z))$ represents the probability, as predicted by the Discriminator, that the sequence produced by the Generator $G(z)$ is a real (or "natural") biological sequence. The Generator is trained to maximize this probability, and this is achieved by minimizing the negative log probability shown above. Minimizing L_G encourages the Generator to create synthetic sequences that the Discriminator cannot easily distinguish from real ones. For example, if the Generator produces a sequence such as "GCCACCATGG," which resembles a natural Kozak consensus sequence, and the Discriminator assigns it a high probability of being real, the Generator receives a lower loss. Crucially, the ultimate goal of the Generator is to match the distribution of real biological sequences – not by generating a single ideal sequence, but by creating multiple synthetic sequences whose collective characteristics (e.g. motifs, nucleotide frequencies, and patterns) closely mimic those observed in the original data. Consider a toy example involving Kozak sequences such as "GCCACCATGG" and "GCCGCCATGG." Initially, the Generator produces diverse random sequences with little biological relevance. Over multiple training iterations, as the Discriminator becomes better at distinguishing authentic sequences, the Generator receives precise feedback on how its outputs deviate from real sequences. Consequently, the Generator's parameter adjustments steer it toward producing sequences that share the same statistical properties as the real dataset.

An example of GANs being applied for DNA sequence generation is ExpressionGAN [25], an approach developed to design synthetic regulatory sequences with a predictable transcriptional output in *S. cerevisiae*. In this framework, the Generator is coupled with a predictive "oracle" model that accurately estimates the gene expression levels corresponding to each generated sequence. During optimization, the predictive model guides the Generator toward sequences achieving specific target expression levels. Notably, despite substantial sequence divergence from natural regulatory DNA (>33%), experimental validations confirmed that sequences generated by ExpressionGAN were functional *in vivo*. Strikingly, 57% of the synthetic regulatory sequences associated with high gene expression surpassed the expression levels corresponding to natural controls [25]. This demonstrates not only the capability of GAN-based models to navigate extensive regulatory sequence-expression landscapes but also their potential for surpassing the limitations of traditional mutagenesis approaches. Thus, ExpressionGAN exemplifies the power of GANs to innovate beyond existing biological constraints, enabling precise control over gene expression levels with significant implications for biotechnology and therapeutic applications.

3.2.2.3 Transformer-Based DNA Models

The introduction of the Transformers architecture in the landmark publication *"Attention Is All You Need"* [81] has revolutionized our ability to train genome-sized sequence models and thus interpret and generate text. The publication also introduced the self-attention mechanism, which allows the model to weigh the relevance of each word in a sentence, achieving a more efficient semantic and syntactic understanding. For example, in the sentence *"She went to the bank to deposit money,"* self-attention allows the model to interpret "bank" as a financial institution rather than the land alongside a river, based on the surrounding words. Importantly, by eliminating any recurrent or convolutional operations and relying solely on self-attention, Transformers are able to process all the positions in an input in parallel, which enables the use of larger models and faster training. Building on this foundational work, models like Generative Pretrained Transformer (GPT) [82] have also demonstrated exceptional performance in generating coherent, context-aware text across a wide range of applications. The success of attention-based models in natural language generative tasks has encouraged their application in other fields. In genomics, where DNA or RNA sequences can be treated as text made up of A, C, G, T, and U nucleotides, genomic language models (GLMs) have proven effective at capturing both local motifs and long-range dependencies. This capability makes them particularly well-suited for synthetic part design, where precise sequence generation and contextual understanding are critical.

Similar to convolution-based models, the training of a GLM requires converting raw DNA sequences into a numeric format. However, instead of applying one-hot

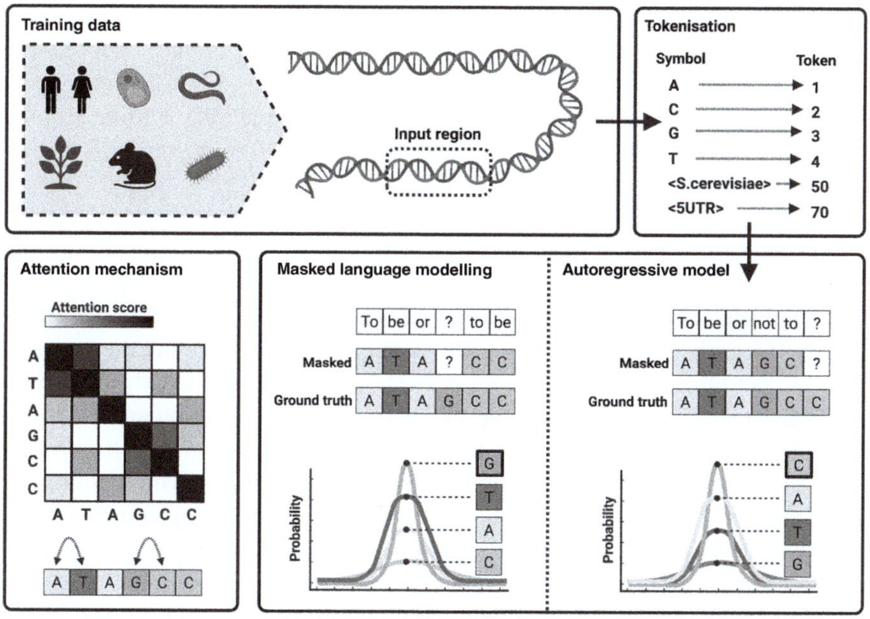

Figure 3.4 Overview of transformer-based language models for DNA sequence understanding. Top panels. DNA sequences from a wide range of organisms are input into a Transformer model and tokenized to convert them to numeric format. Bottom left panel. MLM vs AR training. In MLM (left), the model uses attention scores to infer the identity of masked bases, learning bidirectional dependencies. In AR (right), it generates sequences one base at a time, predicting the most probable nucleotide given prior context. The distribution curves show nucleotide probability outputs at specific positions. Bottom left panel, attention mechanism. The heatmap represents attention scores for each nucleotide in the sequence. Using these scores, Transformers learn context-aware representations of DNA, which can be used for diverse downstream genomic tasks. *Source*: BioRender.

encoding, GLMs rely on tokenization, a process that maps nucleotide letters to numerical representations or tokens (Figure 3.4). In the simplest approach, known as single-character tokenization, each nucleotide is treated as an individual token, similar to how characters are treated in a text corpus (e.g. “A” might be encoded as 1, “C” as 2, and so on). Additional tokens can also be introduced to inform and guide the GLM to design sequences with particular characteristics. For example, in the *regLM* framework [83], synthetic *S. cerevisiae* promoters were generated using a two-token prompt, <X><Y>, prepended to the DNA sequence. Each token represents the desired promoter activity in a specific growth medium, where <X> represents activity in a complex medium and <Y> represents the activity in a defined medium. Token values range from 0 (very low activity) to 4 (very high activity). For instance, the prompt <2><4> instructs the model to generate an *S. cerevisiae* promoter with medium activity in complex medium and very high activity in defined medium. Similarly, for human enhancer generation, *regLM* implemented a three-token prompt, <A><B><C>, where each token specifies enhancer activity in a particular human cell line: <A> for HepG2 (liver), <B> for

K562 (erythroid leukemia), and <C> for SK-N-SH (neuroblastoma). These values range from 0 (inactive) to 3 (very active). For example, the prompt <0><3><1> directs the model to generate an enhancer that is inactive in HepG2, highly active in K562, and exhibits low activity in SK-N-SH. An example of how these prompts are structured is shown below:

$$\langle 2\rangle\langle 4\rangle\text{ATGCGTACGATCGA}\ldots$$

$$\langle 0\rangle\langle 3\rangle\langle 1\rangle\ \text{TCGAGTACCTAGGA}\ldots$$

Alternatively, k-mer tokenization segments the DNA into overlapping subsequences of length k using a sliding window approach. For example, the sequence "ATGCA" would yield the 3-mers "ATG", "TGC," and "GCA." This method preserves local context and introduces a more biologically relevant vocabulary, capturing short motifs such as TFBS. However, k-mer tokenization also increases the vocabulary size exponentially with respect to k (e.g. $4^3 = 64$ for 3-mers), which can impact memory and computational efficiency. The choice between single-nucleotide or k-mer tokenization depends on the model's design goals – whether to prioritize fine-grained base resolution or richer motif-level context.

GLM training typically begins with a pretraining phase, where the model is exposed to large amounts of unlabeled genomic data to learn general representations of DNA syntax. This phase allows the model to capture the underlying patterns and contextual information within the DNA without the need for explicit labels or annotations, enabling it to generalize across different downstream tasks, such as DNA parts' classification or generation. For pretraining, genomes from one or more organisms can be used. For instance, HyenaDNA [84] focuses specifically on human genomic data, enabling it to learn specialized representations tailored to the human genome. In contrast, models like DNABERT-2 [85] aggregate DNA sequences from multiple species – including bacteria, plants, and humans – encouraging broader generalizability across evolutionary contexts.

There are two main pretraining methods: masked language modeling (MLM) and autoregressive modeling (AR) (Figure 3.4 bottom right panel). In MLM, some nucleotides within a sequence are masked, and the model learns to predict the identity of the masked tokens using the unmasked context. In AR modeling, the model is trained to predict the next nucleotide of a sequence based solely on the previous tokens. The key difference between both approaches is that MLM training allows the models to learn bidirectional dependencies from both sides of the masked nucleotide, understanding how upstream and downstream nucleotides influence a given position. For example, if the sequence "ATA_CC" is provided, the model learns to infer that the masked token is likely a "G" given the neighboring nucleotides, using both ATA and CC. On the other hand, in AR training, given "ATAGC," the model estimates the probability distribution over the next base, and selects the most likely candidate of the four nucleotides to extend the sequence.

An important innovation of Transformer-based models is the attention mechanism, illustrated in the lower left panel of Figure 3.4. Attention allows the model to selectively focus on certain parts of the input sequence by assigning importance scores, or attention weights, to each position when predicting nucleotides. Importantly, attention weights are context-dependent, meaning that the relevance assigned to each nucleotide can shift depending on the specific surrounding sequence. For instance, when modeling a promoter region, the model may assign high attention scores to a known regulatory motif such as the TATA box ("TATAAA") in order to predict the transcription start site (TSS) downstream.

Transformer-based DNA language models are trained using cross-entropy loss as the objective function. Cross-entropy compares the predicted probability distribution over nucleotides to the actual correct nucleotide. A lower cross-entropy value means the model is assigning a higher probability to the correct answer. For MLM, the model only predicts the masked positions using full sequence context; for AR modeling, the model predicts each nucleotide from left to right using only previous positions. In both cases, cross-entropy is computed at each relevant position to guide learning. The loss for each setting can be defined as:

$$L_{\mathrm{MLM}} = -\sum_{j=1}^{V} y_{i,j} \log(\hat{y}_{i,j})$$

$$L_{\mathrm{AR}} = -\sum_{i=1}^{N} \sum_{j=1}^{V} y_{i,j} \log(\hat{y}_{i,j})$$

Here, N is the sequence length, and V is the vocabulary size (e.g. 4 for DNA: A, C, G, and T). The term $y_{i,j}$ is a one-hot vector where the correct nucleotide is marked with 1, and all others are 0. Since the loss multiplies each predicted log-probability by $y_{i,j}$, only the correct class contributes (because $1 \times$ log-probability is retained, and $0 \times$ anything else is zero). For example, if the true nucleotide is A (one-hot label = [1,0,0,0]) and the model predicts [0.7,0.1,0.1,0.1], $Loss = -\log(0.7) \approx 0.36$, indicating a confident prediction. In contrast, if the model assigns equal probability to all the nucleotides ([0.25,0.25,0.25,0.25]), $Loss = -\log(0.25) \approx 1.39$, a higher value reflecting uncertainty and poor predictive capabilities.

Typically, MLM is predominantly used by Bidirectional Encoder Representations from Transformers (BERT)-style models [86, 87], where the goal is to predict masked tokens within a sequence by leveraging context from both directions. This bidirectional approach enables BERT-based models to learn contextually rich representations, making them highly effective for downstream tasks such as regression and classification. When applied to DNA sequences, these BERT models often incorporate an additional regression or classification "head" – a small neural network layer placed on top of the pretrained Transformer – to translate the learned representations into specific predictions, such as enhancer activity levels or promoter classifications. In contrast, GPT [82] models adopt an AR approach, predicting nucleotides sequentially from left to right, thus allowing generative capabilities useful for synthesizing new DNA sequences. Consequently, GPT-style AR models are well-suited for tasks involving sequence generation, such as designing novel regulatory elements or generating optimized sequences with desired biological functions [46]. The first DNA-based large language model (LLM), DNABERT [87], was built upon a 12-layer BERT [88] architecture trained on sequences of 512 nucleotides derived from the human genome. Since then, several additional GLMs have emerged, notably Nucleotide Transformer [86] and DNABERT-2 [85]. While all these models use Transformer architectures, they differ considerably in their tokenization strategies. Despite their successes, Transformer models inherently face computational challenges due to their reliance on the self-attention mechanism, resulting in a quadratic computational complexity of $O(n^2)$, where n is the sequence length [81]. This is because self-attention computes pairwise interactions between all positions in the sequence, requiring the model to calculate an attention score for every possible token pair. The quadratic complexity can severely limit the scalability of Transformers, particularly for analyzing very long sequences such as whole chromosomes or performing inference across millions of nucleotides.

To address these scalability challenges, HyenaDNA [84] was introduced. HyenaDNA employs an AR next-token prediction approach at single-nucleotide resolution. Crucially, it scales sub-quadratically with sequence length, achieving a computational complexity of $O(n \log n)$. This results in an up to 160 times faster than quadratic Transformer models. The underlying Hyena architecture supports input sequences of up to one million nucleotides, significantly expanding the range of genomic contexts that can be effectively modeled. By integrating language modeling techniques with supervised sequence-to-function predictive models, HyenaDNA effectively generates realistic synthetic DNA sequences, demonstrated by creating functional yeast promoters and human cell-type-specific enhancers [83].

Building upon these advances, the Evo model [80], based on the StripedHyena architecture, has recently emerged as a versatile approach capable of both sequence generation and functional prediction across DNA, RNA, and proteins. Evo, trained using a large-scale dataset with 7 billion parameters and a context length of 8 kb and 131 kb, exhibits impressive performance. It surpasses specialized models in predicting fitness effects of mutations within diverse biological systems, including *E. coli* proteins, non-coding RNAs, and regulatory DNA sequences. A particularly noteworthy application of Evo is the generative design of synthetic CRISPR-Cas systems, such as Cas9, Cas12, and Cas13 proteins. Evo generates protein-RNA complexes that closely resemble their natural counterparts in both sequence composition and structural integrity. Moreover, Evo extends its generative capabilities to entire transposable elements, exemplified by the IS605 system, producing novel, functionally coherent coding sequences. Remarkably, Evo also enables generation of genome-scale sequences up to 650 kb, significantly expanding the scope of genomic modeling and opening new avenues for large-scale biological engineering.

3.2.2.4 Diffusion Models for the Design of Biological Sequences

Diffusion models have recently emerged as powerful generative architectures, offering an alternative to adversarial and attention-based frameworks for the generation of realistic synthetic regulatory elements [53, 89], long DNA sequences [90] and 5′-UTRs [91]. Unlike GANs, which generate sequences through adversarial training, diffusion models operate by learning to reverse a progressive noising process. As shown in Figure 3.3b, the generative process begins with a real DNA sequence, which is first one-hot encoded into a binary matrix. During the forward diffusion process, Gaussian noise is incrementally added to the one-hot encoded sequence across multiple timesteps, denoted by $x_0 \rightarrow x_1 \rightarrow x_t$. As noise accumulates, the structured patterns of the sequence gradually degrade until the representation resembles pure noise. Mathematically, each noisy sequence x_t is computed using a predefined noise schedule t, with added Gaussian noise $\varepsilon \sim N(0,1)$. This process simulates the loss of biological information, turning a highly structured sequence into a noisy, uninformative distribution where each nucleotide has an equal probability (e.g. ~0.25 for A, C, G, T). This transformation is defined by the equation:

$$x_t = \sqrt{\alpha_t} x_{t-1} + \sqrt{1 - \alpha_t} \varepsilon, \quad \varepsilon \sim \mathcal{N}(0, 1)$$

Here, α_t is a predefined noise schedule that determines how much of the original signal is retained at each timestep, with values that gradually decrease from 1 at each timestep. The term $\sqrt{\alpha_t} x_{t-1}$ preserves the structured information from the previous timestep, while $\sqrt{1 - \alpha_t} \varepsilon$ injects Gaussian noise sampled from a standard normal distribution. Early in the process, the sequence remains mostly intact, but as t increases and α_t shrinks, noise dominates and the sequence becomes increasingly degraded. This gradual corruption process creates the training data for the model, which learns to reverse the noise and recover biologically meaningful sequences during the generative phase.

The reverse diffusion process is where generation occurs. A neural network – often a U-Net or Transformer [92] – is trained to denoise these random sequences, gradually transforming noise back into a meaningful DNA sequence. The model receives a noisy input x_t, along with the current timestep t, and learns to predict the noise $\hat{\varepsilon}_\theta(x_t, t)$ that was added. The training objective is to minimize the difference between the predicted noise and the actual noise used in the forward process.

As shown in Figure 3.3b, denoising begins with a random sequence in which each base has an approximately uniform distribution. At each timestep, the network refines these probabilities, gradually restoring higher confidence in biologically plausible nucleotide choices (e.g. increasing the probability of G at a conserved position in a Kozak motif). Eventually, the model outputs a nearly one-hot encoded matrix, from which the most probable nucleotide is selected at each position to form the final DNA sequence. Notably, the generative process is stochastic, allowing for the creation of diverse but realistic sequences that follow learned biological patterns without replicating any specific sequence from the training data. This iterative refinement process can be thought of like sharpening a blurred image – each step reintroduces critical structural elements such as transcription factor motifs, GC content profiles, or positional constraints, leading to the emergence of high-fidelity synthetic DNA with meaningful biological features.

3.3 Sources of Sequence–Function Data for Deep Learning

The effective training of DL models that translate DNA sequences into functional predictions requires large, well-curated sequence-to-function datasets. These data provide the empirical "ground truth," mapping specific nucleotide patterns to transcriptional outputs, chromatin conformation, or any other cellular state directly modulated by DNA sequences. Therefore, these datasets are indispensable for DL models to learn the complex context-dependent rules that govern gene regulation. In synthetic biology, transcriptional profiling datasets are especially valuable. Available datasets can be broadly divided into two types: native-context datasets derived from genome-wide assays performed in their original chromatin and regulatory context, and synthetic dataset generated by screening the activity of cloned DNA sequences into reporter constructs and introduced into cells via plasmids or genomic integration. In this section, we will describe representative native-context and synthetic datasets, and discuss how each has enabled the training of predictive models suitable for synthetic part design.

3.3.1 Native-Context Genome-Derived Datasets

Native datasets are derived from whole-genome sequences and analyzed in their original biological context. These datasets include raw genomic sequences as well as functional annotations that capture specific biochemical activities across cell types and conditions. Large collections of native-context datasets are publicly available through free-access resources (detailed further in Table 3.1), including ENCODE [37] (Encyclopedia of DNA Elements), GTEx [93] (Genotype-Tissue Expression), the Roadmap Epigenomics Consortium [94], FANTOM [95] (Functional Annotation of the Mammalian Genome), CATlas (Cis-elements Atlas), Expression Atlas [96] RegulonDB [97]. The combination of genomic annotations with whole-genome transcriptional activity datasets has been essential for training DL models that link DNA sequence features, such as the annotation of a promoter sequence, to transcriptional activity. For example, Xpresso is a CNN trained only on promoter sequences and sequence-derived features related to mRNA stability to predict steady-state gene expression levels in humans and mice [55]. To train Xpresso, the promoter sequence for each gene (e.g. the genomic region spanning from ~7 kb upstream to ~3.5 kb downstream of the TSS) was extracted and converted to a one-hot vector to feed the neural network. Each sequence was labeled with its corresponding gene expression level derived from bulk RNA-seq datasets from the Epigenomics Roadmap Consortium and ENCODE. Another CNN-based model, Basenji, predicts cell-type-specific RNA expression profiles using 131 kb genomic regions as inputs and Cap analysis of gene expression (CAGE) datasets as targets [61]. This model applies a dilated convolutional architecture to enable the use of large receptive fields, improving gene expression prediction by modeling long-range enhancer-promoter interactions. The same dataset was used to train Enformer [56] on 196 kb genomic regions. The architecture of Enformer combines convolutional blocks with Transformers to integrate information from longer-range interactions up to 100 kb, outperforming Basenji in gene expression prediction (Pearson's correlation 0.85 vs 0.81). A more recent advancement is Borzoi, a U-Net-style architecture combining convolutional and Transformer layers to predict RNA expression (Pearson's correlation of 0.87) from 524 kb genomic

Table 3.1 Publicly available databases.

Database	Goal	Data types	Organisms
ENCODE	Identification of all functional elements in the genome.	RNA-seq, ChIP-seq, DNase-seq, ATAC-seq, CAGE.	Human Mouse
GTEx	Tissue- and cell-specific gene expression across species and developmental stages.	RNA-seq, eQTL.	Human Nonhuman primates
Roadmap Epigenomics Consortium	Epigenomic profiling of primary tissues and cells.	ChIP-seq, DNAse-seq, MeDIP-seq, MRE-seq, RNA-seq.	Human
FANTOM	Annotation of functional elements in mammalian genomes.	CAGE, RNA-seq.	Human Mouse Mammals
CATlas	Reference maps of cis-regulatory elements from single-cell epigenetics datasets.	ATAC-seq, RNA-seq, snm3-seq, Paired-Tag.	Human Mouse Others
Expression Atlas	Gene and protein expression across tissues, species, and conditions.	RNA-seq, Proteomics, Microarrays.	Animals Plants Fungi
RegulonDB	Comprehensive resource on transcriptional regulation in *Escherichia coli* K-12.	RNA-seq, ChIP-seq, gSELEX-seq.	*E. coli*

windows [98]. Borzoi was trained using over 10,000 human and mouse genomic assays, including CAGE datasets from the FANTOM5 Consortium as well as RNA-seq coverage tracks from ENCODE and GTEx. Using the same datasets, the authors trained Flashzoi, an enhanced version of Borzoi that incorporates rotary positional encoding to reduce training time while preserving or improving accuracy of several predictive tasks, including RNA coverage [99].

3.3.2 Synthetic Datasets

In addition to profiling natural genomic sequences, RNA-seq also serves a crucial role in evaluating the function of synthetically designed biological parts, particularly in combinations with MPRA. In MPRA experiments, large libraries of synthetic DNA sequences – typically encompassing promoter, enhancer, or 5′UTR elements – are cloned upstream of a reporter gene and introduced into cells. Upon transcription, each synthetic construct generates a unique mRNA tagged with a barcode, allowing sequence-specific quantification of expression levels. RNA-seq is then used to measure the relative abundance of each transcript, effectively linking sequence design to transcriptional output.

In a "classic" low-throughput setup, a synthetic dataset of transcriptional activity is created by cloning a set of candidate regulatory DNA sequences besides a reporter gene, such as luciferase [100]. These constructs are then delivered into cells through plasmid transfection or lentiviral integration, and reporter expression is quantified using luminescence-based readouts. The emergence of high-throughput RNA-sequencing, combined with affordable DNA synthesis and advances in laboratory automation, has enabled more efficient screening methods, including MPRA (Figure 3.5). This technique allows the simultaneous analysis of transcriptional activity of thousands to millions of candidate DNA sequences in a single experiment [101]. MPRA exists in several variants, including episomal MPRA, lenti-MPRA [102, 103] and STARR-seq [104], but all share a common framework: libraries of CREs are cloned alongside barcoded reporter genes. These libraries are then

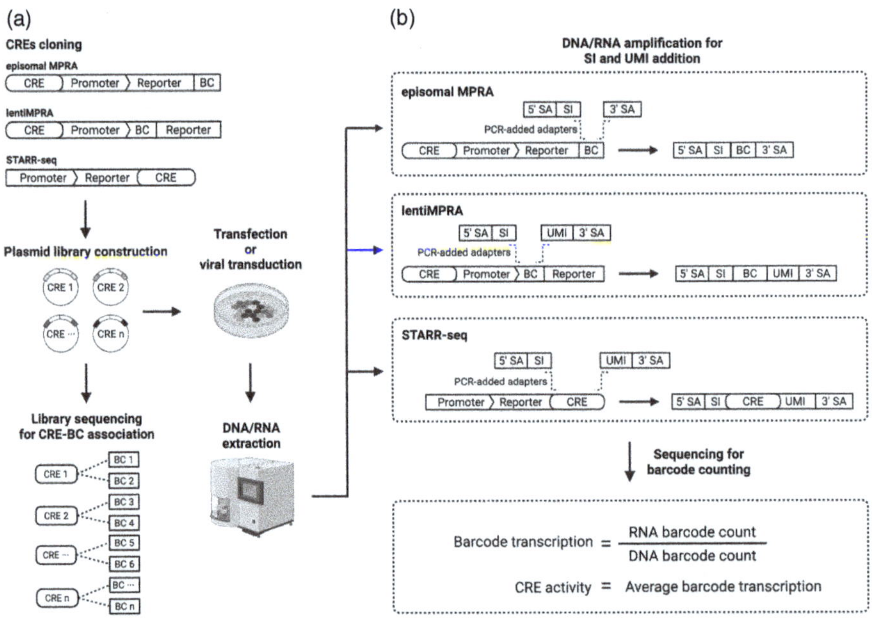

Figure 3.5 Massively parallel reporter assay (MPRA) workflow for validating synthetic regulatory DNA activity. (a) Library design and cloning. Synthetic candidate cis-regulatory elements (CREs) are each linked to multiple unique DNA barcodes (BCs) during oligonucleotide synthesis. Barcode–CRE pairs are cloned into a plasmid backbone (upstream or downstream of a minimal promoter–reporter gene cassette depending on the format) to create a plasmid library in which each CRE is unambiguously associated with its barcodes. (b) CRE activity quantification. Following library transfection (or transduction), both plasmid (or genomic) DNA and total RNA are extracted. Barcode-containing fragments are PCR-amplified from each pool and subjected to high-throughput sequencing. For each CRE, barcode transcription is computed as the barcode counts in the RNA divided by the barcode counts in the DNA. CRE overall activity is defined as the mean barcode transcription across its associated barcodes. *Source*: BioRender.

delivered into cells, and transcriptional activity is quantified in parallel using RNA sequencing. In barcode-based MPRAs, the RNA expression of the associated barcode serves as a proxy for activity, while in STARR-seq, the candidate sequence itself is transcribed and directly measured, eliminating the need for separate barcodes.

One prokaryote MPRA dataset was derived from metagenomic mining of bacterial regulatory sequences [105]. Over 29,000 165-bp regulatory sequences from 184 prokaryotic genomes were synthesized, barcoded, and cloned upstream of a GFP reporter in species-specific plasmids. These constructs were transformed into various bacterial hosts and assayed using RNA-seq and FACS-seq. The resulting dataset was used to train DeepSEED, a DL framework that learned the relationship between promoter sequence and activity and allowed the design of species-specific synthetic promoters [39]. The 2024 DREAM Random Promoter Challenge generated a FACS-based MPRA dataset consisting of over 6.7 million random 80-bp sequences cloned into a promoter context and expressed in *S. cerevisiae.* Fluorescence measurements and high-throughput sequencing quantifications enabled the training of a range of models to predict RNA expression from DNA sequence alone. The participating teams employed a variety of DL architectures, including CNNs, recurrent neural networks (RNNs), and Transformers, to achieve strong sequence-to-expression prediction performance [106]. In an earlier study, Cuperus *et. al.* assayed the activity of 500,000 50-nucleotide-long random 5′-UTRs in *S. cerevisiae* using a histidine-negative growth assay MPRA [107]. A CNN trained using this dataset predicted the effect of both random and native 5′ UTR on protein expression and guided the design of synthetic 5′ UTR by single nucleotide *in silico* evolution. The same group used an MPRA-based approach to quantify the activity of 280,000 random 5′ UTR in human cell lines [57]. The resulting dataset was used to train Optimus 5′-Prime, a CNN that predicts ribosome loading from 5′ UTR sequences. This model was successfully used in conjunction with a genetic algorithm to guide the design of novel 5′ UTRs with precise translational control. The same dataset supported the training of other 5′ UTR activity predictors, such as the CNN-based FramePool series [108], and the Transformer-based UTR-LM [109] and UTR-Insight [110]. MPRA has also been widely used to investigate distal regulatory elements, such as enhancers. For example, UMI-STAR-seq, a variant of STARR-seq, was used to generate high-resolution activity maps of developmental and housekeeping enhancers in Drosophila S2 cells [64]. This dataset was used to train DeepSTARR, a CNN based on Basset architecture [111], to identify novel enhancers and predict their activity. DeepSTARR was used to predict the activity of 1 billion random 249-bp DNA sequences that were not present in the *Drosophila* genome, and the experimental validation of 249 selected sequences showed a Pearson correlation of 0.62, demonstrating its predictive potential. Using an episomal MPRA dataset comprising 776,474 200-bp sequences assayed in K562, HepG2, and SK-N-SH human cell lines, Gosai *et al* [46] trained Malinois, another Basset-based CNN activity predictor. Combining Malinois with sequence design algorithms, they developed CODA (Computational Optimization of DNA Activity), a platform for designing novel cell-type-specific CREs. Importantly, these synthetic CREs were functionally validated *in vitro* using MPRAs and *in vivo* in physiologically related tissues in mice and zebrafish. The same dataset also supported the training of regLM, a workflow that combines HyenaDNA with an oracle-like regressor to generate synthetic cell-type-specific human enhancers [83]. A lentiMPRA approach was used to quantify the transcriptional activity of over 680,000 candidate CREs across three human cell types (HepG2, K562 and WTC11) [112]. The resulting dataset was used to benchmark the predictive potential of four models, with MPRALegNet achieving the highest performance (Pearson $r \approx 0.83$).

Altogether, these examples show how sequence-to-function datasets can be leveraged to train DL models that map DNA sequence to regulatory activity. These approaches are transforming our ability to interpret regulatory codes, design optimized genetic parts, and predict the functional consequences of sequence variation. As experimental and computational technologies continue to evolve, integrating diverse datasets will be key to advancing synthetic biology – from gene circuit design to novel therapies development.

3.4 Evaluating Synthetic Biological Parts Using Motif Analysis and Deep Learning

Designing synthetic biological parts using DL requires not only the generation of plausible sequences but also rigorous evaluation based on biological functionality and context. To streamline assessment across the biological parts, a range of quantitative evaluation strategies has been established. These include distributional similarity metrics (e.g. Fréchet Inception Distance, k-mer correlation), motif syntax evaluations (e.g. spacing and co-occurrence of TFBS), and structural feature checks (e.g. Kozak sequence presence, GC content, or polyadenylation signals). Table 3.2 summarizes these key metrics, offering a compact reference for how synthetic sequences can be quantitatively assessed against natural counterparts, enabling principled design and iterative optimization of synthetic biological systems.

While Table 3.2 provides a structured starting point for evaluating generated sequences, it is important to recognize that the underlying regulatory grammar of DNA is inherently flexible and context-dependent, rather than governed by universally fixed rules. DL models have proven essential in uncovering these complex patterns, learning positional constraints, motif syntax, and higher-order interactions directly from sequence data. For instance, species-aware DNA language models trained on over 800 fungal genomes captured species-specific promoter features, including the preferred position of the TATA box (~30 bp upstream of the TSS in *S. pombe*, 50–120 bp in *S. cerevisiae*) and long-range interactions such as TATA-INR spacing and RNA structural elements [113].

Crucially, DL models can also decode motif syntax – the arrangement, spacing, and combinations of regulatory motifs that influence function. A CNN trained on MPRA data of synthetic B-cell promoters learned that transcriptional activity peaks when the OCT motif lies within 45 bp of the core promoter, and identified synergistic combinations like CArG-box + Runx1 + Arid3a [114]. Similarly, the DeepSTARR model, trained on thousands of Drosophila enhancer sequences, found that only

Table 3.2 Metrics used in generative tasks.

Metric	Objective	Description
Fréchet Inception Distance (FID)	Similarity between datasets	Compares the mean and covariance statistics of the learned latent space corresponding to generated and real images.
k-mer frequency correlation (R^2)	Similarity between sequences	Computes k-mer counts across real and generated sequences, plots natural vs synthetic frequencies, and uses the coefficient of determination from a linear fit.
Levenshtein Distance or Edit Distance	Similarity between sequences	Computes the minimum number of insertions, deletions, or substitutions needed to turn a generated sequence into a real reference sequence.
Jaccard Similarity on k-mer counts	Compositional similarity between sequences	Measures the overlap of k-mer sets between generated and real sequences.
Cosine Similarity on k-mer counts	Compositional similarity between sequences	Treats sequences as k-mer frequency vectors and computes the cosine of the angle between generated vs. real frequency profiles.
TATA-box position	Core promoter positional accuracy	Assesses whether synthetic promoters place the TATA-box at species-specific distances upstream of the TSS (e.g. ~30 bp in *S. pombe*, 50–120 bp in *S. cerevisiae*).
TFBS density	Motif content similarity	Counts known TFBS per unit length in synthetic sequences and compares to natural promoters/enhancers to ensure comparable motif abundance.
Motif spacing consistency	Regulatory grammar consistency	Assesses distances between key motifs – e.g. OCT motif to core promoter – against empirically determined optimal ranges (e.g. ≤45 bp) to ensure spatial arrangement supports activity.
Motif co-occurrence	Regulatory grammar consistency	Measures frequency of specific motif pairs or higher-order combinations (e.g. CArG-box + Runx1 + Arid3a) appearing within defined spatial windows, reflecting synergistic interactions.

(Continued)

Table 3.2 (Continued)

Metric	Objective	Description
GC content	Compositional context fidelity	Compares overall and region-specific GC percentage in synthetic sequences to natural distributions, ensuring organism-specific base composition constraints are respected.
Kozak sequence presence	5′ UTR accuracy	Quantifies matching to the consensus GCCRCCAUGG motif surrounding the start codon in the 5′ UTR (R = purine). Higher scores indicate optimal ribosomal recognition contexts.
Poly(A) signal presence	3′ UTR termination signal accuracy	Detects canonical AAUAAA (and variant) hexamers ~10–30 nt upstream of transcript end in synthetic 3′ UTRs. Ensures correct positioning and sequence context for efficient cleavage and polyadenylation.
Codon Adaptation Index (CAI)	Codon-usage optimization	Measures how well the synonymous codon composition of a coding sequence matches the preferred codon usage of a host organism. Higher CAI indicates better expected translation efficiency.

specific motif instances (e.g. a single GATA motif) drive activity, depending on local context and spacing. These examples show how DL not only captures biologically meaningful features but also formalizes a probabilistic, context-aware grammar of regulatory DNA [64].

3.5 Current Challenges of Generative Part Design

GenAI has proven its capacity to design tailored biological sequences, such as promoters and enhancers, with significant applications in biotechnology – from developing cell-type-specific regulatory elements for therapeutic use to engineering novel enzymes with enhanced catalytic efficiency for industrial purposes [83]. Despite these advancements, current LLMs are limited by their inability to capture long-range interactions in regulatory elements, mainly due to the quadratic complexity ($\mathcal{O}(n^2)$) of existing Transformer models. Emerging models, such as state-space models, like MAMBA [115], address this challenge by achieving linear complexity ($\mathcal{O}(n)$), enabling the training and generation of chromosome-length sequences with reduced memory constraints. The integration of bi-directionality in the Caduceus model [116] can potentially be used for the generation of both upstream and downstream sequences, paving the way for *de novo* chromosome generation.

Another key limitation of LLMs lies in their susceptibility to hallucinations, that is, the generation of biologically implausible or nonfunctional sequences. This issue may stem from limited sequence diversity in training datasets, leading to model overfitting on prevalent patterns while failing to capture rare but functionally important sequence variations. Incomplete or inaccurate genome annotations further compound this problem, often resulting in unreliable outputs. Currently, these challenges remain largely unaddressed in the context of DNA sequence generation. To mitigate such limitations, strategies from natural language processing, such as Reinforcement Learning from Human Feedback (RLHF)) [117], could be adapted to the engineering of synthetic parts. RLHF fine-tunes models based on curated human feedback. In a biological setting, this could be experimental validation feedback, though this will come at significant computational and labor cost.

Despite rapid progress in using DL models to design biologically compatible synthetic DNA sequences, the vast majority of these sequences have not yet been experimentally validated. Current validation approaches often depend on computational metrics like k-mer similarity, which, while useful proxies, are insufficient to fully capture biological function or regulatory effectiveness. Bridging the gap between *in silico* predictions and experimental validation remains a major challenge in the field of generative part design.

3.6 Conclusion

GenAI marks a paradigm shift in synthetic biology, offering a streamlined alternative to the traditional design-build-test-learn cycle. By learning complex sequence–function relationships from large-scale datasets, these models can generate novel, realistic functional sequences with reduced experimental burden. As predictive and generative models continue to improve, we anticipate their integration will become central to the rational programmable design of synthetic elements across diverse organisms. By tailoring the synthesis of biological sequences to specific user needs, GenAI will drive innovation in the bioeconomy and expand the potential of the biotechnology industry.

References

1 Clarke, L. and Kitney, R. (2020). Developing synthetic biology for industrial biotechnology applications. *Biochem. Soc. Trans.* 20 (48): 113–122.

2 Liu, Z., Wang, J., and Nielsen, J. (2022). Yeast synthetic biology advances biofuel production. *Curr. Opin. Microbiol.* 65: 33–39.

3 Georgianna, D.R. and Mayfield, S. (2012). Exploiting diversity and synthetic biology for the production of algal biofuels. *Nature* 15 (488): 329–335.

4 Tan, X., Letendre, J.H., Collins, J.J., and Wong, W.W. (2021). Synthetic biology in the clinic: engineering vaccines, diagnostics, and therapeutics. *Cell* 184 (4): 881–898.

5 Jullesson, D., David, F., Pfleger, B., and Nielsen, J. (2015). Impact of synthetic biology and metabolic engineering on industrial production of fine chemicals. *Biotechnol. Adv.* 33 (7): 1395–1402.

6 Elowitz, M.B. and Leibler, S. (2000). A synthetic oscillatory network of transcriptional regulators. *Nature* 403 (6767): 335–338.

7 Gardner, T., Cantor, C., and Collins, J. (2000). Construction of a genetic toggle switch in Escherichia coli. *Nature* 20 (403): 339–342.

8 Vilanova, C. and Porcar, M. (2014). iGEM 2.0--refoundations for engineering biology. *Nat. Biotechnol.* 32 (5): 420–424.

9 McLaughlin, J.A., Myers, C.J., Zundel, Z. et al. (2018). SynBioHub: a standards-enabled design repository for synthetic biology. *ACS Synth. Biol.* 7 (2): 682–688.

10 Georgescu, R., Bandara, G., and Sun, L. (2003). Saturation mutagenesis. *Methods Mol. Biol.* 231: 75–83.

11 Kretz, K.A., Richardson, T.H., Gray, K.A. et al. (2004). Gene site saturation mutagenesis: a comprehensive mutagenesis approach. *Methods Enzymol.* 388: 3–11.

12 Cirino, P.C., Mayer, K.M., and Umeno, D. (2003). Generating mutant libraries using error-prone PCR. *Methods Mol. Biol.* 231: 3–9.

13 McCullum, E.O., Williams, B.A.R., Zhang, J., and Chaput, J.C. (2010). Random mutagenesis by error-prone PCR. *Methods Mol. Biol.* 634: 103–109.

14 Wang, Y., Xue, P., Cao, M. et al. (2021). Directed evolution: methodologies and applications. *Chem. Rev.* 121 (20): 12384–12444.

15 Arnold, F.H. (2018). Directed evolution: bringing new chemistry to life. *Angew. Chem. Int. Ed. Engl.* 57 (16): 4143–4148.

16 Currin, A., Parker, S., Robinson, C.J. et al. (2021). The evolving art of creating genetic diversity: from directed evolution to synthetic biology. *Biotechnol. Adv.* 50: 107762.

17 Packer, M. and Liu, D.R. (2015). Methods for the directed evolution of proteins. *Nat. Rev. Genet.* (16): 379–394.

18 Wang, B., Liu, Y., Bai, X. et al. (2024). In vitro generation of genetic diversity for directed evolution by error-prone artificial DNA synthesis. *Commun. Biol.* 7 (1): 628.

19 Sarkisyan, K.S., Bolotin, D.A., Meer, M.V. et al. (2016). Local fitness landscape of the green fluorescent protein. *Nature* 533 (7603): 397–401.

20 Romero, P.A. and Arnold, F.H. (2009). Exploring protein fitness landscapes by directed evolution. *Nat. Rev. Mol. Cell Biol.* 10 (12): 866–876.

21 Repecka, D., Jauniskis, V., Karpus, L. et al. (2021). Expanding functional protein sequence spaces using generative adversarial networks. *Nat. Mach. Intell.* 4: 1–10.

22 Yona, A.H., Alm, E.J., and Gore, J. (2018). Random sequences rapidly evolve into de novo promoters. *Nat. Commun.* 9 (1): 1530.

23 de Boer, C.G., Vaishnav, E.D., Sadeh, R. et al. (2020). Deciphering eukaryotic gene-regulatory logic with 100 million random promoters. *Nat. Biotechnol.* 38 (1): 56–65.

24 Whitrow, G.J. (1952). An elementary derivation of Eddington's number 1078. *Nature* 169 (4302): 611–612.

25 Zrimec, J., Fu, X., Muhammad, A.S. et al. (2022). Controlling gene expression with deep generative design of regulatory DNA. *Nat. Commun.* 13 (1): 5099.

26 Tian, R., Rehm, F.B.H., Czernecki, D. et al. (2024). Establishing a synthetic orthogonal replication system enables accelerated evolution in E. coli. *Science* 383 (6681): 421–426.

27 Sellés Vidal, L., Isalan, M., Heap, J.T., and Ledesma-Amaro, R. (2023). A primer to directed evolution: current methodologies and future directions. *RSC Chem. Biol.* 4 (4): 271–291.

28 Salis, H. (2011). The ribosome binding site calculator. *Methods Enzymol.* 498: 19–42.

29 Salis, H.M., Mirsky, E.A., and Voigt, C.A. (2009). Automated design of synthetic ribosome binding sites to control protein expression. *Nat. Biotechnol.* 27 (10): 946–950.

30 Na, D. and Lee, D. (2010). RBSDesigner: software for designing synthetic ribosome binding sites that yields a desired level of protein expression. *Bioinformatics* 26 (20): 2633–2634.

31 Seo, S.W., Yang, J.S., Kim, I. et al. (2013). Predictive design of mRNA translation initiation region to control prokaryotic translation efficiency. *Metab. Eng.* 15: 67–74.

32 Bonde, M.T., Pedersen, M., Klausen, M.S. et al. (2016). Predictable tuning of protein expression in bacteria. *Nat. Methods* 13 (3): 233–236.

33 Zhang, M., Holowko, M.B., Hayman Zumpe, H., and Ong, C.S. (2022). Machine learning guided batched design of a bacterial ribosome binding site. *ACS Synth. Biol.* 11 (7): 2314–2326.

34 Buenrostro, J.D., Wu, B., Chang, H.Y., and Greenleaf, W.J. (2015). ATAC-seq: a Method for assaying chromatin accessibility genome-wide. *Curr. Protoc. Mol. Biol.* (109): 21.29.1–21.29.9.

35 Nagano, T., Lubling, Y., Stevens, T.J. et al. (2013). Single-cell Hi-C reveals cell-to-cell variability in chromosome structure. *Nature* 502 (7469): 59–64.

36 Cusanovich, D.A., Reddington, J.P., Garfield, D.A. et al. (2018). The cis-regulatory dynamics of embryonic development at single-cell resolution. *Nature* 555 (7697): 538–542.

37 Luo, Y., Hitz, B.C., Gabdank, I. et al. (2020). New developments on the encyclopedia of DNA Elements (ENCODE) data portal. *Nucleic Acids Res.* 48 (D1): D882–D889.

38 De Winter, S., Konstantakos, V., and Aerts, S. (2025). Modelling and design of transcriptional enhancers. *Nat. Rev. Bioeng.* 3 (5): 374–389.

39 Zhang, P., Wang, H., Xu, H. et al. (2023). Deep flanking sequence engineering for efficient promoter design using DeepSEED. *Nat. Commun.* 14 (1): 6309.

40 Van Brempt, M., Clauwaert, J., Mey, F. et al. (2020). Predictive design of sigma factor-specific promoters. *Nat. Commun.* 11 (1): 5822.

41 Wang, Y., Wang, H., Wei, L. et al. (2020). Synthetic promoter design in Escherichia coli based on a deep generative network. *Nucleic Acids Res.* 48 (12): 6403–6412.

42 Zrimec, J., Börlin, C.S., Buric, F. et al. (2020). Deep learning suggests that gene expression is encoded in all parts of a co-evolving interacting gene regulatory structure. *Nat. Commun.* 11 (1): 6141.

43 Taskiran, I.I., Spanier, K.I., Dickmänken, H. et al. (2024). Cell-type-directed design of synthetic enhancers. *Nature* 626 (7997): 212–220.

44 de Almeida, B.P., Schaub, C., Pagani, M. et al. (2024). Targeted design of synthetic enhancers for selected tissues in the Drosophila embryo. *Nature* 626 (7997): 207–211.

45 Ventimiglia, L.N. and Zelezniak, A. (2025). Programming of synthetic regulatory DNA for cell-type targeting in humans. *Mol. Cell* 85 (2): 205–207.

46 Gosai, S.J., Castro, R.I., Fuentes, N. et al. (2024). Machine-guided design of cell-type-targeting cis-regulatory elements. *Nature* 634 (8036): 1211–1220.

47 LeCun, Y., Boser, B., Denker, J. et al. (1989). Handwritten digit recognition with a back-propagation network. *Neural. Inf. Process. Syst.* 2: 396–404.

48 Alipanahi, B., Delong, A., Weirauch, M.T., and Frey, B.J. (2015). Predicting the sequence specificities of DNA- and RNA-binding proteins by deep learning. *Nat. Biotechnol.* 33 (8): 831–838.

49 Novakovsky, G., Dexter, N., Libbrecht, M.W. et al. (2023). Obtaining genetics insights from deep learning via explainable artificial intelligence. *Nat. Rev. Genet.* 24 (2): 125–137.

50 Goodfellow, I., Pouget-Abadie, J., Mirza, M. et al. (2020). Generative adversarial networks. *Commun. ACM* 63 (11): 139–144.

51 Killoran, N., Lee, L. J., Delong, A., et al. (2017). Generating and designing DNA with deep generative models [Internet]. *arXiv [cs.LG]*. http://arxiv.org/abs/1712.06148.

52 Ho, J., Jain, A., and Abbeel, P. (2020; abs/2006.11239). Denoising Diffusion Probabilistic Models. *Adv.Neural. Inf. Process. Syst.* 33: 6840–6851.

53 DaSilva LF, Senan S, Patel ZM, Janardhan Reddy A, Gabbita S, Nussbaum Z, et al. DNA-diffusion: leveraging generative models for controlling chromatin accessibility and gene expression via synthetic regulatory elements. *bioRxiv*. 2024. doi:https://doi.org/10.1101/2024.02.01.578352.

54 Radford, A., Narasimhan, K. (2018). Improving language understanding by generative pre-training; https://www.mikecaptain.com/resources/pdf/GPT-1.pdf.

55 Agarwal, V. and Shendure, J. (2020). Predicting mRNA abundance directly from genomic sequence using deep convolutional neural networks. *Cell Rep.* 31 (7): 107663.

56 Avsec, Ž., Agarwal, V., Visentin, D. et al. (2021). Effective gene expression prediction from sequence by integrating long-range interactions. *Nat. Methods* 18 (10): 1196–1203.

57 Sample, P.J., Wang, B., Reid, D.W. et al. (2019). Human 5′ UTR design and variant effect prediction from a massively parallel translation assay. *Nat. Biotechnol.* 37 (7): 803–809.

58 Kshirsagar, M., Yuan, H., Ferres, J.L., and Leslie, C. (2022). BindVAE: Dirichlet variational autoencoders for de novo motif discovery from accessible chromatin. *Genome Biol.* 23 (1): 174.

59 Frazer, J., Notin, P., Dias, M. et al. (2021). Disease variant prediction with deep generative models of evolutionary data. *Nature* 599 (7883): 91–95.

60 Pan X, Rijnbeek P, Yan J, Shen HB. Prediction of RNA-protein sequence and structure binding preferences using deep convolutional and recurrent neural networks. *BMC Genomics* 2018; 19(1). doi:https://doi.org/10.1186/s12864-018-4889-1.

61 Kelley, D.R., Reshef, Y.A., Bileschi, M. et al. (2018). Sequential regulatory activity prediction across chromosomes with convolutional neural networks. *Genome Res.* 28 (5): 739–750.

62 Schwessinger, R., Gosden, M., Downes, D. et al. (2020). DeepC: predicting 3D genome folding using megabase-scale transfer learning. *Nat. Methods* 17 (11): 1118–1124.

63 Fudenberg, G., Kelley, D.R., and Pollard, K.S. (2020). Predicting 3D genome folding from DNA sequence with Akita. *Nat. Methods* 17 (11): 1111–1117.

64 de Almeida, B.P., Reiter, F., Pagani, M., and Stark, A. (2022). DeepSTARR predicts enhancer activity from DNA sequence and enables the de novo design of synthetic enhancers. *Nat. Genet.* 54 (5): 613–624.

65 Jores, T., Tonnies, J., Wrightsman, T. et al. (2021). Synthetic promoter designs enabled by a comprehensive analysis of plant core promoters. *Nat. Plants* 7 (6): 842–855.

66 Géron, A. (2022). *Hands-on Machine Learning with Scikit-learn, Keras, and Tensorflow*, 3e, 864. O'Reilly Media.

67 James, G., Witten, D., Hastie, T., and Tibshirani, R. (2013). *An Introduction to Statistical Learning: with Applications in R*. New York, NY: Springer.

68 Hastie, T., Tibshirani, R., and Friedman, J. (2009). *The Elements of Statistical Learning: Data Mining, Inference, and Prediction*, 2e, 745. Springer Science & Business Media.

69 Benjamini, Y. and Speed, T.P. (2012). Summarizing and correcting the GC content bias in high-throughput sequencing. *Nucleic Acids Res.* 40 (10): e72.

70 Conesa A, Madrigal P, Tarazona S, Gomez-Cabrero D, Cervera A, McPherson A, et al. A survey of best practices for RNA-seq data analysis [Internet]. Vol. 17, *Genome Biol.* 2016. doi:https://doi.org/10.1186/s13059-016-0881-8.

71 Kapoor, S. and Narayanan, A. (2023). Leakage and the reproducibility crisis in machine-learning-based science. *Patterns* 4 (9): 100804.

72 Whalen, S., Schreiber, J., Noble, W.S., and Pollard, K.S. (2022). Navigating the pitfalls of applying machine learning in genomics. *Nat. Rev. Genet.* 23 (3): 169–181.

73 Nair, V. and Hinton, G.E. (2010). Rectified linear units improve restricted Boltzmann machines. *ICML* 21: 807–814.

74 Cui, H., Wang, C., Maan, H. et al. (2024). scGPT: toward building a foundation model for single-cell multi-omics using generative AI. *Nat. Methods* 21 (8): 1470–1480.

75 Lin, Z., Akin, H., Rao, R. et al. (2023). Evolutionary-scale prediction of atomic-level protein structure with a language model. *Science* 379 (6637): 1123–1130.

76 Sumi, S., Hamada, M., and Saito, H. (2024). Deep generative design of RNA family sequences. *Nat. Methods* 21 (3): 435–443.

77 Riley, A.T., Robson, J.M., Ulanova, A., and Green, A.A. (2025). Generative and predictive neural networks for the design of functional RNA molecules. *Nat. Commun.* 16 (1): 4155.

78 Xia, Y., Du, X., Liu, B. et al. (2024). Species-specific design of artificial promoters by transfer-learning based generative deep-learning model. *Nucleic Acids Res.* 52 (11): 6145–6157.

79 Fallahpour, A., Gureghian, V., Filion, G.J. et al. (2025). CodonTransformer: a multispecies codon optimizer using context-aware neural networks. *Nat. Commun.* 16 (1): 3205.

80 Nguyen, E., Poli, M., Durrant, M.G. et al. (2024). Sequence modeling and design from molecular to genome scale with Evo. *Science* 386 (6723): eado9336.

81 Vaswani, A., Shazeer, N., Parmar, N.et al. (2017). Attention is all you need. *arXiv [cs.CL]*. http://arxiv.org/abs/1706.03762.

82 Brown, T.B, Mann, B., Ryder N. et al. (2020). Language models are few-shot learners. *arXiv [cs.CL]*. http://arxiv.org/abs/2005.14165.

83 Lal, A., Garfield, D., Biancalani, T., and Eraslan, G. (2024). Designing realistic regulatory DNA with autoregressive language models. *Genome Res.* 34 (9): 1411–1420.

84 Nguyen, E., Poli, M., Faizi, M., et al. (2023). HyenaDNA: long-range genomic sequence modeling at single nucleotide resolution. *arXiv [cs.LG]*. http://arxiv.org/abs/2306.15794.

85 Zhou, Z., Ji, Y., Li, W., et al. (2023). DNABERT-2: efficient foundation model and benchmark for multi-species genome. *arXiv [q-bio.GN].* http://arxiv.org/abs/2306.15006.

86 Dalla-Torre, H., Gonzalez, L., Mendoza-Revilla, J. et al. (2025). Nucleotide transformer: building and evaluating robust foundation models for human genomics. *Nat. Methods* 22 (2): 287–297.

87 Ji Y, Zhou Z, Liu H, Davuluri R. DNABERT: pre-trained bidirectional encoder representations from transformers model for DNA-language in genome. *bioRxiv* 2020. doi:https://doi.org/10.1101/2020.09.17.301879.

88 Devlin, J., Chang, M. W., Lee, K., Toutanova, K. (2018). BERT: pre-training of deep bidirectional transformers for language understanding. *arXiv [cs.CL].* http://arxiv.org/abs/1810.04805.

89 Sarkar A, Tang Z, Zhao C, Koo P. Designing DNA with tunable regulatory activity using discrete diffusion. *bioRxiv* 2024. doi:https://doi.org/10.1101/2024.05.23.595630.

90 Li, Z., Ni, Y., Huygelen, T. A. B. et al. (2023). Latent diffusion model for DNA sequence generation. *arXiv [cs.LG].* https://github.com/Zehui127/Latent-DNA-Diffusion.

91 Huang, K., Yang, Y., Fu, K.et al. (2024). Latent diffusion models for controllable RNA sequence generation. *arXiv [cs.LG].* http://arxiv.org/abs/2409.09828.

92 Ronneberger, O., Fischer, P., Brox, T. (2015). U-Net: convolutional networks for biomedical image segmentation. *arXiv [cs.CV].* http://arxiv.org/abs/1505.04597.

93 The GTEx Consortium. (2020). The GTEx Consortium atlas of genetic regulatory effects across human tissues. *Science.* doi: https://doi.org/10.1126/science.aaz1776.

94 Roadmap Epigenomics Consortium, Kundaje, A., Meuleman, W. et al. (2015). Integrative analysis of 111 reference human epigenomes. *Nature* 518 (7539): 317–330.

95 Abugessaisa, I., Ramilowski, J.A., Lizio, M. et al. (2021). FANTOM enters 20th year: expansion of transcriptomic atlases and functional annotation of non-coding RNAs. *Nucleic Acids Res.* 49 (D1): D892–D898.

96 George, N., Fexova, S., Fuentes, A.M. et al. (2024). Expression atlas update: insights from sequencing data at both bulk and single cell level. *Nucleic Acids Res.* 52 (D1): D107–D114.

97 Salgado, H., Gama-Castro, S., Lara, P. et al. (2024). RegulonDB v12.0: a comprehensive resource of transcriptional regulation in E. coli K-12. *Nucleic Acids Res.* 52 (D1): D255–D264.

98 Linder, J., Srivastava, D., Yuan, H. et al. (2025). Predicting RNA-seq coverage from DNA sequence as a unifying model of gene regulation. *Nat. Genet.* 57 (4): 949–961.

99 Hingerl JC, Karollus A, Gagneur J. Flashzoi: an enhanced Borzoi model for accelerated genomic analysis [Internet]. *bioRxiv.* 2024. doi:https://doi.org/10.1101/2024.12.18.629121.

100 Li, X., Eastman, E., Schwartz, R., and Draghia-Akli, R. (1999). Synthetic muscle promoters: activities exceeding naturally occurring regulatory sequences. *Nat. Biotechnol.* 17: 241–245.

101 Patwardhan, R.P., Lee, C., Litvin, O. et al. (2009 Dec). High-resolution analysis of DNA regulatory elements by synthetic saturation mutagenesis. *Nat. Biotechnol.* 27 (12): 1173–1175.

102 Gordon, M.G., Inoue, F., Martin, B. et al. (2020). lentiMPRA and MPRAflow for high-throughput functional characterization of gene regulatory elements. *Nat. Protoc.* 15 (8): 2387–2412.

103 Inoue, F., Kircher, M., Martin, B. et al. (2017). A systematic comparison reveals substantial differences in chromosomal versus episomal encoding of enhancer activity. *Genome Res.* 27 (1): 38–52.

104 Arnold, C.D., Gerlach, D., Stelzer, C. et al. (2013). Genome-wide quantitative enhancer activity maps identified by STARR-seq. *Science* 1 (339): 1074–1077.

105 Johns, N.I., Gomes, A.L.C., Yim, S.S. et al. (2018). Metagenomic mining of regulatory elements enables programmable species-selective gene expression. *Nat. Methods* 15 (5): 323–329.

106 Rafi AM, Nogina D, Penzar D, Lee D, Lee D, Kim N, et al. A community effort to optimize sequence-based deep learning models of gene regulation. *Nat. Biotechnol.* 2024; doi: https://doi.org/10.1038/s41587-024-02414-w.

107 Cuperus, J.T., Groves, B., Kuchina, A. et al. (2017). Deep learning of the regulatory grammar of yeast 5′ untranslated regions from 500,000 random sequences. *Genome Res.* 27 (12): 2015–2024.

108 Karollus, A., Avsec, Ž., and Gagneur, J. (2021). Predicting mean ribosome load for 5'UTR of any length using deep learning. *PLoS Comput. Biol.* 17 (5): e1008982.

109 Chu, Y., Yu, D., Li, Y. et al. (2024). A 5′ UTR language model for decoding untranslated regions of mRNA and function predictions. *Nat. Mach. Intell.* 6 (4): 449–460.

110 Pan, S., Wang, H., Zhang, H. et al. (2025). UTR-Insight: integrating deep learning for efficient 5' UTR discovery and design. *BMC Genomics* 26 (1): 107.

111 Kelley, D.R., Snoek, J., and Rinn, J.L. (2016). Basset: learning the regulatory code of the accessible genome with deep convolutional neural networks. *Genome Res.* 26 (7): 990–999.

112 Agarwal, V., Inoue, F., Schubach, M. et al. (2025). Massively parallel characterization of transcriptional regulatory elements. *Nature* 639 (8054): 411–420.

113 Karollus, A., Hingerl, J., Gankin, D. et al. (2024). Species-aware DNA language models capture regulatory elements and their evolution. *Genome Biol.* 25 (1): 83.

114 Fu, Z.H., He, S.Z., Wu, Y., and Zhao, G.R. (2023). Design and deep learning of synthetic B-cell-specific promoters. *Nucleic Acids Res.* 51 (21): 11967–11979.

115 Gu, A., Dao, T. (2023). Mamba: linear-time sequence modeling with selective state spaces [Internet]. *arXiv [cs.LG]*. http://arxiv.org/abs/2312.00752.

116 Schiff, Y., Kao, C. H., Gokaslan, A., et al. (2024). Caduceus: bi-directional equivariant long-range DNA sequence modeling [Internet]. *arXiv [q-bio.GN]*. http://arxiv.org/abs/2403.03234.

117 Ouyang, L., Wu, J., Jiang, X., et al. (2022). Training language models to follow instructions with human feedback [Internet]. *arXiv [cs.CL]*. http://arxiv.org/abs/2203.02155.

4

Machine Learning for Sequence-to-Function Approaches

Rana A. Barghout[1,2], Maxim Kirby[1], Austin Zheng[1], Lya Chinas[1], Marjan Mohammadi[1], Zhiqing Xu[1], Benjamin Sanchez-Lengeling[1,2], and Radhakrishnan Mahadevan[1]

[1] *University of Toronto, Department of Chemical Engineering & Applied Chemistry, 200 College St, M5T 3E5, Toronto, ON, Canada*
[2] *Vector Institute for Artificial Intelligence, 108 College St, M5G 0C6, Toronto, ON, Canada*

4.1 Introduction

Proteins, the molecular machines of life, derive their biochemical and cellular functions from linear sequences of amino acids. They can fold into dynamic three-dimensional structures, ranging from ordered to conformationally flexible states. This relationship can be described using a "language" analogy proposed by Bepler and Berger [1]: amino acids serve as a chemical "alphabet" that combines into structural elements ("words"), which when assembled into sentences, acquire a functional "meaning." However, it's important to note that protein sequences differ fundamentally from human language: they operate through physical and chemical interactions rather than symbolic representation, and their "meaning" is governed by biophysical principles rather than semantic conventions. Deciphering how the sequence of a protein encodes its functional outcomes, known as the sequence-structure–function relationship, contributes to a wider comprehension of evolutionary biology.

The linear arrangement of amino acids in a protein sequence encodes the structural and functional properties with high efficiency [2]. Even minor perturbations to this sequence, such as single mutations, can alter their stability, highlighting the difficulty of inferring function solely from sequence information. Overcoming this challenge is essential for applications in disease research and biotechnology [3]. Inferring the function of a protein directly from its sequence would improve the understanding of genetic diseases [4] and identify novel drug targets [5]. It could also enable the design enzymes with enhanced capabilities for the synthesis of chemicals [6] and degradation of pollutants [7].

Machine Learning and Big Data-enabled Biotechnology, First Edition. Edited by Hal S. Alper.

In recent years, modern sequencing technologies have produced an exponential growth of sequence data. As of 2025, the UniProt database contained over 253 million protein sequences [8], yet fewer than 0.3% had experimentally validated annotations of their function [9]. This is due to the labor-intensive and time-consuming nature of experimental assays, which are also prone to biases and have limited coverage across all proteins [10]. The gap between sequencing and functional annotation creates an urgent need for reliable computational methods that would enable the prediction of protein function from its sequence information.

Machine learning (ML) has enabled major breakthroughs in sequence-to-function mapping. Typically framed as a classification or regression task, ML models are trained on raw amino acid sequences (the input) to accurately predict functional characteristics (the output). Traditional ML approaches require the alignment of protein sequences or manual feature engineering; thus, they are limited in capturing the complex, high-dimensional relationships between a sequence and its corresponding function. In contrast, modern ML models can learn features and nonlinear patterns from large datasets, making them well-suited for protein sequence data. Deep learning (DL) models, using neural networks as building blocks for ML models, employ parameterized transformations of data to learn new representations of data that are optimized for the task at hand. The initial breakthroughs were driven by convolutional neural networks (CNNs), with the identification of sequence motifs, which correspond to catalytic sites or binding domains that drive protein function [11]. These models have also been combined with other architectures, such as recurrent neural networks (RNNs), to capture temporal relationships [12]. Most recent models incorporate transformer architectures [13], which can capture long-range dependencies and recognize amino acid sequences crucial to maintaining protein function throughout evolution.

As ML is increasingly applied to sequence-to-function problems, two key priorities are improving generalizability and integrating multi-omics data. Focusing on these aspects ensures that models remain not only accurate on benchmark datasets but also applicable across diverse biological contexts. A model is considered generalizable if it can make accurate predictions for unseen or diverse sequences beyond its training set. Many models have a significant drop in performance when faced with very dissimilar training data, as demonstrated in validation tasks with evolutionarily distant variants [14]. This highlights the challenge of extrapolating predictions to completely unseen sequence spaces. Methods such as self-supervised pretraining [15] and physics-informed architectures [16] help models move beyond memorization to context-aware predictions. Generalization capabilities go hand in hand with the integration of multi-omics data. Biological function is rarely determined solely by sequence; it emerges from a cascade of molecular interactions. Therefore, the training data for these models requires the integration of genomics, transcriptomics, proteomics, metabolomics, and microbiomics [17]. By prioritizing both generalizability and multi-omics integration, sequence-to-function models can move from narrow, benchmark-specific successes to widely usable tools in research (Figure 4.1).

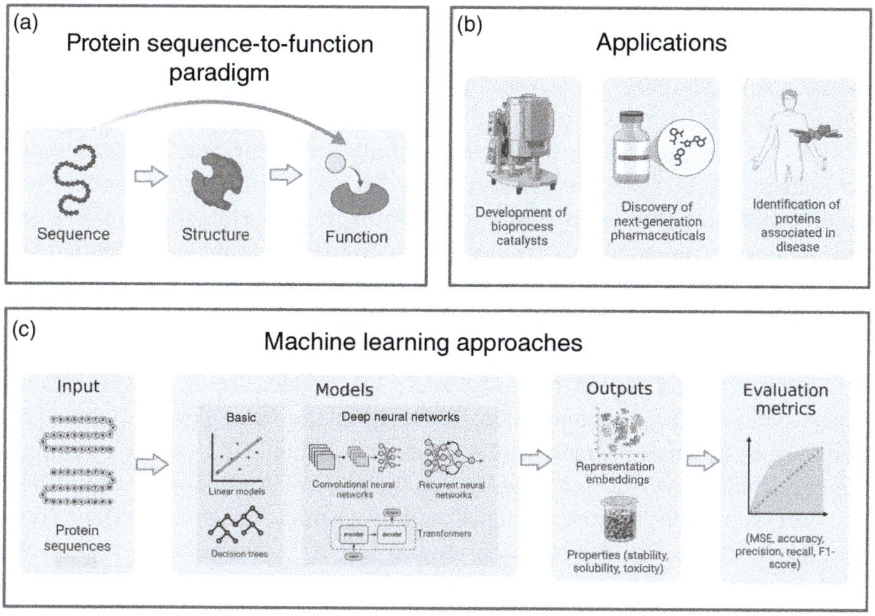

Figure 4.1 Overview of the protein sequence-to-function paradigm and machine learning (ML) approaches. (a) The sequence-to-function paradigm describes how protein sequences determine their structures, which in turn dictate their functions. ML can facilitate direct predictions, bypassing the need for structural determination. (b) Applications of sequence-to-function predictions extend across biotechnology to healthcare sectors. (c) ML approaches involve using protein sequences as inputs, then processed by models ranging from basic methods to advanced deep neural networks. These models produce outputs, including representation embeddings and predictions of biochemical or biophysical characteristics, with performance evaluated through specific metrics.

4.2 Current State of Sequence-to-Function Modeling

4.2.1 Protein Function Prediction: From BLAST to Language Models

Before the now-commonplace ML models were established, many different rule-based methods were proposed. Since 1997, methods such as the basic local alignment search tool (BLAST) and psi-BLAST [18] have been used to predict protein functions by homology by finding sequence similarities in established protein or DNA databases. More recent methods include DIAMOND [19] and HHblits [20], which significantly improve the search throughput for similar sequences.

Early 2000's ML models used algorithms such as k-nearest neighbor (KNN) and support vector machines (SVMs), considering function prediction as a classification problem [21, 22]. Since these ML methods rely on already well-established and annotated data that often comes from biasedly distributed data, they often fail at predicting out-of-distribution protein properties, or proteins with low sequence similarity to existing data.

While supervised learning has advanced protein functionality prediction, self-supervised learning has had a broader impact by exploiting vast unlabeled sequence datasets. Among those techniques, masked language modeling (MLM) has become a dominant strategy. In MLM, models are trained to reconstruct randomly masked amino acids in a sequence using surrounding context, allowing them to learn rich representations.

TAPE [13] was among the first to successfully train a protein language model (pLM) using MLM tasks, demonstrating that a transformer-based model could encode functionality-relevant information from protein sequences. ESM-1b and ESM-2 continuously expanded upon model depth while training on larger datasets, which offered improved performance in both structure and function predictions [23]. Beyond ESM, the ProtTrans suite introduced models like ProtBERT, which retains the bidirectional encoder representation from transformers (BERT)-based MLM approach, and ProtT5 that adapts the T5 encoder-decoder architecture to enable span-masking and enhanced flexibility in sequence representation [24]. In contrast to transformer models, CARP employs dilated CNNs, offering competitive performance with lower computational cost [25]. Adding pLM embeddings have been shown to improve a model's ability to map sequences to their corresponding functions through ablation studies [26, 27], underscoring the efficacy of ML-based self-supervised learning in advancing protein sequence-to-function prediction (Table 4.1).

4.2.2 Gene Ontology

In an effort to standardize the way proteins were described, the community created a vocabulary known as Gene Ontology (GO) in 2000 to provide a consistent way of annotating genes and products in different databases and research groups [28, 29]. GO covers three main aspects of gene/protein functionality: molecular function (MF), biological process (BP), and cellular component/location (CC/CL). The GO structure is hierarchical, with multiple other GO terms branching off from the three parent terms to fully encapsulate gene product functionality. This then provides the perfect groundwork for classification-type ML models to attempt and build GO "trees" for new, unannotated proteins.

In 2017, DeepGO used a CNN to better recognize patterns in sequential data and incorporated multimodal data sources to predict GO annotations [30]. DeepGOPlus [31] and Deep_CNN_LSTM_GO [32] build deeper networks and incorporate long short-term memory (LSTM), leading to predictive improvements and allowing larger sequence input enabling broader protein analysis. In 2021, DeepGraphGO [33] uses a structural GNN protein representation, enabling it to outperform DeepGoPlus when comparing predictions for BP and CC Ontologies. In 2023, DeepGO-SE [34] leveraged pLM embeddings from ESM-2 and allowed the model to generalize and predict GO terms for largely unseen proteins.

4.2.3 Enzyme Commission Numbers

Although GO remains the largest field for classification of sequences, other metrics have also been predicted from other ML/DL models. Within the microbial

Table 4.1 Protein function prediction models: input–output data sources.

Model	Input data	Output data
Gene Ontology		
DeepGO [30]: CNN	Sequences, cross-species PPI networks[a)]	GO terms
DeepGoPlus [31]: CNN	Sequences[b)]	GO terms
Deep_CNN_LSTM_GO [32]: CNN,LSTM	Sequences	GO terms
DeepGraphGO [33]: GNN	Binary feature vectors from sequences from InterPro	GO terms
DeepGO-SE [34]: Neuro-Symbolic Model	ESM-2 (protein embeddings)	GO terms
Enzyme Commission Number		
DEEPre [35]: CNN,RNN,LSTM	Sequences	EC numbers (first 3 digits)
mlDEEPre [36]: CNN,RNN,LSTM	Sequences	EC numbers[d)]
DeepEC [37]: CNN	Sequences	EC numbers
DeepECTransformer [38]: CNN	ProtTrans (pLM embeddings) & Sequences	EC numbers
CLEAN [40]: FFNN	ESM-1b embeddings	EC numbers[c)]
Enzyme Activity		
DLKcat [41]: Attention	CNN (protein embeddings), GNN (substrate embeddings)	k_{cat}
TurNuP [42]: XGBoost	ESM-1b (pLM embeddings) and differential reaction fingerprints	k_{cat}
UniKP [43]: Extra Trees	ProtT5-XL (pLM embeddings) and SMILES transformer (substrate embeddings)	k_{cat}, K_M
CatPred [26]: Gaussian Process	ESM-2 (pLM embeddings) and GNN (substrate embeddings)	k_{cat}, K_M
CPI-Pred [27]: Cross-Attention	Various pLM embeddings and substrate embeddings	k_{cat}, K_M
Thermal Stability		
DeepSTABp [45]: FFNN	FFNN (experimental condition embeddings), ProtT5-XL (pLM embeddings), and FFNN (growth temperature embeddings)	T_m
ProTstab2 [56]: LightGBM	Sequences	T_m
BertThermo [44]: Logistic Regression (LR)	TAPE BERT (pLM embeddings)	Stability
TemProSta [46]: Multiple LR	Sequences	Stability

(Continued)

Table 4.1 (Continued)

Model	Input data	Output data
Toxicity		
ToxinPred [48]: SVM	Compositional/evolutionary/physicochemical features	Toxicity
CSM-Toxin [57]: BERT	ESM-2 (pLM embeddings)	Toxicity
ToxinPred 3.0 [47]: Ensemble of 11 models	Protein sequences	Toxicity
VISH-Pred [49]: Transformer, LightGBM, XGBoost	ESM-2 (pLM embeddings)	Toxicity
Solubility		
PROSO [53]: SVM, Naive Bayes	Expert-curated physicochemical features	Solubility
NetSolP [54]: fine-tuned ESM-1b	Protein sequences	Solubility
RPPSP [55]: XGBoost	Amino acid counts/transitions/distribution features	Solubility
PLM_Sol [52]: LSTM, CNN	ProtT5-XL (pLM embeddings)	Solubility

a) DeepGO has sequence length limitations (<1,002 amino acids) and cannot process proteins with ambiguous amino acids.
b) DeepGoPlus extended the single CNN layer in DeepGO to several layers with different filter lengths and replaced the trigram embedding with one-hot encoding, reducing total parameter count.
c) Instead of directly classifying into EC number labels, CLEAN predicts EC numbers by calculating the distances between EC embeddings.
d) For multi- and monofunctional enzymes.

genome, enzymes constitute one of the most prevalent functional gene classes. Subsequently, their catalytic behavior and functionality are most often represented using an enzyme commission (EC) number. The EC number consists of four digits, representing the class, subclass, sub-subclass, and sub-sub-subclass of the enzyme.

Released in the same time frame as DeepGO, DEEPre [35] utilized raw enzyme sequence data to predict EC numbers. DEEPre uses both CNNs and RNNs, with RNNs extracting sequential, long-term interactions, and was capable of predicting the first three EC digits with a higher degree of accuracy than pre-existing ML models, such as EzyPred, SVM-Prot, COFACTOR, and EFICAz. In 2019, mlDEEPre was released, with additional capabilities to predict functionality for multifunctional enzymes instead of only monofunctional enzymes [36].

DeepEC [37], developed at the same time, consisted of three separate independent CNNs. Additionally, DeepEC implements homology analysis using DIAMOND for EC numbers that cannot be classified by the CNNs. In 2023, DeepECTransformer was released, using transformer layers in its neural network for prediction [38].

It covers more EC numbers compared to DeepEC, and employs a linear layer BERT architecture [39]. The improved architecture allows it to better predict enzymes with low sequence identities in the training set, relying less on the homology analysis than DeepEC utilized.

Finally, the CLEAN model (2023) reframed the EC problem from a multilabel classification to a contrastive learning framework [40]. It uses protein representations obtained from the ESM-1b pLM and a feed-forward neural network (FFNN). Using a distance-based prediction in its learning framework allows it to more accurately and reliably predict EC numbers, particularly for cases like understudied or mislabeled enzymes, outperforming traditional classification models to date. Many of these EC-based learning methods are further detailed in Chapter 5 of the book.

4.2.4 Enzyme Activity

Enzyme activity refers to an enzyme's ability to convert substrate into products. Enzyme activity is often represented by kinetic parameters such as rate of substrate conversion (k_{cat}), substrate binding affinity (K_M), and catalytic efficiency (k_{cat}/K_M). Accurately predicting these parameters is essential for various applications, including enzyme engineering, metabolic modeling, and drug development. Reliable predictions can accelerate the discovery of biocatalysts, optimize pathway design for synthetic biology, and reduce the need for costly and time-consuming experimental characterization.

Recent advances in kinetic parameter prediction have markedly improved k_{cat} and K_M estimates. In 2022, DLKcat aimed to make predictions using protein sequences and substrate representations but struggled with enzymes lacking close homologs [41]. In 2023, TurNuP set out to build a more generalizable k_{cat} predictor by leveraging embeddings from pLMs for sequences and chemical fingerprints for substrates [42]. UniKP expanded this work by not only using pretrained embeddings from pLMs but also from chemical models [43]. With this new framework, UniKP was used to predict different kinds of kinetic parameters, such as K_M (in addition to k_{cat}), resulting in improved performance for all kinetic parameters. In 2024, CatPred utilized newer pretrained embeddings for sequences and substrates, an expanded dataset, and uncertainty predictions, resulting in better K_M predictions [26]. We also recently developed CPI-Pred, a generalizable framework where pretrained models could be interchanged for sequence and substrate embeddings [27]. Interestingly, its best model performed on par with CatPred, suggesting that improvements in model architecture alone may no longer drive substantial gains in prediction accuracy. The key challenge is that future advances may depend more on the availability of higher-quality and more diverse datasets rather than further refinements to deep learning models themselves.

4.2.5 Protein Thermal Stability

Thermal stability refers to a protein's ability to maintain their spatial structures as temperature increases, and stability is critical for BPes. Proteins with low thermal stability tend to aggregate at physiological temperatures, leading to loss of activity, dysfunction, or the formation of toxic protein aggregates [44].

There are two main ways researchers are investigating the thermal stability of proteins. The first involves predicting the melting temperature (T_m) of a protein. T_m is the point at which 50% of the protein loses its native structure and is indicative of how stable a given protein is throughout a temperature range. In 2023, DeepSTABp was developed to predict $\mathrm{T_m}$ for a given protein sequence [44]. The model would extract sequence features and physical features from thermal proteome profiling datasets and run predictions to determine T_m. DeepSTABp was able to achieve state-of-the-art (SOTA) performance, beating out ProTstab2 on this regression task.

The second method treats thermal stability as a classification problem, but does so with a set of binary classification models due to dataset constraints. BertThermo was the first method that would consider whether or not a protein will remain stable above some threshold temperature (analogous to T_m) as a classification task [45]. This way, there are several binary classifiers where the threshold temperature remains static for each model. In 2024, TemProSta used a similar approach, leveraging a larger dataset, and outperformed all other comparable methods on standard benchmarks (iThermo, SAPPHIRE, SCMTPP, and ThermoPred) [46].

4.2.6 Protein Toxicity

Protein toxicity refers to the adverse effects that proteins or peptides can exert on biological systems, potentially leading to harmful physiological responses or even fatal outcomes. Using binary prediction methods to determine whether a protein is toxic is essential in developing protein-based therapeutics [47]. These methods help researchers identify and eliminate candidates that could cause harmful side effects, making the drug discovery process more efficient and ensuring patient safety.

Early toxicity prediction methods were largely template-based, employing sequence alignment tools like BLAST to compare unknown proteins to known toxic ones [48]. These approaches, while pioneering, were limited by the dependency on high-quality template sequences. To overcome this, researchers moved toward feature-based models that extracted compositional, evolutionary, and physicochemical attributes from protein sequences and utilized classical ML techniques. A notable model was ToxinPred [48], developed in 2013, which laid the groundwork for predicting peptide toxicity across various sources.

Recent advances have seen the adoption of alignment-free methods that leverage deep learning architectures such as RNNs, LSTMs, and transformers [47]. Specific models such as CSM-Toxin [48], ToxinPred 3.0 [47], and the more recent VISH-Pred [49] have harnessed the power of pLMs like ProteinBERT [50] and ESM-2 [51] to capture intricate sequence patterns and subtle structural nuances. These SOTA approaches have not only improved predictive accuracy but have also facilitated the creation of accessible web-based tools that integrate into practical workflows, ultimately enhancing the design of safer and more effective protein-based therapeutics.

4.2.7 Protein Solubility

Solubility refers to the proportion of a protein that remains dissolved in a solution and aggregated under physiological conditions. In protein expression, determining whether a protein will remain soluble or form aggregates is crucial for efficient production, which is more common in over-expressed proteins. Classifying proteins by solubility before production helps optimize selection, as insoluble aggregates can reduce yields, increase purification costs, and compromise function. In contrast, soluble proteins are easier to purify and use in downstream applications.

Although there are many model organisms for protein production, most efforts have focused on predicting protein solubility in *Escherichia coli*, as it is widely used in industry and has the most available data [52]. Preliminary methods were built to predict whether a protein is soluble (binary classification task) from its protein sequence through support vector machines and naive-bayes. [53]. PROSO was among the first ML models to manually curate expert features, such as grouping amino acids based on physicochemical properties, to improve predictive generalizability back in 2007. Since then, researchers have explored improved feature engineering and larger datasets. In 2021, NetSolP was an early adopter of pLMs, in the form of a fine-tuned ESM-1b model, to make these predictions and was able to achieve SOTA performance compared to other competitive models at the time [54]. In 2023, researchers identified that features like amino acid counts, transitions between adjacent amino acids, residue distribution along the sequence, and correlations between properties such as charge and hydrophobicity were particularly impactful [55]. This led to the development of the RPPSP model, which leveraged these features to outperform both NetSolP and earlier models. As interest grew in using pLM embeddings, the 2024 model PLM_Sol emerged as the best predictor by passing ProtT5 sequence embeddings through a bidirectional LSTM, and it remains the current SOTA at the time of writing [52].

4.3 Tool Kits and Benchmarks

4.3.1 Overview of Open-Source Tools

Datasets and Databases Large-scale datasets have been utilized extensively over the past decade for a multitude of functional prediction ML tasks. For proteins, UniProt [8] is the most well-known and used database. It contains information on protein sequences, their native host organism, and supplementary data such as 3D structures from the PDB database [58] and/or predictive models such as AlphaFold [59]. UniProt often also hosts other protein functional tags, such as kinetic information, GO, and PPI links, to external databases like STRING [60]. In the enzyme space, BRENDA [61] plays an essential role in providing functional enzymatic information. Many other datasets and databases, as listed in Table 4.2, have also been frequently used in sequence-to-function prediction pipelines.

Table 4.2 Datasets available for sequence-to-function prediction tasks with sources hyperlinked to database/dataset name.

Database	Description	Data size	Predefined splits?
Databases (Comprehensive Resources)			
UniProt [8]	Comprehensive resource for protein sequence and functional information[a)]	240 M+ sequences	No
BRENDA [61]	Comprehensive enzyme information system with manually curated functional and molecular data[b)]	9,000+ enzyme types, 3 M+ data points	No
BFD [62]	Big Fantastic Database for protein language model training[c)]	2.5B+ sequences	No
Pfam [63]	Database of protein families with multiple sequence alignments and HMMs[d)]	19,632 families	No
STRING [60]	PPI network database[e)]	14,000+ organisms, 24.6 M+ proteins	No
Gene Ontology [64]	Structured vocabulary of gene and protein biological attributes[f)]	45,000+ terms across 3 domains	No
InterPro [65]	Integrated database of protein families, domains, and sites[g)]	40,000+ entries	No
Swiss-Prot [66]	Manually annotated subset of UniProt with experimental validation[h)]	570,000+ sequences	No
Datasets (Specialized collections)			
PEER (2022) [67]	Comprehensive benchmark for protein sequence understanding with 21 diverse tasks[j)]	21 tasks across multiple data types	Train/Dev/ Test
ProteinGym (2024) [68]	Fitness prediction benchmark with mutation effect datasets[k)]	87 DMS assays, 139 K fitness values	Train/Test
OpenProteinSet (2023) [69]	Collection of protein sequences, structures, and metrics	470 K structures, 215 M sequences	No
CAFA (2023) [70]	Critical assessment of protein function annotation for community evaluation[l)]	Varies by CAFA challenge round	Target/ Benchmark
ProteinGLUE (2022) [71]	General benchmark for evaluating pLMs on diverse downstream tasks[m)]	13 tasks across 5 categories	Train/Val/ Test

Table 4.2 (Continued)

Database	Description	Data size	Predefined splits?
DeepFRI-Dataset (2021) [72]	Dataset for learning protein function from structure and sequence[n)]	29,513 proteins	Train/Test
TemStaPro-Major-30 (2024) [46]	Non-eukaryotic protein sequences with associated growth temperatures of host microorganism[a)]	1.3 M sequences	Yes
Meltome Atlas (2020) [73]	Mass spectrometry-based thermo-proteome profiling dataset for thermal stability prediction[b)]	48,000 proteins	No
TargetTrack (2017) [74]	Aggregated protein solubility data[c)]	350,000+ sequences	No

a) Contains manually annotated and reviewed (Swiss-Prot) and automatically annotated (TrEMBL) protein records with GO annotations, domains, and functional classifications.
b) Comprehensive resource for enzyme functional data, including substrate specificity, kinetic parameters, inhibition data, and disease-related information.
c) Clustered and filtered metagenomic database created for training pLMs with high sequence diversity.
d) A widely used resource for protein domain annotation with curated seed alignments for each family.
e) Aggregates known and predicted PPI, including direct (physical) and indirect (functional) associations from various sources.
f) Provides controlled vocabulary of gene and protein functions across species with molecular function, biological process, and cellular component domains.
g) Classifies proteins into families and predicts domains and important sites by combining various protein signature databases.
h) Gold-standard manually curated subset with experimental evidence and expert annotation, considered the reference for protein function.
i) Used to train ESM-2 and other large pLMs with self-supervised learning.
j) Tasks span protein classification, regression, generation, mutation effect prediction, and structure prediction.
k) Includes both substitution-based and indel-based fitness prediction tasks.
l) CAFA5 included 100+ research groups, 300 K protein targets, and three ontologies (GO, HPO, and EC) with emphasis on experimental validation.
m) Inspired by GLUE for NLP, it includes tasks in protein function prediction, PPI, protein structure prediction, protein engineering, and protein generation.
n) Used to train deep learning models that predict protein functions directly from structure and sequence using graph neural networks.

Precursor Models Outside of existing databases and datasets, different types of modeling frameworks have also served as essential tool kits for sequence-to-function ML prediction pipelines. These frameworks include ML-based methods (AlphaFold [59], Chai-1 [9]), ESM-2 [51], etc.), as well as non-ML-based methods (molecular dynamics (MD) simulations, quantum-based methods, computational bio methods such as BLAST [18], etc.).

Early protein function prediction relied on sequence homology and domain architecture without the incorporation of any traditional ML methods, as previously discussed in the current state section. BLAST remains a foundational method for identifying homologous sequences through pairwise alignment. Hidden Markov models (HMMs) [75] advanced this through probabilistic profile matching in tools like Pfam [63] and InterProScan [76]. These methods established the paradigm of transferring functional annotations from characterized homologs, achieving moderate accuracy but limited by evolutionary divergence. Outside of these sequence similarity-based methods, sequence-to-function pipelines have also utilized MD simulations to provide physicochemical insights, which can be critical for understanding protein functions like allosteric regulations [77]. Furthermore, density functional theory and quantum mechanical methods can be used as precursor tools in sequence-to-function pipelines as they address electronic effects, which are crucial for metalloenzyme active sites [78].

As for ML-based methods, there are three types of models that typically play a precursor role in these sequence-to-function prediction pipelines. The first class of ML models is one that focuses on predicting protein structures, such as AlphaFold [59] and Chai-1 [79]. Researchers in this field will often dismiss the use of sequence-only function prediction models when comparing their performance against structure-based function models, where they've shown superior performance on monomer, PPI, and antibody–antigen CASP15 tasks [80]. Regardless, sequence-only models provide a big advantage in their lighter computational loads, extensive experimental data availability, and ability to model intrinsically disordered regions. The second class of ML models is pLMs, such as ESM-2 [51], ProtT5 [81], and ProteinBERT [50]. These models utilize different training objectives (masked language modeling, bidirectional sequence reconstruction, GO term prediction) to "learn the protein language," and therefore be able to output embeddings from protein sequences that can be utilized for downstream sequence-to-function prediction tasks. The third class of models is auto-regressive sequence design models, like ProtGPT2 [2] and ProGEN [82], that often serve as a generative surrogate for producing novel sequences that can be treated as synthetic data in these sequence-to-function pipelines [83, 84]. It is worth noting that there are also genomic language models, such as HyenaDNA [85], Evo [86], and gLM2 [87], that have been utilized for protein prediction tasks, although we won't be focusing on that in this chapter.

4.3.2 Importance of Standardized Benchmarks

Standardized benchmarks are crucial for the objective evaluation and comparison of computational methods. While there has been significant progress in the development of these benchmarking datasets (as listed in Table 4.2), studies focusing on benchmarking different models and methods on a wide variety of protein tasks still remain a challenge. This is often due to the lack of open-source models and data, or perhaps a lack of consolidation amongst researchers in this field. In terms of assessment protocols, the field has converged on a few different performance metrics.

For classification tasks, F_{max}, the harmonic mean of precision and recall at optimal threshold; AUPR, area under precision-recall curve accounting for class imbalance; and S_{min}, semantic distance metric for prediction completeness. For regression tasks, the R^2 coefficient of determination, which quantifies the degree of any linear correlation between 2 datasets; Pearson's r, a correlation coefficient that measures the linear correlation between 2 datasets; Spearman's ρ, a correlation coefficient that measures the strength of association between 2 datasets. That being said, it is also important to consider domain-specific evaluation metrics that might be useful for different biological scenarios. For example, CatPred [26] introduces the $p_{1,mag}$ metric, which describes the percent of predictions within one order of magnitude error for prediction of different kinetic parameters. Outside of benchmarking studies and evaluation metrics, it's also crucial to consider data splits. Proteins are really complex when it comes to feature extraction and input space; therefore, we find that models tend to be able to learn and predict things well when they've learned information from sequences that are similar to sequences they've seen during training. A recurring issue with protein datasets is their biased/uneven distribution, where they have very sparse annotations for certain protein classes. It's important to benchmark any sequence-to-function model on a wide variety of sequence-similarity-based splits, which is often done through tools like CDHIT [88] or MMSeqs2 [89] (see 4.2 for datasets with preset splits).

4.4 Emerging ML Methods

ML methods for protein sequence-to-function prediction have rapidly evolved. Increasingly sophisticated approaches have been adopted to overcome different limitations in sequence-to-function learning. Emerging strategies leverage diverse methodologies. Sections below summarize key recent developments in some promising areas.

4.4.1 Contrastive Learning

Contrastive learning utilizes self-supervised strategies to generate robust protein representations by aligning diverse data types, such as sequence embeddings and structural embeddings. This approach significantly enhances generalization, particularly in data-scarce scenarios. An exemplar of contrastive learning's efficacy is the CLEAN model [90], which substantially improves enzyme function prediction. CLEAN excels in annotating understudied enzymes and accurately identifying promiscuous enzymes with multiple EC numbers. It consistently outperforms conventional methods like BLASTp [91], especially in distinguishing closely related enzyme functions and discovering novel enzymatic activities. Experimental validations have demonstrated CLEAN's robustness and reliability in assigning functional annotations to previously uncharacterized enzymes, establishing its value for genomics, synthetic biology, and biocatalysis research.

4.4.2 Meta Learning

Meta learning is a subcategory of ML that trains artificial intelligence (AI) models to understand and adapt to new tasks autonomously. Unlike conventional supervised learning, where models are trained to solve a specific task using a defined training dataset, meta learning involves multiple tasks, each with its own dataset. Models trained via meta learning generalize across tasks. This allows rapid adaptation to novel scenarios with minimal data. The meta learning process consists of two main stages: (1) meta training, where models are exposed to various tasks to discover common patterns and generalizable knowledge, and (2) meta testing, where models are evaluated on new, unseen tasks to assess their adaptive performance. A recent advancement in this area is DeepPFP [92], a multitask-aware model specifically designed for protein function prediction. DeepPFP combines model-agnostic meta-learning (MAML) [93] with ESM-1b to effectively capture universal sequence-function relationships. MAML optimizes a model's initial parameters so that the model can rapidly adapt to new tasks after only a few gradient update steps. ESM, on the other hand, leverages evolutionary information embedded within protein sequences, capturing deep contextual relationships crucial for accurate function prediction. By integrating meta-learning strategies like MAML with powerful sequence representations from ESM, DeepPFP achieves enhanced generalization across diverse enzyme function prediction tasks. Consequently, DeepPFP excels in few-shot learning scenarios, surpassing traditional deep learning models. Evaluations on multiple deep mutational scanning (DMS) datasets highlight its capability to significantly improve predictive performance, especially when sequence similarity alone fails to accurately predict protein functions. This approach effectively addresses the long-tail distribution challenge common in protein function prediction, enabling accurate identification of rare functional variants.

4.5 Case Studies

4.5.1 Prediction of Enzyme Activity and Substrate Specificity

Enzyme activity prediction has been a long-standing functional prediction task. While many models have recently emerged with the goal of predicting enzyme activity, the field has not yet achieved a comprehensive solution that meets key benchmarks, including extrapolating for enzymes outside of the available datasets. Predicting the rate at which enzymes speed up a reaction is arguably the most important prediction task for an enzyme, given that the majority of enzymes function as biological catalysts. Activity prediction can be described through several functional parameters, but the most common term used to describe it is the enzyme turnover number (k_{cat}). These values are essential for understanding kinetic behaviors in the metabolism on a genome-scale for different organisms and also often serve as tunable parameters when looking at enzyme engineering. Several ML models have emerged over the past five years for predicting k_{cat} values.

Although k_{cat} values are one of the most important and commonly used functional descriptors for enzyme assays, ML models that have surfaced over the past five years still struggle to accurately predict these values. One of the works pioneering the k_{cat} prediction task in the ML space is from Heckmann et al. [94]. Heckmann et al. [94] utilized several traditional ML methods to predict enzyme turnover numbers from input fluxomics and proteomics data, including an elastic net, random forest, and an FFNN outputting predictions in an ensembled format. Following the publication of this work, several research groups worked on developing deep learning models for predicting k_{cat} values, as previously mentioned in Section 4.2.4. Although these models utilize protein-compound inputs to predict enzyme turnover numbers, they serve as great examples for enhancing our understanding of protein property prediction. Next, we explore an example of how these pipelines work.

4.5.1.1 Predicting k_{cat} Using CPI-Pred

CPI-Pred, or Compound-Protein Interaction Predictor, is a deep learning framework for inferring kinetic parameters such as k_{cat}, as well as other compound-protein interaction functional descriptors. CPI-Pred takes in two inputs (protein sequence and compound SMILES) and predicts one value (k_{cat}). For the context of this chapter, we can hone in on the sequence side of CPI-Pred. In the computational biology space, proteins typically follow IUPAC nomenclature before they are processed in any downstream process. For a computer to actually understand and process this protein language, it must be able to translate the amino acids into numbers. This is done in a step called tokenization; a protein sequence (e.g. "GCTVE") is converted to numbers by assigning each amino acid a token ID (i.e. "2, 11, 7, 23, 25"). In many of these DL pipelines, pLMs (see Section 1.3.1) are utilized to "featurize" protein sequences of the dataset of interest from a much larger protein dataset (i.e. ESM-2 was trained on 650-million protein sequences). The tokenized protein sequences are passed to the selected pLM and a protein embedding is outputted from the model. In the case of CPI-Pred, the ESM-2 model is identified as the best-performing pLM in the pipeline, although many others were tested. From ESM-2, the protein embedding is outputted with dimension $D_L \times D_E$, with D_L being the protein sequence's length (number of amino acid residues) and D_E being the pLM's embedding dimension; for ESM-2, it ranges from 320 all the way up to 5,120. Often, for protein functional prediction tasks, the proteins of interest don't come in datasets with consistent sequence lengths, meaning that $D_L = \max(D_{l,i})$, or the maximum length out of all of the proteins in our dataset (or batch). Due to the large dimensions, processing these pLM embeddings is a tedious and computationally expensive task. Therefore, different models, including CPI-Pred, have explored different methods of pooling that information in an efficient manner that tries to preserve as much learned residue-specific information as possible. CPI-Pred utilizes three different methods for this task: convolutional 1D pooling, self-attention based pooling, and LSTM variational encoding pooling (Figure 4.2). Following these protein pooling steps, we are left with a fixed-dimension vector that can then be passed to a multilayer perceptron (MLP) for further processing and prediction.

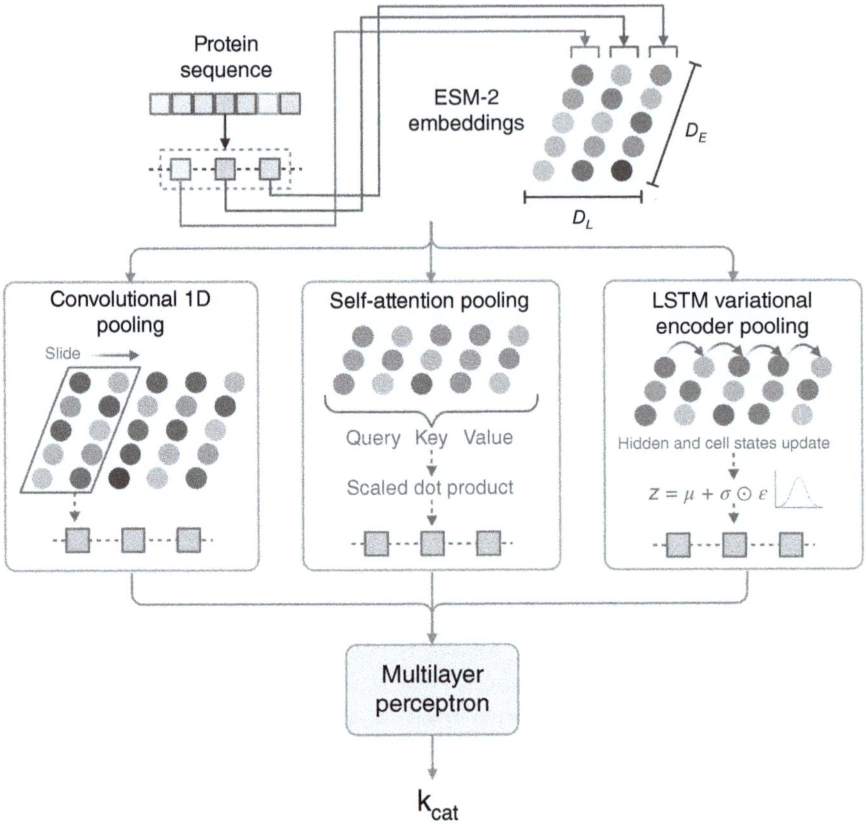

Figure 4.2 A deep learning architecture for predicting enzyme catalytic efficiency (k_{cat}) from protein sequences. The model begins with protein sequence inputs that are converted to ESM-2 embeddings, represented as a high-dimensional space with dimensions D_E and D_L. These embeddings are then processed through three parallel pooling methods: convolutional 1D pooling (which slides across the sequence to extract features), self-attention pooling (using query-key-value mechanisms and scaled dot products), and LSTM variational encoder pooling (which updates hidden and cell states to produce a latent representation (z), from the mean (μ), standard deviation (σ), and some random noise (ε). The outputs from these three pooling methods are fed into a multilayer perceptron, which ultimately predicts k_{cat} of the enzyme.

4.6 Challenges in Sequence-to-Function Mapping

ML approaches show promise in protein engineering by efficiently exploring vast sequence spaces, but face challenges due to complexity and functional sequence sparsity [95, 96]. Protein designs often fail when transitioning from specific assays to biological applications, highlighting the need for better alignment between screening methods and intended outcomes [97–100]. Consequently, candidates from early screening frequently fail in rigorous validation [101–104], while certain protein classes lack suitable high-throughput assays altogether [6, 105–110].

4.6.1 Sparse Experimental Data

High-Dimensional Space and Limited Labels Protein engineering is challenged by vast sequence dimensionality coupled with limited functional annotations (Figure 4.3). Despite high-throughput screening capabilities, obtaining quality labeled datasets requires substantial resources [61, 111–114]. Recent approaches like low-N protein engineering [83] attempt to address this by using ML models such as eUniRep to predict functionality from minimal experimental data, enabling *in silico* screening of millions of sequences.

Imbalanced Datasets and Rare Enzymatic Activities Class imbalance, particularly under-representation of rare enzyme functions, further limits ML performance. This challenge intensifies with high-dimensional sequence embedding methods (Figure 4.3a). Whether developing therapeutic [115, 116], agricultural [117], or industrial proteins [118, 119], researchers face trade-offs between detailed evaluation and broader exploration. While extensive sequence and structure databases exist, functional annotation remains challenging due to diverse catalytic properties and experimental conditions [112]. Most databases weren't designed for ML applications, leading to inconsistencies, missing values, and conflicting annotations [120, 121]. Standardizing formats and adopting FAIR principles [122]

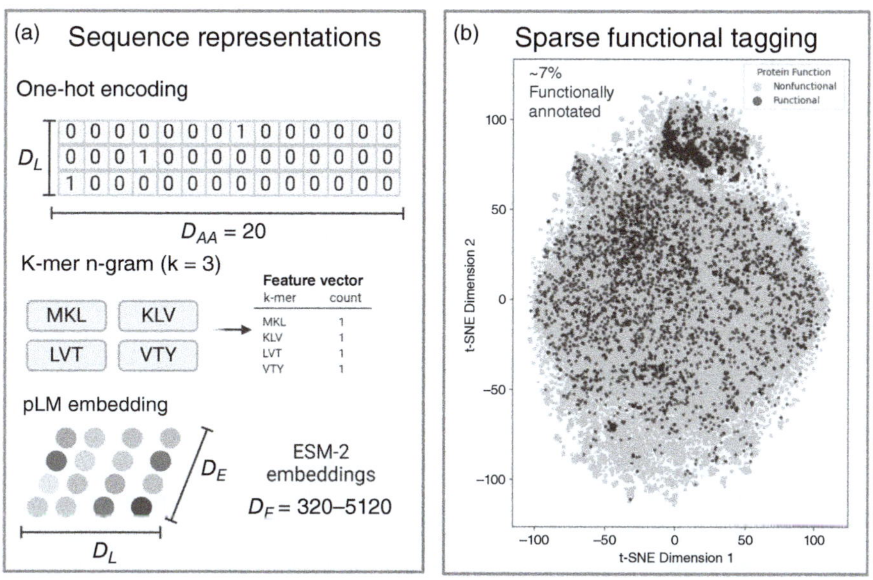

Figure 4.3 Challenges in sequence-to-function datasets. (a) The complexity of sequence representation approaches, comparing simple one-hot encoding (with dimension sequence length (D_L) and number of possible amino acids ($D_{AA} = 20$)) to increasingly sophisticated methods: k-mer n-gram analysis that captures local motifs, and pLMs like ESM-2 that generate high-dimensional embeddings ($D_E = 320-5120$). (b) The sparse annotation problem is demonstrated, visualizing how only 7% of 100 k randomly sampled proteins from Swiss-Prot [66] in the embedding space have functional annotations (red dots), leaving the vast majority (light blue) unlabeled, despite potentially sharing similar structural or functional properties as suggested by their proximity in the t-SNE projection.

and STRENDA guidelines [123] can enhance data quality, while oversampling rare classes or optimizing loss functions helps mitigate bias toward overrepresented categories [112].

4.6.2 Interpretability

The utility of sequence-to-function predictions depends on their biological insights, yet deep learning models often function as "black boxes." Despite pLMs' success in predicting protein structures and functions, their lack of interpretability limits their application in critical domains like drug discovery, where understanding how sequence changes affect function is essential. While some models allow conditional generation by tagging protein features [82, 124, 125], they don't clarify which features influence specific predictions. The concept bottleneck pLM (CB-pLM) improves interpretability by introducing functional tags as biologically meaningful concepts [84]. Each bottleneck neuron corresponds to a specific attribute, enabling sequence generation aligned with desired properties while facilitating interpretation by mapping concepts to outputs. Other approaches include AlphaFold's structure predictions [59] and deep representation learning that generates biologically meaningful embeddings [126].

4.7 Conclusion and Future Directions

This chapter has examined the evolving landscape of ML approaches for protein sequence-to-function prediction, focusing on recent advances in predicting functional annotations (e.g. GO terms, EC numbers, enzyme activity, stability, toxicity, and solubility) directly from amino acid sequences. Key models exemplify how deep learning architectures leverage vast sequence datasets to bridge the annotation gap with the help of pLMs. These models address critical challenges, such as generalizing to evolutionarily distant proteins and integrating multimodal data (e.g. structural, kinetic, and multiomics insights), while highlighting persistent limitations like sparse experimental data, dataset biases, and computational costs. By connecting sequence patterns to functional outcomes, these tools accelerate discoveries in disease mechanisms, drug target identification, and enzyme engineering, though their reliability remains constrained by the quality and diversity of training data.

The progress in ML-driven function prediction underscores its significance in advancing both basic research and applied biotechnology. However, the field must prioritize strategies to enhance model interpretability, incorporate physicochemical constraints, and expand beyond homology-dependent predictions to fully realize its potential. While current models excel in narrow tasks, their broader utility depends on reconciling high-dimensional sequence data with underrepresented functional classes and dynamic biological contexts. These limitations set the stage for emerging methodologies, such as meta-learning and physics-informed architectures, which aim to unify sequence, structure, and environmental variables into predictive frameworks.

References

1 Bepler, T. and Berger, B. (2021). Learning the protein language: evolution, structure, and function. *Cell Syst.* 12 (6): 654–669. ISSN 2405-4712. doi: 10.1016/j.cels.2021.05.017.

2 Ferruz, N., Schmidt, S., and Höcker, B. (2022). ProtGPT2 is a deep unsupervised language model for protein design. 13 (1): 4348. ISSN 2041-1723. doi: 10.1038/s41467-022-32007-7.

3 Blaabjerg, L.M., Jonsson, N., Boomsma, W. et al. (2024). SSEmb: A joint embedding of protein sequence and structure enables robust variant effect predictions. *Nat. Commun.* 15 (1): 9646. ISSN 2041-1723. doi: 10.1038/s41467-024-53982-z.

4 Wekesa, J.S. and Kimwele, M. (2023). A review of multi-omics data integration through deep learning approaches for disease diagnosis, prognosis, and treatment. *Front. Genet.* 14. ISSN 1664–8021: doi: 10.3389/fgene.2023.1199087.

5 Barabási, A.-L., Gulbahce, N., and Loscalzo, J. (2011). Network medicine: a network-based approach to human disease. *Nat. Rev. Genet.* 12 (1): 56–68. ISSN 1471-0064. doi: 10.1038/nrg2918.

6 Zachary, W., Jennifer Kan, S.B., Lewis, R.D. et al. (2019). Machine learning-assisted directed protein evolution with combinatorial libraries. *Proc. Natl. Acad. Sci. U.S.A.* 116 (18): 8852–8858. doi: 10.1073/pnas.1901979116.

7 Jiang, R., Shang, L., Wang, R. et al. (2023). Machine learning based prediction of enzymatic degradation of plastics using encoded protein sequence and effective feature representation. *Environ. Sci. Technol. Lett.* 10 (7): 557–564. doi: 10.1021/acs.estlett.3c00293.

8 UniProt. (2025). UniProtKB statistics, https://www.uniprot.org/uniprotkb/statistics.

9 Chen, J.-Y., Wang, J.-F., Hu, Y. et al. (2025). Evaluating the advancements in protein language models for encoding strategies in protein function prediction: a comprehensive review. 13: 1506508. ISSN 2296-4185. doi: 10.3389/fbioe.2025.1506508.

10 Schnoes, A.M., Ream, D.C., Thorman, A.W. et al. (2013). Biases in the experimental annotations of protein function and their effect on our understanding of protein function space. 9 (5): e1003063. ISSN 1553-734X. doi: 10.1371/journal.pcbi.1003063.

11 Liu, Y., Palmedo, P., Ye, Q. et al. (2018). Enhancing evolutionary couplings with deep convolutional neural networks. 6 (1): 65–74. ISSN 2405–4712. doi: 10.1016/j.cels.2017.11.014.

12 Ekanayake, D., Parackrama, W., Prabodha, K., et al. (2024). Automated protein function prediction using an ensemble of deep neural networks. In *2024 Moratuwa Engineering Research Conference (MERCon)* 260–265. doi: 10.1109/MERCon63886.2024.10688625. ISBN: 9798331529048 Moratuwa, Sri Lanka: IEEE.

13 Rao, R., Bhattacharya, N., Thomas, N. et al. (2019). Evaluating protein transfer learning with TAPE. *Adv. Neural. Inf. Process. Syst.* 32: 9689–9701. ISSN 1049-5258.

14 Ektefaie, Y., Shen, A., Bykova, D., et al. (2024). Evaluating generalizability of artificial intelligence models for molecular datasets. 2024.02.25.581982. doi: 10.1101/2024.02.25.581982.

15 Rives, A., Meier, J., Sercu, T. et al. (2021). Biological structure and function emerge from scaling unsupervised learning to 250 million protein sequences. *Proc. Nat. Acad. Sci. U.S.A.* 118 (15): e2016239118. doi: 10.1073/pnas.2016239118.

16 Doğan, T., Atas, H., Joshi, V. et al. (2021). CROssBAR: comprehensive resource of biomedical relations with knowledge graph representations. *Nucleic Acids Res.* 49 (16): e96. ISSN 0305–1048. doi: 10.1093/nar/gkab543.

17 Xin, L., Huang, C., Li, H., et al. (2024). Artificial intelligence for central dogma-centric multi-omics: Challenges and breakthroughs. http://arxiv.org/abs/2412.12668.

18 Altschul, S.F., Madden, T.L., Schäffer, A.A. et al. (1997). Gapped BLAST and PSI-BLAST: a new generation of protein database search programs. *Nucleic Acids Res.* 25 (17): 3389–3402. ISSN 0305-1048. doi: 10.1093/nar/25.17.3389.

19 Buchfink, B., Xie, C., and Huson, D.H. (2015). Fast and sensitive protein alignment using DIAMOND. *Nat. Methods* 12 (1): 59–60. ISSN 1548-7105. doi: 10.1038/nmeth.3176.

20 Remmert, M., Biegert, A., Hauser, A., and Söding, J. (2012). HHblits: lightning-fast iterative protein sequence searching by HMM-HMM alignment. *Nat. Methods* 9 (2): 173–175. ISSN 1548-7105. doi: 10.1038/nmeth.1818.

21 Cai, C.Z., Han, L.Y., Ji, Z.L. et al. (2003). SVM-prot: webbased support vector machine software for functional classification of a protein from its primary sequence. *Nucleic Acids Res.* 31 (13): 3692–3697. ISSN 0305-1048. https://www.ncbi.nlm.nih.gov/pmc/articles/PMC169006.

22 Liang, L., Djuric, N., Guo, Y., and Vucetic, S. (2013). MSkNN: protein function prediction by integrating multiple data sources. *BMC Bioinf.* 14 (Suppl 3): S8. ISSN 1471-2105. doi: 10.1186/1471-2105-14-S3-S8.

23 Lin, Z., Akin, H., Rao, R. et al. (2023). Evolutionary-scale prediction of atomic level protein structure with a language model. *bioRxiv* doi: 10.1101/2022.07.20.500902.

24 Elnaggar, A., Heinzinger, M., Dallago, C. et al. (2022). Prottrans: towards cracking the language of life's code through self supervised deep learning and high performance computing. *IEEE Trans. Pattern. Anal. Mach. Intell.* 44 (10): 7112–7127. doi: 10.1109/TPAMI.2021.3095381.

25 Yang, K.K., Lu, A.X., and Fusi, N. (2024). Convolutions are competitive with transformers for protein sequence pretraining. *Cell Syst.* 15 (3): 286–294. doi: 10.1016/j.cels.2024.01.008.

26 Boorla, V.S. and Maranas, C.D. (2025). CatPred: a comprehensive framework for deep learning in vitro enzyme kinetic parameters. *Nat. Commun.* 16 (1): 2072, ISSN 2041–1723. doi: 10.1038/s41467-025-57215-9.

27 Xu, Z., Barghout, R.A., Wu, J. et al. (2025). CPI-Pred: A deep learning framework for predicting functional parameters of compound-protein interactions. *bioRxiv* 1 (16): 633372. doi: 10.1101/2025.01.16.633372v1 Section: New Results.

28 Ashburner, M., Ball, C.A., Blake, J.A. et al. (2000). Gene ontology: tool for the unification of biology. The Gene Ontology Consortium. *Nat Genet.* 25 (1): 25–29. doi: 10.1038/75556.

29 The Gene Ontology Consortium (2021). The gene ontology resource: enriching a GOld mine. *Nucleic Acids Res.* 49: D325–D334. ISSN 0305-1048. doi: 10.1093/nar/gkaa1113.

30 Kulmanov, M., Khan, M.A., and Hoehndorf, R. (2018). DeepGO: predicting protein functions from sequence and interactions using a deep ontology-aware classifier. *Bioinformatics* 34 (4): 660–668, ISSN 1367-4803. doi: 10.1093/bioinformatics/btx624.

31 Kulmanov, M. and Hoehndorf, R. (2020). DeepGOPlus: improved protein function prediction from sequence. 36 (2): 422–429. ISSN 1367-4803. doi: 10.1093/bioinformatics/btz595.

32 Elhaj-Abdou, M.E.M., El-Dib, H., El-Helw, A., and El-Habrouk, M. (2021). Deep cnn lstm go: protein function prediction from amino-acid sequences. *Comput. Biol. Chem.* 95: 107584. ISSN 1476-9271. doi: 10.1016/j.compbiolchem.2021.107584.

33 You, R., Yao, S., Mamitsuka, H., and Zhu, S. (2021). Deep-GraphGO: graph neural network for large-scale, multispecies protein function prediction. *Bioinformatics* 37: i262–i271. ISSN 1367-4803. doi: 10.1093/bioinformatics/btab270.

34 Kulmanov, M., Guzmán-Vega, F.J., Roggli, P.D. et al. (2024). Protein function prediction as approximate semantic entailment. *Nat. Mach. Intell.* 6 (2): 220–228, ISSN 2522–5839. doi: 10.1038/s42256-024-00795-w.

35 Yu, L., Wang, S., Umarov, R. et al. (2018). DEEPre: sequence-based enzyme EC number prediction by deep learning. *Bioinformatics* 34 (5): 760–769. ISSN 1367-4803. doi: 10.1093/bioinformatics/btx680.

36 Zou, Z., Tian, S., Gao, X., and Yu, L. (2019). mlDEEPre: multi-functional enzyme function prediction with hierarchical multilabel deep learning. *Front. Genet.* 9. ISSN 1664-8021: doi: 10.3389/fgene.2018.00714.

37 Ryu, J.Y., Kim, H.U., and Lee, S.Y. (2019). Deep learning enables high-quality and high-throughput prediction of enzyme commission numbers. *Proc. Nat. Acad. Sci. U.S.A.* 116 (28): 13996–14001. doi: 10.1073/pnas.1821905116.

38 Kim, G.B., Kim, J.Y., Lee, J.A. et al. (2023). Functional annotation of enzyme encoding genes using deep learning with transformer layers. *Nat. Commun.* 14 (1): 7370. ISSN 2041-1723. doi: 10.1038/s41467-023-43216-z.

39 Devlin, J., Chang, M-W, Lee, K., and Toutanova, K. (2019). BERT: pre-training of deep bidirectional transformers for language understanding. In *Proceedings of the 2019 Conference of the North American Chapter of the Association for Computational Linguistics: Human Language Technologies*, pages 4171–4186. Association for Computational Linguistics, https://aclanthology.org/N19-1423.

40 Yu, T., Cui, H., Li, J.C. et al. (2023). Enzyme function prediction using contrastive learning. *Science* 379 (6639): 1358–1363. doi: 10.1126/science.adf2465.

41 Li, F., Yuan, L., Lu, H. et al. (2022). Deep learning-based k_{cat} prediction enables improved enzyme-constrained model reconstruction. *Nat. Catal.* 5 (8): 662–672, ISSN 2520–1158. doi: 10.1038/s41929-022-00798-z.

42 Kroll, A., Rousset, Y., Xiao-Pan, H. et al. (2023). Turnover number predictions for kinetically uncharacterized enzymes using machine and deep learning. *Nat. Commun.* 14 (1): 4139, ISSN 2041–1723. doi: 10.1038/s41467-023-39840-4.

43 Yu, H., Deng, H., He, J. et al. (2023). UniKP: a unified framework for the prediction of enzyme kinetic parameters. *Nat. Commun.* 14 (1): 8211, ISSN 2041–1723. doi: 10.1038/s41467-023-44113-1.

44 Pei, H., Li, J., Ma, S. et al. (2023). Identification of thermophilic proteins based on sequence-based bidirectional representations from transformer-embedding features. *Applied Sciences* 13 (5): 2858, ISSN 2076–3417. doi: 10.3390/app13052858.

45 Jung, F., Frey, K., Zimmer, D., and Mühlhaus, T. (2023). Deep-STABp: A deep learning approach for the prediction of thermal protein stability. *Int. J. Mol. Sci.* 24 (8): 7444, ISSN 1422–0067. doi: 10.3390/ijms24087444.

46 Pudžiuvelytė, I., Olechnovič, K., Godliauskaite, E. et al. (2024). TemStaPro: protein thermostability prediction using sequence representations from protein language models. *Bioinformatics* 40 (4): btae157, ISSN 1367-4811. doi: 10.1093/bioinformatics/btae157.

47 Rathore, A.S., Choudhury, S., Arora, A. et al. (2024). ToxinPred 3.0: an improved method for predicting the toxicity of peptides. *Comput. Biol. Med.* 179: 108926, ISSN 0010–4825. doi: 10.1016/j.compbiomed.2024.108926.

48 Gupta, S., Kapoor, P., Chaudhary, K. et al. (2013). In Silico approach for predicting toxicity of peptides and proteins. *PLoS One* 8 (9): e73957, ISSN 1932–6203. doi: 10.1371/journal.pone.0073957.

49 Mall, R., Singh, A., Patel, C.N. et al. (2024). VISH-Pred: an ensemble of fine-tuned ESM models for protein toxicity prediction. *Briefings Bioinf.* 25 (4): bbae270, ISSN 1477-4054. doi: 10.1093/bib/bbae270.

50 Brandes, N., Ofer, D., Peleg, Y. et al. (2022). ProteinBERT: a universal deep-learning model of protein sequence and function. *Bioinformatics* 38 (8): 2102–2110, ISSN 1367–4803. doi: 10.1093/bioinformatics/btac020.

51 Lin, Z., Akin, H., Rao, R. et al. (2023). Evolutionary-scale prediction of atomic-level protein structure with a language model. *Science* 379 (6637): 1123–1130. doi: 10.1126/science.ade2574.

52 Zhang, X., Xiaoxuan, H., Zhang, T. et al. (2024). PLM sol: predicting protein solubility by benchmarking multiple protein language models with the updated Escherichia coli protein solubility dataset. *Briefings Bioinf.* 25 (5): bbae404, ISSN 1477-4054. doi: 10.1093/bib/bbae404.

53 Smialowski, P., Martin-Galiano, A.J., Mikolajka, A. et al. (2007). Protein solubility: sequence based prediction and experimental verification. *Bioinformatics* 23 (19): 2536–2542, ISSN 1367-4803. doi: 10.1093/bioinformatics/btl623.

54 Thumuluri, V., Martiny, H.-M., Almagro Armenteros, J.J. et al. (2022). NetSolP: predicting protein solubility in Escherichia coli using language models. *Bioinformatics* 38 (4): 941–946, ISSN 1367-4803. doi: 10.1093/bioinformatics/btab801.

55 Mehmood, F., Arshad, S., and Shoaib, M. (2023). RPPSP: a robust and precise protein solubility predictor by utilizing novel protein sequence encoder. *IEEE Access*, 11:59397–59416, ISSN 2169–3536. doi: 10.1109/ACCESS.2023.3284464. https://ieeexplore.ieee.org/document/10147217. Conference Name: IEEE Access.

56 Yang, Y., Zhao, J., Zeng, L., and Vihinen, M. (2022). ProTstab2 for prediction of protein thermal stabilities. *Int. J. Mol. Sci.* 23 (18): 10798, ISSN 1422-0067. doi: 10.3390/ijms231810798.

57 Morozov, V., Rodrigues, C.H.M., and Ascher, D.B. (2023). CSM-toxin: A web-server for predicting protein toxicity. *Pharmaceutics* 15 (2): 431, ISSN 1999–4923. doi: 10.3390/pharmaceutics15020431.

58 Berman, H.M., Westbrook, J., Feng, Z. et al. (2000). The Protein Data Bank. *Nucleic Acids Res.* 28 (1): 235–242, ISSN 0305–1048. doi: 10.1093/nar/28.1.235.

59 Jumper, J., Evans, R., Pritzel, A. et al. (2021). Highly accurate protein structure prediction with alphafold. *Nature* 596 (7873): 583–589.

60 Szklarczyk, D., Kirsch, R., Koutrouli, M. et al. (2023). The STRING database in 2023: protein-protein association networks and functional enrichment analyses for any sequenced genome of interest. *Nucleic Acids Res.* 51 (D1): D638–D646, ISSN 1362-4962. doi: 10.1093/nar/gkac1000.

61 Jeske, L., Placzek, S., Schomburg, I. et al. (2019). Brenda in 2019: a european elixir core data resource. *Nucleic Acids Res.* 47 (D1): D542–D549.

62 Steinegger, M., Mirdita, M., and Söding, J. (2019). Protein-level assembly increases protein sequence recovery from metagenomic samples manyfold. *Nat. Methods* 16 (7): 603–606, ISSN 1548–7105. doi: 10.1038/s41592-019-0437-4.

63 Mistry, J., Chuguransky, S., Williams, L. et al. (2020). Pfam: the protein families database in 2021. *Nucleic Acids Res.* 49 (D1): D412–D419, ISSN 0305-1048. doi: 10.1093/nar/gkaa913.

64 The Gene Ontology Consortium, Aleksander, S.A., Balhoff, J. et al. (2023). The gene ontology knowledgebase in 2023. *Genetics* 224 (1): iyad031, ISSN 1943-2631. doi: 10.1093/genetics/iyad031.

65 Paysan-Lafosse, T., Blum, M., Chuguransky, S. et al. (2022). InterPro in 2022. *Nucleic Acids Res.* 51 (D1): D418–D427, ISSN 0305-1048. doi: 10.1093/nar/gkac993.

66 Bairoch, A. and Apweiler, R. (2000). The SWISS-PROT protein sequence database and its supplement TrEMBL in 2000. *Nucleic Acids Res.* 28 (1): 45–48. ISSN 0305-1048. https://www.ncbi.nlm.nih.gov/pmc/articles/PMC102476.

67 Xu, M., Zhang, Z., Lu, J., et al. (2022). PEER: A Comprehensive and Multi-Task Benchmark for Protein Sequence Understanding, http://arxiv.org/abs/2206.02096. *arXiv:2206.02096 [cs]*.

68 Notin, P., Kollasch, A., Ritter, D. et al. (2023). Proteingym: large-scale benchmarks for protein fitness prediction and design. In: *Advances in Neural Information Processing Systems*, vol. 36 (ed. A. Oh, T. Naumann, A. Globerson, et al.), 64331–64379. Curran Associates, Inc.

69 Ahdritz, G., Bouatta, N., Kadyan, S., et al. (2023). OpenProteinSet: training data for structural biology at scale, http://arxiv.org/abs/2308.05326. *arXiv:2308.05326 [q-bio]*.

70 Zhou, N., Jiang, Y., Bergquist, T.R. et al. (2019). The CAFA challenge reports improved protein function prediction and new functional annotations for hundreds of genes through experimental screens. *Genome Biol.* 20 (1): 244, ISSN 1474-760X. doi: 10.1186/s13059-019-1835-8.

71 Capel, H., Weiler, R., Dijkstra, M. et al. (2022). ProteinGLUE multi-task benchmark suite for self-supervised protein modeling. *Sci. Rep.* 12 (1): 16047, ISSN 2045–2322. doi: 10.1038/s41598-022-19608-4.

72 Vladimir Gligorijević, P., Renfrew, D., Kosciolek, T. et al. (2021). Structure-based protein function prediction using graph convolutional networks. *Nat. Commun.* 12 (1): 3168, ISSN 2041-1723. doi: 10.1038/s41467-021-23303-9.

73 Jarzab, A., Kurzawa, N., Hopf, T. et al. (2020). Meltome atlas – thermal proteome stability across the tree of life. *Nat. Methods* 17 (5): 495–503, ISSN 1548-7105. doi: 10.1038/s41592-020-0801-4.

74 Micallef, D.I. John Westbrook Margaret J. Berman, G.H.M., et al. (2017). Protein structure initiative – targettrack 2000–2017 – all data files, https://zenodo.org/records/821654.

75 Baum, L.E. and Petrie, T. (1966). Statistical inference for probabilistic functions of finite state markov chains. *Ann. Math. Stat.* 37 (6): 1554–1563.

76 Jones, P., Binns, D., Chang, H.-Y. et al. (2014). InterProScan 5: genome-scale protein function classification. *Bioinformatics* 30 (9): 1236–1240, ISSN 1367-4803. doi: 10.1093/bioinformatics/btu031.

77 Zhu, J., Wang, J., Han, W., and Xu, D. (2022). Neural relational inference to learn long-range allosteric interactions in proteins from molecular dynamics simulations. *Nat. Commun.* 13 (1): 1661, ISSN 2041-1723. doi: 10.1038/s41467-022-29331-3.

78 Wappett, D.A. and Goerigk, L. (2023). Benchmarking density functional theory methods for Metalloenzyme reactions: the introduction of the MME55 set. *J. Chem. Theory Comput.* 19 (22): 8365–8383, ISSN 1549–9618. doi: 10.1021/acs.jctc.3c00558.

79 Chai Discovery (2024). Chai-1: decoding the molecular interactions of life. *bioRxiv*. doi: 10.1101/2024.10.10.615955.

80 Elofsson, A. (2023). Progress at protein structure prediction, as seen in CASP15. *Curr. Opin. Struct. Biol.* 80: 102594, ISSN 0959-440X. doi: 10.1016/j.sbi.2023.102594.

81 Elnaggar, A., Heinzinger, M., Dallago, C. et al. (2020). Prottrans: towards cracking the language of life's code through self-supervised deep learning and high performance computing. *bioRxiv*. doi: 10.1101/2020.07.12.199554.

82 Madani, A., McCann, B., Naik, N., et al. (2020). Progen: language modeling for protein generation. *arXiv preprint arXiv:2004.03497*.

83 Biswas, S., Khimulya, G., Alley, E.C. et al. (2021). Low-n protein engineering with data-efficient deep learning. *Nat. Methods* 18 (4): 389–396.

84 Ismail, A.A., Oikarinen, T., Wang, A., et al. (2024).Concept bottleneck language models for protein design. *arXiv preprint arXiv:2411.06090*.

85 Nguyen, E., Poli, M., Faizi, M., et al. (2023). Hyenadna: Long-range genomic sequence modeling at single nucleotide resolution, https://arxiv.org/abs/2306.15794.

86 Nguyen, E., Poli, M., Durrant, M.G. et al. (2024). Sequence modeling and design from molecular to genome scale with evo. 386 (6723): eado9336. doi: 10.1126/science.ado9336.

87 Cornman, A., West-Roberts, J., Camargo, A.P. et al. (2024). The omg dataset: an open metagenomic corpus for mixed-modality genomic language modeling. *bioRxiv*. doi: 10.1101/2024.08.14.607850.

88 Limin, F., Niu, B., Zhu, Z. et al. (2012). CD-HIT: accelerated for clustering the next-generation sequencing data. *Bioinformatics* 28 (23): 3150–3152, ISSN 1367-4803. doi: 10.1093/bioinformatics/bts565.

89 Steinegger, M. and Söding, J. (2017). MMseqs2 enables sensitive protein sequence searching for the analysis of massive data sets. *Nat. Biotechnol.* 35 (11): 1026–1028, ISSN 1546–1696. doi: 10.1038/nbt.3988.

90 Stojanovic, J., Sattar, S., and Kirov, S. (2022). Enzyme function prediction using contrastive learning. *Nat. Commun.* 13 (1): 3284. doi: 10.1038/s41467-022-30961-9.

91 Altschul, S.F., Gish, W., Miller, W. et al. (1990). Basic local alignment search tool. *J. Mol. Biol.* 215 (3): 403–410. doi: 10.1016/S0022-2836(05)80360-2.

92 Wang, H., Zhang, R., Ni, M. et al. (2025). DeepPFP: a multi-task-aware architecture for protein function prediction. *Briefings Bioinf.* 26 (1): bbae579. doi: 10.1093/bib/bbae579.

93 Finn, C., Abbeel, P., and Levine, S. (2017). Model-agnostic metalearning for fast adaptation of deep networks. In *Proceedings of the 34th International Conference on Machine Learning (ICML)*, volume 70, pages 1126–1135, https://arxiv.org/abs/1703.03400.

94 Heckmann, D., Lloyd, C.J., Mih, N. et al. (2018). Machine learning applied to enzyme turnover numbers reveals protein structural correlates and improves metabolic models. *Nat. Commun.* 9 (1): 5252, ISSN 2041-1723. doi: 10.1038/s41467-018-07652-6.

95 Packer, M.S. and Liu, D.R. (2015). Methods for the directed evolution of proteins. *Nat. Rev. Genet.* 16 (7): 379–394.

96 Romero, P.A. and Arnold, F.H. (2009). Exploring protein fitness landscapes by directed evolution. *Nat. Rev. Mol. Cell Biol.* 10 (12): 866–876.

97 Goldsmith, M. and Tawfik, D.S. (2012). Directed enzyme evolution: beyond the low-hanging fruit. *Curr. Opin. Struct. Biol.* 22 (4): 406–412.

98 Lutz, S. and Patrick, W.M. (2004). Novel methods for directed evolution of enzymes: quality, not quantity. *Curr. Opin. Biotechnol.* 15 (4): 291–297.

99 You, L. and Arnold, F.H. (1996). Directed evolution of subtilisine in bacillus subtilis to enhance total activity in aqueous dimethylformamide. *Protein Eng. Des. Sel.* 9 (1): 77–83.

100 Zhao, H. and Arnold, F.H. (1997). Combinatorial protein design: strategies for screening protein libraries. *Curr. Opin. Struct. Biol.* 7 (4): 480–485.

101 Laverty, H.G., Benson, C., Cartwright, E.J. et al. (2011). How can we improve our understanding of cardiovascular safety liabilities to develop safer medicines? *Br. J. Pharmacol.* 163 (4): 675–693.

102 Scannell, J.W. and Bosley, J. (2016). When quality beats quantity: decision theory, drug discovery, and the reproducibility crisis. *PLoS One* 11 (2): e0147215.

103 Scannell, J.W., Blanckley, A., Boldon, H., and Warrington, B. (2012). Diagnosing the decline in pharmaceutical r&d efficiency. *Nat. Rev. Drug Discovery* 11 (3): 191–200.

104 Silver, L.L. (2011). Challenges of antibacterial discovery. *Clin. Microbiol. Rev.* 24 (1): 71–109.

105 Bedbrook, C.N., Yang, K.K., Rice, A.J. et al. (2017). Machine learning to design integral membrane channelrhodopsins for efficient eukaryotic expression and plasma membrane localization. *PLoS Comput. Biol.* 13 (10): e1005786.

106 Bedbrook, C.N., Yang, K.K., Elliott Robinson, J. et al. (2019). Machine learning-guided channelrhodopsin engineering enables minimally invasive optogenetics. *Nat. Methods* 16 (11): 1176–1184.

107 Lutz, S. (2010). Beyond directed evolution – semi-rational protein engineering and design. *Curr. Opin. Biotechnol.* 21 (6): 734–743.

108 Marcandalli, J., Fiala, B., Ols, S. et al. (2019). Induction of potent neutralizing antibody responses by a designed protein nanoparticle vaccine for respiratory syncytial virus. *Cell* 176 (6): 1420–1431.

109 Romney, D.K., Murciano-Calles, J., Wehrmuller, J.E., and Arnold, F.H. (2017). Unlocking reactivity of trpb: a general biocatalytic platform for synthesis of tryptophan analogues. *JACS* 139 (31): 10769–10776.

110 Silva, D.-A., Shawn, Y., Ulge, U.Y. et al. (2019). De novo design of potent and selective mimics of il-2 and il-15. *Nature* 565 (7738): 186–191.

111 Fowler, D.M. and Fields, S. (2014). Deep mutational scanning: a new style of protein science. *Nat. Methods* 11 (8): 801–807.

112 Mazurenko, S., Prokop, Z., and Damborsky, J. (2019). Machine learning in enzyme engineering. *ACS Catal.* 10 (2): 1210–1223.

113 Wang, C.Y., Chang, P.M., Ary, M.L. et al. (2018). Protabank: A repository for protein design and engineering data. *Protein Sci.* 27 (6): 1113–1124.

114 Wittmann, B.J., Johnston, K.E., Zachary, W., and Arnold, F.H. (2021). Advances in machine learning for directed evolution. *Curr. Opin. Struct. Biol.* 69: 11–18.

115 Daniel Lagassé, H.A., Alexaki, A., Simhadri, V.L. et al. (2017). Recent advances in (therapeutic protein) drug development. *F1000Research* 6: 113.

116 Marshall, S.A., Lazar, G.A., Chirino, A.J., and Desjarlais, J.R. (2003). Rational design and engineering of therapeutic proteins. *Drug Discovery Today* 8 (5): 212–221.

117 Gururaj, A. and Rao. (2008). The outlook for protein engineering in crop improvement. *Plant Physiol.* 147 (1): 6–12.

118 Schmid, A., Dordick, J.S., Hauer, B. et al. (2001). Industrial biocatalysis today and tomorrow. *Nature* 409 (6817): 258–268.

119 Sheldon, R.A. and Pereira, P.C. (2017). Biocatalysis engineering: the big picture. *Chem. Soc. Rev.* 46 (10): 2678–2691.

120 Bastian, F.B., Chibucos, M.C., Gaudet, P. et al. (2015). The confidence information ontology:a step towards a standard for asserting confidence in annotations. *Database* 2015: bav043.

121 Yang, Y., Urolagin, S., Niroula, A. et al. (2018). Pon-tstab: protein variant stability predictor. Importance of training data quality. *Int. J. Mol. Sci.* 19 (4): 1009.

122 Wilkinson, M.D., Dumontier, M., Aalbersberg, I.J.J. et al. (2016). The fair guiding principles for scientific data management and stewardship. *Sci. Data* 3 (1): 1–9.

123 Tipton, K.F., Armstrong, R.N., Bakker, B.M. et al. (2014). Standards for reporting enzyme data: the strenda consortium: what it aims to do and why it should be helpful. *Perspect. Sci.* 1 (1–6): 131–137.

124 Hayes, T., Rao, R., Akin, H., et al. (2024).Simulating 500 million years of evolution with a language model. https://www.biorxiv.org/content/10.1101/2024.07,1:v1,2024.

125 Ruffolo, J.A., Bhatnagar, A., Beazer, J. et al. (2024). Adapting protein language models for structure-conditioned design. *bioRxiv*. doi: 10.1101/2024.08.03.606485.

126 Alley, E.C., Khimulya, G., Biswas, S. et al. (2019). Unified rational protein engineering with sequence-based deep representation learning. *Nat. Methods* 16 (12): 1315–1322.

5

Prediction of Enzyme Functions by Artificial Intelligence

Ha Rim Kim[1,2], *Hongkeun Ji*[1,2], *Gi Bae Kim*[1,2,4], *and Sang Yup Lee*[1,2,3,4,5]

[1] *KAIST Institute for BioCentury, Korea Advanced Institute of Science and Technology (KAIST), Metabolic and Biomolecular Engineering National Research Laboratory, Department of Chemical and Biomolecular Engineering (BK21 four), 291 Daehak-ro, Yuseong-gu, Daejeon 34141, Republic of Korea*
[2] *KAIST Institute for BioCentury, Korea Advanced Institute of Science and Technology (KAIST), Systems Metabolic Engineering and Systems Healthcare Cross-Generation Collaborative Laboratory, 291 Daehak-ro, Yuseong-gu, Daejeon 34141, Republic of Korea*
[3] *KAIST Institute for BioCentury, Korea Advanced Institute of Science and Technology (KAIST), Graduate School of Engineering Biology, 291 Daehak-ro, Yuseong-gu, Daejeon 34141, Republic of Korea*
[4] *KAIST Institute for BioCentury, Korea Advanced Institute of Science and Technology (KAIST), BioProcess Engineering Research Center, 291 Daehak-ro, Yuseong-gu, Daejeon 34141, Republic of Korea*
[5] *KAIST Institute for BioCentury, Korea Advanced Institute of Science and Technology (KAIST), Center for Synthetic Biology, 291 Daehak-ro, Yuseong-gu, Daejeon 34141, Republic of Korea*

5.1 Introduction

Enzymes play a crucial role in maintaining cellular physiological processes, effectively catalyzing metabolic reactions for operating energy metabolism, biosynthesis, and homeostasis. Thus, to thoroughly understand cell metabolism, it is essential to identify which proteins are enzymes and analyze their specific functions. While whole-genome sequencing technologies have significantly enhanced our capacity to elucidate the genetic underpinnings of metabolism by identifying putative enzyme-coding genes, the precise functional annotation of these genes has remained a persistent challenge. Traditional experimental approaches demand substantial time, labor, and cost, limiting high-throughput analysis. Consequently, many enzyme-encoding genes have been sequenced but remain functionally uncharacterized.

The development of computational enzyme function classification has complemented the limitations of traditional experimental approaches. These computational methods have enabled enzyme function prediction to transition from a localized scale to a genome-wide analysis, significantly expanding its scope and applicability. As described in Chapter 4, among the earliest computational strategies for enzyme function prediction are methods based on sequence homology. These homology-based approaches compare unknown sequences against well-characterized enzymes to identify regions of similarity, where similar

Machine Learning and Big Data-enabled Biotechnology, First Edition. Edited by Hal S. Alper.

sequences often suggest analogous biological functions. The effectiveness of these methods is widely acknowledged. However, they rely heavily on previous knowledge, and there are many cases where sequence similarity does not necessarily correspond to functional similarity. Thus, the importance of leveraging the intrinsic properties in enzyme sequences or structures for functional prediction has become increasingly emphasized. The emergence of artificial intelligence has facilitated accurate function prediction by learning and capturing the complex relationships between enzyme features and their functions [1, 2]. In this chapter, we review recent advancements in enzyme function prediction using computational approaches, encompassing conventional methods to advanced deep learning models. We focus on providing a comprehensive analysis of the field, emphasizing fundamental principles and methodological developments. Finally, we discuss ongoing challenges and future directions for further developing artificial intelligence-based enzyme functional annotation.

5.2 Conventional Computational Approaches for Predicting Enzyme Function

To predict whether an unknown protein functions as an enzyme and to determine its catalytic activity, the most fundamental and widely adopted computational approach involves transferring knowledge from already characterized enzymes to uncharacterized proteins by analyzing the similarities between them. This strategy has been widely employed since the early stages of computer-based enzyme characterization [3]. From an evolutionary perspective, proteins with high sequence similarity are presumed to have evolved from a common ancestor, thereby conserving their functions [4, 5].

As protein sequence availability increases with advances in whole genome sequencing technology, amino acid sequences and biochemical features derived from these sequences have become the main resources for enzyme functional classification. The basic local alignment search tool (BLAST) has been considered the gold standard tool for biological sequence similarity analysis [6]. It heuristically measures the local similarity for a pair of biological sequences, rapidly identifying sequences that are similar to a query from the large size of biological databases. Enzyme functions can be assigned to query proteins with statistically significant matches, making BLAST one of the most extensively utilized tools in protein and enzyme function prediction [7–10].

Although BLAST enables effective searching of protein sequences with global sequence similarity, it is challenging to detect sequences of remote homology that might affect enzyme functions. To address this limitation, position-specific iterated BLAST (PSI-BLAST) [11], a protein sequence search method that utilizes a position-specific score matrix (PSSM), was developed. PSI-BLAST iteratively performs sequence alignment to update PSSM, which is used for the next round of iteration, enabling it to better detect distant homologs and transfer knowledge of enzyme functions to the query protein. For example, ModFun successfully predicted the function of CYP7 – a peptidyl-prolyl cis-trans isomerase – by utilizing PSI-BLAST and protein–protein interaction data, which was not predicted by performing BLAST alone [8].

Advances in bioinformatics have further enabled enzyme function analysis at the fine-grained level of functional domains, expanding beyond sequence-level analysis. Profile-HMM is a hidden Markov model that represents the position-specific frequencies of the amino acids and their probabilities of state transitions (i.e. match, deletion, and insertion) derived from a multiple sequence alignment (MSA) of a set of proteins [12]. This approach enabled the accurate recognition of conserved sequences or functional domains that are undetectable through conventional sequence alignments [13, 14]. Profiles derived from enzyme families with the same function provide functional motifs, domains, and families that characterize the enzyme functions. Aligning a query protein to these profiles facilitates the identification and analysis of its enzyme functions [14–16]. Thus, sequence similarity-based approaches have progressed from simple global sequence alignment to the precise detection of functional domains linked to enzyme activity through locally conserved sequences, enabling computer-aided high-throughput enzyme function prediction.

5.3 Prediction of Enzyme Functions Using Machine Learning

The homology-based approach contributed to shifting the trend in enzyme function prediction from experimental to *in silico* approaches. However, the diversity of evolutionary processes often significantly reduces their effectiveness [17]. An example is convergent evolution, which allows proteins to have similar functions despite low sequence similarity [18]. Another example is divergent evolution, which induces proteins to acquire different functions even with high sequence similarity [19]. Therefore, there has been a high demand for a new prediction methodology that does not rely on similarity. Machine learning, which learns the intrinsic features of data to make predictions, has addressed this need in enzyme function prediction quite effectively [20].

Two critical factors must be considered for the development of high-performance machine learning models. The first is feature extraction, which involves determining meaningful properties to extract from the raw data and how to preprocess them into a machine-readable format. Even when utilizing the same raw dataset and model architecture, model performance can vary substantially depending on the selection and processing of features for training. The second factor is the construction of machine learning models for effectively processing given features. Because various machine learning algorithms have distinct advantages and limitations, it is crucial to design models with a comprehensive consideration of the inherent properties of these algorithms and the training data. In this section, we explore different methods for extracting features from amino acid sequences and highlight machine learning algorithms frequently used for predicting enzyme functions.

5.3.1 Extraction of Enzyme Features from Amino Acid Sequences

The amino acid sequence is the fundamental determinant of enzyme characteristics, inherently encoding its chemical, structural, and evolutionary information. Enzyme features derived from amino acid sequences can be broadly categorized into direct and derived features. Directed features are defined as characteristics that can be immediately extracted from enzyme sequences based on statistical analyses of the physicochemical properties of amino acids (Figure 5.1a). In contrast, derived features are properties that require additional processing of raw sequence data to extract nonintuitive information such as higher-order structure, functional domains, and evolutionary profiles (Figure 5.1b).

Amino acid composition (AAC) is a direct feature that straightforwardly represents the entire physicochemical properties of enzymes by calculating the frequency of each amino acid sequence [21]. Due to its simplicity, it is widely used in the early stages of machine learning-based approaches [22–26]. For instance, AAC was used for training an enzyme function prediction model based on the k-nearest neighbor (kNN) algorithm [25]. Similarly, a support vector machine (SVM)-based model was designed to co-utilize AAC and physicochemical properties to classify enzyme functions [26]. However, AAC is insufficient to comprehensively reflect the detailed characteristics of a protein molecule.

To address this limitation, various physicochemical features derived from specific properties of each amino acid residue have been employed in enzyme function prediction [23, 24, 27–30]. Physicochemical properties are highly correlated with protein function, as they encompass critical information regarding molecular interactions and catalytic activity within the cellular environment. Several computational tools have been developed to extract and vectorize physicochemical properties from enzyme sequences based on statistical analysis. Representative examples are Pepstats, ProtParam, and PROFEAT [31–33]. Pepstats calculates molecular weight, isoelectric point, AAC, and the composition of aliphatic, aromatic, polar, nonpolar, charged, basic, and acidic residues [32]. ProtParam returns statistical values for negative and positive charge residues, and the count of each atom [31]. PROFEAT offers a comprehensive set of physicochemical features, including AAC, dipeptide composition, autocorrelation, and amino acid distribution patterns, calculating 1,447 descriptor values [33]. When integrated with machine learning models, these computed features are effective in classifying enzyme functions [28–30]. For example, ECPred leverages Pepstats to extract protein features, which are subsequently used to train an SVM for Enzyme Commission number (EC number) prediction [29]. However, both AAC and physicochemical features cannot provide the positional information of each amino acid in the enzyme sequence. The use of pseudo-amino acid composition (PseAAC), an extension of AAC that incorporates correlations between adjacent amino acids, has addressed this problem. Because physicochemical properties of amino acids were used for calculating these correlations, PseAAC can represent both physicochemical properties and local sequence order information of enzymes [34, 35].

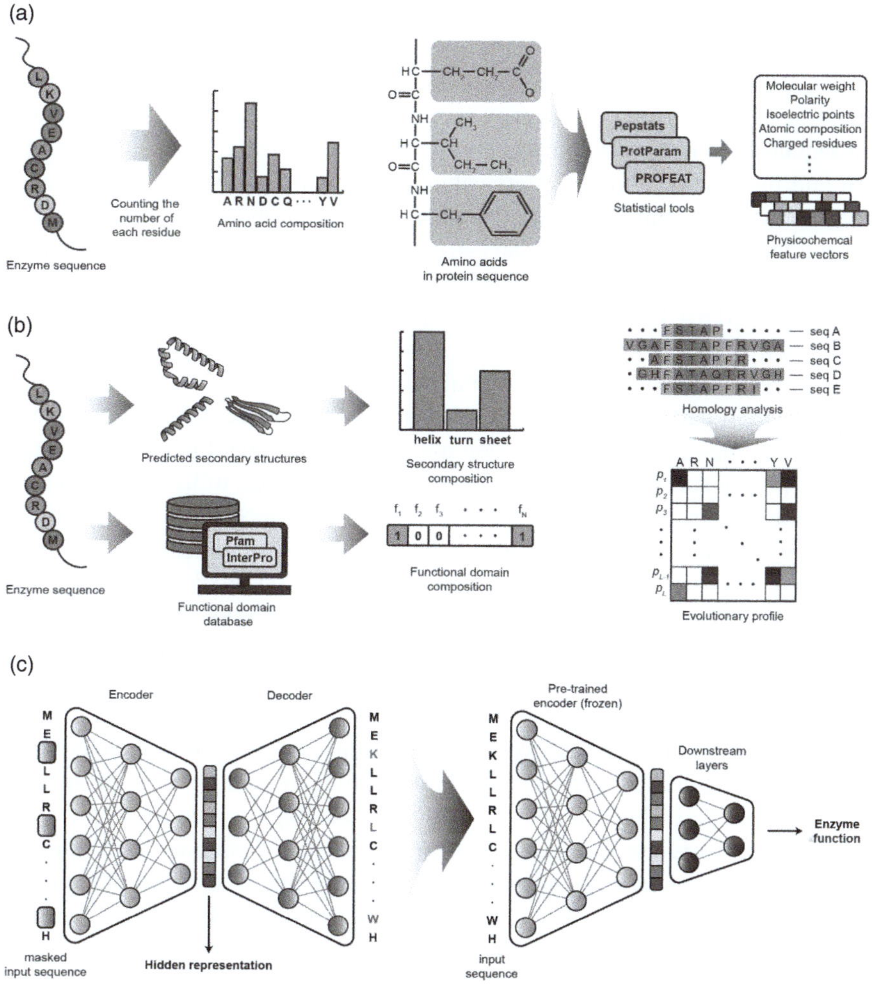

Figure 5.1 The schematic illustration of feature extraction from amino acid sequences for artificial intelligence-based enzyme function prediction. (a) Graphical scheme of extraction of the direct features. AAC represents the frequency of each amino acid in the enzyme sequence. The physicochemical features of a protein are determined based on various intrinsic properties of each amino acid (e.g. molecular weight, polarity, isoelectric point, atomic composition, and charged residues). Feature vector extraction can be performed using statistical tools such as Pepstats, ProtParam, and PROFEAT. (b) Graphical scheme of the extraction of derived features. Secondary structure composition quantifies the frequency of local folding patterns in a protein, such as helices, turns, and sheets. Functional domain composition is represented as a vector indicating the presence or absence of specific functional domains within a protein. Several databases, such as Pfam and InterPro, can be used for the extraction of functional domain composition. (c) Graphical scheme of construction and utilization of pLM for downstream tasks. In the pretraining stage, input sequences are masked at the residue level and fed into encoder layers to be embedded into hidden representations. The model trains to predict masked residues by employing features of unmasked residues, which allows the pLM to learn the general sequence context of training data. After training, the hidden representations extracted from the encoder of the pLM can reflect intrinsic characteristics of proteins, serving as a useful starting point for downstream tasks.

Although direct features are straightforward to extract and interpret, relying solely on these features may be insufficient for accurately capturing the functional properties of enzymes. Derived features include nonintuitive information such as structural and evolutionary properties, in which additional preprocessing of amino acid sequences is necessary for extraction. For instance, the composition of structural and functional elements offers valuable insights into enzyme functionality, though additional computational processing is required and often complicated to extract from raw sequences. Secondary structure composition, which represents the local conformation of polypeptide backbones, can be extracted using computational tools such as STRIDE [36] and PSIPRED [37]. For example, secondary structure composition obtained from STRIDE is utilized to train SVM model to classify enzyme types, incorporating additional structural features (e.g. surface area, disulfide bonds, surface AAC) and physicochemical properties (e.g. AAC, cofactor types, metal ions) [27]. Functional domain composition quantifies the frequency of occurrence of specific domains within a protein sequence, utilizing information from protein domain databases such as Pfam and InterPro [38, 39]. Because functional domains are closely associated with the biological functions of proteins, this feature has been widely employed in enzyme function prediction [40–42]. For example, ECS is an SVM-based model trained on Pfam-based functional domain composition, successfully classifying protein sequences into six enzyme classes [40]. Each sequence was represented as a 2,657-dimensional binary vector corresponding to the presence or absence of functional domains from the Pfam database.

Evolutionary features, which were derived from conventional homology-based approaches, were also used to train machine learning models for accurate enzyme function prediction. For example, in ECPred, homology scores derived from a BLAST search were used to train the kNN module, predicting enzyme function based on the functional class of sequences that were highly scored [29]. Another example is EzyPred, in which PSSMs were used to provide detailed information related to evolutionarily conserved regions in protein sequences [43].

5.3.2 Machine Learning-Based Approaches Algorithms for Enzyme Function Prediction

Various machine learning algorithms process input data using distinct strategies. Therefore, selecting appropriate machine learning algorithms to effectively analyze the characteristics of given features is crucial. In enzyme function classification, the three most frequently used types of machine learning approaches are instance-based, boundary-based, and tree-based algorithms. Instance-based algorithms derive predictions directly from existing data without parameter optimization of explicit models. kNN is a representative example of this type, inferring the predominant enzyme functional class from the nearest neighbors within the corresponding feature space [44]. For instance, feature vectors derived from InterPro-based functional domain composition can be used to calculate the distance of each enzyme for kNN-based prediction [41]. In another example, feature vectors derived from PSSM were used to compute the distance of each protein sequence for predicting six enzyme classes of EC number (i.e. the first digit of EC numbers) with kNN [45]. Several advanced kNN algorithms have also been employed to better capture ambiguous relationships between biological features and labels. For instance, optimized evidence-theoretic kNN was used to consider the reliability of adjacent data affecting the decision-making process [43].

Instance-based approaches facilitate the straightforward identification of data points that significantly influence predictions. However, the need to store the full dataset and compute pairwise distances for each query leads to high computational costs, limiting its scalability for whole-genome scale analysis. Boundary-based approaches have addressed these limitations by defining specific regions in the feature space corresponding to each class [46]. SVM is a boundary-based algorithm that is most frequently used for enzyme functional classification [23, 24, 26, 27, 29, 30, 40]. For example, the classification of EC number and protein family was conducted with separate SVM models trained to identify each family, utilizing various physicochemical features extracted from protein sequences [23]. However, the effectiveness of both instance-based and boundary-based approaches is constrained by their sensitivity to feature distribution and the inherent challenges in interpreting their predictive outcomes. Tree-based approaches recursively split data according to well-defined criteria derived from feature attributes, enhancing the interpretability of predictions and offering robustness against variations in feature distribution [47]. However, relying on a single decision tree is prone to overfitting. Therefore, random forest [48], an ensemble tree-based method, was predominantly used for enzyme function prediction to improve generalization and predictive robustness [28, 42, 49, 50]. Multiple studies have demonstrated the effectiveness of random forest algorithms in the classification of enzyme functions. For example, PredictEFC, a random forest model for classification of seven EC classes trained on InterPro-based functional domain features, outperformed an SVM-based model on the same dataset [42]. EFPrf utilized random forests trained on CATH domain profiles derived from input protein sequences, exhibiting robust performance on proteins with low sequence identity relative to those in the training dataset. Analyzing feature importance scores facilitated the identification of specificity-determining residues that play a critical role in enzyme functionality, thereby improving the interpretability and reliability of predictions [50]. However, as the number of input features increases, tree-based approaches would become less effective in identifying important features, reducing their interpretability.

Since each machine learning approach has distinct advantages and limitations, integrating multiple algorithms could compensate for their respective shortcomings. Furthermore, it facilitates the systematic integration of diverse features into model prediction, further enhancing the performance of machine learning-based enzyme function prediction. For example, EFICAz^2 employed a tree-based approach to integrate SVM with functionally discriminating residue detection modules, enabling the prediction of EC numbers using protein features derived from MSAs and the physicochemical properties of amino acids [51]. ECPred combined kNN and SVM-based modules to simultaneously utilize homology scores and physicochemical features of protein sequence, respectively [29].

The application of machine learning has enabled accurate enzyme function predictions even in cases with low sequence-function similarity, utilizing intrinsic properties of proteins. However, traditional machine learning models are typically designed to learn linear relationships or simple nonlinear patterns, making it challenging to capture the intricate and highly complex relationships inherent in biological data. The necessity of hand-crafted extraction of enzyme features is also a significant constraint in machine learning-based approaches. Although manually extracted features are based on an in-depth understanding of enzymology, they are often inefficient in reflecting the correlation between protein sequences and functions. Recent advancements in deep learning have addressed these limitations by enabling automatic feature learning, eliminating the need for manual feature extraction. Deep learning models facilitate inherently capturing hierarchical and complex relationships within biological data, leading to more accurate enzyme function predictions compared to traditional machine learning approaches.

5.4 Prediction of Enzyme Functions Using Deep Learning

Artificial neural networks (ANNs), developed by imitating the transmission of information between neurons via synapses in the human brain, serve as the fundamental algorithm underlying all deep neural networks [52]. For decades, ANNs have been employed to capture complex and nonlinear data patterns, enabling the analysis of nonlinear relationships between protein sequences and their functions [53, 54]. For instance, ANNs trained on sequence similarity to known domain groups can predict functional domain groups in input proteins, even for those with low sequence similarity, which was challenging for traditional BLAST searches [55]. Similarly, ProtFun was developed as an ensemble of ANNs to predict enzyme functional classes from various physicochemical properties, including hydrophobicity, number of charged residues, and secondary structures [56]. Although early use of ANNs was constrained to shallow network architectures due to limited computational capabilities, their emergence is highly significant as they form the foundation for all subsequent advanced deep neural networks.

In deep learning-based enzyme prediction models, EC numbers and GO terms are commonly used as target labels for model training. The EC number is one of the useful classification schemes for enzyme functions that assigns general functions of enzymes to specific digits [57]. The EC number consists of four digits separated by periods in a hierarchical system, with more detailed functional information toward the end of the EC number. The first digit is the enzyme class, assigned according to the primary reactions performed by the enzyme, classified as oxidoreductase (EC:1), transferase (EC:2), hydrolase (EC:3), lyase (EC:4), isomerase (EC:5), ligase (EC:6), and translocase (EC:7). The enzyme functions are further described by the second and third digits, known as subclass and sub-subclass, providing more detailed functional roles. The final digit indicates the type of substrates the enzyme reacts with. For example, EC:3.4.11.4 denotes a hydrolase that acts on peptide bonds, cleaving off the amino-terminal residue from polypeptides or proteins. Gene ontology (GO) terms, a gene functional classification system, also describe protein properties in terms of molecular function, cellular component, and biological process. Each category of GO terms represents molecular-level activities, cellular locations where these functions occur, and large-scale biological processes carried out by multiple molecular-level activities. By tagging multiple GO terms and linking them hierarchically, this system provides more detailed descriptions and complex relationships, structured as a directed acyclic graph.

With the breakthroughs in computing power, the development of novel neural network architectures, and the advent of big data, various deep neural network architectures have been developed to improve learning efficiency. Such neural networks have shown powerful performance in predicting enzyme functions. Basically, deep neural networks learn from amino acid sequences and/or features derived from sequences, identifying patterns inherent in an enzyme [1]. In this section, we discuss the recent advances in deep learning models used for predicting EC numbers and GO terms, focusing on their architecture for effectively processing enzyme data.

5.4.1 Convolutional Neural Network

Convolutional neural networks (CNNs) were initially developed to learn patterns of image inputs by using channels and convolutional kernels, where kernels compress nearby pixels according to the kernel size [58]. These convolutional layers effectively capture local feature patterns through multiplication and summation between convolutional kernels and local regions of the input (Figure 5.2). These characteristics of CNNs can also be applied to learn local contexts of enzyme sequences, thereby enhancing performance in enzyme function prediction. By utilizing CNNs, various deep learning-based tools have been developed to predict EC numbers and/or GO terms [59–66]. A notable example is DeepEC, a model that predicts enzyme EC numbers using both CNN-based modules and a homology analysis module [63]. DeepEC operates in two steps: firstly, three CNN sub-modules predict whether the input sequence is an enzyme or not, the EC number up to the third digit, and the fourth digit in parallel. Then, if the EC numbers cannot be predicted consistently (i.e. the EC numbers predicted up to the third digit and the fourth digit are inconsistent), DeepEC assigns the EC number by employing homology analysis based on sequence similarity. Another example is ECPICK, which also utilizes three parallel convolutional layers alongside four hierarchical layers predicting each digit of the EC number, respectively [67]. The hierarchical layers perform two types of prediction: global prediction of all four digits at once from the last latent representation, and local prediction of each digit sequentially from the same level's latent representation. The final predicted probabilities are a weighted average of these two predictions.

Models for predicting GO terms as enzyme functions have also been developed using amino acid sequences with CNNs. For example, DeepGOPlus takes an amino acid sequence as input and embeds it into a one-hot encoded vector with dimensions of (21, 2000), representing the number of amino acid types and sequence length, respectively [65]. One-hot encoding simplifies information derived from the

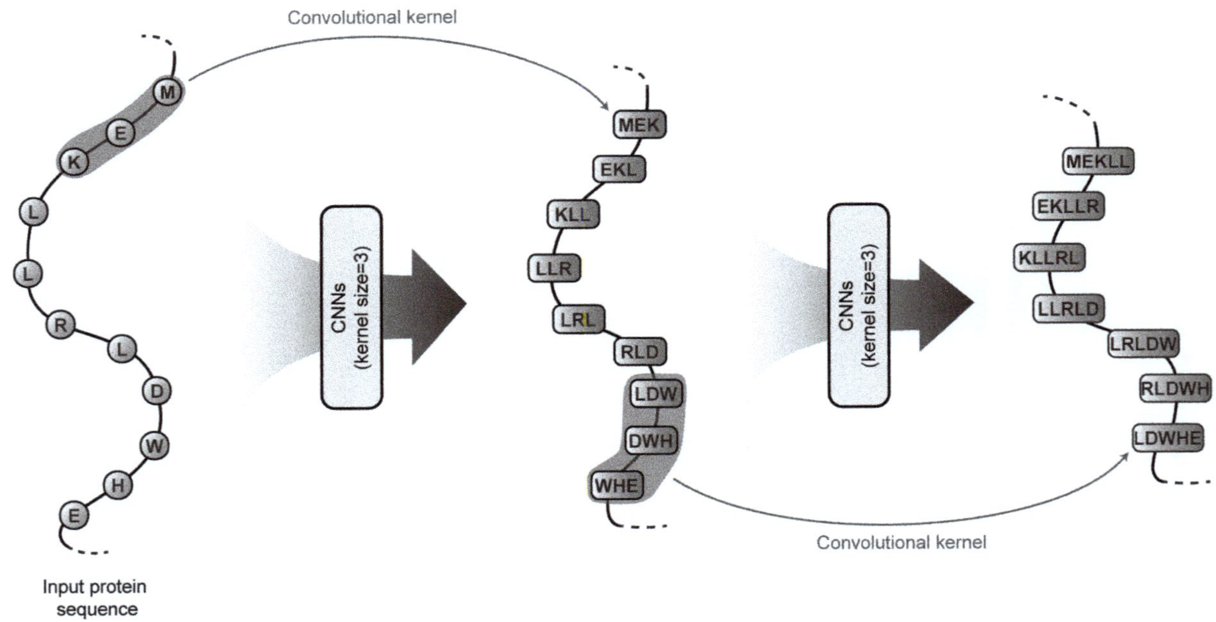

Figure 5.2 The CNN scheme for capturing local features of a protein at the residue level. The residues of the input sequence are condensed together with sequentially adjacent residues based on filter size by passing a learnable convolution kernel. The condensed features involve information localized from residues that pass the kernel together.

sequence without requiring complex calculations or learnable parameters in the embedding process. For input proteins longer than 2,000 amino acid residues, only the first 2,000 residues are retained, while shorter sequences are zero-padded in the one-hot encoded vectors. The one-hot encoded vector is then fed into 16 convolutional layers, followed by max-pooling layers and a fully connected layer, to predict the likelihood of each GO term. Convolutional layers are effective at extracting latent features from two-dimensional inputs like images due to their translational invariance, making them widely used in developing deep neural networks for GO term prediction from one-hot encoded vectors. Another model, ProteInfer, predicts both GO terms and EC numbers by employing dilated convolutional layers to identify a broader range of contexts with low computational cost and residual layers to ensure the stability of neural networks during training [66]. ProteInfer leverages these advantages to predict multi-label sequences (i.e. multifunctional enzymes) effectively.

5.4.2 Recurrent Neural Network

Unlike CNNs for capturing local patterns in images, recurrent neural networks (RNNs) were devised to process sequential data, such as natural language. RNNs read each token (e.g. words in natural language and amino acid residues in proteins) one at a time and retain information from previous tokens using internal memory. Inspired by the fact that, similar to natural language, the properties of proteins are predominantly specified by the order of amino acids, RNNs have been used to extract global contexts from protein sequences (Figure 5.3) [60, 64, 68–70]. For instance, DEEPre was developed by integrating CNNs with a subsequent long short-term memory (LSTM), a type of RNN [60]. DEEPre takes sequence-length-dependent features as input for CNNs to capture local contexts of enzyme sequences. The global representation is then derived from compressed convolutional features through LSTM. This global representation is concatenated with sequence-length-independent features to predict the EC number of the input sequence. DEEPre hierarchically predicts each digit of the EC number using a

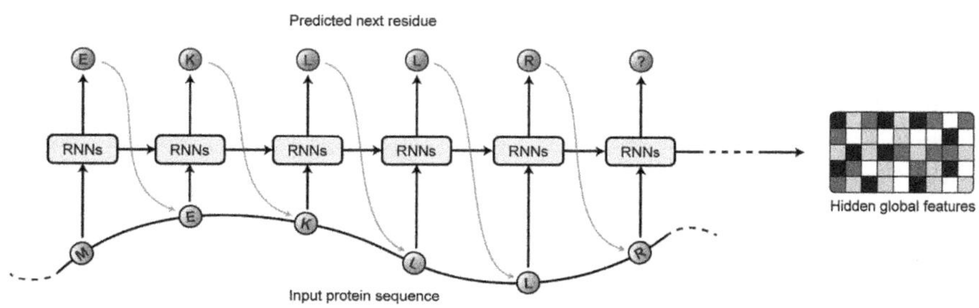

Figure 5.3 The RNN scheme for extracting global features of a protein at the residue level. The features of the first residue are input into the first layer. The output of those features is then sequentially fed into the next layer, merging with features of the second residue. This mechanism proceeds until the end of the sequence is reached, and global representation is obtained from the final layer of the RNNs.

specifically trained submodel for each digit prediction. mlDEEPre, an extension of DEEPre, predicts multi-label enzymes by estimating the probability of all digits in EC numbers [61]. mlDEEPre performs a four-stage prediction: (1) determining if an input protein is an enzyme, (2) identifying enzyme promiscuity, (3) predicting the first digit of the EC number, and (4) predicting the full four-digit EC number.

RNNs are effective neural architectures for predicting not only EC numbers but also GO terms. For instance, UDSMProt, a model composed of LSTM that applies self-supervised pretraining, predicts both EC numbers and GO terms [68]. UDSMProt was first trained on an unlabeled protein sequence database, predicting next residue based on previous residues, allowing the model to learn general protein sequence contexts. Following such self-supervised pretraining, fully connected layers are added at the end of LSTM layers frozen with pretrained parameters and trained again on a labeled dataset to classify enzyme functions based on EC numbers and GO terms of the input sequences.

5.4.3 Transformer and Protein Language Models

While RNNs are valuable for processing sequential data, this architecture suffers from a fundamental drawback: initial information gradually fades as the sequence progresses through RNNs. However, the emergence of transformer architectures, designed to capture long-range dependencies between tokens in natural language, addresses this limitation by simultaneously identifying all relationships and influences among pairs of tokens [71]. Transformer architectures have also pushed the boundaries of protein science, drawing inspiration from the parallels between natural language and protein sequences. In proteins, transformer architectures are particularly effective for learning relationships between residues (Figure 5.4) [70, 72–81].

To utilize these strengths of transformer architectures with the tremendous amount of protein data deposited in various databases [82, 83], protein language models (pLMs) have emerged, mimicking BERT-like models pretrained with masked language modeling (Figure 5.1c) [84–96]. For example, ESM was constructed with 650 million parameters and trained on sequences in the UniRef50 database by applying masked language models to learn evolutionary relationships between residues (e.g. coevolution, contact, etc.) [85]. Recently, ESM has been updated to ESM3, which was trained on various aspects of proteins, such as sequence, structure, and other residue-level characteristics [96]. Another model, ProtT5, was designed as a pretrained pLM for downstream tasks, employing span masking to learn local contextual patterns by masking consecutive residue domains rather than single residues [87]. ProstT5, while maintaining the underlying architecture of ProtT5, incorporates structural information and supports bidirectional conversion (i.e. from structure to sequence features, or *vice versa*) using 3Di tokens, which encode structural characteristics of residues and facilitate sequence-structure transformations [95, 97]. Embeddings extracted from these pLMs serve as a robust foundation for downstream applications, effectively representing inherent patterns and characteristics. For instance, DeepECtransformer utilizes transformer

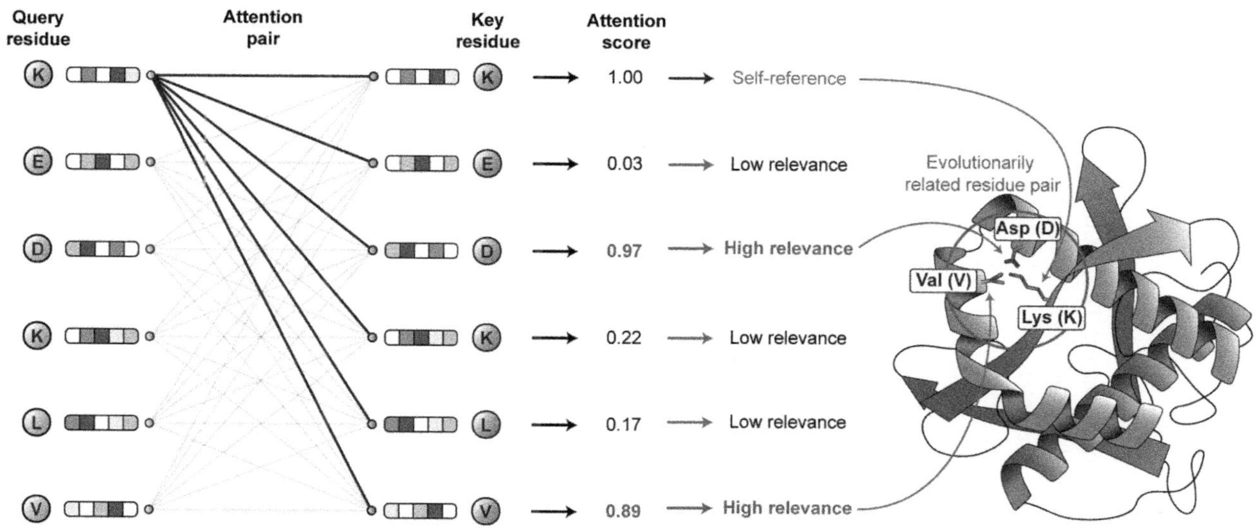

Figure 5.4 Visualization of transformer mechanism using attention algorithm at the residue level. The features of each residue (i.e. query residue) in the input sequence are used to compute attention scores with other residues (i.e. key residues) based on their features. The attention scores represent predicted contextual relevance for each residue pair, which may indicate evolutionary relationships such as residue contacts or coevolution patterns. A transformer can also capture long-range dependencies because the attention algorithm computes each query-key score at once.

layers and CNNs to predict EC numbers [72]. The transformer layers consist of two subsequent self-attention encoders from the ProtBERT model to embed the input amino acid sequences of enzymes into hidden representations [87]. The resulting hidden representation transformed through transformer layers is fed into two consecutive CNN layers, effectively learning local and global features of the input enzyme sequence. DIAMOND, a tool for a homology search algorithm, supplements DeepECtransformer in cases where neural networks fail to predict the EC number of an input enzyme, assigning EC numbers based on homologous enzymes [98]. Thanks to transformer layers, DeepECtransformer has demonstrated superior performance compared to previous EC number prediction tools.

The advent of transformer architectures and pLMs has also affected the prediction of GO terms as well as EC numbers. PFresGO, for instance, first uses protein sequences and GO term embeddings as inputs. The protein sequence is converted into two types of representations: features encoded from one-hot encoding of the input sequence and embeddings extracted from the encoder of the ProtT5 model [73, 87]. These representations are summed and fed into the multi-head cross-attention layer of a transformer along with GO term embeddings, learning which GO term is relatively more associated with each residue in the input sequence. Another example, PhiGnet, utilizes the ESM-1b model as the initial stage of neural networks to convert input enzyme sequences into hidden representations [85, 99]. The hidden representations are processed using two types of graph convolutional networks (GCNs): one captures evolutionary couplings representing relationships between pairwise residues at two covariant sites, and the other identifies residue communities as hierarchical interactions among residues. Based on these relationships (i.e. edges in the graph), corresponding features in the representations are updated and used to predict EC numbers and GO terms.

Similarly, protein structure information has been utilized for predicting GO terms, as a protein's three-dimensional structure often determines its roles in catalysis, molecular transport, and other functions. A recent study developed ESM-S, a pLM that incorporates structural information [81]. ESM-S builds upon ESM2, which was pretrained on sequence databases using masked language modeling (i.e. predicting masked tokens in a sequence) and further trained on a remote homology detection dataset to learn structural information. The learned representation from ESM-S has shown increased performance not only in GO term prediction but also in EC number prediction tasks, underscoring the importance of structural information for enzyme functional classification. PredGO combines representations learned from a pLM, protein-protein interaction networks, and three-dimensional structure information to predict GO terms, highlighting how exploiting heterogeneous data sources enhances the performance of GO term prediction [100]. To reflect structural information in predicting EC numbers, MAPred, a state-of-the-art model for EC number prediction, exploits ESM and ProstT5 as pretrained models to extract embeddings based on sequence and structure, respectively [80, 93, 95]. MAPred also employs cross-attention mechanisms to interlace sequence and 3Di representations for extracting global features, alongside convolutional layers using diverse kernel sizes for local feature extraction. Finally, MAPred concatenates global and local features, feeding them into multilayer perceptrons to predict digits of the EC number hierarchically.

Advances in machine learning training schemes have also transformed enzyme functional classification using deep learning. Supervised learning, a machine learning approach where a model is trained on labeled data to make predictions or classify new inputs, has been used for enzyme function prediction. However, the type of supervised model that predicts labels of the input itself has limitations when the sequence data used for training includes significant noise or the model encounters unseen labels during training. To complement this weakness, CLEAN was trained on EC numbers by using contrastive learning, which was devised to train the model to ensure that representations with the same EC number are more similar rather than focusing solely on inferring labels of the input itself [77]. This approach clusters enzymes with similar functional groups closely together in the latent space, while functionally distinct enzymes are spaced farther apart. Despite being composed entirely of simple neural networks, CLEAN has achieved superior performance.

5.4.4 Graph Neural Network

In line with the advancement in deep learning approaches, the development of AlphaFold and RoseTTAFold has pioneered protein structure prediction with high confidence, facilitating an exponential increase in the accumulation of protein structure data [101, 102]. With the increased accessibility of protein structure data, such as the AlphaFold DB, protein structure information has become instrumental in predicting EC numbers and GO terms, demonstrating significant potential [103]. Graph neural networks (GNNs) are well-suited deep learning architectures for processing protein structure data, capable of generating diverse, information-rich graphs based on how nodes (e.g. atoms or residues) and edges (e.g. spatial distance, sequential order, extrinsic/intrinsic distance, etc.) are defined (Figure 5.5) [92, 99, 100, 104–111]. For instance, GraphEC employs ProtTrans to extract protein embeddings and ESMFold to generate predicted three-dimensional protein structures from input sequences [87, 93, 111]. The nodes and edges are defined by residues and relationships between near residues, respectively. Using this GNN, the model predicts three types of outputs for the input enzyme: active sites, optimal pH, and EC number. For EC number prediction, the training of GraphEC-AS is preceded by the prediction of active sites of an input enzyme, whose trained parameters are used to emphasize which residues in the input sequence are more influential for EC number prediction. The hidden representation embedded by the GNN is compressed by using pooling based on weighted residues, predicting the EC number of the input enzyme. GraphEC also uses label diffusion to balance the prediction and labels derived from homologs of the input sequence. GraphEC has shown superior performance by utilizing evolutionary and structural information with suitable neural network architectures.

The increasing accessibility of three-dimensional protein structure data has also provided useful information for predicting GO terms. DeepFRI employs both

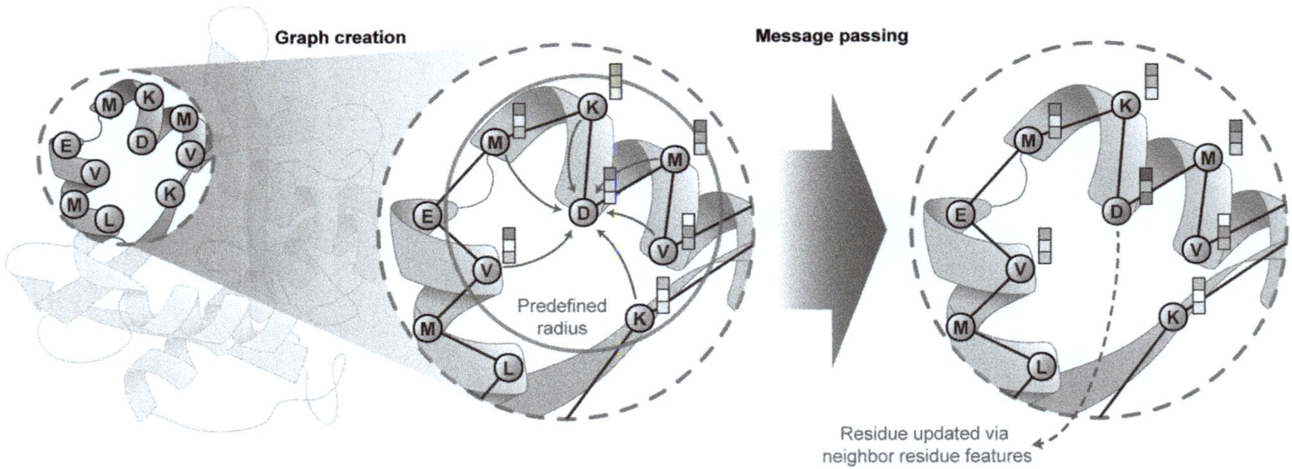

Figure 5.5 Graph creation from protein structure and GNN process with message passing at the residue level. A protein graph is constructed using the atomic coordinates of a protein structure by defining residues as nodes and relationships between neighboring residues as edges, where neighboring residues are considered to be those within a predefined distance. During the message passing stage, each residue updates its representation by aggregating features from neighboring residues so that the residue can contain information from its neighbors. It allows GNNs to efficiently learn structural information, emphasizing spatial characteristics. Furthermore, according to how the nodes and edges are defined, GNNs can capture diverse geometric and spatial features from protein graphs.

sequence and structural information of enzymes to predict GO terms [106]. For this task, LSTMs are used to embed the sequence of an input enzyme, trained to predict the next amino acid based on preceding residues. Three-dimensional protein structures generate contact maps to identify spatially proximate residue pairs, forming protein graphs at the residue level based on these contact maps. GCN, a type of GNN, updates the features of each residue by reflecting features of neighboring residues using the representation from LSTMs and edges from the contact map. Combining sequence information with protein structure data enhances GO term classification for input enzymes, while also allowing for the interpretation and identification of residue-level contributions to GO term prediction. In another approach, TransFun employed protein structures predicted by AlphaFold2, which create adjacency matrices indicating neighboring residues of each residue in proteins [109]. Equivariant GNNs were used to predict GO terms, with node and edge features defined by embeddings from ESM-1b and relationships between neighboring residues.

With abundant protein three-dimensional structure data available, pretraining deep learning models on protein structures similar to protein sequences has been employed to explicitly capture spatial information crucial for protein function determination. GearNet, for example, a GNN pretrained on AlphaFoldDB, predicts EC numbers and GO terms [90, 94]. Unlike pretraining on sequence data using masked language modeling, pretraining on structural data is more complex. To address this challenge, GearNet employed multiview contrastive learning, an approach that distinguishes views (i.e. subsequence and substructure) of the same protein from those of different proteins, embedding biologically related substructures closely within the latent space. Although the pretraining dataset contained fewer than one million data points, much smaller than the millions to billions of protein sequences used for pLMs, GearNet showed high prediction performance in downstream tasks of EC number and GO term prediction, highlighting the potential of structure-based protein function analysis.

GNNs can also be utilized to learn information about protein-protein interaction networks, in which relationships between proteins serve as an important factor, as it is difficult to determine the role of a protein independently, such as in the context of GO terms. DeepGOGAT-SE was designed using a graph attention network, a type of GNN, to apply graph structure derived from protein-protein interaction, employing the ESM2 model to extract sequence-derived embeddings as inputs [78]. DeepGOGAT-SE achieved high performance in GO term classification by effectively integrating representations learned from ESM2 with functional contexts derived from protein-protein interaction networks.

5.5 Concluding Remarks

In this chapter, we introduce the enzyme functional classification with the development and progression of various prediction schemes using computational learning algorithms (Table 5.1). The conventional machine learning algorithms were initially

Table 5.1 Machine learning- and deep learning-based tools for enzyme function prediction.

Name	Characteristics	Reference
SVM-Prot	– AAC, van der Waals volume, polarity, polarizability, charge, surface tension, secondary structure, and solvent accessibility as inputs – Binary classification for each family using SVM	[23]
Syed et al., 2003	– Protein sequence, secondary structure, and database features as inputs – The use of subcellular location information as database features	[119]
Cai et al., 2004	– AAC, van der Waals volume, polarity, polarizability, charge, surface tension, secondary structure, and solvent accessibility as inputs – Employment of principal component analysis (PCA) for reduction of overlapping features	[24]
Paul et al., 2004	– Secondary structure content, amino acid propensity, and surface property as inputs – One-versus-Rest approach with 15 submodels	[27]
EzyPred	– Functional domain composition and Pse-PSSM as inputs – Independent modules for processing functional domain composition and Pse-PSSM	[43]
Concu et al., 2009	– Tertiary structure as inputs – The use of electrostatic, hydrophobic, and surface potentials	[120]
EFICAz²	– Protein sequence as inputs – The use of decision tree for data classification	[51]
Kumar et al., 2009	– Pepstats and ProtParam as inputs – The use of physicochemical features	[49]
Kumar et al., 2012	– Pepstats and ProtParam as inputs – Hierarchical architecture with modules for 1st digits and 2nd digits	[28]
EFPrf	– Protein sequence as inputs – Employment of prediction modules for each CATH	[50]
Che et al., 2016	– The use of Pse-PSSM and AAC together	[45]
Amidi et al., 2017	– Torsion angle in peptide chain of each residue, homology score as inputs – The use of feature-level fusion for prediction	[121]
Predhan et al., 2017	– Molecular weight, number of amino acids, isoelectric point, AAC, charge of residues, atomic composition, aliphatic index, and hydrophobicity as inputs – Hierarchical architecture with binary SVM and multiclass SVM	[26]

Table 5.1 (Continued)

Name	Characteristics	Reference
Watanabe et al., 2020	– Protein sequence and reactants as inputs – Employment of PCA for dimension reduction	[30]
DeepGO	– The prediction of GO terms as outputs – The use of convolutional layers with protein-protein interaction networks and the GO graph hierarchy	[59]
DEEPre	– The prediction of EC numbers as outputs – The use of convolutional layers and LSTM layers for hierarchical classification using models specialized in each EC number	[60]
mlDEEPre	– The prediction of EC numbers as outputs – The use of convolutional layers for hierarchical classification with DEEPre after the second digit and multi-label classification based on the first digit of the enzyme	[61]
DeepEC	– The prediction of EC numbers as outputs – The use of convolutional layers combined with a homologous search module	[63]
DeepGOA	– The prediction of GO terms as outputs – The use of convolutional layers and graph convolutional layers specialized for maize proteins	[105]
HECNet	– The prediction of EC numbers as outputs – The use of convolutional layers and LSTM layers with a Siamese network employing triplet loss	[64]
UDSMProt	– The prediction of EC numbers and GO terms as outputs – The use of LSTM layers through pretraining and fine-tuning of the classifier	[68]
DeepGOPlus	– The prediction of GO terms as outputs – The use of convolutional layers combined with sequence homology search	[65]
DeepGraphGO	– The prediction of GO terms as outputs – The use of graph convolutional layers integrated into the NetGO framework	[107]
NetGO 2.0	– The prediction of GO terms as outputs – The use of BiLSTM layers with a learn-to-rank framework	[69]
DeepFRI	– The prediction of EC number and GO terms as outputs – The use of LSTM layers and graph convolutional layers with protein structure information	[106]

(Continued)

Table 5.1 (Continued)

Name	Characteristics	Reference
GearNet	– The prediction of EC numbers and GO terms as outputs – The use of graph convolutional layers and edge message passing layers with three kinds of edge types and multiview contrastive learning for pretraining	[90]
DeeProtGO	– The prediction of GO terms as outputs – The use of fully connected layers for hierarchical classification of GO terms	[54]
DeepGOZero	– The prediction of GO terms as outputs – The use of fully connected layers with zero-shot learning and embedding proteins into a space where GO axioms are embedded by the ELEmbeddings method	[53]
GAT-GO	– The prediction of GO terms as outputs – The use of convolutional layers and graph attention networks with protein structure information and a protein language model	[108]
DCNN	– The prediction of EC numbers as outputs – The use of convolutional layers with the kNN algorithm for integrating each feature and inferring output	[62]
CaLM	– The prediction of GO terms as outputs – The use of transformer layers for learning patterns of synonymous codon usage and pretraining on cDNA sequences	[79]
HDMLF	– The prediction of EC numbers as outputs – The use of transformer layers, BiGRU layers, and attention layers with multiple sequence alignment for fine-tuning prediction, multitask learning for optimizing predicted outputs, and prediction of promiscuous enzyme functions	[74]
DeepECtransformer	– The prediction of EC numbers as outputs – The use of transformer layers and convolutional layers combined with a homologous search module	[72]
CLEAN	– The prediction of EC numbers as outputs – The use of transformer layers for EC number prediction using contrastive learning	[77]
CFAGO	– The prediction of GO terms as outputs – The use of transformer layers with pretraining by self-supervised learning, fine-tuning for GO term prediction, and utilization of protein-protein interaction networks	[76]

Table 5.1 (Continued)

Name	Characteristics	Reference
Chemical-SA-BiLSTM	– The prediction of GO terms as outputs – The use of self-attention layer and bidirectional LSTM layers specialized for proteins from grain	[70]
ProteInfer	– The prediction of GO terms as outputs – The use of dilated convolutional layers for providing coarse-grained functional localization	[66]
NetGO 3.0	– The prediction of GO terms as outputs – The use of transformer layers with a protein language model and a learn-to-rank framework	[75]
PFresGO	– The prediction of GO terms as outputs – The use of transformer layers with a protein language model and the GO graph hierarchy	[73]
DeepGO-SE	– The prediction of GO terms as outputs – The use of transformer layers for embedding proteins into a space where GO axioms are embedded by the ELEmbeddings method	[78]
HNetGO	– The prediction of GO terms as outputs – The use of a graph attention network with SeqVec-extracted embeddings, protein-protein interaction networks, and the GO graph hierarchy	[110]
TransFun	– The prediction of GO terms as outputs – The use of transformer layers and equivariant graph neural networks combined with sequence homology search and protein structure information	[109]
PredGO	– The prediction of GO terms as outputs – The use of transformer layers and geometric vector perceptron graph neural networks with protein-protein interaction networks and protein structure information	[100]
PhiGnet	– The prediction of EC numbers and GO terms as outputs – The use of transformer layers and graph convolutional layers with evolutionary couplings and residue community information	[99]
MAPred	– The prediction of EC numbers as outputs – The use of transformer layers, convolutional layers, and cross-attention layers with embeddings extracted by ESM-1b and ProtT5, utilizing global and local features	[80]

(Continued)

Table 5.1 (Continued)

Name	Characteristics	Reference
GraphEC	- The prediction of EC numbers, active sites, and optimal pH as outputs - The use of transformer layers and a graph convolutional network with a protein language model, structural features extracted by ESMFold, and label diffusion to reflect homologs information	[111]
ESM-S	- The prediction of GO terms as outputs - The use of transformer layers pretrained on the remote homology detection task	[81]

used to predict enzyme functions utilizing computational resources. As various classification schemes are devised, such as kNN, SVM, and random forest, the machine learning-based models have become more reliable with protein-derived features as inputs, yet encountered limitations, including data quality, feature extraction, computational resources, and parametric constraints of traditional algorithms. To address these challenges, deep learning approaches have emerged alongside advances in computing devices such as graphics processing units, significantly expanding the field of computational biology. Thus, we discuss various types of deep neural networks initially designed for other fields but now applied to biology due to shared characteristics, such as spatial locality in images and the sequential nature of natural language.

Additionally, we review how deep neural networks have been used to classify enzyme functions in terms of well-defined classification nomenclature schemes – EC numbers and GO terms – illustrating deep learning-based models categorized by representative deep neural architectures. However, despite these numerous efforts, the fundamental limitation remains in predicting EC numbers and GO terms using deep learning, leading to constraints on predictive accuracy caused by data imbalance in underrepresented EC number and GO term classes. These challenges are unavoidable but must be addressed to achieve more robust and reliable predictions of enzyme functions in real-world applications. Moreover, with the growing availability of biological big data, trends in enzyme functional prediction using artificial intelligence have shifted toward integrating diverse biological resources and proteins. This approach has proven efficient for predicting various properties, including functions, of proteins at the omics levels [79]. Furthermore, by leveraging the characteristics of generative models such as the hallucination effect, enzyme engineering strategies are expanding their area into non-natural enzyme space, where engineered enzymes exhibit low sequence similarity to natural ones but retain the same or similar structural and functional properties [112–117]. These innovative changes are propelling biology into a new phase, accelerating the advancement of enzyme engineering through enzyme functional predictions.

Acknowledgments

This chapter was prepared by expanding our recent review paper published in Trends in Biotechnology (doi: 10.1016/j.tibtech.2025.03.003) [118]. We have extended the coverage from classical methodologies for enzyme function prediction to state-of-the-art deep learning models, emphasizing the evolution and future directions of prediction methods driven by deep learning advancements. This work was supported by the Development of advanced synthetic biology source technologies for leading the biomanufacturing industry project (RS-2024-00399424) from National Research Foundation supported by the Korean Ministry of Science and ICT. This research was also supported by a grant of the Korea-US Collaborative Research Fund (KUCRF), funded by the Ministry of Science and ICT and Ministry of Health & Welfare, Republic of Korea (grant number: RS-2024-00468410).

References

1 LeCun, Y., Bengio, Y., and Hinton, G. (2015). Deep learning. *Nature* 521 (7553): 436–444.

2 Zou, J., Huss, M., Abid, A. et al. (2019). A primer on deep learning in genomics. *Nat. Genet.* 51 (1): 12–18.

3 Lee, D., Redfern, O., and Orengo, C. (2007). Predicting protein function from sequence and structure. *Nat. Rev. Mol. Cell Biol.* 8 (12): 995–1005.

4 Pellegrini, M., Marcotte, E.M., Thompson, M.J. et al. (1999). Assigning protein functions by comparative genome analysis: protein phylogenetic profiles. *PNAS* 96 (8): 4285–4288.

5 Loewenstein, Y., Raimondo, D., Redfern, O.C. et al. (2009). Protein function annotation by homology-based inference. *Genome Biol.* 10 (2): 207.

6 Altschul, S.F., Gish, W., Miller, W. et al. (1990). Basic local alignment search tool. *J. Mol. Biol.* 215 (3): 403–410.

7 Shah, I. and Hunter, L. (1997). Predicting enzyme function from sequence: a systematic appraisal. *Proc. Int. Conf. Intell. Syst. Mol. Biol.* 5: 276–283.

8 Espadaler, J., Eswar, N., Querol, E. et al. (2008). Prediction of enzyme function by combining sequence similarity and protein interactions. *BMC Bioinf.* 9: 249.

9 Hung, S.S., Wasmuth, J., Sanford, C. et al. (2010). DETECT--a density estimation tool for enzyme classification and its application to Plasmodium falciparum. *Bioinformatics* 26 (14): 1690–1698.

10 Pathak, A. and Jayaram, B. (2022). Seq2Enz: An application of mask BLAST methodology with a new chemical logic of amino acids for improved enzyme function prediction. *Biochim. Biophys. Acta, Proteins Proteomics* 1870 (1): 140721.

11 Altschul, S.F., Madden, T.L., Schaffer, A.A. et al. (1997). Gapped BLAST and PSI-BLAST: a new generation of protein database search programs. *Nucleic Acids Res.* 25 (17): 3389–3402.

12 Eddy, S.R. (1998). Profile hidden Markov models. *Bioinformatics* 14 (9): 755–763.
13 Soding, J. (2005). Protein homology detection by HMM-HMM comparison. *Bioinformatics* 21 (7): 951–960.
14 Finn, R.D., Clements, J., and Eddy, S.R. (2011). HMMER web server: interactive sequence similarity searching. *Nucleic Acids Res.* 39: W29–W37.
15 Tian, W., Arakaki, A.K., and Skolnick, J. (2004). EFICAz: a comprehensive approach for accurate genome-scale enzyme function inference. *Nucleic Acids Res.* 32 (21): 6226–6239.
16 Remmert, M., Biegert, A., Hauser, A. et al. (2011). HHblits: lightning-fast iterative protein sequence searching by HMM-HMM alignment. *Nat. Methods* 9 (2): 173–175.
17 Glasner, M.E., Gerlt, J.A., and Babbitt, P.C. (2006). Evolution of enzyme superfamilies. *Curr. Opin. Chem. Biol.* 10 (5): 492–497.
18 Omelchenko, M.V., Galperin, M.Y., Wolf, Y.I. et al. (2010). Non-homologous isofunctional enzymes: a systematic analysis of alternative solutions in enzyme evolution. *Biol. Direct* 5: 31.
19 Gerlt, J.A. and Babbitt, P.C. (2001). Divergent evolution of enzymatic function: mechanistically diverse superfamilies and functionally distinct suprafamilies. *Annu. Rev. Biochem.* 70: 209–246.
20 Jordan, M.I. and Mitchell, T.M. (2015). Machine learning: trends, perspectives, and prospects. *Science* 349 (6245): 255–260.
21 Jukes, T.H., Holmquist, R., and Moise, H. (1975). Amino acid composition of proteins: selection against the genetic code. *Science* 189 (4196): 50–51.
22 M., d.J., Karp, P.D., Krummenacker, M. et al. (1997). Prediction of enzyme classification from protein sequence without the use of sequence similarity. *Proc. Int. Conf. Intell. Syst. Mol. Biol.* 5: 92–99.
23 Cai, C.Z., Han, L.Y., Ji, Z.L. et al. (2003). SVM-Prot: Web-based support vector machine software for functional classification of a protein from its primary sequence. *Nucleic Acids Res.* 31 (13): 3692–3697.
24 Cai, C.Z., Han, L.Y., Ji, Z.L. et al. (2004). Enzyme family classification by support vector machines. *Proteins* 55 (1): 66–76.
25 Nasibov, E. and Kandemir-Cavas, C. (2009). Efficiency analysis of KNN and minimum distance-based classifiers in enzyme family prediction. *Comput. Biol. Chem.* 33 (6): 461–464.
26 Pradhan, D., Padhy, S., and Sahoo, B. (2017). Enzyme classification using multiclass support vector machine and feature subset selection. *Comput. Biol. Chem.* 70: 211–219.
27 Dobson, P.D. and Doig, A.J. (2005). Predicting enzyme class from protein structure without alignments. *J. Mol. Biol.* 345 (1): 187–199.
28 Kumar, C. and Choudhary, A. (2012). A top-down approach to classify enzyme functional classes and sub-classes using random forest. *EURASIP J. Bioinform. Syst. Biol.* 2012 (1): 1.
29 Dalkiran, A., Rifaioglu, A.S., Martin, M.J. et al. (2018). ECPred: a tool for the prediction of the enzymatic functions of protein sequences based on the EC nomenclature. *BMC Bioinf.* 19 (1): 334.

30 Watanabe, N., Murata, M., Ogawa, T. et al. (2020). Exploration and Evaluation of Machine Learning-Based Models for Predicting Enzymatic Reactions. *JCIM* 60 (3): 1833–1843.

31 Wilkins, M.R., Gasteiger, E., Bairoch, A. et al. (1999). Protein identification and analysis tools in the ExPASy server. *Methods Mol. Biol.* 112: 531–552.

32 Rice, P., Longden, I., and Bleasby, A. (2000). EMBOSS: the european molecular biology open software suite. *Trends Genet.* 16 (6): 276–277.

33 Li, Z.R., Lin, H.H., Han, L.Y. et al. (2006). PROFEAT: a web server for computing structural and physicochemical features of proteins and peptides from amino acid sequence. *Nucleic Acids Res.* 34: W32–W37.

34 Chou, K.C. (2001). Prediction of protein cellular attributes using pseudo-amino acid composition. *Proteins* 43 (3): 246–255.

35 Zhou, X.B., Chen, C., Li, Z.C. et al. (2007). Using Chou's amphiphilic pseudo-amino acid composition and support vector machine for prediction of enzyme subfamily classes. *J. Theor. Biol.* 248 (3): 546–551.

36 Frishman, D. and Argos, P. (1995). Knowledge-based protein secondary structure assignment. *Proteins* 23 (4): 566–579.

37 McGuffin, L.J., Bryson, K., and Jones, D.T. (2000). The PSIPRED protein structure prediction server. *Bioinformatics* 16 (4): 404–405.

38 Sonnhammer, E.L., Eddy, S.R., and Durbin, R. (1997). Pfam: a comprehensive database of protein domain families based on seed alignments. *Proteins* 28 (3): 405–420.

39 Apweiler, R., Attwood, T.K., Bairoch, A. et al. (2001). The InterPro database, an integrated documentation resource for protein families, domains and functional sites. *Nucleic Acids Res.* 29 (1): 37–40.

40 Lu, L., Qian, Z., Cai, Y.D. et al. (2007). ECS: an automatic enzyme classifier based on functional domain composition. *Comput. Biol. Chem.* 31 (3): 226–232.

41 De Ferrari, L., Aitken, S., van Hemert, J. et al. (2012). EnzML: multi-label prediction of enzyme classes using InterPro signatures. *BMC Bioinf.* 13: 61.

42 Chen, L., Zhang, C., and Xu, J. (2024). PredictEFC: a fast and efficient multi-label classifier for predicting enzyme family classes. *BMC Bioinf.* 25 (1): 50.

43 Shen, H.B. and Chou, K.C. (2007). EzyPred: a top-down approach for predicting enzyme functional classes and subclasses. *Biochem. Biophys. Res. Commun.* 364 (1): 53–59.

44 Cover, T. and Hart, P. (1967). Nearest neighbor pattern classification. *IEEE Trans. Inf. Theory* 13 (1): 21–27.

45 Che, Y., Ju, Y., Xuan, P. et al. (2016). Identification of multi-functional enzyme with multi-label classifier. *PLoS One* 11 (4): e0153503.

46 Cortes, C. and Vapnik, V. (1995). Support-vector networks. *Mach. Learn.* 20 (3): 273–297.

47 Quinlan, J.R. (1986). Induction of decision trees. *Mach. Learn.* 1 (1): 81–106.

48 Breiman, L. (2001). Random forests. *Mach. Learn.* 45 (1): 5–32.

49 Kumar C., Li G., and Choudhary A. (2009). Enzyme function classification using protein sequence features and random forest. 2009 3rd International Conference on Bioinformatics and Biomedical Engineering. p 1–4.

50 Nagao, C., Nagano, N., and Mizuguchi, K. (2014). Prediction of detailed enzyme functions and identification of specificity determining residues by random forests. *PLoS One* 9 (1): e84623.

51 Arakaki, A.K., Huang, Y., and Skolnick, J. (2009). EFICAz2: enzyme function inference by a combined approach enhanced by machine learning. *BMC Bioinf.* 10: 107.

52 Rosenblatt, F. (1958). The perceptron: a probabilistic model for information storage and organization in the brain. *Psychol. Rev.* 65 (6): 386–408.

53 Kulmanov, M. and Hoehndorf, R. (2022). DeepGOZero: improving protein function prediction from sequence and zero-shot learning based on ontology axioms. *Bioinformatics* 38 (Suppl 1): i238–i245.

54 Merino, G.A., Saidi, R., Milone, D.H. et al. (2022). Hierarchical deep learning for predicting GO annotations by integrating protein knowledge. *Bioinformatics* 38 (19): 4488–4496.

55 Murvai, J., Vlahovicek, K., Szepesvari, C. et al. (2001). Prediction of protein functional domains from sequences using artificial neural networks. *Genome Res.* 11 (8): 1410–1417.

56 Jensen, L.J., Gupta, R., Blom, N. et al. (2002). Prediction of human protein function from post-translational modifications and localization features. *J. Mol. Biol.* 319 (5): 1257–1265.

57 Martinez, C.S., Rahman, S.A., Furnham, N. et al. (2015). The classification and evolution of enzyme function. *Biophys. J.* 109 (6): 1082–1086.

58 Krizhevsky, A., Sutskever, I., and Hinton, G.E. (2012). *Imagenet classification with deep convolutional neural networks*, 25. Syst: Adv. Neural Inf. Process.

59 Kulmanov, M., Khan, M.A., Hoehndorf, R. et al. (2018). DeepGO: predicting protein functions from sequence and interactions using a deep ontology-aware classifier. *Bioinformatics* 34 (4): 660–668.

60 Li, Y., Wang, S., Umarov, R. et al. (2018). DEEPre: sequence-based enzyme EC number prediction by deep learning. *Bioinformatics* 34 (5): 760–769.

61 Zou, Z., Tian, S., Gao, X. et al. (2018). mlDEEPre: multi-functional enzyme function prediction with hierarchical multi-label deep learning. *Front. Genet.* 9: 714.

62 Gao, R., Wang, M., Zhou, J. et al. (2019). Prediction of Enzyme Function Based on Three Parallel Deep CNN and Amino Acid Mutation. *Int. J. Mol. Sci.* 20 (11): 2845.

63 Ryu, J.Y., Kim, H.U., and Lee, S.Y. (2019). Deep learning enables high-quality and high-throughput prediction of enzyme commission numbers. *PNAS* 116 (28): 13996–14001.

64 Memon, S.A., Khan, K.A., and Naveed, H. (2020). HECNet: a hierarchical approach to enzyme function classification using a Siamese Triplet Network. *Bioinformatics* 36 (17): 4583–4589.

65 Kulmanov, M. and Hoehndorf, R. (2020). DeepGOPlus: improved protein function prediction from sequence. *Bioinformatics* 36 (2): 422–429.

66 Sanderson, T., Bileschi, M.L., Belanger, D. et al. (2023). ProteInfer, deep neural networks for protein functional inference. *Elife* 12: e80942.

67 Han, S.R., Park, M., Kosaraju, S. et al. (2023). Evidential deep learning for trustworthy prediction of enzyme commission number. *Briefings Bioinf.* 25 (1): bbad401.

68 Strodthoff, N., Wagner, P., Wenzel, M. et al. (2020). UDSMProt: universal deep sequence models for protein classification. *Bioinformatics* 36 (8): 2401–2409.

69 Yao, S., You, R., Wang, S. et al. (2021). NetGO 2.0: improving large-scale protein function prediction with massive sequence, text, domain, family and network information. *Nucleic Acids Res.* 49 (W1): W469–W475.

70 Liu, J., Tang, X., and Guan, X. (2023). Grain protein function prediction based on self-attention mechanism and bidirectional LSTM. *Briefings Bioinf.* 24 (1): bbac493.

71 Vaswani, A., Shazeer, N., Parmar, N. et al. (2017). Attention is all you need. *Adv. Neural Inf. Process. Syst.* 30: 6000–6010.

72 Kim, G.B., Kim, J.Y., Lee, J.A. et al. (2023). Functional annotation of enzyme-encoding genes using deep learning with transformer layers. *Nat. Commun.* 14 (1): 7370.

73 Pan, T., Li, C., Bi, Y. et al. (2023). PFresGO: an attention mechanism-based deep-learning approach for protein annotation by integrating gene ontology inter-relationships. *Bioinformatics* 39 (3): btad094.

74 Shi, Z., Deng, R., Yuan, Q. et al. (2023). Enzyme commission number prediction and benchmarking with hierarchical dual-core multitask learning framework. *Research (Wash D C)* 6: 0153.

75 Wang, S., You, R., Liu, Y. et al. (2023). NetGO 3.0: protein language model improves large-scale functional annotations. *Genomics Proteomics Bioinf.* 21 (2): 349–358.

76 Wu, Z., Guo, M., Jin, X. et al. (2023). CFAGO: cross-fusion of network and attributes based on attention mechanism for protein function prediction. *Bioinformatics* 39 (3): btad123.

77 Yu, T., Cui, H., Li, J.C. et al. (2023). Enzyme function prediction using contrastive learning. *Science* 379 (6639): 1358–1363.

78 Kulmanov, M., Guzmán-Vega, F.J., Duek, R.P. et al. (2024). Protein function prediction as approximate semantic entailment. *Nat. Mach. Intell.* 6 (2): 220–228.

79 Outeiral, C. and Deane, C.M. (2024). Codon language embeddings provide strong signals for use in protein engineering. *Nat. Mach. Intell.* 6 (2): 170–179.

80 Rong D., Zheng W., Zhong B., et al. (2024). Autoregressive enzyme function prediction with multi-scale multi-modality fusion. *arXiv preprint arXiv:2408.06391.*

81 Zhang Z., Lu J., Chenthamarakshan V., et al. (2024). Structure-informed protein language model. *arXiv preprint arXiv:2402.05856.*

82 ww P. D. B. c. (2019). Protein Data Bank: the single global archive for 3D macromolecular structure data. *Nucleic Acids Res.* 47 (D1): D520–D528.

83 UniProt, C. (2023). UniProt: the universal protein knowledgebase in 2023. *Nucleic Acids Res.* 51 (D1): D523–D531.

84 Devlin J., Chang M.-W., Lee K., et al. (2019). BERT: pre-training of deep bidirectional transformers for language understanding. *Proceedings of the 2019 conference of the North American chapter of the association for computational linguistics: human language technologies*, volume 1 (long and short papers). p 4171-86.

85 Rives, A., Meier, J., Sercu, T. et al. (2021). Biological structure and function emerge from scaling unsupervised learning to 250 million protein sequences. *PNAS* 118 (15): e2016239118.

86 Brandes, N., Ofer, D., Peleg, Y. et al. (2022). ProteinBERT: a universal deep-learning model of protein sequence and function. *Bioinformatics* 38 (8): 2102–2110.

87 Elnaggar, A., Heinzinger, M., Dallago, C. et al. (2022). ProtTrans: toward understanding the language of life through self-supervised learning. *IEEE Trans. Pattern Anal. Mach. Intell.* 44 (10): 7112–7127.

88 Ferruz, N., Schmidt, S., and Hocker, B. (2022). ProtGPT2 is a deep unsupervised language model for protein design. *Nat. Commun.* 13 (1): 4348.

89 Weissenow, K., Heinzinger, M., and Rost, B. (2022). Protein language-model embeddings for fast, accurate, and alignment-free protein structure prediction. *Structure* 30 (8): 1169–1177.

90 Zhang Z., Xu M., Jamasb A., et al. (2022). Protein representation learning by geometric structure pretraining. *arXiv preprint arXiv:2203.06125*.

91 Mansoor, S., Baek, M., Juergens, D. et al. (2023). Zero-shot mutation effect prediction on protein stability and function using RoseTTAFold. *Protein Sci.* 32 (11): e4780.

92 Buton, N., Coste, F., and Le Cunff, Y. (2023). Predicting enzymatic function of protein sequences with attention. *Bioinformatics* 39 (10).

93 Lin, Z., Akin, H., Rao, R. et al. (2023). Evolutionary-scale prediction of atomic-level protein structure with a language model. *Science* 379 (6637): 1123–1130.

94 Zhang Z., Wang C., Xu M., et al. (2023). A systematic study of joint representation learning on protein sequences and structures. *arXiv preprint arXiv:2303.06275*.

95 Heinzinger, M., Weissenow, K., Sanchez, J.G. et al. (2024). Bilingual language model for protein sequence and structure. *NAR Genom. Bioinform.* 6 (4): lqae150.

96 Hayes, T., Rao, R., Akin, H. et al. (2025). Simulating 500 million years of evolution with a language model. *Science* 387 (6736): 850–858.

97 van Kempen, M., Kim, S.S., Tumescheit, C. et al. (2024). Fast and accurate protein structure search with Foldseek. *Nat. Biotechnol.* 42 (2): 243–246.

98 Buchfink, B., Reuter, K., and Drost, H.G. (2021). Sensitive protein alignments at tree-of-life scale using DIAMOND. *Nat. Methods* 18 (4): 366–368.

99 Jang, Y.J., Qin, Q.Q., Huang, S.Y. et al. (2024). Accurate prediction of protein function using statistics-informed graph networks. *Nat. Commun.* 15 (1): 6601.

100 Zheng, R., Huang, Z., and Deng, L. (2023). Large-scale predicting protein functions through heterogeneous feature fusion. *Briefings Bioinf.* 24 (4).

101 Baek, M., DiMaio, F., Anishchenko, I. et al. (2021). Accurate prediction of protein structures and interactions using a three-track neural network. *Science* 373 (6557): 871–876.

102 Jumper, J., Evans, R., Pritzel, A. et al. (2021). Highly accurate protein structure prediction with AlphaFold. *Nature* 596 (7873): 583–589.

103 Varadi, M., Anyango, S., Deshpande, M. et al. (2022). AlphaFold protein structure database: massively expanding the structural coverage of protein-sequence space with high-accuracy models. *Nucleic Acids Res.* 50 (D1): D439–D444.

104 Hermosilla P., Schäfer M., Lang M., et al. (2020). Intrinsic-extrinsic convolution and pooling for learning on 3d protein structures. *arXiv preprint arXiv:2007.06252.*

105 Zhou, G., Wang, J., Zhang, X. et al. (2020). Predicting functions of maize proteins using graph convolutional network. *BMC Bioinf.* 21 (Suppl 16): 420.

106 Gligorijevic, V., Renfrew, P.D., Kosciolek, T. et al. (2021). Structure-based protein function prediction using graph convolutional networks. *Nat. Commun.* 12 (1): 3168.

107 You, R., Yao, S., Mamitsuka, H. et al. (2021). DeepGraphGO: graph neural network for large-scale, multispecies protein function prediction. *Bioinformatics* 37 (Suppl_1): i262–i271.

108 Lai, B. and Xu, J. (2022). Accurate protein function prediction via graph attention networks with predicted structure information. *Briefings Bioinf.* 23 (1): bbab502.

109 Boadu, F., Cao, H., and Cheng, J. (2023). Combining protein sequences and structures with transformers and equivariant graph neural networks to predict protein function. *Bioinformatics* 39 (39 Suppl 1): i318–i325.

110 Zhang, X., Guo, H., Zhang, F. et al. (2023). HNetGO: protein function prediction via heterogeneous network transformer. *Briefings Bioinf.* 24 (6).

111 Song, Y., Yuan, Q., Chen, S. et al. (2024). Accurately predicting enzyme functions through geometric graph learning on ESMFold-predicted structures. *Nat. Commun.* 15 (1): 8180.

112 Dauparas, J., Anishchenko, I., Bennett, N. et al. (2022). Robust deep learning-based protein sequence design using ProteinMPNN. *Science* 378 (6615): 49–56.

113 Goverde, C.A., Wolf, B., Khakzad, H. et al. (2023). De novo protein design by inversion of the AlphaFold structure prediction network. *Protein Sci.* 32 (6): e4653.

114 Watson, J.L., Juergens, D., Bennett, N.R. et al. (2023). De novo design of protein structure and function with RFdiffusion. *Nature* 620 (7976): 1089–1100.

115 Yeh, A.H., Norn, C., Kipnis, Y. et al. (2023). De novo design of luciferases using deep learning. *Nature* 614 (7949): 774–780.

116 Goverde, C.A., Pacesa, M., Goldbach, N. et al. (2024). Computational design of soluble and functional membrane protein analogues. *Nature* 631 (8020): 449–458.

117 Vazquez, T.S., Leung, P.J.Y., Venkatesh, P. et al. (2024). De novo design of high-affinity binders of bioactive helical peptides. *Nature* 626 (7998): 435–442.

118 Kim, H.R., Ji, H., Kim, G.B. et al. (2025). Enzyme functional classification using artificial intelligence. *Trends Biotechnol.* 43 (9): 2214–2231.

119 Syed U., and Yona G. (2003). Using a mixture of probabilistic decision trees for direct prediction of protein function. *Proceedings of the seventh annual international conference on Research in computational molecular biology*. p. 289–300.

120 Concu, R., Dea-Ayuela, M.A., Perez-Montoto, L.G. et al. (2009). 3D entropy and moments prediction of enzyme classes and experimental-theoretic study of peptide fingerprints in Leishmania parasites. *Biochim. Biophys. Acta* 1794 (12): 1784–1794.

121 Amidi, S., Amidi, A., Vlachakis, D. et al. (2017). Automatic single- and multi-label enzymatic function prediction by machine learning. *PeerJ* 5: e3095.

6

Design of Biochemical Pathways via AI/ML-Enabled Retrobiosynthesis

Hongxiang Li[1,2,3], *Xuan Liu*[1,3,4,5], *and Huimin Zhao*[1,2,3,4,5]

[1] *University of Illinois Urbana-Champaign, NSF Molecule Maker Lab Institute, Urbana, IL, 61801, USA*
[2] *University of Illinois Urbana-Champaign, Department of Chemistry, Urbana, IL, 61801, USA*
[3] *University of Illinois Urbana-Champaign, Carl R. Woese Institute for Genomic Biology, Urbana, IL, 61801, USA*
[4] *University of Illinois Urbana-Champaign, Department of Chemical and Biomolecular Engineering, Urbana, IL, 61801, USA*
[5] *University of Illinois Urbana-Champaign, NSF iBioFoundry, Urbana, IL, 61801, USA*

6.1 Introduction

6.1.1 Computer-Aided Synthesis Planning

In the 1960s, the concept of computer-aided synthesis planning (CASP) was first proposed by E. J. Corey [1], aiming to learn synthetic logic from reaction databases and generate synthesis routes for target molecules. CASP involves systematically breaking down target molecules using retrosynthesis tools until reaching commercially available compounds [2–5]. This is a complex process that includes multiple tasks, such as single-step synthesis, multistep search algorithms, and ranking systems. The retrosynthesis process begins with selecting suitable databases for training. The USPTO [6] database is a commonly used resource for reaction training and benchmarking, while the Reaxys [7] database is a commercially available database containing reactions from the literature. For single-step synthesis, two main approaches are used: template-based and template-free methods. These approaches determine how precursors are predicted from target molecules in a single-step reaction.

In template-based tools, reaction templates can be constructed either by extracting reaction patterns from reaction databases or by curating reaction rules with input from domain experts. When extracting templates, the reaction center is identified to capture structural changes between reactants and products, forming a set of reaction rules. These rules are then applied to predict precursors for new reactions. In contrast, template-free tools train a model to predict precursors directly from target molecules without relying on predefined reaction templates. Template-based tools generally yield more accurate precursors due to the structural similarity between query reactions and existing templates, whereas template-free tools enable out-of-domain predictions, expanding the range of possible synthetic pathways.

Machine Learning and Big Data-enabled Biotechnology, First Edition. Edited by Hal S. Alper.

A summary of the ML-based approaches discussed in this chapter is provided in Table 6.1.

6.1.2 Retrobiosynthesis

The development of retrobiosynthesis tools has grown rapidly in recent years [2–5]. Generally, retrobiosynthesis follows the same principles as retrosynthesis but differs in certain aspects due to the unique characteristics of biosynthetic pathways.

The first key difference lies in the availability of biological reaction databases, which are significantly smaller than chemical reaction databases, containing only about 10% of the data points found in chemical databases. Commonly used biological reaction databases include BRENDA [28], KEGG [29], MetaCyc [30], SABIO-RK [31], and various integrated databases [32]. Several approaches have been developed to address the issue of limited data availability. One solution is data augmentation, as demonstrated in BioNavi-NP [11], where chemical reactions with products structurally like natural products are incorporated to expand the biological training dataset. Another strategy involves using expert-curated templates to bypass the problem of limited data points, as exemplified by the RetrobioCat [8] tool.

Another key difference is that biological reactions involve metabolites, which are small molecules involved in the chemical reactions of metabolism, which impact both termination methods and reaction template selection. In retrosynthesis, synthetic routes typically terminate when a commercially available compound is reached, often based on assigned price limits. However, in retrobiosynthesis, some algorithms use predefined metabolites as termination points to facilitate the biosynthesis of target molecules. Additionally, NovoStoic [33] utilizes *in vivo* reaction templates to construct complete biosynthetic pathways.

6.2 Retrobiosynthesis Tools

Pathway discovery via retrobiosynthesis is a complex process that involves single-step reaction prediction, multistep searching, and a ranking system. In this section, we will discuss various AI/ML tools for pathway discovery. For single-step transformations, there are two main approaches: template-based and template-free methods. Representative tools for both approaches will also be introduced. After predicting single-step reactions, multistep searching algorithms are integrated to form complete synthesis pathways. Various search algorithms are compared to achieve shorter and more reliable pathways while minimizing computational costs. Finally, the ranking system provides a list of pathways based on user-defined criteria such as cost, compound availability, and other relevant factors.

6.2.1 Template-Based Tools

In template-based tools, reaction templates can be extracted from databases or curated by human experts (Figure 6.1a). When extracting templates, a suitable database is first selected. For enzymatic reactions, the database entries typically include reactants, products, enzymes, and sometimes cofactors. Reactants, products,

Table 6.1 Summary of key computational approaches and tools for pathway discovery, enzyme selection, and optimization in AI–/ML-enabled retrobiosynthesis.

Approach/Algorithm	Description	Reference
RetrobioCat	A template-based retrobiosynthesis tool using expert-curated enzymatic templates for pathway exploration.	Finnigan et al. [8]
ASKCOS (chemoenzymatic mode)	A computational synthesis planning tool that integrates enzymatic templates extracted from BKMS database with chemical synthesis templates.	Levin et al. [9]
Rxn4chemistry	A template-free machine learning-based retrobiosynthesis tool trained on enzymatic and chemical reactions for pathway prediction.	Probst et al. [10]
BioNavi-NP	A machine learning model for retrobiosynthesis focusing on natural products, using data augmentation strategies to expand biological reaction datasets.	Zheng et al. [11]
AiZynthFinder	A retrosynthesis tool applying Monte Carlo Tree Search (MCTS) to construct multistep synthetic pathways efficiently.	Segler et al. [12]
SCScore	A synthetic complexity scoring system that ranks synthesis routes based on the estimated difficulty of molecule construction.	Coley et al. [13]
SAscore	A score estimating synthetic accessibility of compounds based on fragment contributions and molecular complexity.	Ertl and Schuffenhauer [14]
Selenzyme	Enzyme selection tool based on sequence homology, reaction similarity, and phylogenetic distance.	Carbonell et al. [15]
ESP (Enzyme-Substrate Prediction)	Transformer-based model predicting enzyme-substrate relationships using sequence embeddings.	Kroll et al. [16]
MEI (Molecule-Enzyme Interaction)	Hierarchical neural network integrating enzyme sequences and molecule atomic environments.	Qian et al. [17]

(Continued)

Table 6.1 (Continued)

Approach/Algorithm	Description	Reference
AutoDock	Molecular docking tool for evaluating enzyme-ligand binding.	Morris et al. [18]
CatPred	Deep learning framework predicting turnover number, Michaelis constant, and inhibition constants from enzyme sequences and structures.	Boorla and Maranas [19]
DLKcat	CNN and GNN-based model for kcat prediction across species and mutant enzymes.	Li et al. [20]
UniKP	Ensemble learning framework for predicting enzyme kinetic parameters with pretrained models.	Yu et al. [21]
TurNuP	Transformer model predicting turnover numbers and handling low-similarity enzymes.	Kroll et al. [22]
RFdiffusion	Diffusion model generating protein backbones to host designed catalytic sites.	Watson et al. [23]
ProteinMPNN	Deep learning model assigning amino acid identities to protein backbones.	Dauparas et al. [24]
ProteinGAN	GAN trained to produce functional enzyme sequences.	Repecka et al. [25]
ProGen	Large language model generating protein sequences based on enzyme classifications.	Madani et al. [26]
ProtGPT2	Protein language model generating novel protein sequences.	Nijkamp et al. [27]

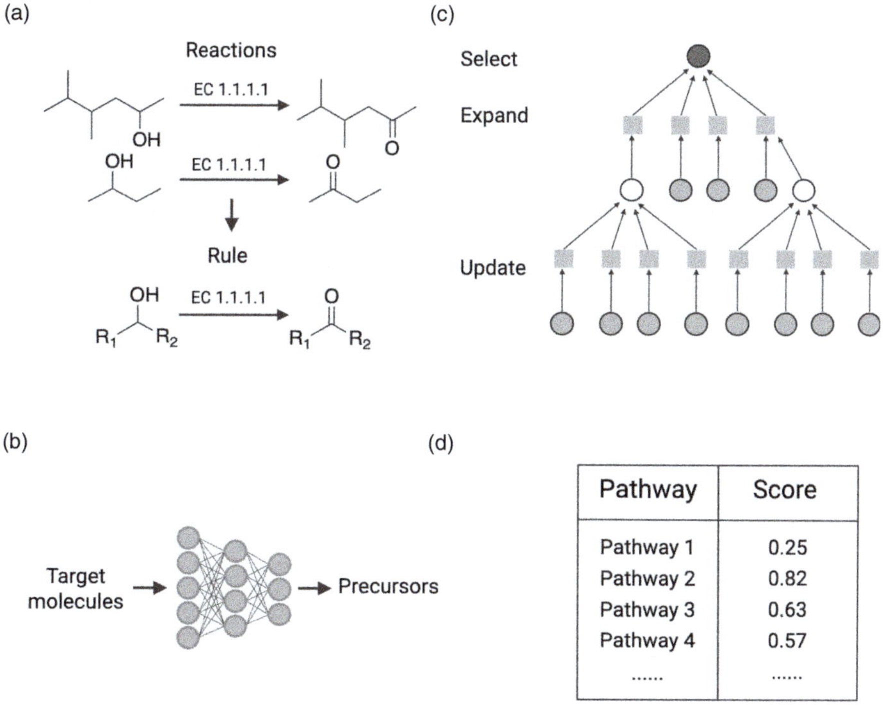

Figure 6.1 Retrobiosynthesis tools. (a) Template-based method: Reaction templates are extracted from a database to capture changes in the reaction center and are then applied to new target molecules. (b) Template-free method: Machine learning (ML) algorithms are trained to predict precursors from target molecules. (c) Multistep searching: A search algorithm is used to construct multistep synthesis routes following the select, expand, and update procedure. (d) Pathway ranking: A scoring system is applied to synthesis routes to evaluate their reliability and effectiveness, ranking them accordingly.

and cofactors are represented as SMILES [34], while enzymes are classified using Enzyme Commission number (EC number). Next, the differences between reactants and products are identified to form reaction templates. This process begins with locating the reaction center – the atoms or bonds that undergo changes during the reaction – followed by defining the template radius to balance prediction accuracy and broad applicability. Reaction templates are stored as SMARTS [35], and when a new target molecule is input, the precursor can be predicted using the template. Additionally, the enzymes associated with the template reaction can serve as references. One challenge in enzymatic reactions is the lack of a unified database and the presence of redundant or similar reactions across different databases. A possible solution is to use integrated databases such as BKMS [36], which consolidates data from BRENDA [28], KEGG [29], MetaCyc [30], and SABIO-RK [31]. Another challenge is the generalizability of reaction templates. Unlike chemical reactions, enzymatic reactions require enzyme-substrate binding, meaning factors such as substrate bulkiness and accessibility to the enzyme's active site must be considered when applying templates to new reactions. As a result, targets for retrobiosynthesis often share structural similarity with metabolites, and selecting suitable enzymes requires specialized tools.

RetrobioCat [8] is a template-based retrobiosynthesis tool that utilizes human expert-curated templates. The tool compiles enzymatic reactions from the literature into an in-house database, categorizing reactions based on enzyme types to facilitate template extraction and enzyme identification. Through a community-driven platform, users can explore synthesis routes for target molecules and contribute to expanding the reaction template database.

ASKCOS [9] is an example of a tool that extracts reaction templates from databases. Built upon a previous chemical retrosynthesis tool, ASKCOS [9] extracts enzymatic reaction templates from the BKMS [36] database and integrates them into a chemoenzymatic synthesis tool. A prioritization mechanism selects templates for each step from a combined chemical and biological reaction library. In a comparison between chemical and biological synthesis approaches, some pathways were found exclusively through enzymatic reactions. Additionally, chemoenzymatic synthesis planning predicted shorter synthetic routes compared to traditional chemical synthesis planning methods.

6.2.2 Template-Free Tools

In contrast to template-based methods, template-free tools utilize machine learning (ML) algorithms to predict precursors for a given target molecule (Figure 6.1b). While the output of template-based methods is deterministic after designing and selecting reaction templates, ML-based methods offer greater tunability. The representation of molecules, selection of algorithms, and techniques used in the training process all contribute to prediction performance. The reaction database is typically divided into training, validation, and test subsets, with prediction performance evaluated based on the model's ability to reproduce reactions in the test set. Molecules can be represented as SMILES [34] strings or drawn as graphs to encode structural information. Since retrosynthesis outputs are chemical structures, several factors must be considered to ensure their validity, including bond formation rules, structural thermodynamic stability, and isomerization effects. To address these issues, filters can be applied to eliminate invalid structures before generating the final output. Beyond that, due to the presence of recurring reaction types in the database, ML models may develop biases toward certain reaction types or specific transformations. Proper categorization and weighting of the database can help achieve a more balanced reaction output.

Rxn4chemistry [10], developed by IBM, is the first ML-based retrobiosynthesis tool. It uses reactions from the ECREACT database [10] for training. Chemicals are represented as SMILES and tokenized into an end-to-end transformer architecture for both forward and backward synthesis predictions. Notably, to expand the dataset and incorporate chemical knowledge into model training, reactions from the USPTO [6] chemical database were included using multitask transfer learning. On the Rxn4chemistry website, if a retrosynthesis route could not be found using enzymatic pathways, a supplementary chemical route is suggested instead. As the first ML-based retrosynthesis tool, Rxn4chemistry [10] demonstrated promising results while also identifying incorrect predictions in its published work, highlighting limitations and areas for further improvement.

Another ML-based tool, BioNavi-NP [11], focuses on natural product synthesis. Recognizing the issue of limited biological reaction data, BioNavi-NP implemented data augmentation techniques. Specifically, it uses the DNP database to represent natural products and calculates molecular similarities between reactions in the USPTO [6] dataset and compounds in the DNP dataset. Reactions with high similarity are added to the training set, and enhanced performance has been observed with the expanded database.

6.2.3 Searching Algorithm

In retrosynthesis, some tools focus on single-step synthesis, aiming to achieve high reproducibility and accuracy. However, designing a complete synthesis pathway for practical applications and industrial-scale synthesis requires multistep synthesis routes (Figure 6.1c). Since each target molecule can have multiple predicted precursors, the number of possible synthesis routes grows exponentially with each additional step. Therefore, rather than simply linking single-step reactions together, a multistep synthesis algorithm must balance route quality and computational efficiency. Starting from the target molecule, various precursor candidates can be generated. To reduce computational costs, the most promising precursor is selected for expansion at each step. During this process, the reliability and effectiveness of each reaction are evaluated and compared. Reliability is often determined by the degree of similarity to known reaction templates or established reaction rules, while effectiveness can be assessed using synthetic complexity scores, which will be discussed in detail in Section 2.4. After these evaluations, the next node(s) are selected for expansion, and the process continues until commercially available compounds are reached.

The MCTS [37] is a widely used decision-making algorithm that determines the optimal next move based on probabilistic exploration. In retrosynthesis, AiZynthFinder [12] was the first tool to apply MCTS for retrosynthesis planning, following a process of selection, expansion, simulation, and backpropagation. In this framework, an initial molecule is selected for expansion, generating potential precursors. The search tree is explored through simulations, where each branch is evaluated to determine whether it leads to viable starting materials or requires further expansion. The calculation results help guide the search, allowing pathways to be ranked based on likelihood and feasibility.

Retro* [11] is another multistep retrosynthesis search method that follows a best-first search strategy with a neural network-guided bias. It structures the retrosynthesis problem as an AND-OR search tree, where molecules represent OR nodes and reaction steps represent AND nodes. Retro* utilizes a learned value function to guide node selection, prioritizing high-quality and low-cost reaction pathways. The cost function is derived from a transformer-based neural network, which assigns confidence scores to reaction feasibility. A pathway is expanded iteratively, and at each step, the most statistically favorable precursor is selected based on reaction likelihood and synthetic accessibility. Once a complete pathway to available starting materials is found, it is recorded, and the search continues to identify alternative routes.

6.2.4 Ranking

Appropriate ranking of synthesis routes is crucial both in the process of expanding the search tree and in presenting the final list of synthesis routes. Since multiple synthesis routes can be predicted, a reliable scoring system helps users select the best prediction, reducing time spent on experimental validation and applications (Figure 6.1d). In synthesis route ranking, reliability represents the confidence in the predicted reaction and is often measured by the similarity between templates and reactions. Additionally, since forward reaction prediction generally has higher accuracy than retrosynthesis planning, a forward reaction predictor can be used to verify the predicted results and identify potential side reactions. Effectiveness in retrosynthesis can be defined by whether the predicted routes lead to commercially available compounds, and several scoring metrics have been developed to evaluate this aspect. The most well-known metric is the SCScore [13], which introduces the concept of synthetic complexity for ranking synthesis routes. In practical applications, additional factors must be considered, including the cost of starting materials, the overall cost of the synthesis route, the yield of each step, and the toxicity of reagents used. To enable a more comprehensive evaluation, relevant data can be extracted from online literature databases and assigned different ranking weights.

The SCScore [13] is widely used in retrosynthesis, and both RetrobioCat [8] and Rxn4chemistry [10] utilize it for route ranking. This scoring system is based on the concept of synthetic complexity. Specifically, reactants and products in a reaction are analyzed, and products are considered synthetically more complex than reactants. All compounds are classified on a gradient from 1 to 5 according to their synthetic complexity. In route ranking, reactions that proceed from a synthetically more complex molecule to a simpler one are preferred, and scores are accumulated to determine the overall synthetic complexity of a given route.

The SAscore [14], on the other hand, evaluates whether a compound is easy to synthesize, recognizing that some complex structures can be formed through simple reactions. This approach involves fragmenting molecules from the PubChem database and calculating fragmentation scores under the assumption that easily synthesized substructures appear more frequently in the database. Additionally, complex molecular features, such as rings and stereocenters, are considered by applying penalty scores to these structures in the SAscore calculation.

6.3 Enzyme Selection and Optimization

Once a biochemical pathway is designed for a target molecule via retrobiosynthesis, selecting appropriate enzymes for the pathway becomes crucial to ensuring that each step in the pathway is feasible and efficient [38]. Unlike traditional synthetic chemistry, where chemical catalysts may provide broad reactivity, enzymes exhibit high specificity, regioselectivity, and stereoselectivity, making them particularly valuable for synthesizing complex molecules with minimal side reactions. Moreover,

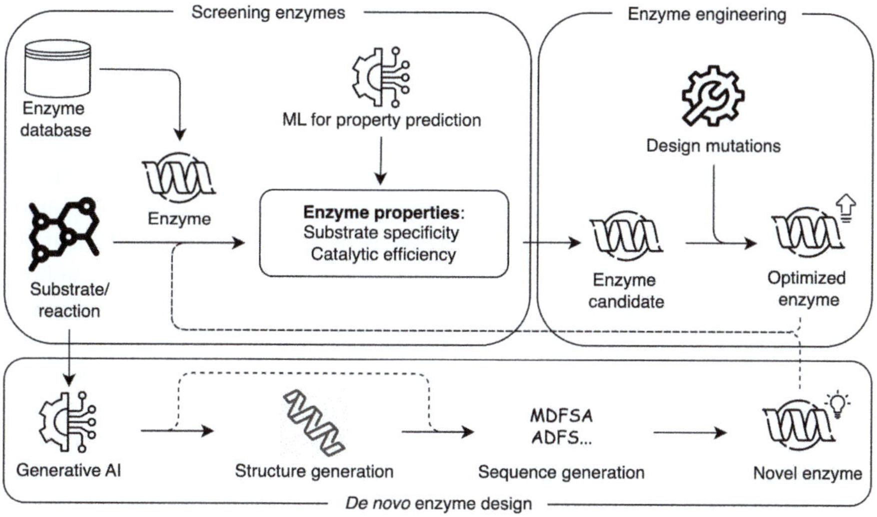

Figure 6.2 Overview of the enzyme selection and optimization. Enzyme selection begins with a database of known enzymes, which are assessed for substrate specificity and catalytic efficiency. In enzyme engineering, candidate enzymes undergo refinement through optimization to enhance their properties. Concurrently, *de novo* enzyme design leverages generative AI to create novel enzyme structures and sequences tailored for specific substrates or reactions.

identifying the suitable enzyme for a given transformation ensures overall process efficiency. In this section, we provide a high-level overview and in-depth technical details on how AI/ML tools can be used for enzyme selection and optimization. We focus on four key aspects: predicting enzyme substrate specificity and catalytic efficiency, engineering enzyme variants with improved properties, and designing novel enzymes *de novo* (Figure 6.2). These approaches complement the enzyme discovery tools that are described in Chapters 4 and 5.

6.3.1 Enzyme Substrate Specificity

Enzyme substrate specificity refers to an enzyme's ability to recognize and act on particular substrates. In retrobiosynthesis, selecting an enzyme is a critical step to ensure the feasibility of predicted reactions. Traditionally, enzyme selection relied on sequence or substrate similarity searches in databases (e.g. using tools likc BLAST [39] against BRENDA [28] or UniProt [40]). While useful, similarity-based methods can miss enzymes with low sequence identity but suitable activity. AI/ML enhances the predictive power by learning complex patterns that dominate enzyme–substrate interactions. By training on known enzyme–substrate pairs, AI/ML models can generalize and predict whether a given enzyme will accept a novel substrate, enabling *in silico* screening of enzymes for predicted reactions in retrobiosynthesis.

Moving beyond EC number-based enzyme classification, which indicates what type of reaction an enzyme can conduct, enzyme substrate specificity prediction models are required to predict enzyme compatibility with a given substrate. Several AI/ML-based approaches have been proposed to generalize enzyme-substrate interactions across diverse enzyme families. Selenzyme is an enzyme selection tool that recommends candidates for a reaction based on sequence homology, reaction similarity, and phylogenetic distance [15]. It is integrated into synthetic biology workflows for metabolic pathway optimization. Mellor et al. built a Gaussian process for enzyme search [41]. Their work introduced a semi-supervised Gaussian process model that assigns probability scores for enzyme-catalyzed reactions based on reaction fingerprints and enzyme sequence similarity. The model successfully identified novel *E. coli* enzymes involved in metabolite transformations. Overall, sequence-based AI/ML predictors provide a rapid way to screen candidate enzymes for a novel substrate, representing an invaluable tool in retrosynthesis planning.

Recently, protein language models, like evolutionary scale modeling (ESM) [42], have significantly advanced sequence-based models to predict enzyme substrate specificity. The enzyme-substrate prediction (ESP) model employs a modified transformer architecture to predict enzyme-substrate relationships [16]. Unlike traditional models that rely on explicit structural information, ESP derives sequence-based embeddings of enzymes, capturing functional features from protein language models. It addresses the challenge of imbalanced datasets by augmenting training with randomly sampled small molecules as non-substrates, achieving 91% accuracy across various enzyme families. In addition, the molecule-enzyme interaction (MEI) model integrates both enzyme sequence data and small-molecule atomic environments using hierarchical neural networks with attention mechanisms [17]. This approach significantly outperforms conventional compound-protein interaction (CPI) models, which struggle with enzyme promiscuity. The model successfully predicts CYP450 enzyme activity and enzymatic degradation of complex plastics with an accuracy of 90.5%.

When high-resolution enzyme structures (experimental or predicted) are available, structure-based methods can refine specificity predictions. Molecular docking simulations such as AutoDock [18] have long been used to evaluate enzyme–ligand binding. Mou et al. integrated ligand docking, structural modeling, and physicochemical descriptors with AI/ML models (random forests, support vector machines, and gradient-boosted trees) [43]. The approach yielded an 82% accuracy in predicting the substrate scope of bacterial nitrilases.

Overall, combining sequence-derived patterns, structural modeling (docking or AlphaFold-predicted structures), and modern AI/ML yields powerful predictors of enzyme–substrate specificity. These tools enable retrosynthesis planners to select enzymes likely to catalyze a given reaction *in silico*, narrowing down candidates before any lab experiments.

6.3.2 Enzyme Catalytic Efficiency

Once the candidate enzyme is identified for a reaction, the next challenge is to ensure the reaction proceeds at a desirable rate. Catalytic efficiency encompasses the kinetic parameters (turnover number k_{cat}, Michaelis constant K_m, and catalytic efficiency k_{cat}/K_m) and how readily an enzyme converts substrates to products. Improving catalytic activity often involves understanding the reaction's energy barrier and the enzyme's dynamics. Computational chemistry methods like quantum mechanics/molecular mechanics (QM/MM) can explicitly calculate transition state energies for enzyme-catalyzed reactions, and molecular dynamics (MD) simulations reveal how enzyme conformational changes affect catalysis [44]. However, these physics-based methods are time-consuming and not feasible for screening large mutant libraries or many enzyme–substrate combinations. AI/ML offers faster predictive models by learning from either experimental kinetic data or high-level theoretical calculations. In this way, AI/ML can guide enzyme optimization by predicting which mutations or conditions will increase catalytic efficiency, all before actual wet-lab testing.

A growing area of AI/ML research is the prediction of enzyme kinetic parameters directly from enzyme and substrate features. Recent deep learning models have achieved promising results in this area. CatPred is a comprehensive deep learning framework designed for predicting k_{cat}, K_m, and inhibition constants (K_i) values of enzymes [19]. It employs pretrained protein language models (pLMs) and three-dimensional structural features to improve generalization, particularly on out-of-distribution samples. Unlike previous methods that struggle with generalization, CatPred incorporates uncertainty quantification techniques to assess prediction reliability. In addition, DLKcat employs a combination of convolutional neural networks (CNNs) for enzyme sequence processing and graph neural networks (GNNs) for substrate structure representation [20]. It has been applied for large-scale k_{cat} predictions across 343 yeast species and successfully parameterized enzyme-constrained metabolic models. DLKcat achieves high predictive accuracy and even captures the impact of enzyme mutations on catalytic turnover, making it useful for enzyme engineering applications. Similarly, Kroll et al. developed a deep learning model for predicting K_m values using a combination of a graph neural network for molecular fingerprints and a deep representation of enzyme amino acid sequences [45]. Their approach successfully generalizes to genome-scale predictions for 47 model organisms, making it a valuable tool for metabolic modeling and systems biology.

Ensemble learning techniques, particularly gradient-boosted trees, have demonstrated strong performance in predicting enzyme kinetic parameters by leveraging pretrained language models and biochemical features. UniKP integrates multiple AI/ML models and pretrained language models to predict k_{cat}, K_m, and k_{cat}/K_m from enzyme sequences and substrate structures [21]. It employs an ensemble-based regression approach, outperforming previous models by systematically considering environmental factors such as pH and temperature. Additionally, UniKP incorporates re-weighting methods to improve prediction accuracy for high-value kinetic parameters, addressing a key limitation in many AI/ML models. On the other hand, TurNuP predicts k_{cat} by utilizing transformer neural networks to extract enzyme sequence features and employs differential reaction fingerprints to encode complete reaction information [22]. Compared to traditional models, TurNuP shows superior generalization to enzymes with low sequence similarity to training data. The model has been successfully applied to parameterize metabolic models, improving predictions of proteome allocation.

These studies highlight that, given enough training data, AI/ML models can serve as surrogates for laborious enzyme kinetics experiments. For retrosynthesis, such models can rank enzyme candidates not just on whether they *can* catalyze a reaction, but how efficiently they will likely do so, focusing attention on the most catalytically competent options.

6.3.3 Enzyme Engineering

Enzyme engineering involves modifying an enzyme to enhance its performance. For instance, improving activity, selectivity, stability, or altering its substrate scope. In AI/ML-enabled retrosynthesis, once an enzyme is chosen for a step, one might engineer it to better fit the process requirements (e.g. higher turnover, tolerance to solvents, or acceptance of a non-natural substrate). AI/ML has emerged as a powerful tool to navigate the immense sequence space of possible enzyme variants. Instead of random or exhaustive mutagenesis, AI/ML-guided enzyme engineering uses data-driven models to prioritize mutations or designs likely to yield improvement.

The general workflow of ML-guided directed evolution (MLDE) involves: (1) generating an initial diverse library and measuring activities (the experimental data), (2) training an AI/ML model (such as a regression model or classifier, potentially a neural network) on that data to learn sequence–function relationships, and (3) using the model to propose the next set of variants expected to have even better activity. The proposed variants are then tested experimentally, new data are gathered, and the cycle repeats. By prioritizing beneficial mutations, AI/ML-guided approaches can explore sequence space more broadly rather than greedily following just the best current variant. This mitigates the risk of missing global optima. Yang et al. demonstrated the promise of this paradigm, showing that ML models could guide directed evolution to improve enzyme activity while using fewer screening iterations [46]. As a concrete example, an AI/ML-guided evolution of Sortase A was able to find improved variants more efficiently than traditional methods by training a Bayesian model [47]. Similarly, a recent study used an MLDE approach to evolve a ketoreductase with enhanced diastereoselectivity for a drug intermediate, successfully identifying highly selective variants [48]. These successes illustrate how data-driven models can accelerate enzyme optimization, effectively reducing the experimentalburden while achieving equal or better improvements.

6.3.4 *De Novo* Enzyme Design and Discovery

Despite the expanding repertoire of natural enzymes, retrosynthetic plans sometimes require reactions for which no known enzymes exist or no enzymes are efficient enough. In such cases, the ability to design new enzymes *de novo* – essentially inventing biocatalysts for "new-to-nature" reactions – is a game-changer. Traditional enzyme engineering starts from an existing protein scaffold, but *de novo* enzyme design aims to create an enzyme from scratch to catalyze a desired reaction. This is an enormously complex task: it requires imagining a protein structure that can bind the substrates in the correct orientation, stabilize the transition state of the reaction, and perform necessary catalytic chemistry, all while folding into a stable, expressible protein. Early efforts in *de novo* enzyme design (before the deep learning era) showed that it is possible in principle. For example, Röthlisberger et al. designed a novel enzyme for the Kemp elimination reaction by computationally inserting a theoretical catalytic motif into stable protein scaffolds, and after subsequent laboratory evolution, the enzyme achieved measurable activity [49]. However, those efforts had very low success rates – tens of thousands of designs might yield one weakly active enzyme – underscoring the difficulty of the problem [50].

Recent advances in AI and computational modeling are dramatically improving the success of *de novo* enzyme design. One breakthrough has come from applying generative models, where AI models are trained to produce new protein sequences or structures with certain desired properties. Instead of manually specifying an active site geometry and painstakingly designing a scaffold (as in classical approaches), researchers can now leverage deep learning to "imagine" new proteins. A striking example is the "hallucination" approach developed by the Baker lab, in which a neural network is used to generate novel protein structures that satisfy certain design objectives [51]. This method was used to create light-emitting enzymes (luciferases) that are unlike any natural protein [50]. The team defined the chemical reaction they wanted (oxidation of a luciferin substrate to emit photons) and then used a deep-learning algorithm to hallucinate thousands of possible protein folds that could catalyze this reaction. Crucially, these hallucinated proteins were not modifications of known enzymes – they were built upon *de novo*-designed scaffolds generated by AI. After incorporating a putative active site (for example, positioning an arginine in the model to stabilize the transition state's negative charge) and optimizing sequences, several designs were synthesized and tested. Remarkably, at least three designs showed luciferase activity, and one named "LuxSit" was highly efficient – emitting blue light with an intensity visible to the naked eye. LuxSit is only 117 amino acids in size (smaller than any known natural luciferase) and proved to be quite robust (partially folded even near boiling temperatures). This case opens the door to custom enzymes for almost any chemical reaction in principle.

Another revolutionary tool in *de novo* enzyme design is diffusion-based protein models. RFdiffusion adopted the principles of diffusion models (which have been used to generate images and text) to protein structures [23]. RFdiffusion was used to design a complex serine hydrolase enzyme from scratch, achieving catalytic efficiency on par with a naturally occurring enzyme [52]. The design pipeline began by specifying the desired reaction mechanism: in this case, ester hydrolysis via a classic serine-histidine-aspartate catalytic triad and an oxyanion hole for stabilizing intermediates. The 3D coordinates of the ideal active site (the arrangement of those key residues and the substrate binding geometry) were fed into RFdiffusion, which then generated entire protein structures that could hold that active site. The output of RFdiffusion is a novel protein backbone (a fold) that positions the catalytic residues correctly. Then, a second AI algorithm (such as a sequence design network like ProteinMPNN [24]) assigns amino acid identities to produce a viable sequence, and a third program evaluates how well each design might stabilize all steps of the reaction mechanism. Using this approach, the team generated about 1,000 *de novo* designs and screened them in wet lab, finding several that successfully catalyzed the full four-step hydrolysis mechanism with efficiencies comparable to natural enzymes. This level of catalytic performance was unprecedented for an enzyme designed *de novo*, underscoring the capacity of generative AI methods to capture intricate catalytic requirements associated with multistep reactions, which have historically challenged conventional design approaches. It is important to recognize that this achievement still required iterative experimental validation, where AI-generated candidates were synthesized, tested, and optimized. Nonetheless, the initial selection of enzyme designs was systematically guided by AI, representing a significant advancement over traditional random or trial-and-error strategies.

Beyond these specific techniques, a variety of generative models for protein design have flourished. These include variational autoencoders (VAEs) that learn a latent representation space of protein sequences [53], generative adversarial networks (GANs) like ProteinGAN [25] that are trained to produce plausible enzyme sequences, and large-scale language models such as ProGen [26] and ProtGPT2 [27] which can generate novel protein sequences character by character, having learned grammar from millions of natural proteins. These models can be used to generate proteins with specific properties. For example, ProGen has been used to create enzyme sequences based on a particular enzyme classification (EC number) or protein family, resulting in proteins that closely match the intended function or group. Such AI-generated enzymes still need experimental validation, but they expand the search beyond what evolution has sampled. In the context of retrosynthesis, one can imagine using generative models to propose an enzyme for a "missing reaction." For example, designing an enzyme active site for a carbon–silicon bond formation (an unnatural reaction) and letting the model create a protein to catalyze it. While still an emerging area, there have been proof-of-concept successes where AI-designed enzymes enabled entirely new biocatalytic steps.

Despite these advances, *de novo* enzyme design remains challenging. Success often requires generating a huge number of candidates and depends on subtle factors that are not fully captured by current models (such as protein dynamics, solvent effects, and the need for cofactors or metal ions in some reactions). However, the rapid improvement in AI techniques is cause for optimism. The luciferase and hydrolase examples show that each year brings orders-of-magnitude improvements in success rate and enzyme performance. As generative models are further refined (for instance, incorporating physics-based feedback or training on enzyme mechanism data), we can expect bespoke enzymes for virtually any reaction to become achievable. For retrosynthetic planning, this means that even if a proposed biosynthetic pathway has "gaps" (steps with no known enzyme), one day it may be routine to fill those gaps by designing a new enzyme *in silico*, thus truly expanding biocatalysis to new frontiers of chemistry.

6.4 Perspectives

AI/ML is transforming retrosynthesis and biocatalysis, but there are still significant opportunities and challenges ahead. Examples include the integration of biocatalysis with traditional chemocatalysis in synthetic route design, the development of more powerful AI models (including foundation models and large language models) for synthesis planning, advances in predicting enzyme properties and designing enzymes for new-to-nature reactions, the standardization of data resources for model training, and improvements in the interpretability of deep learning models. Addressing these will push the boundaries of what's possible in computational retrosynthesis and accelerate the discovery of sustainable synthetic routes.

6.4.1 Integrating Biocatalysis with Chemocatalysis

In the future, we envision retrosynthetic planners that seamlessly combine biocatalysis and chemocatalysis, capitalizing on the strengths of each. For example, an early step in a synthetic route might be best done by a highly selective enzyme, while a later step might require a metal-catalyzed coupling that no enzyme can do. AI models should be developed to consider such trade-offs and propose hybrid pathways [54]. Recent discussions highlight that current retrosynthesis algorithms seldom incorporate biocatalytic steps alongside traditional steps. To change this, more data on chemo-enzymatic sequences of reactions will be needed so that planners can learn patterns of when to insert an enzyme-catalyzed step for optimal results. There have been successful examples of chemoenzymatic synthesis, such as routes to certain pharmaceuticals where an enzyme step improved yield or selectivity compared to purely chemical routes [55].

In practical terms, a retrosynthesis AI may propose a multistep route where an engineered enzyme converts a cheap starting material into an intermediate, which is then coupled chemically to another fragment, followed by an enzymatic resolution step, etc. Such hybrid routes can yield shorter, more economical syntheses. Moving forward, the community will benefit from efforts to bridge the gap between organic chemistry and biochemistry datasets. Initiatives might include expanding retrosynthetic rule sets to cover enzymatic transformations, and teaching models to evaluate sequences of steps by combined metrics (e.g. overall yield considering both chemo and bio steps, compatibility of reaction conditions). Successfully combining biocatalysis and chemocatalysis *in silico* will mirror what synthetic chemists often do in the lab – using "the best tool for the job" at each step – but with the speed and breadth of exploration that AI offers. Ultimately, this integration is expected to produce more creative and greener synthetic routes for complex molecules, transcending the boundaries of traditional retrosynthesis.

6.4.2 Toward Next-Generation AI for Retrobiosynthesis Planning

While significant progress has been made in retrosynthetic route prediction, there is a clear need for more powerful AI models dedicated to biosynthetic planning. Many current retrosynthesis algorithms (whether template-based or template-free) were developed for organic synthesis and then adapted to include biotransformation in a limited way. Future models should be built from the ground up with biochemistry in mind. One promising direction is the development of foundation models for chemistry and biology – large AI models trained on massive amounts of reaction data or biochemical knowledge that can then be fine-tuned for specific tasks. In the chemistry domain, we already see large transformer models that have "learned" chemistry from millions of reaction examples. Analogous models could be trained on networks of metabolic pathways and enzymatic reactions [10]. A foundation model could encode general principles of retrosynthetic logic (e.g. recognizing strategic bonds to break in molecules) while also understanding enzyme constraints (like recognizing which functional groups enzymes commonly act upon).

Additionally, the advent of large language models (LLMs) like ChatGPT opens the door to interactive retrosynthesis planning. One can imagine a ChatBot for retrosynthesis that allows chemists to dialog with the AI: the user might ask for a synthetic plan for a target, the AI proposes a route (mixing chemical and enzymatic steps) and explains its reasoning, and the user can ask follow-up "what if" questions (such as swapping in an available enzyme or considering an alternative starting material). To achieve this, LLMs would need to be imbued with retrosynthetic knowledge – possibly by training on texts (papers, patents) about synthetic methods and pathways. Early experiments in this vein are appearing, where LLMs have been used to suggest chemical reactions or annotate pathways, but much development is needed for reliable performance.

6.4.3 Enhancing Enzyme Prediction and Design Capabilities

As retrosynthetic planning models become more advanced, a parallel need is for more powerful AI tools to predict enzyme properties and to design new enzymes. Having a pathway plan is only half the battle – one must ensure the chosen enzymes can actually carry out their steps efficiently. We discussed earlier how AI/ML can predict certain properties and how generative models can create new enzymes. Going forward, these capabilities need to be expanded and made more accurate to truly enable on-demand biocatalyst development for any reaction.

One critical challenge is accurately predicting enzyme-substrate pairs: given a candidate enzyme, can we predict if it will accept a proposed substrate and at what rate? Current approaches, like docking simulations or simple sequence similarity, have limited reliability, especially for novel substrates. Moreover, current AI models primarily rely on protein-level embeddings and are often unable to reliably predict catalytic efficiency across diverse enzyme families. Given that even single mutations can drastically alter catalytic properties, existing models frequently struggle with transferability and accuracy, severely limiting their trustworthiness. Improving existing models will likely require more training data – e.g. high-throughput screening results or kinetics data for various enzyme-substrate combinations and enzymes with single-mutation resolution – which calls for community efforts to share and standardize such data. If successful, such models would greatly aid retrosynthesis: whenever a model suggests a reaction, the enzyme-activity predictor could suggest a feasible enzyme or the need for engineering.

On the enzyme engineering front, better AI/ML models for mutation effects will make it easier to optimize enzymes *in silico*. We foresee more use of large pretrained protein models (like those that produce residue embeddings) to predict the effect of mutations on function. Currently, there are computational methods that leverage transformer models to predict beneficial mutations by analyzing patterns within multiple sequence alignments. However, current methods remain highly dependent on the initial diversity of mutation libraries and struggle significantly with capturing epistatic interactions between mutations, which are critical but often overlooked in current modeling approaches. Future tools might allow a scientist to engineer enzymes with a bulkier substrate, and the AI model outputs a set of mutations predicted to enlarge the binding pocket without disrupting the enzyme's fold or activity. More powerful AI/ML models, possibly incorporating physics-informed neural networks, could even estimate how a mutation affects the transition state stabilization – a key piece for catalysis that is currently almost impossible to know without experiments.

Importantly, to unlock truly new-to-nature reactions, *de novo* enzyme design using generative models will continue to advance. In the near future, generative models might design enzymes for reactions that are completely artificial (for instance, forming chemical bonds not found in any metabolic pathway). Some research groups are already working on training generative models conditioned on reaction chemistry descriptors, essentially asking the model to "hallucinate" an enzyme given a description of the reaction. We expect the quality of AI-designed enzymes to improve to the point that fewer design-build-test cycles are required. As generative models become more adept, enzyme design could become a standard step in retrosynthesis: whenever no natural enzyme is available for a desired transformation, the retrosynthesis program could invoke a generative model to propose a new enzyme, effectively adding a "virtual enzyme inventory" to the planner's toolkit.

6.4.4 Data Standardization and Model Interpretability

The progress of ML in retrobiosynthesis is tightly linked to the availability of high-quality data. One pressing need is to standardize and expand databases for enzymes, reactions, and pathways, in formats conducive to AI/ML model development. Currently, information is scattered across resources: enzyme sequences and basic functions in UniProt [40], detailed kinetic data in BRENDA [28] or SABIO-RK [31], metabolic pathways in KEGG [29] or MetaCyc [56], and experimental protein engineering data in literature or niche databases like ProtaBank [57]. Many of these databases were not designed with ML in mind – they may lack negative examples, have inconsistent naming, or miss links between datasets (e.g. which sequence in UniProt [40] corresponds to which kinetics entry in BRENDA [28]). To build robust AI/ML models, we need comprehensive datasets with consistent annotations. For example, a model that predicts enzyme specificity would benefit from a dataset that, for each enzyme, lists substrates it acts on and does not act on, under standard conditions. Achieving this requires the community to agree on data schemas and share experimental results even when they are "failures" (to avoid biasing models). Going forward, concerted initiatives to link these databases (so that a model can easily fetch all relevant info about an enzyme: sequence, structure, reactions, kinetics, mutants, etc.) will greatly facilitate AI/ML development.

Finally, as AI models for retrobiosynthesis become more complex (deep neural networks with millions of parameters), there is a growing call for interpretable and explainable models. Currently, many deep learning models are "black boxes" – they might predict which enzyme to use or which mutation to make, but not why. Increasing interpretability can take several forms. One approach is to integrate mechanistic knowledge or constraints into the model, so that its outputs follow known rules of chemistry. For instance, a retrosynthesis model might be constrained by atom-mapping rules that ensure mass balance, or an enzyme activity predictor might highlight which amino acid positions in the active site contributed most to its decision (perhaps via an attention mechanism that can be visualized). Another approach is post-hoc interpretation: analyzing a trained model to extract human-comprehensible insights. For example, if a neural network predicts enzyme function from sequence, tools like saliency maps could identify motifs or residues that the network finds important – which might correspond to known catalytic motifs or suggest new ones. In one study, an AI/ML classifier for enzyme families was designed to be "biochemically interpretable," meaning it associated certain learned features with known biochemical properties [58]. Such interpretability not only builds trust but can lead to scientific discovery (e.g. uncovering new sequence patterns that correlate with specificity).

In conclusion, the future of AI/ML-enabled retrobiosynthesis is bright. We anticipate a positive feedback loop where better data leads to better models, and better models enable experiments that generate even more valuable data. In this synergy of computation and biotechnology, enzymes – nature's catalysts – will play a starring role, guided by algorithms that can learn from nature's examples and imagination that can go beyond them. With integrated chemoenzymatic planning, powerful AI-driven design tools, and a focus on interpretable and trustworthy AI, the vision of on-demand biosynthesis of any molecule is becoming increasingly tangible. The coming years will likely witness retrosynthesis AI evolving from a clever assistant to an indispensable creative partner in synthetic chemistry and biotechnology.

Acknowledgments

This work was supported by the Molecule Maker Lab Institute: An AI Research Institute program supported by the US National Science Foundation under grant no. 2019897 and 2505932 (HZ), the NSF iBioFoundry (DBI- 2400058), the NSF Global Center for Reliable and Scalable Biofoundries (OISE-2435374), and the IBM-Illinois Accelerated Discovery Institute.

References

1 Corey, E.J. and Wipke, W.T. (1969). Computer-assisted design of complex organic syntheses. *Science* 166 (3902): 178–192.

2 Boob, A.G., Chen, J., and Zhao, H. (2024). Enabling pathway design by multiplex experimentation and machine learning. *Metab. Eng.* 81: 70–87.

3 Gricourt, G., Meyer, P., Duigou, T., and Faulon, J.-L. (2024). Artificial intelligence methods and models for retro-biosynthesis: a scoping review. *ACS Synth. Biol.* 13 (8): 2276–2294.

4 Volk, M.J., Tran, V.G., Tan, S.-I. et al. (2023). Metabolic engineering: methodologies and applications. *Chem. Rev.* 123 (9): 5521–5570.

5 Yu, T., Boob, A.G., Volk, M.J. et al. (2023). Machine learning-enabled retrobiosynthesis of molecules. *Nat. Catal.* 6 (2): 137–151.

6 Coley Connor, W., Jin, W., Rogers, L. et al. (2019). A graph-convolutional neural network model for the prediction of chemical reactivity. *Chem. Sci.* 10 (2): 370–377.

7 Segler, M.H.S. and Waller, M.P. (2017). Neural-symbolic machine learning for retrosynthesis and reaction prediction. *Chem. Eur. J.* 23 (25): 5966–5971.

8 Finnigan, W., Hepworth, L.J., Flitsch, S.L., and Turner, N.J. (2021). RetroBioCat as a computer-aided synthesis planning tool for biocatalytic reactions and cascades. *Nat. Catal.* 4 (2): 98–104.

9 Levin, I., Liu, M., Voigt, C.A., and Coley, C.W. (2022). Merging enzymatic and synthetic chemistry with computational synthesis planning. *Nat. Commun.* 13 (1): 7747.

10 Probst, D., Manica, M., Nana Teukam, Y.G. et al. (2022). Biocatalysed synthesis planning using data-driven learning. *Nat. Commun.* 13 (1): 964.

11 Zheng, S., Zeng, T., Li, C. et al. (2022). Deep learning driven biosynthetic pathways navigation for natural products with BioNavi-NP. *Nat. Commun.* 13 (1): 3342.

12 Segler, M.H.S., Preuss, M., and Waller, M.P. (2018). Planning chemical syntheses with deep neural networks and symbolic AI. *Nature* 555 (7698): 604–610.

13 Coley, C.W., Rogers, L., Green, W.H., and Jensen, K.F. (2018). SCScore: synthetic complexity learned from a reaction corpus. *JCIM* 58 (2): 252–261.

14 Ertl, P. and Schuffenhauer, A. (2009). Estimation of synthetic accessibility score of drug like molecules based on molecular complexity and fragment contributions. *J. Cheminf.* 1 (1): 8.

15 Carbonell, P., Wong, J., Swainston, N. et al. (2018). Selenzyme: enzyme selection tool for pathway design. *Bioinformatics* 34 (12).

16 Kroll, A., Ranjan, S., Engqvist, M.K.M., and Lercher, M.J. (2023). A general model to predict small molecule substrates of enzymes based on machine and deep learning. *Nat. Commun.* 14 (1): 2787.

17 Qian, W., Wang, X., Huang, Y. et al. (2025). Deep learning-driven insights into enzyme-substrate interaction discovery. *JCIM* 65 (1): 187–200.

18 Morris, G.M., Huey, R., and Olson, A.J. (2008). Using autodock for ligand-receptor docking. *Curr. Protoc. Bioinformatics* 24 (1): 8–14.

19 Boorla, V.S. and Maranas, C.D. (2025). CatPred: a comprehensive framework for deep learning in vitro enzyme kinetic parameters. *Nat. Commun.* 16 (1): 2072.

20 Li, F., Yuan, L., Lu, H. et al. (2022). Deep learning-based k_{cat} prediction enables improved enzyme-constrained model reconstruction. *Nat. Catal.* 5 (8): 662–672.

21 Yu, H., Deng, H., He, J. et al. (2023). UniKP: a unified framework for the prediction of enzyme kinetic parameters. *Nat. Commun.* 14 (1): 8211.

22 Kroll, A., Rousset, Y., Hu, X.P. et al. (2023). Turnover number predictions for kinetically uncharacterized enzymes using machine and deep learning. *Nat. Commun.* 14 (1): 4139.

23 Watson, J.L., Juergens, D., Bennett, N.R. et al. (2023). De novo design of protein structure and function with RFdiffusion. *Nature* 620 (7976): 1089–1100.

24 Dauparas, J., Anishchenko, I., Bennett, N. et al. (2022). Robust deep learning–based protein sequence design using ProteinMPNN. *Science* 378 (6615): 49–56.

25 Repecka, D., Jauniskis, V., Karpus, L. et al. (2021). Expanding functional protein sequence spaces using generative adversarial networks. Nature. *Machine Intelligence* 3 (4): 324–333.

26 Madani, A., Krause, B., Greene, E.R. et al. (2023). Large language models generate functional protein sequences across diverse families. *Nat. Biotechnol.* 41 (8): 1099–1106.

27 Nijkamp, E., Ruffolo, J.A., Weinstein, E.N. et al. (2023). ProGen2: exploring the boundaries of protein language models. *Cell Syst.* 14 (11): 968–978.

28 Chang, A., Jeske, L., Ulbrich, S. et al. (2021). BRENDA, the ELIXIR core data resource in 2021: new developments and updates. *Nucleic Acids Res.* 49 (D1): D498–D508.

29 Kanehisa, M. and Goto, S. (2000). KEGG: Kyoto encyclopedia of genes and genomes. *Nucleic Acids Res.* 28 (1): 27–30.

30 Karp, P.D., Billington, R., Caspi, R. et al. (2017). The BioCyc collection of microbial genomes and metabolic pathways. *Briefings Bioinf.* 20 (4): 1085–1093.

31 Wittig, U., Kania, R., Golebiewski, M. et al. (2012). SABIO-RK – database for biochemical reaction kinetics. *Nucleic Acids Res.* 40 (D1): D790–D796.

32 Yuan, L., Tian, Y., Ding, S. et al. (2019). PrecursorFinder: a customized biosynthetic precursor explorer. *Bioinformatics* 35 (9): 1603–1604.

33 Kumar, A., Wang, L., Ng, C.Y., and Maranas, C.D. (2018). Pathway design using de novo steps through uncharted biochemical spaces. *Nat. Commun.* 9 (1): 184.

34 Weininger, D. (1988). SMILES, a chemical language and information system. 1. Introduction to methodology and encoding rules. *J. Chem. Inf. Comput. Sci.* 28 (1): 31–36.

35 Coley, C.W., Green, W.H., and Jensen, K.F. (2019). RDChiral: an RDKit wrapper for handling stereochemistry in retrosynthetic template extraction and application. *JCIM* 59 (6): 2529–2537.

36 Lang, M., Stelzer, M., and Schomburg, D. (2011). BKM-react, an integrated biochemical reaction database. *BMC Biochem.* 12 (1): 42.

37 Browne, C.B., Powley, E., Whitehouse, D. et al. (2012). A survey of monte carlo tree search methods. *IEEE Trans. Comput. Intell. AI Games* 4 (1): 1–43.

38 Rittié, L., Perbal, B., Rittié, L., and Perbal, B. (2008). Enzymes used in molecular biology: a useful guide. *J. Cell Commun. Signaling* 2 (1): 25–45.

39 Ye, J., McGinnis, S., and Madden, T.L. (2006). BLAST: improvements for better sequence analysis. *Nucleic Acids Res.* 34: W6–W9.

40 Consortium TU (2019). Uni Prot: a worldwide hub of protein knowledge. *Nucleic Acids Res.* 47 (D1): D506–D515.

41 Mellor, J., Grigoras, I., Carbonell, P., and Faulon, J.L. (2016). Semisupervised Gaussian process for automated enzyme search. *ACS Synth. Biol.* 5 (6): 518–528.

42 Lin, Z., Akin, H., Rao, R. et al. (2023). Evolutionary-scale prediction of atomic-level protein structure with a language model. *Science* 379 (6637): 1123–1130.

43 Mou, Z., Eakes, J., Cooper, C.J. et al. (2021). Machine learning-based prediction of enzyme substrate scope: application to bacterial nitrilases. *Proteins* 89 (3): 336–347.

44 Zhou, J., Huang, M., Zhou, J., and Huang, M. (2024). Navigating the landscape of enzyme design: from molecular simulations to machine learning. *Chem. Soc. Rev.* 53 (16): 8202–8239.

45 Kroll, A., Engqvist, M.K.M., Heckmann, D., and Lercher, M.J. (2021). Deep learning allows genome-scale prediction of Michaelis constants from structural features. *PLoS Biol.* 19 (10): e3001402.

46 Yang, K.K., Wu, Z., and Arnold, F.H. (2019). Machine-learning-guided directed evolution for protein engineering. *Nat. Methods* 16 (8): 687–694.

47 Saito, Y., Oikawa, M., Sato, T. et al. (2021). Machine-learning-guided library design cycle for directed evolution of enzymes: the effects of training data composition on sequence space exploration. *ACS Catal.* 11 (23): 14615–14624.

48 Huang, C., Zhang, L., Tang, T. et al. (2024). Application of directed evolution and machine learning to enhance the diastereoselectivity of ketoreductase for dihydrotetrabenazine synthesis. *JACS Au* 4 (7): 2547–2556.

49 Rothlisberger, D., Khersonsky, O., Wollacott, A.M. et al. (2008). Kemp elimination catalysts by computational enzyme design. *Nature* 453 (7192): 190–195.

50 Yeh, A.H.-W., Norn, C., Kipnis, Y. et al. (2023). De novo design of luciferases using deep learning. *Nature* 614 (7949): 774–780.

51 Anishchenko, I., Pellock, S.J., Chidyausiku, T.M. et al. (2021). De novo protein design by deep network hallucination. *Nature* 600 (7889): 547–552.

52 Lauko, A., Pellock, S.J., Sumida, K.H. et al. (2025). Computational design of serine hydrolases. *Science* 388: eadu2454.

53 Ziegler, C., Martin, J., Sinner, C., and Morcos, F. (2023). Latent generative landscapes as maps of functional diversity in protein sequence space. *Nat. Commun.* 14 (1): 2222.

54 Li, H., Liu, X., Jiang, G., and Zhao, H. (2024). Chemoenzymatic synthesis planning guided by reaction type score. *JCIM* 64 (24): 9240–9248.

55 Kaspar, F. and Schallmey, A. (2022). Chemo-enzymatic synthesis of natural products and their analogs. *Curr. Opin. Biotechnol.* 77: 102759.

56 Caspi, R., Billington, R., Keseler, I.M. et al. (2020). The MetaCyc database of metabolic pathways and enzymes - a 2019 update. *Nucleic Acids Res.* 48 (D1): D445–D453.

57 Wang, C.Y., Chang, P.M., Ary, M.L. et al. (2018). ProtaBank: a repository for protein design and engineering data. *Protein Sci.* 27 (6): 1113–1124.

58 Kavvas, E.S., Yang, L., Monk, J.M. et al. (2020). A biochemically-interpretable machine learning classifier for microbial GWAS. *Nat. Commun.* 11 (1): 2580.

7

Machine Learning to Accelerate the Discovery of Therapeutic Peptides

Nicole Soto-Garcia[1], *Mehdi D. Davari*[2], *and David Medina-Ortiz*[1,2]

[1] *Universidad de Magallanes, Departamento de Ingeniería en Computación, 6210427, Avenida Bulnes 01855, Punta, Arenas, Chile*
[2] *Leibniz Institute of Plant Biochemistry, Department of Bioorganic Chemistry, 06120, Weinberg 3, Halle, Germany*

7.1 Introduction

In this chapter, we present a comprehensive overview of current machine learning (ML)-based methods and strategies that are transforming and accelerating the discovery and design of therapeutic peptides. We begin by introducing key biological concepts related to peptides, including their therapeutic relevance, mechanisms of action, and the advantages they offer over traditional small-molecule drugs – such as higher specificity, lower toxicity, and the ability to target challenging biological interfaces. We also address their limitations, including stability, bioavailability, and susceptibility to degradation.

Next, we describe the main data sources available for extracting relevant information about peptides, covering aspects such as biological activity, molecular function, pharmacokinetics, and potential toxic effects. We then introduce traditional data-driven strategies used in developing ML models for peptide property prediction, focusing on supervised learning pipelines that incorporate sequence encoding, model training, and validation.

Following this foundation, we explore both classical and state-of-the-art ML approaches across various tasks, including bioactivity classification, prediction of toxic adverse effects (such as cytotoxicity, hemolysis, and immunogenicity), and estimation of key pharmacological properties such as half-life, IC_{50}, or stability in physiological environments. In addition, we examine recent advances in generative modeling and *de novo* peptide design, highlighting how approaches such as variational autoencoders (VAEs), generative adversarial networks (GANs), and diffusion models enable the creation of novel peptide sequences with optimized properties.

Finally, we discuss the emerging field of autonomous peptide design, driven by the integration of ML-based agents capable of iterative learning, optimization, and decision-making. We evaluate current frameworks that incorporate reinforcement

Machine Learning and Big Data-enabled Biotechnology, First Edition. Edited by Hal S. Alper.

learning, active learning, and closed-loop design pipelines, and we assess their potential to enable self-improving platforms for therapeutic peptide discovery. The chapter concludes with a critical analysis of the advantages, limitations, and open challenges in building fully autonomous systems, emphasizing how ML can drive the next generation of intelligent, data-guided peptide therapeutics.

7.2 Peptides: Definitions and Main Characteristics

In contrast to the full-length enzymes discussed in Chapters 5 and 6, peptides are short chains of amino acids linked by peptide bonds, typically ranging from 2 to 100 amino acid residues, and with molecular weights between 0.5 and 10 kDa [1].

Structurally, peptides can adopt a variety of secondary conformations, including α-helices, β-sheets, and random coils [2]. However, their stability can be compromised by enzymatic degradation and conformational flexibility, limiting their bioavailability and therapeutic efficacy [3].

Peptides are involved in a wide array of biological processes, including hormone regulation, immune response, cell signaling, and tissue repair, making them key molecular tools across both scientific and industrial domains [4].

Due to their structural and functional versatility, peptides play essential roles in biological systems and hold high value in biotechnology, medicine, and various industrial applications [5]. Figure 7.1 summarizes the most common applications of peptides.

In the food industry, peptides function as bio-active compounds with antioxidant [6] and food preservative properties [7].

In biotechnology, peptides are utilized in the development of biosensors, where they act as recognition elements for detecting specific molecules [8]. They are also applied in plant protection, acting as biopesticides or disease resistance enhancers [9], in immune modulation [10] and in protein purification workflows through affinity-tag-based techniques [11].

Finally, in biomedical research and therapeutics applications, peptides play a central role in drug design and delivery. Their uses include the development of antimicrobial peptides (AMPs) with antifungal, antiviral, and antibacterial activity, as well as tumor-targeting agents and components in cell signaling pathways [4].

In industrial contexts, peptides play a vital role in the food industry, particularly as natural antioxidants and food preservative agents. Beyond their applications in food technology, peptides are widely utilized in biotechnology. Peptides serve important functions in plant protection, acting as bio-based insecticides and herbicides, as well as components in biosensors and the development of disease-resistant crops through genetic engineering. Moreover, peptides are crucial in the pharmaceutical and medical fields. They function as therapeutic agents with antimicrobial, antiviral, and anticancer properties, and are also employed in drug delivery systems due to their biocompatibility, specificity, and low toxicity. Their versatility and efficacy make them indispensable tools across multiple industrial sectors.

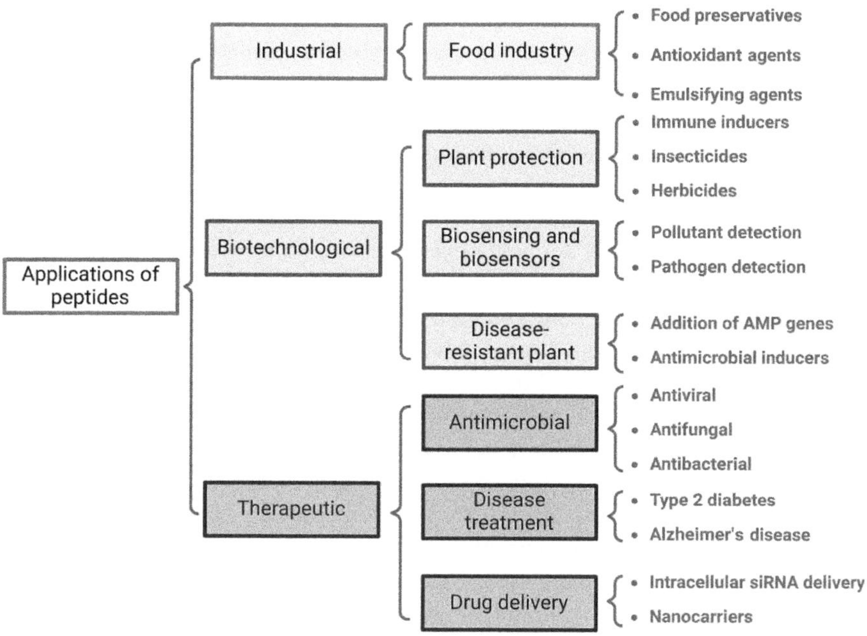

Figure 7.1 Peptides exhibit diverse applications, including industrial, biotechnological, and therapeutic.

7.3 Benefits and Limitations of Therapeutic Peptides

The therapeutic potential of peptides stems from their ability to bind with high affinity and specificity to cell surface receptors, triggering precise intracellular signaling pathways [12]. This selectivity makes them particularly effective against a variety of pathogens, including viruses, bacteria, fungi, and parasites [13]. Therapeutic peptides exhibit enhanced tissue penetration, improved bioavailability, and increased potency—making them promising candidates for targeted therapies with reduced off-target effects and minimized side effects [14].

Compared to small-molecule drugs, therapeutic peptides offer several advantages. Their synthesis is often simpler and more cost-effective, particularly through solid-phase peptide synthesis or recombinant production methods [15]. Their high target specificity not only boosts efficacy but also significantly reduces the likelihood of adverse effects [16]. Furthermore, peptides generally exhibit lower immunogenicity, improving their clinical safety profile. Being biodegradable, they are less likely to accumulate in tissues, thus reducing systemic toxicity and aligning well with the principles of precision medicine [17].

Despite all described advantages, therapeutic peptides face several clinical and pharmacological challenges. Their short half-life, due to rapid enzymatic degradation and renal clearance, limits their therapeutic window [18]. Oral administration is typically ineffective due to poor gastrointestinal stability and low permeability, often necessitating parenteral delivery routes (e.g. intravenous, subcutaneous, or intramuscular), which can reduce patient adherence and introduce risks such as discomfort and infection [19]. Moreover, peptides are vulnerable to denaturation under extreme pH conditions and exhibit low membrane permeability, limiting intracellular access and bioefficacy [20].

To address these limitations, ongoing research focuses on improving peptide stability, bioavailability, and delivery strategies. Techniques such as peptide cyclization, PEGylation, lipidation, nanoparticle encapsulation, and the use of cell-penetrating peptides are being explored to enhance pharmacokinetics and therapeutic performance [21].

In parallel, drug discovery and drug repurposing strategies are accelerating therapeutic peptide development. Drug discovery efforts aim to identify novel bioactive peptides through high-throughput screening (HTS), computational modeling, and peptide libraries. In contrast, drug repurposing investigates new therapeutic indications for already approved peptides, significantly reducing development time and cost [22]. These approaches have broadened the clinical landscape of peptide therapeutics, providing new treatment options for diseases that were previously challenging to manage with traditional small-molecule drugs [23].

As advances in peptide engineering, delivery technologies, and computational biology continue to evolve, therapeutic peptides are poised to become even more powerful tools in modern medicine—offering personalized, targeted, and safer alternatives for a wide array of conditions [24].

7.4 Computational Design of Therapeutic Peptides

The integration of computational strategies has revolutionized the discovery and design of therapeutic peptides, significantly accelerating the identification of candidates with high bioactivity, specificity, and stability [25]. Techniques such as ML, peptide-protein docking, and bioinformatics enable the rapid screening of vast peptide libraries and the prediction of essential properties, including binding affinity, solubility, stability, and toxicity [26]. By leveraging large-scale biological datasets and predictive modeling, these tools reduce reliance on time-consuming and costly wet-lab experiments, thereby streamlining and expediting drug development pipelines [27].

Among computational methodologies, deep learning models and sequence-based algorithms have proven particularly effective in predicting peptide bioactivities and interactions with biological targets [28]. Trained on comprehensive datasets, these models uncover intricate sequence–function relationships, supporting the design of highly potent and selective therapeutic peptides [29]. Additionally, molecular dynamics simulations provide atomic-level insights into peptide conformational stability and target binding, facilitating the optimization of physicochemical and pharmacological properties [30].

The advent of generative models, including transformers, GANs, and VAEs, has further advanced *de novo* peptide design. These models are capable of generating novel peptide sequences with tailored pharmacokinetic and pharmacodynamic profiles, expanding the chemical and functional space beyond naturally occurring sequences [5].

Moreover, structure-based drug design (SBDD) and computational docking refine peptide–target interactions by predicting binding affinities, conformational flexibility, and energetically favorable poses, thereby supporting the selection and optimization of lead candidates [31]. AI-driven screening pipelines also integrate bioinformatics resources to identify natural peptides with therapeutic potential, analyzing protein–peptide interaction networks, sequence motifs, and 3D structural features [25].

To ensure the clinical viability of therapeutic peptide candidates, multi-objective optimization algorithms are employed to simultaneously balance key parameters such as stability, potency, bioavailability, and toxicity [32]. The application of these strategies enables the rational prioritization of candidates for preclinical validation.

As computational methodologies continue to evolve, their synergy with experimental validation is critical for achieving effective and reliable peptide design [33]. Hybrid approaches – which combine *in silico* predictions with HTS, structural biology, and biophysical characterization – enhance both the accuracy and efficiency of the discovery process [34]. Furthermore, AI-driven peptide engineering platforms enable rational modifications aimed at improving enzymatic stability, membrane permeability, and systemic bioavailability [35].

By merging computational power with empirical validation, modern strategies are reshaping the landscape of therapeutic peptides, offering innovative solutions for complex and previously untreatable diseases, and unlocking new avenues for precision medicine [26].

7.5 Data Sources for Peptide Discovery

The advancement of drug design, driven by AI and the increasing availability of biological data, has led to the development of specialized databases and computational methodologies that are now fundamental to biomedical research [5]. Core resources such as the Protein Data Bank [36] and UniProt [37] provide comprehensive structural and sequence data for proteins, supporting tasks such as molecular modeling, protein annotation, and SBDD. In parallel, dedicated peptide databases – such as Peptipedia v2.0 or DRAMP – provide specialized information on peptide sequences (see Table 7.1 for a list of the most relevant databases used to facilitate the study of therapeutic peptides). These platforms play a crucial role in the identification and optimization of therapeutic peptide candidates, centralizing large-scale datasets and integrating computational tools that accelerate peptide discovery, design, and refinement for clinical and industrial applications.

Peptide-related databases encompass a wide range of resources, from general repositories that catalog diverse peptide types to specialized databases focused on particular peptide families, biological functions, or pharmacological properties [25, 49, 57]. These repositories support therapeutic and biotechnological applications by providing access to peptide sequences, structural data, and functional annotations [58].

General-purpose peptide databases such as StraPep and PepBank compile extensive datasets to support research into structure–function relationships and

Table 7.1 Summary of the most relevant databases to facilitate the study of therapeutic peptides.

Database	Description	Records
Peptipedia v2.0 [4]	Database containing peptides with annotated biological activities and predictive models.	3.983.654
LAMP [38]	Platform for antimicrobial peptides with tools for therapeutic optimization.	22.164
dbAMP [39]	Curated repository of natural, synthetic, and patented antimicrobial peptides.	21.834
PepBank [40]	Text mining-based database focused on peptides for therapeutic and diagnostic applications.	19.792
DBAASP [41]	Database storing antimicrobial peptides along with activity spectra and structural properties.	16.722
CAMPR [42]	Repository of antimicrobial peptides with experimental data and predictive tools.	6.695
DRAMP [43]	Antimicrobial peptide database integrating structural and functional information.	5.551
CancerPPD [44]	Repository focused on anticancer peptides and their biological properties.	3.491
AVPdb [45]	Database of antiviral peptides targeting viruses such as HIV, RSV, and SARS.	1.817
Hemolytik [46]	Database of hemolytic peptides relevant for toxicity studies.	1.530
StraPep [47]	Repository of natural bioactive peptides with data on their structural features.	1.201
BioDADPep [48]	Database of antihypertensive and antidiabetic peptides with IC50 values.	1.136
PEPlife [49]	Database of peptide half-life under various *in vivo* conditions.	938
THPdb [50]	Platform of approved therapeutic peptides and proteins, including pharmacokinetics and mechanisms of action.	852
InverPep [51]	Specialized database of inverted antimicrobial peptides for clinical applications.	774
B3Pdb [52]	Platform compiling peptides capable of crossing the blood–brain barrier.	465
Quorumpeps [53]	Database of peptides involved in quorum-sensing communication.	354
PLifePred [54]	Predictive tool and repository for peptide half-life estimation.	234
AntiAngioPred [55]	Repository of antiangiogenic peptides for biomedical applications.	202
PepTherDia [56]	Compilation of 105 FDA-approved peptide drugs, detailing administration routes and clinical profiles.	105

biological activities. StraPep focuses on natural bioactive peptides, documenting their structural features, while PepBank uses text mining techniques to extract and curate peptide data relevant to therapeutic and diagnostic development [40, 47]. Within functional classifications, AMPs have attracted considerable interest due to the growing threat of antibiotic resistance [59]. Several AMP-specific databases – including Peptipedia v2.0, dbAMP, DBAASP, LAMP, DRAMP, InverPep, and CAMPR – provide curated information on natural, synthetic, and patented peptides, covering their antimicrobial spectra, structural characteristics, and optimization tools for therapeutic development [4, 38, 39, 41–43, 51].

Beyond AMPs, specialized databases focused on peptides with particular biomedical applications have been implemented. For example, AVPdb catalogs antiviral peptides (AVPs) targeting pathogens such as HIV, RSV, and SARS [45], while CancerPPD is dedicated to anticancer peptides (ACPs) [44]. Other focused repositories include Hemolytik (hemolytic peptides) [46], B3Pdb (blood–brain barrier-penetrating peptides) [52], Quorumpeps (quorum-sensing peptides) [53], and AntiAngioPred (antiangiogenic peptides) [55]. In addition, databases like PEPlife and PLifePred provide data on peptide half-life and *in vivo* stability, which are critical for pharmacokinetic optimization [49, 54, 60]. Other repositories track bioactivity metrics such as IC_{100} values for AVPs [61] and IC_{50} values for antihypertensive and antidiabetic peptides [48].

Understanding peptide–target interactions is essential for the rational design of bioactive peptides. Databases focusing on these interactions are typically categorized into large-scale general repositories, reference datasets, and specialized collections derived from specific experimental studies [25]. In addition to research-oriented resources, databases documenting regulatory-approved peptides are vital for drug repurposing and clinical translation [62]. For instance, THPdb provides curated information on therapeutic peptides and proteins, including sequences, physicochemical properties, mechanisms of action, and pharmacodynamics [50]. Similarly, PepTherDia compiles detailed data on 105 FDA-approved peptide drugs, covering administration routes, pharmacokinetics, and therapeutic indications [56].

Collectively, these repositories not only support fundamental research but also accelerate drug discovery by enabling data-driven peptide optimization and expanding the therapeutic potential of peptides across various disease areas.

7.6 ML-Based Strategies to Accelerate the Discovery of Therapeutic Peptides

The design of ML-based predictive models has become a fundamental strategy for classifying bioactivities and estimating the physicochemical or pharmacological properties of therapeutic peptides. These data-driven approaches follow a systematic pipeline that includes the collection and curation of peptide datasets, numerical encoding of peptide sequences, model training using suitable ML algorithms, performance evaluation through established metrics, and application of the trained model to predict properties or activities of novel peptide sequences [63].

This section outlines the methodologies used to develop sequence-based predictive models, emphasizing data preprocessing, feature representation strategies, and model training workflows. Moreover, the applicability of ML-based strategies for classifying bioactive functions, predicting toxicological risks, and estimating relevant pharmacological properties of therapeutic peptides is highlighted.

7.6.1 Data-Driven Approaches

The application of data-driven approaches for developing sequence-based predictive models aimed at estimating peptide properties or bioactivities is illustrated in Figure 7.2. This process generally involves two main stages: the preparation of data for model training and the training and evaluation of the predictive model itself.

The first stage begins with the collection and preprocessing of peptide sequences, which must be carefully aligned with the specific objective of the model – whether it is to classify AMPs, predict half-life values, or estimate solubility. These sequences are typically obtained from curated datasets and public peptide databases, as described in Section 7.5. Once collected, the dataset undergoes a cleaning process to remove redundant entries, correct inconsistencies, and filter out sequences that contain noncanonical amino acids or incomplete annotations. This step ensures the quality and reliability of the data used for training.

After curation, the peptide sequences must be transformed into numerical representations to make them compatible with ML algorithms. Several encoding strategies can be employed for this purpose, ranging from classical approaches such as one-hot encoding and physicochemical property descriptors to frequency-based methods like *k*-mers and signal transformations. In recent years, more advanced techniques based on pretrained protein language models – such as ESM, Mistral, and ProT5 – have gained popularity, as they generate embeddings that capture the contextual and functional features of peptide sequences with remarkable accuracy [1].

Once the numerical representations are ready, the dataset is typically divided into training, validation, and testing subsets, often using a 70:20:10 ratio. In cases where datasets are relatively small, an 80:20 split between training and testing is preferred, supplemented with cross-validation techniques to minimize overfitting and enhance the model's generalization capabilities.

With the data partitioned, the training phase begins. During this stage, ML models are constructed using the training subset. A variety of algorithms can be applied depending on the problem and the nature of the data, including distance-based, rule-based, regression-based, kernel-based, and ensemble-based methods. In some cases, particularly when working with large datasets or rich sequence embeddings, deep learning architectures such as convolutional or recurrent neural networks (RNNs) may be used to capture complex, high-dimensional patterns in the data.

Following training, the performance of the model is evaluated using the validation and test datasets. Classification tasks are typically assessed using metrics such as accuracy, precision, recall, and F1-score, while regression tasks are evaluated through correlation-based metrics like Pearson and Spearman correlation

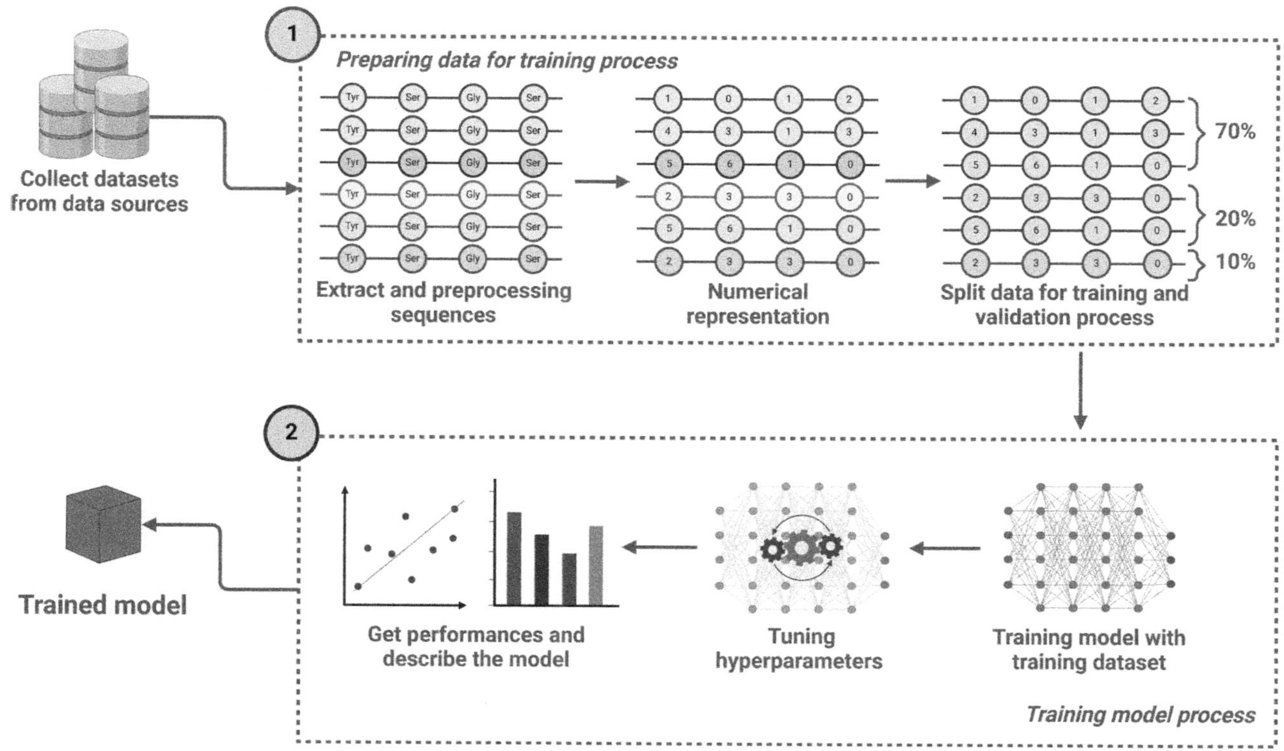

Figure 7.2 Overview of the workflow for developing sequence-based predictive models.

coefficients, as well as error-based metrics such as mean absolute error (MAE) and root mean squared error (RMSE). The validation dataset is also essential for tuning hyperparameters, which is commonly done using optimization strategies such as grid search, genetic algorithms, or Bayesian optimization to identify the most effective configuration for the model.

Finally, the independent test dataset is used to evaluate the model's generalization ability and to report final performance metrics. This dataset also serves as a benchmark for comparison with alternative methods or previously published results. Once validated, the trained model can be exported and is ready for deployment in practical applications, including peptide screening, drug discovery, and the rational design of therapeutic candidates.

The process begins with the construction of a dataset by gathering peptide or protein sequences from public databases or repositories. Preprocessing steps are then applied to remove redundancy, correct inconsistencies, and ensure data quality. The curated sequences are transformed into numerical representations using encoding techniques, making them suitable for ML algorithms. These encoded datasets are used to train predictive models through supervised learning. Finally, the model's performance is evaluated using validation and testing datasets to ensure accuracy, reliability, and generalization.

7.6.2 ML Strategies for Peptide Bioactivity Classification

The growing number of experimentally characterized peptides has accelerated the adoption of computational biology and data-driven techniques for sequence analysis, functional prediction, and design optimization. In particular, ML approaches have emerged as powerful tools for developing predictive models capable of classifying and characterizing peptide bioactivities, including antimicrobial, antifungal, antiviral, and anticancer properties, among others [64]. These models not only facilitate the identification of promising therapeutic candidates but also reduce experimental costs and time by enabling *in silico* screening prior to synthesis and validation.

In this section, we present an overview of the current state of the art in ML-based predictive modeling for therapeutic peptides. We describe the main components of the modeling pipeline, including data collection, sequence encoding strategies, model training, and performance evaluation, and discuss recent advancements leveraging deep learning and protein language models. Special attention is given to tools and frameworks designed to support the discovery and development of peptides with specific bioactivities.

7.6.2.1 Classification Models for Antimicrobial Peptides

Recent advances in AMP classification models have been driven by the integration of cutting-edge ML and deep learning techniques. These models have become increasingly sophisticated, leveraging diverse architectures and learning strategies to improve the accuracy and generalizability of AMP prediction and design.

One notable example is Diff-AMP, a generative framework that integrates attention mechanisms, diffusion models, and reinforcement learning to automate AMP generation, activity prediction, and sequence optimization. By incorporating pretraining and transfer learning, Diff-AMP achieves enhanced performance and adaptability across diverse peptide datasets [65]. In parallel, the representation of peptide sequences has become more varied and nuanced. For instance, AMPActiPred relies on traditional feature engineering approaches, extracting a comprehensive set of compositional and physicochemical descriptors to construct predictive features [66].

Other models take hybrid approaches. AntiBP3 combines binary profiles of terminal residues with distributed representations (embeddings), achieving outstanding performance with an AUC of 0.98 in predicting Gram-negative antibacterial peptides [67]. Meanwhile, Deep-ABPpred utilizes bidirectional long short-term memory (BiLSTM) networks in combination with word2vec embeddings to capture sequential and contextual information from peptide sequences, demonstrating over 94% accuracy on independent test datasets [68].

More recently, large-scale models have embraced the power of pretrained protein language models. AMP-Detector, for example, integrates representations from protein language models with ML classifiers to screen large peptide libraries. This framework successfully identified over 190,000 putative AMPs and facilitated the design of more than 500 novel candidate peptides with predicted antimicrobial activity [1].

7.6.2.2 Classification Models for Antiviral Peptides

AVP prediction has evolved considerably over the past decade, progressing from early ML classifiers to more complex deep learning architectures and virus-specific frameworks. AVPpred was the first predictive model specifically designed for AVPs. It employed a support vector machine (SVM) classifier and achieved an accuracy of 86% with a Matthews correlation coefficient (MCC) of 0.71, laying the groundwork for subsequent computational methods in the field [69].

Later models introduced ensemble learning techniques, such as Random Forests and AdaBoost, which improved predictive performance, reaching accuracies as high as 93.26% [70]. Ensemble-based meta-predictors like Meta-iAVP and PePred-Suite further enhanced classification performance by integrating multiple base classifiers and feature types, demonstrating the value of combining diverse model outputs for robust AVP prediction [71, 72].

With the rise of deep learning, architectures incorporating bidirectional RNNs (BiRNNs) and convolutional neural networks (CNNs) have been developed to automatically extract high-level sequence patterns, improving the generalization and scalability of AVP predictors [73]. One innovative example is AI4AVP, which combines CNNs with GANs to synthetically augment training datasets, addressing the limited availability of labeled AVP sequences and improving model robustness [74].

More recent models have extended their scope beyond binary classification to include multi-label prediction of target viral families, as seen in FFM-AVP [75].

In parallel, virus-specific predictors have been developed to identify AVPs active against particular pathogens, including coronaviruses [76], dengue virus [77], hepatitis C virus [78], and HIV [79]. For instance, iDVIP focuses on peptides that inhibit HIV integrase, while the DBAASP platform includes predictive tools for linear peptides that block viral entry into host cells [80, 81].

7.6.2.3 Classification Models for Antifungal Peptides

Computational models for antifungal peptide (AFP) classification have evolved rapidly, integrating increasingly sophisticated ML and deep learning techniques to enhance predictive performance and model interpretability. Early models like AntiFP employed a support vector machine classifier using simple amino acid composition features, representing one of the first attempts at *in silico* AFP prediction [82].

As the field progressed, models began to incorporate more informative descriptors. Deep-AntiFP, for example, combined physicochemical and quasi-sequential features with deep neural networks, enabling improved discrimination between AFPs and non-AFPs [83]. Further advancement came with Deep-AFPpred, which introduced a hybrid deep learning framework based on one-dimensional CNNs (1D-CNNs) and BiLSTM networks, enhanced by transfer learning. This architecture demonstrated strong generalization capabilities on independent test datasets [84].

Building on these architectures, DeepAFP achieved even higher predictive accuracy (93.29%) by integrating evolutionary information – such as position-specific scoring matrices (PSSMs) – alongside physicochemical features within a CNN-BiLSTM hybrid model, showcasing the benefits of combining sequence conservation with structural properties [85].

More recently, cutting-edge models have adopted transformer-based encodings and attention mechanisms. iAFPs-Mv-BiTCN represents a state-of-the-art model that fuses skip-gram and transformer-derived sequence embeddings, SHAP-based feature selection, and a bidirectional temporal convolutional network (BiTCN), offering enhanced sequence modeling capabilities and interpretability [86]. Similarly, MLAFP-XN employed eight different feature extraction and selection strategies in conjunction with XGBoost, achieving a remarkable accuracy of 99.48%. Notably, it incorporates SHAP (SHapley Additive exPlanations) to interpret model predictions and identify key features driving classification decisions [87].

These advances reflect the growing complexity and refinement of AFP prediction frameworks. From traditional handcrafted features to embedding-based and explainable models, the field continues to push the boundaries of what is possible in peptide-based antifungal drug discovery. Continued integration of evolutionary, structural, and contextual sequence information promises to further improve the reliability and applicability of these tools in therapeutic design.

7.6.2.4 Additional Classification Models for Therapeutic Peptides

In addition to antimicrobial, antiviral, and AFPs, a variety of ML-based predictive models have been developed to classify other therapeutic peptide categories with significant clinical relevance. These include antiparasitic, anticancer, tumor-homing, wound-healing, and antiaging peptides, among others. This section highlights representative computational approaches used for the classification of these specialized peptides, underscoring the diversity of techniques and applications in the field.

For antiparasitic peptides (APPs), models such as i2APP adopt a two-stage approach that integrates LightGBM and SVM classifiers, achieving an accuracy of 91.3% and an AUC of 0.935, showcasing the benefit of hybrid ensemble strategies [88]. Similarly, PredAPP employs a combination of feature selection techniques and ensemble classifiers to improve predictive robustness and generalization performance [89].

In the domain of ACP prediction, recent models have embraced deep learning and dimensionality reduction for enhanced performance. ACPPfel, for instance, combines multiple feature selection and embedding techniques with deep learning models, reaching an impressive accuracy of 98.53% and an AUC of 0.997 [90]. Likewise, cACP-DeepGram integrates FastText-based skip-gram embeddings with deep neural networks, effectively capturing local sequence patterns relevant to anticancer activity [91]. MLACP 2.0 takes a meta-learning approach, combining CNNs with diverse encoding schemes to support multi-view learning and further improve ACP classification accuracy [92].

For tumor-homing peptides (THPs), THPep uses a Random Forest classifier trained on compositional and physicochemical descriptors to distinguish peptides with tumor-targeting potential, offering a useful tool for peptide-based drug delivery research [93].

In other therapeutic categories, models continue to expand the scope of peptide discovery. WHP-Pred leverages XGBoost to predict wound-healing peptides (WHPs) based on a curated feature set derived from sequence and biochemical properties [94]. Meanwhile, Antiaging-FL focuses on identifying antiaging peptides by incorporating feature-level fusion and ML classifiers, enabling the exploration of peptides with potential regenerative or protective effects in age-related biological processes [95].

These models illustrate the growing application of ML in peptide therapeutics beyond conventional categories. As curated datasets grow and representation methods improve, predictive tools tailored to specific therapeutic contexts will become increasingly valuable for accelerating *in silico* peptide discovery.

7.6.3 Strategies for Building Toxicity Classification Models

ML techniques have greatly facilitated the development of predictive models aimed at the early detection of adverse toxic effects in therapeutic peptides. These toxic responses can severely limit the clinical applicability of peptide-based drugs and typically include cytotoxicity, hemolysis, and allergenicity. Cytotoxicity refers to a peptide's potential to damage or destroy host cells, leading to tissue injury or compromised organ function. Hemolysis involves the rupture of red blood cell membranes, which can result in the release of hemoglobin into the bloodstream and cause serious complications such as anemia or renal failure. Allergenicity denotes the ability of a peptide to provoke immune hypersensitivity reactions, ranging from mild inflammatory responses to severe and potentially life-threatening allergic reactions.

Given the complexity and variability of these toxicological outcomes, computational models have emerged as valuable tools for systematically assessing safety risks during the early stages of peptide development. In particular, ML-based approaches can learn intricate sequence–function relationships from experimental data, enabling them to identify peptides likely to cause adverse effects with high accuracy.

This section highlights a variety of ML frameworks designed to predict toxicity in therapeutic peptides. These include traditional classification models based on physicochemical descriptors, as well as more advanced deep learning architectures that leverage sequence embeddings and transfer learning to enhance predictive performance. By integrating these tools into the peptide discovery pipeline, researchers can effectively filter out unsafe candidates prior to synthesis and biological testing, thereby improving both the safety profile and the efficiency of therapeutic peptide development.

7.6.3.1 Toxicity Prediction

ToxinPred was one of the pioneering tools developed for predicting peptide toxicity, employing SVMs trained on sequence-derived features to classify toxic and nontoxic peptides with reasonable accuracy [96]. Since then, the field has evolved significantly, incorporating more powerful ML and deep learning techniques to capture the complex relationships underlying toxic effects.

Recent models have leveraged deep neural networks to improve generalization and predictive accuracy. For instance, extensions of ToxinPred now incorporate deep learning frameworks that can automatically extract hierarchical features from peptide sequences, providing more robust toxicity assessments [97]. Other approaches, such as ToxDL, utilize CNNs to model local residue-level patterns associated with toxicity, enabling the detection of subtle sequence motifs that traditional models may overlook [98].

Hybrid architectures have also shown strong performance in toxicity prediction tasks. Models like ToxIBTL combine CNNs with BiLSTM networks, capturing both spatial and sequential dependencies within peptide sequences [99]. Furthermore, the integration of protein language models, such as CSM-Toxin, has marked a significant advancement by using large-scale unsupervised pretraining to generate contextualized embeddings, which are then fine-tuned for toxicity prediction [100].

More recently, the incorporation of structural information into toxicity models has opened new avenues for enhancing prediction accuracy. Tools such as ToxGIN integrate both sequence and structural data using Graph Isomorphism Networks (GINs), enabling the model to capture complex topological and physicochemical interactions at the molecular level [101]. This sequence-structure fusion represents a promising direction for developing more biologically informed and explainable toxicity predictors.

7.6.3.2 Immunogenicity Identification

The prediction of peptide immunogenicity – an essential step in therapeutic peptide design – has advanced considerably with the integration of ML and deep learning techniques. POPI was the first computational model developed specifically to predict immunogenic peptides. It utilized SVMs trained on physicochemical properties of amino acids to distinguish peptides capable of eliciting immune responses from those that do not [102]. Building on this foundation, PAAQD introduced a more refined feature space by incorporating amino acid pairwise contact parameters and quantum topological molecular similarity descriptors, allowing the model to better capture structural and spatial characteristics relevant to immunogenicity [103].

Subsequent models have embraced ensemble learning and advanced neural network architectures to enhance prediction performance. INeo-Epp, for example, employed Random Forest classifiers and demonstrated improved accuracy in predicting T-cell epitope immunogenicity, benefiting from both feature diversity and ensemble robustness [104]. Meanwhile, deep learning approaches such as DeepHLApan integrated RNNs to model peptide–MHC binding affinities and immunogenic potential, effectively capturing sequential dependencies in the amino acid chains [105].

Other models, like DeepImmuno, utilized CNNs to learn local residue patterns associated with immunogenicity, offering a powerful approach for epitope identification and neoantigen prediction [106]. More recently, the incorporation of transfer learning strategies has further advanced the field. For example, TLImmuno2 combines pretrained embeddings from protein language models with downstream classifiers to improve learning efficiency and predictive performance, particularly in data-scarce scenarios [107].

7.6.3.3 Hemolysis Evaluation

The prediction of hemolytic activity – a critical safety concern in peptide therapeutics – has been the focus of several computational models aimed at identifying peptides with potential hemotoxic effects. Early tools such as HemoPI, HemoPred, and HemoNet laid the groundwork for *in silico* hemotoxicity assessment. These models primarily relied on sequence-derived features, physicochemical descriptors, and structural properties to classify peptides based on their likelihood of causing red blood cell lysis [108–110]. Such tools have been instrumental in filtering out hemolytic peptides during the early stages of drug discovery, improving safety profiles prior to experimental validation.

Building upon these foundations, more recent approaches have introduced advanced learning paradigms to further improve prediction accuracy and model interpretability. Quantum ML (QML) has emerged as a novel strategy for identifying nonhemolytic peptides, leveraging the principles of quantum computing to capture complex sequence relationships that traditional models may overlook [111]. Additionally, transfer learning has been employed to enhance hemotoxicity prediction in the context of AMPs. For example, models like AMPDeep use pretrained representations and domain-specific fine-tuning to more accurately distinguish between hemolytic and nonhemolytic variants, even when training data is limited [112].

A particularly promising advancement involves the combination of QML with protein language models, enabling simultaneous prediction of binary hemolytic activity and continuous toxicity metrics such as HC_{50} concentration, a quantitative measure of hemolytic potency. These integrative approaches not only improve the granularity of toxicity assessments but also support the rational design of safer therapeutic peptides by providing both classification and regression outputs for hemolytic risk evaluation [113].

7.6.3.4 Other Toxic Adverse Effect Predictions

In addition to models developed for general toxicity, immunogenicity, and hemolytic activity, specialized predictive frameworks have been created to address other biologically relevant adverse effects associated with therapeutic peptides. These include neurotoxicity, allergenicity, and autoimmune responses, each of which presents unique challenges in peptide-based drug development.

Deep learning techniques have recently been applied to predict neurotoxin-related adverse effects, particularly in the context of peptide toxins such as conotoxins. These neurotoxic peptides, often derived from marine organisms, can interfere with neuronal signaling and ion channel function. Emerging models trained on conotoxin datasets use deep neural architectures to classify neurotoxic peptides and evaluate their potential neurological risks, supporting safer design of peptide therapeutics targeting the nervous system [114].

Allergenicity prediction has also received considerable attention, especially due to its relevance in peptide-based vaccines, food-derived peptides, and biologics. Tools like AllerTOP v.2 and AllerCatPro 2.0 apply ML to analyze amino acid sequences and physicochemical features, enabling the identification of peptides with high allergenic potential [115, 116]. These models have been validated using curated allergen databases and offer user-friendly platforms for allergen screening during peptide design.

Furthermore, the prediction of autoimmune responses to specific peptide epitopes – such as those associated with gluten intolerance – has led to the development of more specialized models. In the context of celiac disease, tools like CDpred and CICERON predict T-cell epitopes that trigger immune responses in genetically predisposed individuals. These models analyze sequence motifs associated with HLA-DQ2/DQ8 binding and immune activation, facilitating the identification and exclusion of immunopathogenic peptides in therapeutic or nutritional applications [117, 118].

7.6.4 Data-Driven Strategies for Modeling Pharmacological Profiles

The computational prediction of peptide properties – such as half-life, stability, and pharmacological activityhas become increasingly important due to its direct impact on therapeutic efficacy and safety profiles [49]. Accurate *in silico* models allow researchers to assess and optimize peptide candidates before experimental validation, saving both time and resources in the drug development pipeline.

A variety of computational approaches have been developed to estimate peptide half-life, each leveraging different types of input data and modeling strategies. ProtParam, for example, predicts peptide half-life based on the N-terminal rule, which relates the stability of a peptide to its N-terminal amino acid residue [119]. Other models utilize multivariable regression frameworks to incorporate additional sequence features and biochemical properties into half-life prediction algorithms [120].

Some tools are designed for specific peptide classes. PyAMPA, for instance, focuses on predicting half-life for AMPs using tailored descriptors and modeling strategies suited to their structural features [121]. Other models account for specific physiological conditions; HLP, for example, uses SVMs to predict peptide half-life in a digestive-tract-like environment, reflecting the challenges of oral peptide delivery [122].

Advanced models like PepMSND integrate a wide array of features – including molecular descriptors, physicochemical properties, SMILES strings, and 3D structural data – to improve predictions of blood stability [123]. Similarly, Plifepred combines SVMs, linear regression, and Gaussian process models, using molecular descriptors and binary residue profiles to predict half-life in both natural and chemically modified peptides [54]. More comprehensive tools such as PepAnalyzer go beyond half-life estimation by predicting multiple peptide attributes, including *in vitro* stability, net charge, and isoelectric point, providing a versatile platform for early-stage peptide characterization [124].

In parallel with stability modeling, there has been substantial progress in the prediction of pharmacological properties, particularly regarding peptide potency and toxicity. AVP-IC50Pred estimates the IC_{50} values of AVPs, achieving a correlation coefficient of up to 0.74 [125]. Nath's meta-classifier employs a stacking-based ensemble approach to predict both IC_{50} and pIC_{50} values with improved accuracy [126]. For AVPs targeting the Flaviviridae family, a proteochemometric modeling (PCM) approach achieved a correlation coefficient of 0.85 in pIC_{50} prediction, highlighting the utility of integrating chemical and biological descriptors [127].

In the antimicrobial domain, CNNs have been used for simultaneous classification and regression to estimate the minimum inhibitory concentration (MIC), offering insight into both potency and spectrum of activity [128]. Finally, tools like xDeep-AcPEP incorporate structural and compositional features to predict a wide range of pharmacological endpoints – including EC_{50}, LC_{50}, IC_{50}, and LD_{50} – for ACPs, supporting both efficacy and toxicity assessment in a single framework [129].

7.6.5 *De novo* Design of Therapeutic Peptides

Generative models are a class of ML algorithms that learn to model the underlying distribution of data to generate new, previously unseen samples with similar characteristics. In the context of therapeutic peptides, generative models can propose novel sequences with optimized bioactivity, reduced toxicity, enhanced stability, or target specificity – thus accelerating rational peptide design [5].

Several types of generative architectures have been adapted for peptide discovery, including VAEs, GANs, transformer-based language models, and, more recently, diffusion models. Each strategy offers unique advantages in capturing peptide sequence representations and guiding the generation process based on biological, chemical, or structural constraints.

These models differ not only in architecture but also in the type of control they offer over the design process. For instance, VAEs are well-suited for navigating continuous latent spaces and enabling conditional generation, GANs excel at producing highly realistic sequences through adversarial learning, transformers provide context-aware generation informed by global sequence dependencies, and diffusion models offer a powerful framework for generating peptides while incorporating multimodal biological constraints.

In the following section, we focus specifically on the application of ML strategies for *de novo* design of therapeutic peptides, detailing the algorithms, architectures, and design principles that underpin these next-generation generative frameworks.

7.6.5.1 Variational Autoencoder-Based Approaches

VAEs are probabilistic generative models that encode input data into a latent space and then decode it to reconstruct or generate new samples. They are particularly effective at learning smooth, continuous representations of peptide sequences, which can be manipulated to generate novel variants with desired properties.

Different VAE-based strategies have been developed for peptide discovery. For AMPs, models such as PepVAE [130], PepCVAE [131], and GM-Pep [132] have shown strong potential in generating active peptide sequences. [133] proposed a deep autoencoder model for AMP design, which achieved a 10% hit rate within 48 days of generation and experimental validation.

A notable development in this area is HydrAMP, a conditional VAE (cVAE) that learns a continuous latent representation of peptide space, enabling guided and controllable generation of sequences. This approach led to the creation of Hydraganan-1, a novel and highly active AMP derived from the parent peptide Pexiganan, with confirmed efficacy against *E. coli* strains [134].

Beyond AMPs, VAE-based models have been used to generate target-specific inhibitors [135], peptides with membrane permeability and bioactivity optimization [136], inhibitors of β-catenin [137], and peptides with high-affinity MHC binding potential [138].

7.6.5.2 Generative Adversarial Networks

GANs represent another widely used generative architecture. They consist of two neural networks – a generator and a discriminator – trained in competition. GANs are particularly useful in generating biologically plausible sequences due to their ability to model complex, high-dimensional distributions.

One of the most well-known peptide design tools based on GANs is GANDALF (Generative Adversarial Network Drug-tArget Ligand Fructifier), a deep convolutional GAN that incorporates drug-target interaction data, 3D protein structures from PDB, and protein pocket information from CASTp. Instead of relying on protein–protein interaction datasets, GANDALF is trained on FDA-approved peptides from the THP database and has been used to generate peptides targeting cancer-related proteins such as PD-1, PDL-1, and CTLA-4 [50, 139–141].

Another GAN-based approach, PandoraGAN, focuses on generating bioactive AVPs. It uses LeakGAN, a modified GAN architecture originally developed for text generation, to produce sequences that are then validated using physicochemical filters like net charge, instability index, and motif patterns [142].

Other notable GAN implementations include AMPGAN v2 [143], which supports controlled generation of AMP sequences, AI4AVP for AVP design [74], and HelixGAN, which focuses on generating peptides with helical conformations often associated with membrane activity [144].

7.6.5.3 Transformer-Based Language Models

Inspired by advances in natural language processing (NLP), transformer-based language models have also been applied to peptide generation. These models treat peptides as sequences of tokens (amino acids) and learn contextual representations that preserve structural and functional motifs.

AMPTrans-LSTM is a hybrid language model that combines a long short-term memory sampler with a transformer-based converter to generate AMPs. This dual architecture enables the preservation of essential AMP features while enhancing generative creativity. The model reports success rates ranging from 30% to 50% for generating active peptides [145].

7.6.5.4 Diffusion Models

Diffusion models have emerged as a powerful alternative to traditional generative architectures, especially in molecular and structural biology. These models work by iteratively denoising random noise to generate samples that resemble the training data distribution.

In peptide design, diffusion-based methods such as AMP-Diffusion combine latent diffusion models with protein language embeddings, allowing for high-quality generation of AMPs with tunable properties [146]. Watson et al. [147] demonstrated that the combination of structure prediction networks and diffusion generative frameworks can facilitate the rational design of therapeutic peptides with broad-spectrum antiviral activity.

Other innovative methods include Geometric Latent Diffusion, which incorporates 3D geometric constraints into the generation process [148], and multimodal contrastive diffusion models, which align peptide generation with both structural and functional modalities for improved biological relevance [149].

7.7 Next-Generation Peptide Design Through Multi-Agent Systems

In recent years, peptide and protein designs have undergone a significant transformation, driven by data-driven methods and ML, which have emerged as fundamental tools in the discovery of novel proteins with specific properties [150]. These strategies have greatly improved our ability to predict protein structure, properties, and functions directly from amino acid sequences. A landmark achievement in this area is AlphaFold, a DL model that revolutionized protein structure prediction and set a new standard in the field [151].

Despite these advances, several challenges and open questions remain – particularly regarding how to develop intelligent systems capable of addressing complex tasks by integrating diverse sources of knowledge and methodologies. One promising solution lies in the implementation of multi-agent systems powered by large language models (LLMs) [152]. LLMs have been pivotal in enabling conversational AI agents and have facilitated a shift from human–AI interaction to AI–AI and AI–tool collaboration, allowing greater autonomy in task execution [153].

At present, there is no single, universally accepted definition of an agent. However, some authors describe an agent as a system capable of perceiving its environment, making decisions, and acting in response to external stimuli [154]. Others define it as an autonomous system that continuously receives environmental input and performs actions to accomplish a specific task [155].

Autonomous agent models are an emerging topic in scientific research and have found applications across multiple domains [155]. Initially, their development focused on automating tasks such as literature reading and article analysis. Examples include PaperQA, an agent that retrieves articles from databases, analyzes their content, and answers queries about them [156]; and WikiCrow, an extension of PaperQA that additionally generates summaries of scientific publications [157].

In parallel, new agents have been developed to assist researchers in experimental design and data analysis within laboratory environments. These include Context-Aware Language Models for Science (CALMS) [158], BioPlanner [159], CRISPR-GPT [160], LLM-RDF [161], and Organa, which provides feedback to researchers and generates comprehensive reports based on experimental results [162].

From a chemical perspective, specialized models such as Chemist-X, ChemCrow, AtomAgents, and LIDDIA have been developed to serve as virtual assistants in various tasks, including organic synthesis, drug discovery, materials design, and reaction planning [163–166].

While the application of autonomous agents in peptide and protein design remains relatively underexplored, the development of ProtAgent represents a significant step forward. ProtAgent is a multi-agent system capable of handling a wide range of protein design tasks – from sequence retrieval and functional analysis to *de novo* protein generation. Its structured workflow is composed of multiple specialized agents: the Planner defines strategies tailored to each task; the Assistant executes those plans using various computational tools; the Critic monitors performance, provides feedback, and refines the outputs; and the Chat Manager coordinates communication and task synchronization across the system [153].

Looking ahead, the implementation of multi-agent systems such as ProtAgent has the potential to revolutionize protein design by minimizing the need for manual intervention and accelerating the discovery process. This approach could significantly facilitate the generation of peptides and proteins with customized properties for biomedical, biotechnological, and pharmaceutical applications.

7.8 Developing AI-Agent for Autonomous Therapeutic Peptide Design

The discovery of therapeutic peptides is a multifaceted challenge that requires the integration of biological, chemical, and computational knowledge. To streamline and accelerate this complex process, we propose a modular and adaptive agent-based methodology designed to manage the complete peptide design pipeline – from user-defined prompts to *in silico* validation and experimental refinement (see Figure 7.3). This system harnesses the power of transformer-based deep learning models for both sequence generation and predictive evaluation, while reinforcement learning serves as a mechanism to optimize the entire design process based on experimental feedback.

The workflow begins with a user inputting a structured or natural language prompt that defines the desired characteristics of the target peptide. For example, a prompt might specify antiviral activity against a protein target like GP41, a half-life range between three and five hours, and the absence of antimicrobial activity or known toxic effects such as immunogenicity or hemolysis. This prompt is processed by an interpretation agent, which translates it into a formalized set of computational objectives and constraints. This step ensures that all downstream processes – generation, evaluation, and validation – are driven by clear, quantifiable design goals.

Next, a generative agent invokes a transformer-based model pretrained on large-scale peptide or protein datasets to produce novel sequences. Depending on the available domain-specific data, this model may be fine-tuned or conditioned through the input prompt to increase specificity. The generated peptides are designed to exhibit high internal diversity, minimizing redundancy and increasing the likelihood of identifying functionally novel candidates.

The candidate peptides are then assessed by a suite of predictive agents, which oversee two key types of models: classifiers and regressors. The classification agents evaluate discrete properties such as antiviral, antimicrobial, and antifungal activities, as well as safety-related attributes like hemolysis, allergenicity, and immunogenicity. These models are primarily transformer-based, benefiting from transfer learning and fine-tuning on high-quality, curated datasets. In parallel, regression agents use deep learning models to predict quantitative pharmacokinetic parameters such as half-life, IC_{50}, and pIC_{50}. Together, these predictive agents provide a comprehensive assessment of each peptide, with results synthesized into a ranked list based on overall fit to the initial design criteria.

Peptides that meet the evaluation thresholds are passed to a validation agent, which performs *in silico* analyses using a range of bioinformatic tools. This includes molecular docking to estimate binding affinities, structural modeling

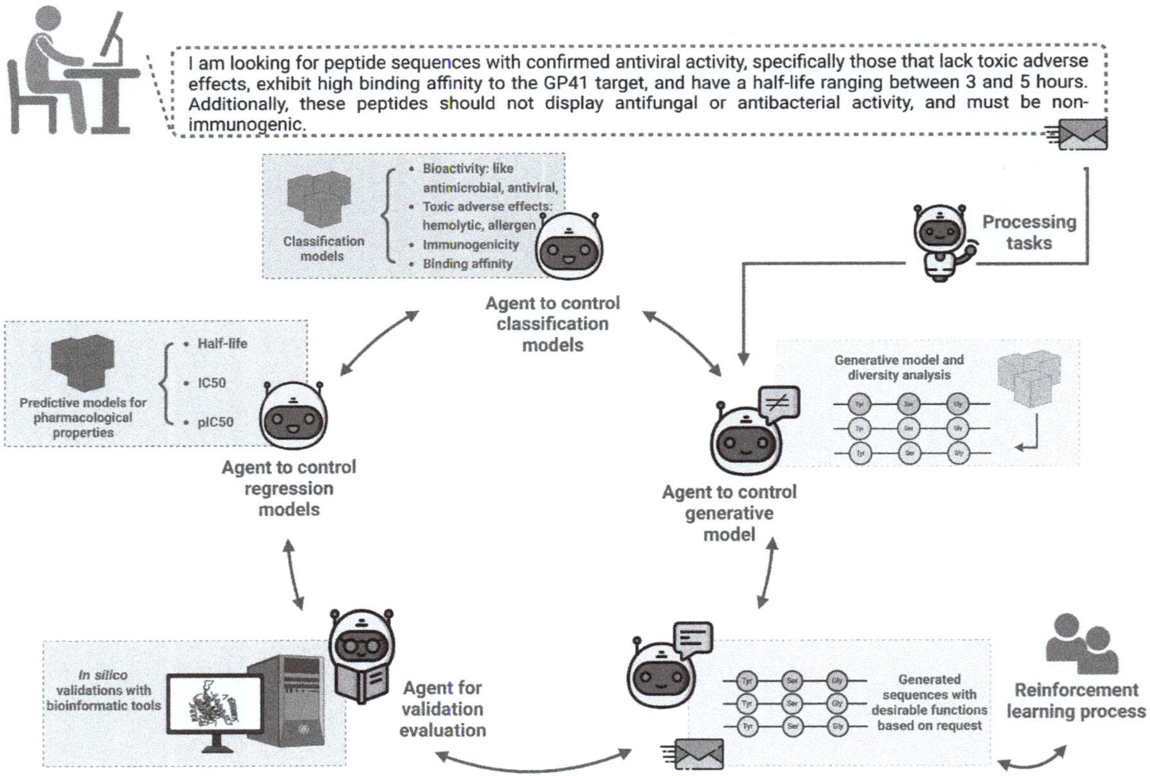

Figure 7.3 Agent-based framework for therapeutic peptide design and discovery. The system receives a user-defined prompt specifying desired biological and pharmacological properties. An agent-controlled generative model produces diverse peptide sequences, which are evaluated by classification models for activity (e.g. antiviral), safety (e.g. nonhemolytic, non-immunogenic), and specificity. Additional agents oversee regression models predicting pharmacokinetic properties such as half-life and binding affinity. Candidate sequences undergo *in silico* validation using bioinformatic tools. Throughout the process, agent coordination enables task distribution and information flow. A reinforcement learning mechanism integrates experimental feedback to iteratively improve the system's performance and guide future generations toward optimal candidates.

to assess conformational stability, and physicochemical profiling to evaluate solubility, charge, and other essential characteristics. The integration of these validations helps ensure that only biologically viable and computationally promising candidates are selected for experimental testing.

To continuously improve the quality of generated peptides, the system incorporates a reinforcement learning agent that implements a feedback loop. Experimentally tested peptides serve as the foundation for reward signals that are fed back into the generative and predictive models. Successful candidates reinforce generation and prediction strategies, while poor outcomes penalize ineffective patterns. Importantly, this adaptive learning process not only updates the generative model but also refines the classification and regression models, ensuring that the entire predictive framework evolves in response to empirical data and converges toward increasingly accurate and functional outputs.

The entire system operates through a modular, agent-based architecture, where each agent is responsible for a specific task – ranging from prompt interpretation, sequence generation, classification, regression, and structural validation to learning from experimental feedback. These agents interact within a shared environment using standardized communication protocols, enabling seamless data flow and coordinated decision-making.

One of the core strengths of this approach is its modularity. Individual components can be independently improved or replaced as new models, algorithms, or datasets become available, without disrupting the integrity of the pipeline. This flexibility allows the system to adapt to novel therapeutic targets, new evaluation metrics, or evolving biological constraints. The incorporation of reinforcement learning adds another layer of adaptability, allowing the platform to refine its performance over time based on real-world outcomes.

Despite its many strengths, the methodology is not without limitations. Running multiple transformer-based models and conducting large-scale simulations in parallel demands significant computational resources. Moreover, coordinating heterogeneous models – each potentially trained on different input features and objectives – requires careful integration to maintain consistency and interpretability. Additionally, the reinforcement learning component is dependent on high-quality experimental data, which can be costly and time-intensive to obtain. Finally, building and maintaining a multi-agent architecture introduces initial complexity, requiring robust orchestration logic and well-defined interfaces to ensure effective collaboration among agents.

7.9 Conclusion and Perspectives

The study and application of therapeutic peptides have rapidly evolved, supported by their biochemical versatility, high specificity, and broad therapeutic relevance. From their foundational roles in natural biological systems to their tailored synthesis for biomedical use, peptides have demonstrated exceptional utility in domains ranging from diagnostics and biosensing to antimicrobial and anticancer therapies. Nevertheless, their inherent limitations – such as low stability, enzymatic degradation, and delivery challenges – continue to pose significant hurdles for clinical translation. To address these limitations, recent research has turned toward computational approaches, including the use of artificial intelligence and ML, which offer promising avenues to rationally design, optimize, and predict peptide functionality with greater efficiency and precision.

The integration of ML techniques has marked a paradigm shift in peptide science. By leveraging curated datasets, numerical representations, and advanced learning algorithms, researchers can now build highly accurate classification and regression models that predict biological activity, toxicity profiles, and pharmacokinetic properties. Pretrained protein language models and deep learning architectures have proven particularly effective in capturing the subtle sequence–function relationships that govern peptide behavior. Furthermore, the use of generative learning has enabled the exploration of entirely new regions of peptide chemical space, allowing the *de novo* design of sequences with desirable and custom-tailored properties.

Looking toward the future, the implementation of intelligent agent-based systems is expected to further transform the field by enabling autonomous peptide discovery pipelines. These multi-agent systems, composed of specialized modules that oversee generation, classification, validation, and feedback integration, allow for flexible and scalable design workflows. By coordinating predictive and generative models through shared communication protocols and reinforcement learning, these platforms minimize the need for manual intervention while continuously learning from experimental feedback. Such frameworks are particularly suited for addressing the complexity of therapeutic peptide development, as they can rapidly respond to new design constraints and integrate heterogeneous sources of biological knowledge.

Therapeutic peptide research stands at the intersection of molecular biology, computational science, and artificial intelligence. The integration of ML-driven methodologies, supported by accessible databases and experimental feedback loops, is not only accelerating discovery but also expanding the functional and therapeutic landscape of peptides. As this chapter has outlined, the adoption of data-driven and autonomous systems holds great promise for next-generation peptide engineering. These approaches are likely to play a central role in the development of personalized and precision-based therapeutics, ultimately transforming how peptide-based drugs are discovered, optimized, and brought to the clinic.

Acknowledgments

DM-O and NS-G acknowledge ANID for the project "SUBVENCIÓN A INSTALACIÓN EN LA ACADEMIA CONVOCATORIA AÑO 2022," Folio 85220004. DM-O acknowledges support from the Centre for Biotechnology and Bioengineering – CeBiB (PIA project FB0001 and AFB240001, ANID, Chile). MDD acknowledges funding by the Deutsche Forschungsgemeinschaft (DFG, German Research Foundation) – within the Priority Program Molecular Machine Learning SPP2363 (Project Number 497207454). MDD acknowledges EU COST Action CA21160 (ML4NGP).

References

1 D. Medina-Ortiz, Contreras, S., Fernández, D. et al. (2024). Protein language models and machine learning facilitate the identification of antimicrobial peptides, *International Journal of Molecular Sciences*, 25 (16): 8851.

2 Trier, N., Hansen, P., and Houen, G. (2019). Peptides, antibodies, peptide antibodies and more. *International journal of molecular sciences* 20 (24): 6289.

3 Bozovičar, K. and Bratkovič, T. (2021). Small and simple, yet sturdy: conformationally constrained peptides with remarkable properties. *International journal of molecular sciences* 22 (4): 1611.

4 Cabas-Mora, G., Daza, A., Soto-García, N. et al. (2024). Peptipedia v2. 0: a peptide sequence database and user-friendly web platform. A major update. *Database* 2024: baae113.

5 Goles, M., Daza, A., Cabas-Mora, G. et al. (2024). Peptide-based drug discovery through artificial intelligence: towards an autonomous design of therapeutic peptides. *Briefings in Bioinformatics* 25 (4): bbae275.

6 Lorenzo, J.M., Munekata, P.E., Gomez, B. et al. (2018). Bioactive peptides as natural antioxidants in food products–a review. *Trends in food science & technology* 79: 136–147.

7 Morales-Camacho, J.I., Espinosa-Hernández, E., Rosas-Cárdenas, F.F. et al. (2019). Insertions of antihypertensive peptides and their applications in pharmacy and functional foods. *Applied microbiology and biotechnology* 103: 2493–2505.

8 Wang, M., Li, L., Zhang, L. et al. (2021). Peptide-derived biosensors and their applications in tumor immunology-related detection. *Analytical Chemistry* 94 (1): 431–441.

9 Zhang, Y.-M., Ye, D.X., Liu, Y. et al. (2023). Peptides, new tools for plant protection in eco-agriculture. *Advanced Agrochem* 2 (1): 58–78.

10 Keymanesh, K., Soltani, S., and Sardari, S. (2009). Application of antimicrobial peptides in agriculture and food industry. *World Journal of Microbiology and Biotechnology* 25: 933–944.

11 Chu, W., Prodromou, R., Day, K.N. et al. (2021). Peptides and pseudopeptide ligands: a powerful toolbox for the affinity purification of current and next-generation biotherapeutics. *Journal of Chromatography A* 1635: 461632.

12 Wang, L., Wang, N., Zhang, W. et al. (2022). Therapeutic peptides: current applications and future directions. *Signal transduction and targeted therapy* 7 (1): 48.

13 Bin Hafeez, A., Jiang, X., Bergen, P.J., and Zhu, Y. (2021). Antimicrobial peptides: an update on classifications and databases. *International journal of molecular sciences* 22 (21): 11691.

14 Rizvi, S.F.A. and Zhang, H. (2021). Emerging trends of receptor-mediated tumor targeting peptides: a review with perspective from molecular imaging modalities. *European Journal of Medicinal Chemistry* 221: 113538.

15 Fosgerau, K. and Hoffmann, T. (2015). Peptide therapeutics: current status and future directions. *Drug discovery today* 20 (1): 122–128.

16 Pereira, A.J., de Campos, L.J., Xing, H., and Conda-Sheridan, M. (2024). Peptide-based therapeutics: challenges and solutions. *Medicinal Chemistry Research* 33, 1275 (8): -1280.

17 Kim, H., Taslakjian, B., Kim, S. et al. (2024). Therapeutic peptides, proteins and their nanostructures for drug delivery and precision medicine. *ChemBioChem* 25 (8): e202300831.

18 Wu, J., Sahoo, J.K., Li, Y. et al. (2022). Challenges in delivering therapeutic peptides and proteins: a silk-based solution. *Journal of Controlled Release* 345: 176–189.

19 Bajracharya, R., Song, J.G., Back, S.Y., and Han, H.-K. (2019). Recent advancements in non-invasive formulations for protein drug delivery. *Computational and structural biotechnology journal* 17: 1290–1308.

20 Tan, X., Liu, Q., Fang, Y. et al. (2024). Predicting peptide permeability across diverse barriers: a systematic investigation. *Molecular Pharmaceutics* 21 (8): 4116–4127.

21 Fetse, J., Kandel, S., Mamani, U.-F., and Cheng, K. (2023). Recent advances in the development of therapeutic peptides. *Trends in Pharmacological Sciences* 44 (7): 425–441.

22 Pintado-Grima, C. and Ventura, S. (2025). The role of amphipathic and cationic helical peptides in Parkinson's disease. *Protein Science* 34 (1): e70020.

23 Hua, Y., Dai, X., Xu, Y. et al. (2022). Drug repositioning: Progress and challenges in drug discovery for various diseases. *European Journal of Medicinal Chemistry* 234: 114239.

24 Hashemi, S., Vosough, P., Taghizadeh, S., and Savardashtaki, A. (2024). Therapeutic peptide development revolutionized: harnessing the power of artificial intelligence for drug discovery. *Heliyon* 10 (22): e40265.

25 Zhai, S., Liu, T., Lin, S. et al. (2025). Artificial intelligence in peptide-based drug design. *Drug Discovery Today* 30: 104300.

26 Gupta, R., Srivastava, D., Sahu, M. et al. (2021). Artificial intelligence to deep learning: machine intelligence approach for drug discovery. *Molecular diversity* 25: 1315–1360.

27 W. Porto, A. Pires, O. Franco (2017). Computational tools for exploring sequence databases as a resource for antimicrobial peptides, *Biotechnology advances* 35 (3): 337–349.

28 Fernández, D., Olivera-Nappa, Á., Uribe-Paredes, R., Medina-Ortiz, D. (2023). Exploring machine learning algorithms and protein language models strategies to develop enzyme classification systems, International Work-Conference on Bioinformatics and Biomedical Engineering, Springer, pp. 307–319.

29 Craig, M., Caicedo, J., Das, P. et al. (2022). AI and drug discovery. *Cell Reports Physical Science* 3 (11): 101142.

30 Ulmschneider, J.P. and Ulmschneider, M.B. (2018). Molecular dynamics simulations are redefining our view of peptides interacting with biological membranes. *Accounts of chemical research* 51 (5): 1106–1116.

31 Wang, X., Song, K., Li, L., and Chen, L. (2018). Structure-based drug design strategies and challenges. *Current Topics in Medicinal Chemistry* 18 (12): 998–1006.

32 Angelo, J.S., Guedes, I.A., Barbosa, H.J., and Dardenne, L.E. (2023). Multi-and many-objective optimization: present and future in de novo drug design. *Frontiers in Chemistry* 11: 1288626.

33 Chang, L., Mondal, A., and Perez, A. (2022). Towards rational computational peptide design. *Frontiers in Bioinformatics* 2: 1046493.

34 Goshisht, M.K. (2024). Machine learning and deep learning in synthetic biology: key architectures, applications, and challenges. *ACS omega* 9 (9): 9921–9945.

35 Dara, S., Dhamercherla, S., Jadav, S.S. et al. (2022). Machine learning in drug discovery: a review. *Artificial intelligence review* 55 (3): 1947–1999.

36 Berman, H.M., Westbrook, J., Feng, Z. et al. (2000). The protein data bank. *Nucleic acids research* 28 (1): 235–242.

37 Consortium, T.U. (2025). UniProt: the universal protein knowledgebase in 2025. *Nucleic Acids Research* 53 (D1): D609–D617.

38 Ye, G., Wu, H., Huang, J. et al. (2020). LAMP2: a major update of the database linking antimicrobial peptides. *Database* 2020: baaa061.

39 Jhong, J.-H., Yao, L., Pang, Y. et al. (2022). dbAMP 2.0: updated resource for antimicrobial peptides with an enhanced scanning method for genomic and proteomic data. *Nucleic acids research* 50 (D1): D460–D470.

40 Shtatland, T., Guettler, D., Kossodo, M. et al. (2007). PepBank-a database of peptides based on sequence text mining and public peptide data sources. *BMC bioinformatics* 8: 1–10.

41 Pirtskhalava, M., Amstrong, A.A., Grigolava, M. et al. (2021). DBAASP v3: database of antimicrobial/cytotoxic activity and structure of peptides as a resource for development of new therapeutics. *Nucleic acids research* 49 (D1): D288–D297.

42 Waghu, F.H., Barai, R.S., Gurung, P., and Idicula-Thomas, S. (2016). CAMPR3: a database on sequences, structures and signatures of antimicrobial peptides. *Nucleic acids research* 44 (D1): D1094–D1097.

43 Shi, G., Kang, X., Dong, F. et al. (2022). DRAMP 3.0: an enhanced comprehensive data repository of antimicrobial peptides. *Nucleic acids research* 50 (D1): D488–D496.

44 Tyagi, A., Tuknait, A., Anand, P. et al. (2015). CancerPPD: a database of anticancer peptides and proteins. *Nucleic acids research* 43 (D1): D837–D843.

45 Qureshi, A., Thakur, N., Tandon, H., and Kumar, M. (2014). AVPdb: a database of experimentally validated antiviral peptides targeting medically important viruses. *Nucleic acids research* 42 (D1): D1147–D1153.

46 Gautam, A., Chaudhary, K., Singh, S. et al. (2014). Hemolytik: a database of experimentally determined hemolytic and non-hemolytic peptides. *Nucleic acids research* 42 (D1): D444–D449.

47 Wang, J., Yin, T., Xiao, X. et al. (2018). StraPep: a structure database of bioactive peptides. *Database* 2018: bay038.

48 Roy, S. and Teron, R. (2019). BioDADPep: a bioinformatics database for anti diabetic peptides. *Bioinformation* 15 (11): 780.

49 Mathur, D., Prakash, S., Anand, P. et al. (2016). PEPlife: a repository of the half-life of peptides. *Scientific reports* 6 (1): 36617.

50 Usmani, S.S., Bedi, G., Samuel, J.S. et al. (2017). THPdb: database of FDA-approved peptide and protein therapeutics. *PloS one* 12 (7): e0181748.

51 E. A. Gómez, P. Giraldo, y S. Orduz (2017). InverPep: a database of invertebrate antimicrobial peptides, *Journal of global antimicrobial resistance*, 8: 13–17.

52 Kumar, V., Patiyal, S., Kumar, R. et al. (2021). B3Pdb: an archive of blood–brain barrier-penetrating peptides. *Brain Structure and Function* 226: 2489–2495.

53 Wynendaele, E., Bronselaer, A., Nielandt, J. et al. (2013). Quorumpeps database: chemical space, microbial origin and functionality of quorum sensing peptides. *Nucleic acids research* 41 (D1): D655–D659.

54 Mathur, D., Singh, S., Mehta, A. et al. (2018). In silico approaches for predicting the half-life of natural and modified peptides in blood. *PloS one* 13 (6): e0196829.

55 Ettayapuram Ramaprasad, A. S., Singh, S., Gajendra, P.S.R., Venkatesan, S. (2015). AntiAngioPred: a server for prediction of anti-angiogenic peptides, *PloS one*, 10 (9): e0136990.

56 D'Aloisio, V., Dognini, P., Hutcheon, G.A., and Coxon, C.R. (2021). PepTherDia: database and structural composition analysis of approved peptide therapeutics and diagnostics. *Drug Discovery Today* 26 (6): 1409–1419.

57 Ramazi, S., Mohammadi, N., Allahverdi, A. et al. (2022). A review on antimicrobial peptides databases and the computational tools. *Database* 2022: baac011.

58 Dong, F., Zhao, G., Tong, H. et al. (2021). The prospect of bioactive peptide research: a review on databases and tools. *Current Bioinformatics* 16 (4): 494–504.

59 Hassan, M., Kjos, M., Nes, I. et al. (2012). Natural antimicrobial peptides from bacteria: characteristics and potential applications to fight against antibiotic resistance. *Journal of applied microbiology* 113 (4): 723–736.

60 Jenssen, H. and Aspmo, S.I. (2008). *Serum Stability of Peptides, en Peptide-Based Drug Design*, 177–186. Springer.

61 Zhang, Q., Chen, X., Li, B. et al. (2022). A database of anti-coronavirus peptides. *Scientific data* 9 (1): 294.

62 Park, K. (2019). A review of computational drug repurposing. *Translational and clinical pharmacology* 27 (2): 59–63.

63 Asim, M.N., Asif, T., Mehmood, F., and Dengel, A. (2025). Peptide classification landscape: an in-depth systematic literature review on peptide types, databases, datasets, predictors architectures and performance. *Computers in Biology and Medicine* 188: 109821.

64 Lv, H., Yan, K., and Liu, B. (2023). TPpred-LE: therapeutic peptide function prediction based on label embedding. *BMC biology* 21 (1): 238.

65 Wang, R., Wang, T., Zhuo, L. et al. (2024). Diff-AMP: tailored designed antimicrobial peptide framework with all-in-one generation, identification, prediction and optimization. *Briefings in Bioinformatics* 25 (2): bbae078.

66 Yao, L. et al. (2024). AMPActiPred: a three-stage framework for predicting antibacterial peptides and activity levels with deep forest. *Protein Science* 33 (6): e5006.

67 Bajiya, N., Choudhury, S., Dhall, A., and Raghava, G.P. (2024). AntiBP3: a method for predicting antibacterial peptides against gram-positive/negative/variable bacteria. *Antibiotics* 13 (2): 168.

68 Sharma, R., Shrivastava, S., Kumar Singh, S. et al. (2021). Deep-ABPpred: identifying antibacterial peptides in protein sequences using bidirectional LSTM with word2vec. *Briefings in Bioinformatics* 22 (5): bbab065.

69 Thakur, N., Qureshi, A., and Kumar, M. (2012). AVPpred: collection and prediction of highly effective antiviral peptides. *Nucleic acids research* 40 (W1): W199–W204.

70 Zare, M., Mohabatkar, H., Faramarzi, F.K. et al. (2015). Using Chou's pseudo amino acid composition and machine learning method to predict the antiviral peptides. *The open bioinformatics journal* 9 (1).

71 Schaduangrat, N., Nantasenamat, C., Prachayasittikul, V., and Shoombuatong, W. (2019). Meta-iAVP: a sequence-based meta-predictor for improving the prediction of antiviral peptides using effective feature representation. *International journal of molecular sciences* 20 (22): 5743.

72 Wei, L., Zhou, C., Su, R., and Zou, Q. (2019). PEPred-suite: improved and robust prediction of therapeutic peptides using adaptive feature representation learning. *Bioinformatics* 35 (21): 4272–4280.

73 Li, J., Pu, Y., Tang, J. et al. (2020). DeepAVP: a dual-channel deep neural network for identifying variable-length antiviral peptides. *IEEE Journal of Biomedical and Health Informatics* 24 (10): 3012–3019.

74 Lin, T.-T., Sun, Y.Y., Wang, C.T. et al. (2022). AI4AVP: an antiviral peptides predictor in deep learning approach with generative adversarial network data augmentation. *Bioinformatics Advances* 2 (1): vbac080.

75 Cao, R., Hu, W., Wei, P. et al. (2023). FFMAVP: a new classifier based on feature fusion and multitask learning for identifying antiviral peptides and their subclasses. *Briefings in Bioinformatics* 24 (6): bbad353.

76 H. Kurata, S. Tsukiyama, y B. Manavalan (2022). iACVP: markedly enhanced identification of anti-coronavirus peptides using a dataset-specific word2vec model, *Briefings in Bioinformatics*, 23 (4): bbac265.

77 Muthuvel, E.E.S.K. (2022). Characterization and prediction of dengue virus targeting peptides based on three class of descriptors using k-NN and random Forest algorithm. *Research Square [Preprint]* [En línea]. Disponible en: http://doi.org/10.21203/rs.3.rs-1652354/v1.

78 Saraswat, A., Sharma, U., Gandotra, A. et al. (2024). Pred-AHCP: robust feature selection enabled sequence specific prediction of anti-hepatitis C peptides via machine learning. *Journal of Chemical Information and Modeling* 64: 9111–9124.

79 Poorinmohammad, N. and Mohabatkar, H. (2015). A comparison of different machine learning algorithms for the prediction of anti-HIV-1 peptides based on their sequence-related properties. *International Journal of Peptide Research and Therapeutics* 21: 57–62.

80 Huang, K.-Y., Kao, H.-J., Weng, T.-H. et al. (2022). iDVIP: identification and characterization of viral integrase inhibitory peptides. *Briefings in Bioinformatics* 23 (6): bbac406.

81 Vishnepolsky, B., Grigolava, M., Gabrielian, A. et al. (2023). Analysis, Modeling, and target-specific predictions of linear peptides inhibiting virus entry. *ACS omega* 8 (48): 46218–46226.

82 Agrawal, P., Bhalla, S., Chaudhary, K. et al. (2018). In silico approach for prediction of antifungal peptides. *Frontiers in microbiology* 9: 318353.

83 Ahmad, A., Akbar, S., Khan, S. et al. (2021). Deep-AntiFP: prediction of antifungal peptides using distanct multi-informative features incorporating with deep neural networks. *Chemometrics and Intelligent Laboratory Systems* 208: 104214.

84 Sharma, R., Shrivastava, S., Kumar Singh, S. et al. (2022). Deep-AFPpred: identifying novel antifungal peptides using pretrained embeddings from seq2vec with 1DCNN-BiLSTM. *Briefings in Bioinformatics* 23 (1): bbab422.

85 Yao, L., Zhang, Y., Li, W. et al. (2023). DeepAFP: an effective computational framework for identifying antifungal peptides based on deep learning. *Protein Science* 32 (10): e4758.

86 S. Akbar, Q. Zou, A. Raza, F. K. Alarfaj, iAFPs-mv-BiTCN: predicting antifungal peptides using self-attention transformer embedding and transform evolutionary based multi-view features with bidirectional temporal convolutional networks, *Artificial Intelligence in Medicine*, vol. 151, p. 102860, 2024.

87 Sultan, M.F., Shaon, M.S.H., Karim, T. et al. (2024). MLAFP-XN: leveraging neural network model for development of antifungal peptide identification tool. *Heliyon* 10 (18).

88 Jiang, M., Zhang, R., Xia, Y. et al. (2022). i2APP: a two-step machine learning framework for antiparasitic peptides identification. *Frontiers in Genetics* 13: 884589.

89 Zhang, W., Xia, E., Dai, R. et al. (2022). PredAPP: predicting anti-parasitic peptides with undersampling and ensemble approaches, Interdiscip Sci Comput. *Life Sciences* 14: 258–268.

90 Liu, M., Wu, T., Li, X. et al. (2024). ACPPfel: explainable deep ensemble learning for anticancer peptides prediction based on feature optimization. *Frontiers in Genetics* 15: 1352504.

91 Akbar, S., Hayat, M., Tahir, M. et al. (2022). cACP-DeepGram: classification of anticancer peptides via deep neural network and skip-gram-based word embedding model. *Artificial intelligence in medicine* 131: 102349.

92 H. W. Park, T. Pitti, T. Madhavan, Y.-J. Jeon, B. Manavalan, y others, MLACP 2.0: An updated machine learning tool for anticancer peptide prediction, Computational and Structural Biotechnology Journal, vol. 20, pp. 4473–4480, 2022.

93 W. Shoombuatong, N. Schaduangrat, R. Pratiwi, y C. Nantasenamat, THPep: a machine learning-based approach for predicting tumor homing peptides, *Computational biology and chemistry*, vol. 80, pp. 441–451, 2019.

94 Gunaseelan, S.K., Khandelwal, Y., Dutta, A. et al. (2024). Machine learning-aided screening framework for wound healing peptides. *Bulletin of Materials Science* 47 (4): 263.

95 Z. Zhang, Y. Chen, G. Chen, M. Wang, Y. Pan, E. Li, Prediction and analysis of anti-aging peptides using machine learning algorithms, *Researchsquare*, 2024, doi: https://doi.org/10.21203/rs.3.rs-4621278/v1.

96 Gupta, S., Kapoor, P., Chaudhary, K. et al. (2013). In silico approach for predicting toxicity of peptides and proteins. *PloS one* 8 (9): e73957.

97 Rathore, A.S., Choudhury, S., Arora, A. et al. (2024). ToxinPred 3.0: an improved method for predicting the toxicity of peptides. *Computers in Biology and Medicine* 179: 108926.

98 Pan, X., Zuallaert, J., Wang, X. et al. (2020). ToxDL: deep learning using primary structure and domain embeddings for assessing protein toxicity. *Bioinformatics* 36 (21): 5159–5168.

99 Wei, L., Ye, X., Sakurai, T. et al. (2022). ToxIBTL: prediction of peptide toxicity based on information bottleneck and transfer learning. *Bioinformatics* 38 (6): 1514–1524.

100 V. Morozov, C. H. Rodrigues, D. B. Ascher, CSM-toxin: a web-server for predicting protein toxicity, *Pharmaceutics*, vol. 15, 2, p. 431, 2023.

101 Yu, Q., Zhang, Z., Liu, G. et al. (2024). ToxGIN: an in silico prediction model for peptide toxicity via graph isomorphism networks integrating peptide sequence and structure information. *Briefings in Bioinformatics* 25 (6): bbae583.

102 Tung, C.-W. and Ho, S.-Y. (2007). POPI: predicting immunogenicity of MHC class I binding peptides by mining informative physicochemical properties. *Bioinformatics* 23 (8): 942–949.

103 Saethang, T., Hirose, O., Kimkong, I. et al. (2013). PAAQD: predicting immunogenicity of MHC class I binding peptides using amino acid pairwise contact potentials and quantum topological molecular similarity descriptors. *Journal of Immunological Methods* 387 (1–2): 293–302.

104 Wang, G., Wan, H., Jian, X. et al. (2020). INeo-Epp: a novel T-cell HLA class-I immunogenicity or Neoantigenic epitope prediction method based on sequence-related amino acid features. *BioMed research international* 2020 (1): 5798356.

105 Wu, J., Wang, W., Zhang, J. et al. (2019). DeepHLApan: a deep learning approach for neoantigen prediction considering both HLA-peptide binding and immunogenicity. *Frontiers in immunology* 10: 2559.

106 Li, G., Iyer, B., Prasath, V.S. et al. (2021). DeepImmuno: deep learning-empowered prediction and generation of immunogenic peptides for T-cell immunity. *Briefings in bioinformatics* 22 (6): bbab160.

107 Wang, G., Wu, T., Ning, W. et al. (2023). TLimmuno2: predicting MHC class II antigen immunogenicity through transfer learning. *Briefings in Bioinformatics* 24 (3): bbad116.

108 Chaudhary, K., Kumar, R., Singh, S. et al. (2016). A web server and mobile app for computing hemolytic potency of peptides. *Scientific reports* 6 (1): 22843.

109 Win, T.S., Malik, A.A., Prachayasittikul, V. et al. (2017). HemoPred: a web server for predicting the hemolytic activity of peptides. *Future medicinal chemistry* 9 (3): 275–291.

110 Yaseen, A., Gull, S., Akhtar, N. et al. (2021). HemoNet: predicting hemolytic activity of peptides with integrated feature learning. *Journal of Bioinformatics and Computational Biology* 19 (05): 2150021.

111 Zhuang, S., Tanner, J., Wu, Y. et al. (2024). Non-hemolytic peptide classification using a quantum support vector machine. *Quantum Information Processing* 23 (11): 379.

112 M. Salem, A. Keshavarzi Arshadi, J. S. Yuan, AMPDeep: hemolytic activity prediction of antimicrobial peptides using transfer learning, *BMC bioinformatics*, vol. 23, 1, p. 389, 2022.

113 Rathore, A.S., Kumar, N., Choudhury, S. et al. (2025). Prediction of hemolytic peptides and their hemolytic concentration. *Communications Biology* 8 (1): 176.

114 Li, R., Yu, J., Ye, D. et al. (2025). Conotoxins: classification, prediction, and future directions in bioinformatics. *Toxins* 17 (2): 78.

115 I. Dimitrov, I. Bangov, D. R. Flower, I. Doytchinova, AllerTOP v. 2—a server for in silico prediction of allergens, *Journal of molecular modeling*, vol. 20, pp. 1–6, 2014.

116 Nguyen, M.N., Krutz, N.L., Limviphuvadh, V. et al. (2022). AllerCatPro 2.0: a web server for predicting protein allergenicity potential. *Nucleic Acids Research* 50 (W1): W36–W43.

117 R. Tomer, S. Patiyal, A. Dhall, G. P. Raghava, Prediction of celiac disease associated epitopes and motifs in a protein, *Frontiers in immunology*, vol. 14, p. 1056101, 2023.

118 Bizzotto, E., Zampieri, G., Treu, L. et al. (2024). Classification of bioactive peptides: a systematic benchmark of models and encodings. *Computational and Structural Biotechnology Journal* 23: 2442–2452.

119 Gasteiger, E., Hoogland, C., Gattiker, A. et al. (2005). Protein identification and analysis tools on the ExPASy server. In: *The Proteomics Protocols Handbook* (ed. J.M. Walker), 571–607. Totowa, NJ: Humana Press doi: 10.1385/1-59259-890-0:571.

120 Glassman, P.M. (2024). Development of a predictive algorithm for the efficacy of half-life extension strategies. *International Journal of Pharmaceutics* 660: 124382.

121 Ramos-Llorens, M., Bello-Madruga, R., Valle, J. et al. (2024). PyAMPA: a high-throughput prediction and optimization tool for antimicrobial peptides. *MSystems* 9 (7): e01358-23.

122 A. Sharma, D. Singla, M. Rashid, G. P. S. Raghava, Designing of peptides with desired half-life in intestine-like environment, *BMC bioinformatics*, vol. 15, pp. 1–8, 2014.

123 Haomeng, H., Zhang, C., Zhenyu, X., Duan, H. (2024). PepMSND: Integrating Multi-level Feature Engineering and Comprehensive Databases to Enhance in vivo/in vitro Peptide Blood Stability Prediction, *bioRxiv,* pp. 2024–12,

124 Rustagi, V., Gupta, S.R., Bajaj, M. et al. (2023). PepAnalyzer: predicting peptide properties using its sequence. *Amino acids* 55 (10): 1371–1379.

125 Qureshi, A., Tandon, H., and Kumar, M. (2015). AVP-IC50Pred: multiple machine learning techniques-based prediction of peptide antiviral activity in terms of half maximal inhibitory concentration (IC50). *Peptide Science* 104 (6): 753–763.

126 Nath, A. (2021). Prediction for understanding the effectiveness of antiviral peptides. *Computational Biology and Chemistry* 95: 107588.

127 D. Singh, A. Mahadik, S. Surana, P. Arora, Proteochemometric method for pic50 prediction of flaviviridae, *BioMed Research International*, vol. 2022, 1, p. 7901791, 2022.

128 Witten, J. and Witten, Z. (2019). Deep learning regression model for antimicrobial peptide design. *BioRxiv* 692681.

129 Chen, J., Cheong, H.H., and Siu, S.W. (2021). xDeep-AcPEP: deep learning method for anticancer peptide activity prediction based on convolutional neural network and multitask learning. *Journal of chemical information and modeling* 61 (8): 3789–3803.

130 Dean, S.N., Alvarez, J.A.E., Zabetakis, D. et al. (2021). PepVAE: variational autoencoder framework for antimicrobial peptide generation and activity prediction. *Frontiers in microbiology* 12: 725727.

131 Das, P., Wadhawan, K., Chang, O. et al. (2018). Pepcvae: Semi-supervised targeted design of antimicrobial peptide sequences, *arXiv preprint arXiv:1810.07743,*

132 Chen, Q., Yang, C., Xie, Y. et al. (2022). GM-pep: a high efficiency strategy to de novo design functional peptide sequences. *Journal of Chemical Information and Modeling* 62 (10): 2617–2629.

133 P. Das, Sercu, T., Wadhawan, K. et al. (2021). Accelerated antimicrobial discovery via deep generative models and molecular dynamics simulations *Nature Biomedical Engineering*, vol. 5, 6, pp. 613–623. https://doi.org/10.1038/s41551-021-00689-x.

134 Szymczak, P., Możejko, M., Grzegorzek, T. et al. (2023). Discovering highly potent antimicrobial peptides with deep generative model HydrAMP. *Nature Communications* 14 (1): 1453.

135 Chen, S., Lin, T., Basu, R. et al. (2024). Design of target specific peptide inhibitors using generative deep learning and molecular dynamics simulations. *Nature Communications* 15 (1): 1611.

136 Fukunaga, I., Matsukiyo, Y., Kaitoh, K., and Yamanishi, Y. (2024). Automatic generation of functional peptides with desired bioactivity and membrane permeability using Bayesian optimization. *Molecular Informatics* 43 (4): e202300148.

137 Chen, S., Lin, T., Basu, R. et al. (2022). Design of peptide inhibitors targeting β-catenin using generative deep learning and molecular dynamics simulations, *Research Square.* https://doi.org/10.21203/rs.3.rs-2019700/v1.

138 Bell, D.R., Domeniconi, G., Yang, C.-C. et al. (2021). Dynamics-based peptide–mhc binding optimization by a convolutional variational autoencoder: a use-case model for Castelo. *Journal of Chemical Theory and Computation* 17 (12): 7962–7971.

139 Rossetto, A. Zhou, W. sep. (2020). GANDALF: Peptide Generation for Drug Design using Sequential and Structural Generative Adversarial Networks, en *Proceedings of the 11th ACM International Conference on Bioinformatics, Computational Biology and Health Informatics*, ACM, pp. 1–10. doi: 10.1145/3388440.3412487.

140 Dundas, J., Ouyang, Z., Tseng, J. et al. (2006). CASTp: computed atlas of surface topography of proteins with structural and topographical mapping of functionally annotated residues. *Nucleic acids research* 34 (suppl_2): W116–W118.

141 Hoos, A. (2016). Development of immuno-oncology drugs—from CTLA4 to PD1 to the next generations. *Nature reviews Drug discovery* 15 (4): 235–247.

142 Surana, S., Arora, P., Singh, D. (2022). PandoraGAN: Generating antiviral peptides using Generative Adversarial Network, *bioRxiv*, p. 2021.02.15.431193, ene. doi: 10.1101/2021.02.15.431193.

143 Van Oort, C.M., Ferrell, J.B., Remington, J.M. et al. (2021). AMPGAN v2: machine learning-guided design of antimicrobial peptides. *Journal of chemical information and modeling* 61 (5): 2198–2207.

144 Xie, X., Valiente, P.A., Kim, P.M. (2023). HelixGAN a deep-learning methodology for conditional de novo design of α-helix structures, *Bioinformatics*, 39 (1): btad036.

145 J. Mao, Guan, S., Chen, Y. et al. (2023). Application of a deep generative model produces novel and diverse functional peptides against microbial resistance, *Computational and Structural Biotechnology Journal*, 21: 463–471. https://doi.org/10.1016/j.csbj.2022.12.029.

146 Chen, T., Vure, P., Pulugurta, R., Chatterjee, P. (2024). AMP-diffusion: Integrating latent diffusion with protein language models for antimicrobial peptide generation, *bioRxiv*, pp. 2024–03.

147 Watson, J.L., Juergens, D., Bennett, N.R. et al. (2022). Broadly applicable and accurate protein design by integrating structure prediction networks and diffusion generative models, *BioRxiv*, pp. 2022–12.

148 Kong, X., Huang, W., and Liu, Y. (2024). Full-atom peptide design with geometric latent diffusion. *Advances in Neural Information Processing Systems* 37: 74808–74839.

149 Wang, Y., Liu, X., Huang, F. (2024). A Multi-Modal Contrastive Diffusion Model for Therapeutic Peptide Generation, en *Proceedings of the AAAI Conference on Artificial Intelligence*, pp. 3–11.

150 Eguchi, R.R., Choe, C.A., and Huang, P.-S. (2022). Ig-VAE: generative modeling of protein structure by direct 3D coordinate generation. *PLoS computational biology* 18 (6): e1010271.

151 Jumper, J., Evans, R., Pritzel, A. et al. (2021). Highly accurate protein structure prediction with AlphaFold. *Nature* 596 (7873): 583–589.

152 Lu, Y., Aleta, A., Du, C. et al. (2024). LLMs and generative agent-based models for complex systems research. *Physics of Life Reviews* 51: 283–293.

153 Ghafarollahi, A. and Buehler, M.J. (2024). ProtAgents: protein discovery via large language model multi-agent collaborations combining physics and machine learning. *Digital Discovery* 3 (7): 1389–1409.

154 Xi, Z., Chen, W., Guo, X. et al. (2025). The rise and potential of large language model based agents: a survey. *Science China Information Sciences* 68 (2): 121101.

155 Ramos, M.C., Collison, C.J., and White, A.D. (2025). A review of large language models and autonomous agents in chemistry. *Chemical Science* 16 (6): 2514–2572.

156 Láreceived, J., O'Donoghue, O., Shtedritski, A. et al. (2023). Paperqa: retrieval-augmented generative agent for scientific research, *arXiv preprint arXiv:2312.07559*,

157 Cox, S., Hammerling, M., Lála, J. et al. (2023). Wikicrow: Automating synthesis of human scientific knowledge. In *WikiCrow: Automating synthesis of human scientific knowledge.*

158 Prince, M.H., Chan, H., Vriza, A. et al. (2024). Opportunities for retrieval and tool augmented large language models in scientific facilities. *npj Computational Materials* 10 (1): 251.

159 O'Donoghue, O., Shtedritski, A., Ginger, J. et al. (2023). BioPlanner: automatic evaluation of LLMs on protocol planning in biology, *arXiv preprint arXiv:2310.10632.*

160 Qu, Y., Huang, K., Cousins, H. et al. (2024). CRISPR-GPT: an LLM agent for automated design of gene-editing experiments, *bioRxiv*, pp. 2024–04.

161 Ruan, Y., Lu, C., Xu, N. et al. (2024). An automatic end-to-end chemical synthesis development platform powered by large language models. *Nature communications* 15 (1): 10160.

162 Darvish, K., Skreta, M., Zhao, Y. et al. (2025). ORGANA: a robotic assistant for automated chemistry experimentation and characterization. *Matter* 8 (2): 101897.

163 Chen, K., Li, J., Wang, K. et al. (2023). Chemist-X: large language model-empowered agent for reaction condition recommendation in chemical synthesis, *arXiv preprint arXiv:2311.10776.*

164 Bran, A.M., Cox, S., Schilter, O. et al. (2024). Augmenting large language models with chemistry tools. *Nature Machine Intelligence* 6 (5): 525–535.

165 Ghafarollahi, A., Buehler, M. J. (2024). AtomAgents: alloy design and discovery through physics-aware multi-modal multi-agent artificial intelligence, *arXiv preprint arXiv:2407.10022.*

166 Averly, R., Baker, F. N., Ning, X., (2025). LIDDIA: language-based intelligent drug Discovery Agent, *arXiv preprint arXiv:2502.13959.*

8

Machine Learning Approaches for High-Throughput Microbial Identification/Culturing

Mohamed Mastouri[1] *and Yang Zhang*[1,2,3]

[1] *Harbin Institute of Technology (Shenzhen), School of Biomedical Engineering, Taoyuan Street, Taoyuan, Guangdong, 518055, China*
[2] *Tianjin University, School of Pharmaceutical Science and Technology, No. 92, Weijin Road, Nankai District, Tianjin 300072, China*
[3] *Shenzhen Loop Area Institute, 1068 Xueyuan Avenue, Shenzhen University Town, Shenzhen, Guangdong, China*

8.1 Introduction

Microbial identification and culturing are essential techniques in microbiology for understanding the diversity, structure, and function of microbial communities. Culturing involves growing microorganisms under controlled laboratory conditions, while identification aims to determine the taxonomic classification of these organisms [1, 2]. The importance of these methods lies in their ability to reveal previously unknown microorganisms and provide insights into their ecological roles. For instance, culturomics has enabled the cultivation of hundreds of new microorganisms associated with humans, offering new perspectives on host-bacteria relationships [1, 2]. Similarly, coupling traditional cultivation with advanced sequencing techniques has revealed that some cultured soil bacteria were rare or absent in culture-independent community analyses, demonstrating unexpected accessibility to the rare biosphere [2]. However, it's crucial to note that traditional culturing methods have limitations. Only a small fraction (0.1–1%) of microorganisms in natural ecosystems can be cultured under standard laboratory conditions [3, 4]. This has led to the development of culture-independent approaches like metagenomics and single-cell genomics, which allow for the analysis of uncultured microorganisms and provide a more comprehensive understanding of microbial diversity and ecology [5].

Traditional methods for microbial identification have long relied on culture-based and biochemical techniques. These approaches involve isolating microorganisms in culture media, followed by enzymatic reactions and morphological analyses [6]. While these methods remain the core basis and standard practice in most laboratories, they are known to be time-consuming, labor-intensive, and may not detect all microorganisms present in a sample. Interestingly, molecular techniques have revolutionized microbial identification, offering faster and more accurate results.

Machine Learning and Big Data-enabled Biotechnology, First Edition. Edited by Hal S. Alper.

Methods such as 16S ribosomal RNA gene sequencing based on polymerase chain reaction (PCR) have become increasingly popular for comprehensive bacterial studies. Other molecular approaches include fluorescent *in situ* hybridization (FISH), denaturing gradient gel electrophoresis (DGGE), and terminal-restriction fragment length polymorphism (T-RFLP) [7]. While traditional culture-based and biochemical methods remain important, especially for antibiotic sensitivity testing [8], molecular techniques have significantly enhanced our ability to identify and study microorganisms. These newer methods offer advantages such as higher sensitivity, specificity, and the ability to detect unculturable microorganisms. However, it's important to note that each method has its own strengths and limitations, and the choice of technique should be based on the specific research objectives and sample types [7].

Traditional approaches in various fields face significant challenges that limit their effectiveness and efficiency. In microbial diagnostics, conventional methods struggle with sample handling, culture difficulties, misidentification, and delays in determining susceptibility [9]. These limitations can lead to inaccurate or delayed diagnoses, potentially impacting patient care.

High-throughput (HTP) techniques have revolutionized microbiology by enabling rapid and comprehensive analysis of microbial genomes, transcriptomes, and communities. These techniques, particularly high-throughput sequencing (HTS), have significantly advanced our understanding of microbial diversity, function, and interactions in various environments, including food systems [10, 11]. HTS technologies allow researchers to generate vast amounts of genetic data at an unprecedented pace, facilitating genome-wide analysis and the elucidation of molecular mechanisms in biological systems [12]. Interestingly, HTP techniques have found applications beyond traditional microbiology. In food microbiology, HTS has been used to predict the prevalence of microorganisms in food samples, study microbial responses to food-associated conditions, and develop genome-assisted approaches for correlating genotype and phenotype. Moreover, the combination of single-cell genomics and metagenomics has improved the efficiency and accuracy of obtaining whole genome information from complex microbial communities, enabling researchers to link function to species and study host-virus interactions. HTP techniques, particularly HTS, have become indispensable tools in microbiology, offering unprecedented insights into microbial communities and their functions. These technologies have applications in food safety, optimization of food fermentations, understanding probiotic mechanisms, and investigating the effects of ingested microbes on the human gut microbiota [13]. As sequencing and computational methods continue to advance, the role of HTP techniques in microbiology is expected to expand further, providing deeper insights into microbial ecosystems and their interactions with the environment and human health.

ML and artificial intelligence (AI) are revolutionizing the field of microbiology, offering powerful tools for data analysis, prediction, and decision-making. These technologies are being applied to various aspects of microbiology, including clinical diagnostics, AMR detection, and laboratory automation [14]. In clinical microbiology, ML algorithms are being used to reduce diagnostic workload through laboratory automation, detect and identify clinically relevant microorganisms, and detect AMR [14]. These applications have the potential to significantly improve the efficiency and accuracy of microbiological analyses. For instance, AI-based systems can analyze digital data from medical imaging and diagnostic techniques in a comprehensive, rapid, and noninvasive manner, offering unprecedented insights into diseases [15]. However, it's important to note that despite the promising potential of ML and AI in microbiology, their integration into routine clinical laboratory practice is still limited.

This chapter explores the synergy of MI with various HTP platforms such as genomic and metagenomic sequencing, mass spectrometry (MS), flow cytometry (FCM), and imaging technologies. The integrated strategies enable precise microbial species identification, AMR prediction, and phenotypic characterization on an unprecedented scale. Moreover, ML models are being applied for automated strain selection, process optimization in bioreactors, and even synthetic biology applications like CRISPR-based genome editing. We also revisit the fundamental ideas of ML, compare different paradigms of learning (supervised, unsupervised, reinforcement), and briefly mention their application to microbial research. Issues like standardization of data, model interpretability, and computational overheads are critically analyzed. Finally, we also hint at directions ahead, such as the potential of multi-omics integration, eXplainable AI (XAI), and newer technologies like quantum computing and federated learning for advancing microbial research and diagnostics. This chapter will attempt to give an in-depth overview of how ML is revolutionizing HTP microbial research, providing both seminal context and current examples to inform future scientific inquiry.

8.2 High-Throughput (HTP) Techniques in Microbial Research

8.2.1 Definition and Scope of HTP Microbial Techniques

HTP microbial technology refers to cutting-edge strategies and platforms for efficient, rapid, and large-scale processing or analysis of microbial samples. HTP technology has applications in many areas, including genomics, biofuel generation, and environmental monitoring. In microbial genomics, HTP sequencing generates huge levels of high-quality sequence data, allowing for rapid accumulation of evidence and robust identification of microbial-related evidence in forensic science [16]. This technology has highly improved the rate and accuracy of microbial analysis and identification, crucial in solving bio-crimes and responding to bio-terrorism attacks. HTP technologies have also found application in microbial fuel cells, where microbial metabolism is utilized in biofuel production, energy, and bioremediation [17]. These systems interest various disciplines and offer scope for sustainable applications in various environments. In the production of recombinant proteins, HTP screening techniques have been developed for increased efficiency at initial process design. For instance, the BioLector micro-bioreactor platform has been used as an HTP cultivation system to screen *E. coli* expression clones with high transferability of results to large-scale bioreactor systems [18]. This approach largely obviates the need for experiments in fully controlled bioreactors, accelerating process development at lower costs. HTP technology has also been applied to produce membrane proteins, which have been considered in the past a formidable challenge. A parallel HTP approach to cloning and expression screening of membrane proteins has shown unexpectedly high success rates, with 71% of *E. coli* integral membrane proteins that are expressed at the levels needed for purification [19]. HTP approaches have revolutionized microbial research, delivering novel and powerful systems for rapid analysis of microorganisms and microbial interactions. These approaches have also been applied to all areas of microbiology, from food microbiology to environmental microbiology and clinical diagnostics [10, 20, 21]. HTP sequencing has been utilized for microbial communities and metagenome studies with insights into microbial diversity and functionality [20, 22]. Even as microbial research became overrun with HTP sequencing, there are other HTP approaches which have emerged as powerful tools. For instance, MALDI-TOF MS bio-typing has been promising in the rapid identification of microbes, particularly in medical settings [23]. Droplet microfluidics has also opened new avenues for HTP microbial culture and analysis with the ability to explore single-cell heterogeneity and microbial interactions [18, 24]. HTP methodologies have immensely facilitated us in exploring microbial systems and their applications. From sequencing-based approaches to novel cultivation methods, these approaches are transforming various fields of microbiology. As technology continues to evolve, we can expect new advances in HTP approaches, which may open the door to more comprehensive and efficient microbial research strategies (Figure 8.1) [22].

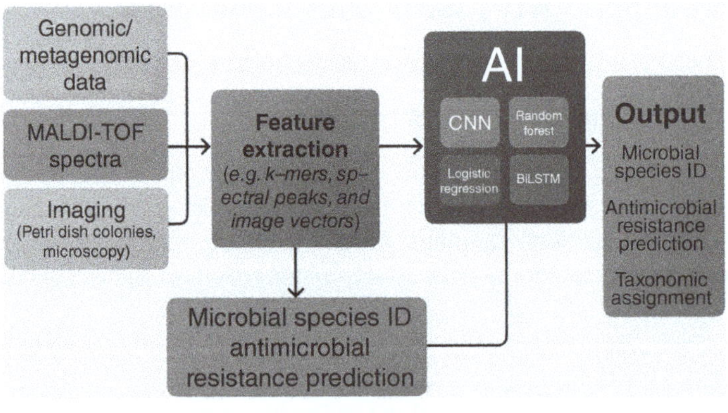

Figure 8.1 Workflow of AI-integrated high-throughput microbial identification and antimicrobial resistance prediction: This diagram illustrates the integration of AI models into high-throughput microbial identification workflows. Input data from genomic/metagenomic sequencing, MALDI-TOF mass spectrometry (MS), and microscopy-based imaging undergo feature extraction (e.g. k-mers, spectral peaks, and image vectors). Extracted features are analyzed using machine learning algorithms such as CNN, Random Forest (RF), Logistic Regression, and BiLSTM. The outputs include microbial species identification, antimicrobial resistance (AMR) prediction, and taxonomic assignment, demonstrating the enhanced speed and accuracy provided by AI approaches in microbial diagnostics.

8.2.2 Metagenomics and Next-Generation Sequencing (NGS)

Metagenomics, the analysis of DNA from microbial communities in environmental samples without prior culturing, has been revolutionized by advances in Next-Generation Sequencing (NGS) technologies [25]. NGS enables rapid analysis of microbial community composition and diversity in various habitats, providing HTP sequencing of millions of DNA fragments simultaneously [26, 27].

Interestingly, NGS has expanded beyond traditional genomics to include metagenomics, allowing for the detection of unexpectedly disease-associated viruses and the discovery of novel human viruses, including cancer-related viruses [28]. Moreover, NGS has facilitated the exploration of the human virome in both healthy and disease conditions through metagenomics-based strategies [28].

NGS has significantly impacted metagenomics research, offering comprehensive insights into microbial diversity, community structure, and functional activities [29]. This technology has been applied across various fields, including clinical microbiology, public health, food and agriculture, energy, and environment [30]. As NGS continues to evolve, it promises to further enhance our understanding of microbial-host interactions and provide valuable insights into the complex world of microorganisms [26, 27].

8.2.3 Mass Spectrometry-Based Proteomics (MALDI-TOF MS)

Matrix-assisted laser desorption/ionization time of flight mass spectrometry (MALDI-TOF MS) is a groundbreaking technology used in clinical microbiology that identifies microorganisms immediately and accurately. MALDI-TOF MS is more convenient, quicker, and more precise than conventional biochemical tests [31, 32]. This technology allows for the direct identification of bacterial species from newly grown colonies in minutes, a decrease of approximately one day in microbial identification time in typical workflows [32, 33].

While initially designed for bacterial identification, follow-up studies have witnessed the use of MALDI-TOF MS being expanded to cover the identification of yeasts, molds, and mycobacteria [34]. Besides, the technique is explored for various other leading-edge applications such as strain typing, antibiotic resistance detection, and antifungal susceptibility [34, 35]. Despite its multiple uses, one of the limitations of MALDI TOF MS is analyzing small molecular compounds. To overcome the limitation, newer nanomaterial-based matrices have been developed that have enhanced the sensitivity and accuracy of the technique to analyze these compounds [36].

With its ever-wider variety of applications, MALDI-TOF MS has emerged as an asset in clinical microbiology, which could be used to improve patient outcomes by facilitating the speed of diagnoses and reducing hospital stays [32, 37]. With its ongoing innovation in application, the technique is poised to continue developing microbial diagnostics even further and further.

8.2.4 Flow Cytometry

FCM is a high-sensitivity technology widely used for HTP sorting and single-cell analysis of cells, based on their physical and biochemical properties. It combines microfluidics, optics, and data collection to provide quantitative data about populations of cells [38]. This technology offers the provision for the simultaneous measurement of various parameters such as cell size, shape, complexity, and fluorescence to obtain accurate characterization and separation of well-defined subsets of cells [38].

High-end systems with enhanced capability have been a product of recent FCM advancements. For example, imaging flow cytometry (IFC) combines digital microscopy and traditional FCM to enable high-content cellular function and morphology analysis [39]. Moreover, breakthroughs in all-silica fiber microflow cytometers have enabled portable HTP devices with the same performance as commercial hardware [40]. Furthermore, label-free techniques, such as Raman FCM, are utilized to study the chemical composition of living cells without fluorescent tagging [39].

In short, FCM continues to advance, with advancements in microfluidic technologies being one of the driving factors behind this [41]. Such advancements have remarkably widened the scope of application of FCM in fields ranging from biomedical research to clinical diagnostics and drug development. With advancing technological progress, we can anticipate increasingly higher resolution, sensitivity, and throughput, hence converting FCM into a valuable resource for sorting and analyzing individual cells.

8.2.5 Microfluidics and Lab-on-a-Chip Systems

Microfluidics and Lab-on-a-Chip (LOC) devices have transformed healthcare diagnostics and analytical laboratory methods by miniaturizing, integrating, and automating complex processes on a single chip, thus enabling precise manipulation of extremely small fluid volumes typically in the order of 10^{-9} to 10^{-18} L via microchannels of 10 to 100 μm [42]. This development has immensely expanded the applications of LOC devices in many biomedical fields, such as integrated bioassays, cell analysis, metabolomics, drug discovery, and personalized medicine [43]. Such developments have the potential to revolutionize diagnostic activities through more efficient, portable, and cost-effective means compared to traditional laboratory techniques.

However, while microfluidic systems have developed significantly from an engineering point of view, the development of chromatographic separation systems has stalled in the last few years. Initially, the tremendous promises of microfluidics for chromatographic use created high enthusiasm, but the challenges of miniaturizing the systems effectively, particularly in terms of reproducibility, sensitivity, and interfacing to other analytical techniques, have kept them from becoming generally established [44].

One such potential development to break through these constraints is the integration of electrowetting-on-dielectric (EWOD) actuators with other microfluidic manipulation techniques. The integration has been observed to hold potential in elevating the capability of digital microfluidics (DMF) for sophisticated biomedical assays, enabling more precise regulation of fluid manipulation and the development of more complex and multifunctional LOC devices [45]. Despite such advances, the integration of various operational steps within a microfluidic system, fluidic handling, molecular recognition, and signal amplification remains a key limitation for achieving complete, functional, and robust point-of-care diagnostic devices [46].

While the promise of microfluidics and LOC systems to transform point-of-care diagnostics, environmental monitoring, and drug discovery remains enormous, the translation from prototype to complete and ruggedized systems requires additional advances in engineering, materials science, and biological integration. The mastery of the remaining technical challenges, particularly in system integration and reproducibility, will hold the answer to unlocking the full potential of such systems for biomedical research and healthcare diagnostics [46]. The technologies are highly promising, yet their more extensive adoption will depend on overcoming the complexities of integrating multiple components into streamlined, functional platforms.

8.2.6 High-Content Imaging and Phenotyping

High-content imaging and phenotyping have greatly enhanced plant and cell biology research by enabling large-scale, automated screening of complex biological systems. The techniques combine sophisticated microscopy with sophisticated image analysis algorithms to extract quantitative information on several cellular or organismal features automatically at the same time [47]. In plant biology, HTP phenotyping facilities that utilize various imaging modalities have been set up to nondestructively measure growth-related characters, stress response, and crop performance, which deliver information previously challenging to acquire with traditional methods [48]. Since phenotype screening experiments began earlier on with model Arabidopsis, it has gradually begun shifting toward establishing approaches for high-content imaging technologies for economically important crops such as sorghum to extend the platform for high-content imaging within agrarian research [48].

The integration of multiple imaging techniques like visible light, thermal imaging, chlorophyll fluorescence, and hyperspectral imaging has boosted the potential for completely phenotyping morphological and physiological traits and biochemical responses in horticultural plants [49]. The holistic approach enables researchers to realize plant performance, hence making crop improvement methods more accurate. At the same time, in the field of cell biology, high-content screening has become an extremely powerful tool to profile drug action, investigate gene function, and investigate cellular response to various perturbations at single-cell resolution [50]. One of the key contributors to success with high-content imaging is the development of advanced image analysis pipelines and MI approaches, which have played an essential role in extracting useful biological information from the large datasets generated by these technologies [51].

These computational tools allow researchers to quantify complex phenotypes, detect nuanced treatment effects, and uncover hidden biological mechanisms.

With advances in the field, the integration of high-content imaging with other omics technologies, as well as the development of standardized data repositories, holds enormous potential to further propel biological discovery and crop breeding. While these advances are full of promise, innovation in computational technologies and data integration will be needed to fully realize the potential of high-content imaging for plant biology at both the basic and applied levels.

8.3 Fundamentals of Machine Learning

8.3.1 Definition of Machine Learning and AI

ML is a branch of AI that enables computers to make conclusions and improve performance based on learning from experience, without explicit programming for specific tasks [52]. AI, however, describes the overall concept of machines performing tasks typically associated with human intelligence, such as reasoning, solving problems, and decision-making [52, 53].

While ML is widely regarded as a subfield of AI, some researchers differentiate ML as a narrow AI, focusing on its ability to learn through data and prediction, in contrast to AI, which encompasses various types of machine intelligence [53]. This differentiation emphasizes the specificity of ML in its application, where the goal is to design algorithms and statistical models that enable machines to improve task performance through experience, data patterns, and learning.

AI, as a general discipline, aims to create intelligent systems that can mimic the cognitive abilities of humans. On the other hand, ML specifically targets algorithm development to enhance machine performance through learning from data and is hence an integral component of AI success. Both AI and ML hold tremendous potential across a wide array of applications in medicine, optics, photonics, and industry by automating complex decision-making, function optimization, and enhanced prediction accuracy [52–55]. These technologies are transforming industries by enabling more efficient and accurate analysis, with the potential to deliver improvements in diagnostics, manufacturing, and other fields.

8.3.2 Supervised vs. Unsupervised vs. Reinforcement Learning

ML is a branch of AI which allows computers to make inferences and enhance performance through learning from data without direct programming for the purpose [52]. AI is the more extensive idea of machines executing tasks conventionally related to human intelligence like reasoning, problem-solving, and decision-making [52, 53].

While ML is commonly viewed as a subfield of AI, other scholars differentiate ML as a type of narrow AI, highlighting its specialization in data-driven learning and predictive function, in contrast to the general scope of AI, which encompasses other types of machine intelligence [53]. This distinction emphasizes the uniqueness of ML in its application, where the goal is to develop algorithms and statistical models that facilitate machines to improve task fulfilment through experience, data patterns, and inferences.

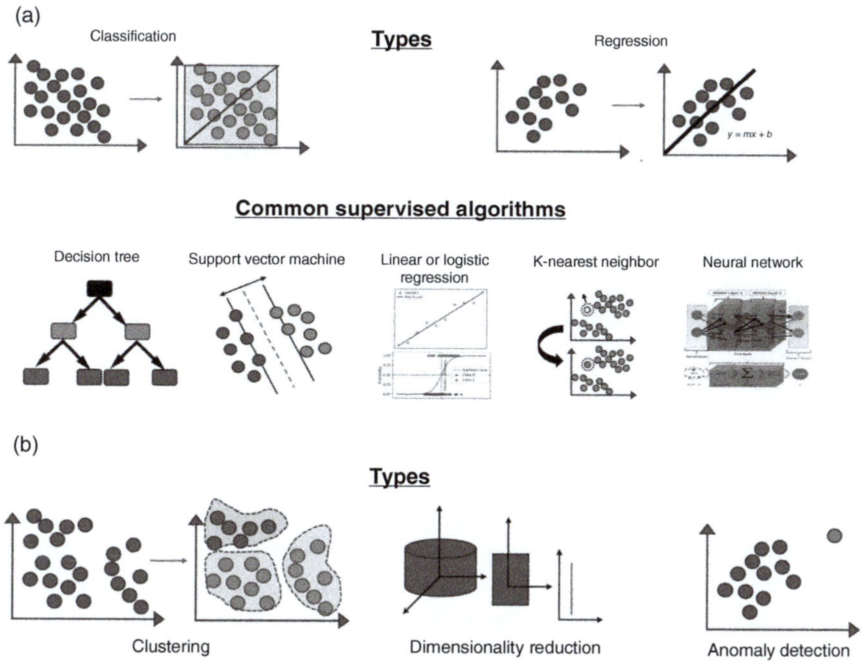

Figure 8.2 Visual overview of supervised and unsupervised machine learning approaches: (a) Supervised learning techniques use labeled data to perform classification and regression tasks. Common algorithms include decision trees, support vector machines (SVMs), linear/logistic regression, K-nearest neighbors (KNN), and neural networks. Classification involves learning decision boundaries between categories, while regression fits continuous trends to data. (b) Unsupervised learning methods work with unlabeled data to uncover hidden patterns. Major tasks include clustering, dimensionality reduction, and anomaly detection. These approaches are used to identify natural groupings, reduce data complexity, or find rare, outlier data points. [62] / with permission of Elsevier.

AI, in the widest sense, focuses on creating smart systems that can mimic human intellectual functions. On the other hand, supervised, unsupervised, and RL are three different ML methodologies based on their strategy and area of application.

Supervised learning (Figure 8.2a) is the training of a model on labeled data, for which input features and corresponding output labels are provided [56]. The algorithm is trained to map input values to target outputs, most typically for classification or regression issues. For example, in classification, the model is trained to make an estimate of a function that maps input vectors to previously given classes based on labeled examples [56]. Supervised learning works best when there is an abundance of high-quality labeled data available, where the algorithm can adequately predict or classify. However, the significant limitation here is the acquisition cost and time of labeled data, which might be prohibitively costly in most actual applications.

Unsupervised learning (Figure 8.2b), conversely, processes unlabeled data with the aim of identifying inherent patterns or organization without target output labels [56]. This approach is particularly handy when having labeled data is difficult or scarce because it allows discovering the patterns that are implicit in the data. Perhaps the greatest difficulty with unsupervised learning is discovering and analyzing these underlying structures within the data [57]. Unsupervised learning tends to utilize methods like clustering and dimensionality reduction. For example, unsupervised feature learning can be used to learn discriminative features from high-dimensional data, for example, in hyperspectral image analysis, where the goal is to reduce data complexity without losing useful information [58].

Though supervised and unsupervised learning are the most developed paradigms, RL is another important one not covered in this book. RL concerns training models by interacting with an environment, where the algorithm gets feedback (reward or penalty) for actions taken, rather than employing labeled data. The paradigm is best applied to decision-making issues in dynamic and uncertain environments.

ML framework-supervised learning, unsupervised learning, and RL have their own strengths and are optimal for a particular kind of problem. Supervised learning shines when labeled data exists, allowing for very accurate models for classification and regression tasks. Unsupervised learning, on the other hand, is used particularly when it is used to reveal hidden patterns or structures in unlabeled data and is thus particularly well-suited for clustering and dimensionality reduction. RL is distinct from the others in that it focuses on decision making in changing circumstances where an agent learns by interacting with its environment and receiving rewards or penalties accordingly. It is particularly useful for sequential decision-making problems [56].

ML contributes directly to the success of AI through the development of algorithms that optimize performance through data-driven learning. Both AI and ML have immense potential in many fields like healthcare, optics, photonics, and industry because they can be employed to automate complex decision-making processes, optimize performance, and increase predictive capability [55]. Both are transforming industries because they enable more analyses that lead to diagnostics, manufacturing, and other category breakthroughs.

RL differs from supervised and unsupervised learning in that it focuses on learning optimal actions based on feedback from the environment [56]. Here, an agent learns to maximize the cumulative sum of rewards by trial and error, and it has had success in, for example, video summarization [59].

There are also hybrid approaches that embed elements of such learning paradigms. Semi-supervised learning, for example, makes use of both unlabeled and labeled samples to improve the performance of a model, and is thus practical when there isn't much labeled data but an abundance of unlabeled data. Techniques such as self-training, co-training, and deep learning (DL)-based methods are generally applied to semi-supervised learning [60].

While each learning paradigm has its merits, they each have demerits too. Supervised learning can be quite demanding when it requires large amounts of labeled data, unsupervised learning does not necessarily translate to tangible task objectives, and RL is challenging to implement in intricate environments. Emerging breakthroughs in hybrid approaches have reported encouraging outcomes. For instance, contrastive learning methods have bridged the gap between unsupervised and supervised learning in computer vision applications [61].

The use of unsupervised feature learning within supervised tasks has also been seen to improve performance, particularly where there is a lack of sufficient labeled data [58].

Lastly, choosing an adequate ML paradigm is contingent on the problem one is facing, the existence of labeled data, and the objectives of the task. As more research continues, new techniques and hybrid approaches are being developed, broadening the field of ML to meet the demands of diverse areas.

8.3.3 ML Algorithms Commonly Used in Microbial Identification

ML algorithms have become increasingly important in microbial identification and classification. Several commonly used ML algorithms have shown promising results in this field. Decision Trees and Random Forests (RFs) are popular choices for microbial identification. Decision Trees create a flowchart-like structure to classify samples based on features, while RFs combine multiple decision trees to improve accuracy and reduce overfitting. These algorithms are particularly useful for handling complex datasets with many features [63]. SVMs have demonstrated high performance in microbial classification tasks. SVMs work by finding the optimal hyperplane to separate different classes in high-dimensional space. They are effective in handling both linear and nonlinear relationships in the data [64]. K-nearest neighbors (KNN) is another widely used algorithm in microbial identification. It classifies samples based on the majority class of their KNN in the feature space. KNN has shown good results in various studies, particularly when dealing with smaller datasets [64]. Neural Networks and DL approaches have gained significant attention in recent years. These methods can automatically learn complex patterns from large datasets, making them suitable for analyzing high-dimensional microbial data. DL models have shown remarkable performance in image-based microbial identification tasks [63, 64]. Feature selection and dimensionality reduction techniques play a crucial role in improving the performance of ML models for microbial identification. These methods help identify the most relevant features and reduce the complexity of the dataset, leading to more efficient and accurate models. Techniques such as Principal Component Analysis (PCA) and recursive feature elimination are commonly employed [65]. The role of big data in microbial ML applications cannot be overstated. With the increasing availability of large-scale genomic and metagenomic datasets, ML algorithms can leverage this wealth of information to improve their predictive capabilities. Big data approaches enable the discovery of novel patterns and relationships in microbial communities that may not be apparent through traditional analysis methods [64]. Model validation and performance metrics are essential for assessing the reliability and effectiveness of ML models in microbial identification. Common metrics include accuracy, precision, recall, and F1-score. Accuracy measures the overall correctness of predictions, while precision and recall focus on the model's ability to correctly identify positive cases. The F1-score provides a balanced measure of precision and recall. Receiver Operating Characteristic (ROC) curves and Area Under the Curve (AUC) are valuable tools for evaluating the performance of binary classification models.

They illustrate the trade-off between true positive rate and false positive rate across different classification thresholds [64]. It's important to note that the performance of ML algorithms can vary depending on the specific microbial identification task and the nature of the dataset. For instance, in a study on cervical cancer prediction, RF, Decision Tree, Adaptive Boosting, and Gradient Boosting algorithms achieved 100% accuracy, while SVM reached 99% accuracy [63]. In another study on brain tumor classification, KNN and Neural Networks outperformed other models with accuracies of 98.5% and 98.4%, respectively [64]. The choice of data scaling methods can also significantly impact model performance. For example, in a heart disease diagnosis study, the CART algorithm combined with Robust Scaler or Quantile Transformer outperformed other combinations, achieving 100% accuracy.

8.4 Machine Learning Approaches for High-Throughput Microbial Identification

ML techniques are increasingly applied to search through intricate datasets of diverse provenances, including genomic information, microbiota information, microscopy images, and spectroscopy information [66]. The objectives of these techniques are to improve the accuracy and speed of microbial identification, which transcends traditional techniques.

Some of the algorithms applied by ML to identify microbes are SVMs, Artificial Neural Networks (ANN), RF, and logistic regression [67, 68]. The algorithms work efficiently to manage large volumes of data and determine complex patterns that are difficult to detect by human experts [69]. For instance, in MALDI-TOF MS analysis, ML methods have been employed to enable improved identification of species and improved determination of AMR [69].

Surprisingly, while more sophisticated models like RF may offer improved performance under certain circumstances, even extremely simple models like L2-regularized logistic regression can offer comparable performance in most cases. The advantage of simpler models is that they can be learned quickly and are easier to explain, which is particularly valuable where such transparency considerations and computational speed are of crucial importance [68]. This then gives rise to the need to select an appropriate ML model depending on the specific research objectives, making compromises on the model's complexity, its accuracy, and interpretability.

In general, ML is revolutionizing microbial identification by offering the possibility to process complex microbial data quickly and precisely. Sufficient attention should be paid to the trade-offs between model complexity, computational power, and interpretability, though, to ensure that the chosen ML approach aligns with the research study or clinical practice goal.

8.4.1 Genomic and Metagenomic Data Processing

Metagenomic and genomic data analysis have become more and more important in microbial ecology and environmental science. Sequence alignment and feature detection are central activities involved in the interpretation of such intricate datasets, and ML tools have emerged as powerful methods for species identification and classification for metagenomic samples.

Sequence alignment is most often the first step when working with metagenomic data. Traditional methods, such as 16S rRNA gene amplicon (AMP) sequencing, are popular but can be subject to amplification bias and only look at a restricted genomic range. To break through such restraints, shotgun metagenomic sequencing has been the preferred choice, with an extended view of microbial communities. This is at a greater expense and is computationally more demanding [70].

Feature extraction from metagenomic data is necessary for downstream analysis. A new approach is the Operational Genomic Unit (OGU) method, which uses sequence alignment hits to individual reference genomes as the minimal unit to quantify microbial community diversity [71]. This method provides high-resolution community composition information and frames features into a phylogenomic tree hierarchy, in which phylogeny-aware tools not ordinarily applied to shotgun metagenomics can be utilized [71].

ML use in species recognition has revolutionized metagenomic data analysis. Various ML approaches, including SVMs, decision trees, RF, neural networks (NN), and deep learning (DL), are used to classify, regress, or cluster complex metabolomic data [72]. The methods have shown great potential in improving disease modeling and diagnosis through metabolomic profiling [72]. DL has been particularly successful in 16S rRNA classification. For example, DL-TODA, a program based on a convolutional neural network (CNN) architecture, has been demonstrated to classify metagenomic reads with high accuracy without the need to construct a reference database [73]. On synthetic data simulated with 2,454 genomes from 639 species, DL-TODA achieved over 0.98 classification accuracy, ranks above the genus level and 0.97 at the species level, outperforming state-of-the-art tools like Kraken2 and Centrifuge.

Another new solution for 16S rRNA classification is RiboDetector, a software package that utilizes a bidirectional long short-term memory (BiLSTM) neural network [74]. RiboDetector rapidly and precisely identifies rRNA reads from various types of sequence data, including transcriptomic, metagenomic, and metatranscriptomic datasets. It showed superior performance to existing methods, with fewer misclassifications and great generalizability for identifying novel rRNA sequences [74].

An interesting case study of ML-based metagenomic species profiling is DeepVirFinder, a reference-free and alignment-free method for viral sequence identification from metagenomic data [75]. DeepVirFinder was trained using viral RefSeq sequences and evaluated using recently discovered sequences, and was shown to be better than the existing tools with varying contig lengths. In application to human gut metagenomic samples, it identified over 50,000 viral sequences from colorectal carcinoma patients and suggested possible viral presence vs cancer status correlations [75].

Wang *et al.* constructed a novel ML framework in a 2015 paper titled MetaBoot, whose goal was to facilitate the discovery of nonredundant taxonomical biomarkers in metagenomic data. MetaBoot integrates the minimum Redundancy Maximum Relevance (mRMR) method with bootstrapping resampling to enhance robustness, accuracy, and biological relevance of biomarker selection. The system works through data normalization, initial feature selection, repeated bootstrapping for redundancy and noise reduction, and final ranking based on the frequency of selection of biomarkers. MetaBoot was compared with synthetic data of varying statistical distributions (normal, gamma, and mixed) and with actual oral and soil microbiome datasets. MetaBoot, compared to traditional methods like LEfSe, Metastats, and Wilcoxon tests, performed superiorly in detection of biologically relevant and nonredundant biomarkers with enhanced classification accuracy. The study is called MetaBoot as a reliable platform for biomarker discovery in taxonomical applications, with significant application in microbial community profiling, environmental monitoring, and clinical diagnosis [76].

During their review, Fiannaca *et al.* [77] highlight the growing significance of ML, particularly DL architectures, in achieving accurate bacterial taxonomic assignment within metagenomic studies. They note that traditional approaches, alignment-based methods, and the widely used RDP classifier pose challenges in terms of computational costs and reduced accuracy, especially in handling short-read sequences from NGS data. To address these constraints, the authors suggested and tested two DL models, CNNs and deep belief networks (DBNs), in combination with k-mer-based sequence representation. Their models surpassed, particularly in handling AMP sequencing data, with more than 91% accuracy at the genus level, surpassing the 83.8% of the RDP classifier.

ML, more so DL, plays a critical role in tackling sequence noise, hypervariable regions of the 16S rRNA gene, and the vast quantities of metagenomic data. This study points out that DL offers a precise and scalable solution for the detection of bacteria, paving the way for its use in diagnostic devices, environmental sampling, and microbiome research, where precise microbial community profiling is critical [77]. The study introduces ANNA16 (Figure 8.3), a novel DL-based estimator of the 16S rRNA gene copy number (GCN) from DNA sequence, addressing a significant gap in 16S rRNA-based microbial community profiling. Bacteria and archaea possess variable 16S GCN between 1 and 21, and traditional read count-based methods are prone to biasing correct cell abundances in microbial communities. Existing methods, by being taxonomy- or phylogeny-based, are constrained by database heterogeneity, high computational complexity, and reduced accuracy.

ANNA16 outvies all these constraints by utilizing a Stacked Ensemble Model (SEM), combining multilayer perceptron (MLP) and SVM architectures. The algorithm was trained against over 19,000 full-length 16S rRNA gene sequences of the rrnDB database. By 6-mer representation-based prediction, ANNA16 enables direct regression-based prediction of 16S GCN without phylogenetic placement. Performance comparison suggests that ANNA16 performs much better than the traditional tools, such as rrnDB, CopyRighter, PAPRICA, and PICRUSt2 for both full-length sequences and universally amplified hypervariable regions (V1–V9). The model performed best in root mean squared error (RMSE), suggesting that it possesses the best accuracy and generalizability.

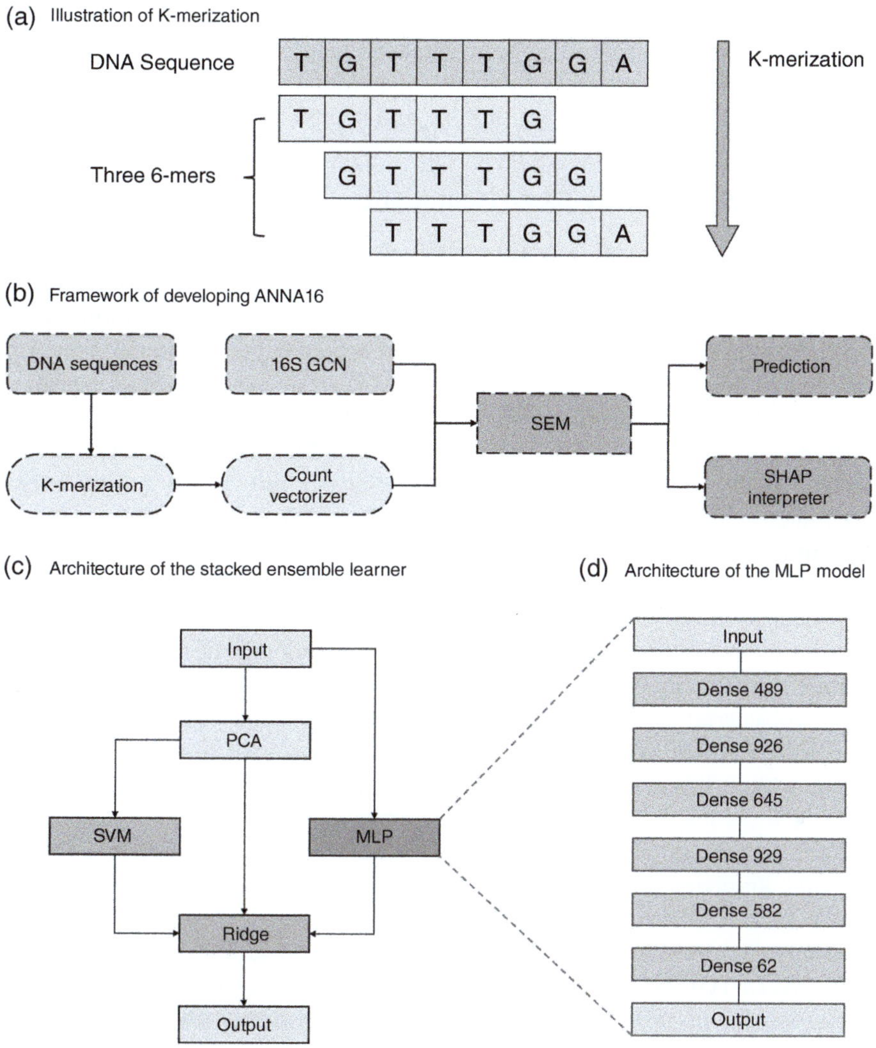

Figure 8.3 Illustration of the framework and architecture of the deep learning models. (a) DNA sequence: K-merization (K = 6). (b) Framework for ANNA16. (c) Architecture of the MLP-based Stacked Ensemble Model (SEM). (d) Architecture of the MLP module inside the SEM. [78] / Springer Nature / CC BY 4.0.

Additionally, by applying SHAP (SHapley Additive exPlanations) analysis, the authors demonstrate that ANNA16 can identify very informative sequence positions, such as noncanonical conserved positions that are normally overlooked by traditional methods. The observation suggests that DL algorithms like ANNA16 can detect subtle evolutionary signals beyond the established hypervariable sites. The paper also describes potential future improvements, such as the implementation of transfer learning and addressing data imbalance for species with very high GCN values.

In short, ANNA16 illustrates the power of DL in microbiome research by providing a scalable, accurate, and interpretable program for resolving 16S rRNA GCN biases. The program plays a key role in quantitative microbiome profiling in clinical diagnostics, environmental studies, and microbial ecology [78].

8.4.2 Mass Spectrometry-Based Identification

MALDI-TOF MS has transformed microbial identification in clinical microbiology laboratories by offering ease, speed, accuracy, and affordability compared to traditional biochemical methods [32, 75]. Technology can gain around a day in everyday practice for microbial identification, and it has become an essential clinical microbiology tool [32]. The MALDI-TOF MS method is based on whole-cell fingerprinting, in conjunction with dedicated bioinformatic software programs for spectral analysis and identification [79]. ML algorithms, and particularly those integrated into the MALDI Biotyper software, are what enable processing of spectral data and matching with reference databases for accurate identification [79].

A case study in clinical microbiology demonstrated that MALDI-TOF MS was very accurate in the identification of staphylococcal strains, with 99.3% of isolates correctly identified to the species level [79]. This technology has also shown great promise in the identification of other medically relevant microorganisms, including mycobacterial species, yeasts, and a variety of pathogens [34]. Additionally, MALDI-TOF MS is also promising for application in AMR detection and strain typing, with utility in epidemiological investigations [80]. Despite its rapid development, there are still challenges in optimizing sample preparation, expanding reference databases, and refining identification algorithms for certain microbial taxa [81]. Resolving these challenges will be instrumental to the further evolution of MALDI-TOF MS in clinical diagnostics.

Systematic review by Weis *et al.* [69] considers the integration of ML techniques with MALDI-TOF MS for microbe identification and antimicrobial susceptibility testing (AST). In reviewing 36 studies, the review details the most common ML algorithms employed, i.e. SVM, Genetic Algorithms (GA), ANN, and RFs. These techniques significantly enhance discrimination among species, particularly among closely related microbes such as *Staphylococcus*, *Streptococcus*, and *Klebsiella*, and facilitate proper prediction of AMR, e.g. methicillin-resistant *Staphylococcus aureus* (MRSA) and vancomycin-intermediate strains.

Despite these results, the review reports several important limitations: small sample sizes, limited external validation (a paltry 11% of studies), and poor reproducibility due to dependence on commercial software and lack of open data or shared code. The authors propose collaborative steps to generate publicly available datasets with resistance profiles, adopt transparent ML workflows, and incorporate external validation to improve clinical utility of such methods.

This study underscores the potential of ML-augmented MALDI-TOF MS to further speed up fast microbial diagnosis but points to the need for standardization and reproducibility to gain broader clinical approval [69]. The research by

De Bruyne *et al.* [14] created a benchmark protocol for the identification of bacterial species using the method of MALDI-TOF MS followed by data processing and ML algorithms. The researchers targeted *Leuconostoc*, *Fructobacillus*, and *Lactococcus* species and optimized sample preparation protocols, like cell smears and cell extracts and utilized rigorous preprocessing protocols like normalization and background correction.

PCA, multidimensional scaling (MDS), and kNN analysis were conducted to reveal species-specific spectral patterns. Identification accuracy of 84.6% for *Leuconostoc* and *Fructobacillus*, and 94.2% for *Lactococcus* species was achieved. Upon application of ML models in the form of SVMs and RFs, classification accuracy was significantly enhanced, with SVM achieving a high 98.4% accuracy and 96.8% F-score.

The study found that MALDI-TOF MS and ML-based identification is an efficient, economic, and reliable process of bacterial species identification and hence has a very vast field of application in microbial taxonomy and diagnosis [14].

The study by Feucherolles *et al.* [82] investigates the combination of MALDI-TOF MS and ML for rapid AMR screening in *Campylobacter jejuni* and *Campylobacter coli*, two major foodborne pathogens. A total of 224 *C. jejuni* and 116 *C. coli* isolates from humans, animals, and the environment were tested for resistance to seven antibiotics, including ciprofloxacin and tetracycline. Protein spectra were analyzed using RF and Logistic Regression models, achieving high sensitivity (up to 92.3%) and precision (up to 81.2%) in detecting resistant strains, particularly for ciprofloxacin and tetracycline.

The study also assessed species-specific and extraction method impacts, finding no significant differences between extraction types. Furthermore, potential AMR biomarkers were identified, contributing to the understanding of resistance mechanisms in these pathogens.

The authors conclude that the integrated MALDI-TOF MS and ML pipeline represents a promising, fast, and cost-effective tool for early AMR screening, facilitating faster and targeted antibiotic treatments in clinical and food safety contexts. However, they recommend larger datasets and expanded pathogen coverage for clinical validation to further refine and validate the approach [82].

This study established an end-to-end pipeline of MALDI-TOF MS and supervised ML models to enhance three related *Enterococcus* species' (*E. faecalis*, *E. faecium*, and *E. lactis*) identification at the species level. With 410 strains and 1,640 spectra, the authors compared KNN, SVMs, and RF models. Traditional MALDI-TOF MS and commercial Biotyper software incorrectly identified *E. lactis* as *E. faecium*. Conversely, the Random Forest model performed better, with 99.1% accuracy and AUC values greater than 0.98.

The study also detected species-specific mass spectral biomarkers that were validated by whole-genome sequencing, including peaks for proteins like the Type I toxin-antitoxin system and isoleucyl-tRNA synthetase. The authors concluded that ML-enhanced MALDI-TOF MS significantly improves classification accuracy and discrimination of very similar *Enterococcus* species, providing a rapid and cost-effective tool for clinical diagnostics and food safety monitoring.

8.4.3 Imaging-Based Identification

AI has revolutionized the study of microbial colony morphology into quick and precise identification of microorganisms. The Annotated Germs for Automated Recognition (AGAR) dataset of 18,000 images of five microorganisms was developed to aid in the automation of microbial colony analysis [83]. The dataset includes single and mixed culture images under various light conditions, which are annotated as countable, uncountable, or empty. The application of deep neural network architectures, such as Faster R-CNN and Cascade R-CNN, to the AGAR dataset, has demonstrated the vast potential of applying DL methods in automatically localizing microbe and classifying it from Petri dish images [83]. Label-free single-cell IFC, when combined with digital holography microscopy, has emerged as a powerful imaging technique for life sciences and clinical diagnostics [84]. DL applied to holographic imaging has enabled the effective use of point-of-care devices. A combined approach of Mask R-CNN for cell detection, a convolutional autoencoder for feature extraction, and a feedforward neural network for single-cell classification has been proposed to avoid biases from data derived under varying experimental conditions [84]. The approach has already been used successfully for the challenging problem of drug-resistant endometrial cancer cell recognition. Real-time visualization and automatic detection and classification of untreated cancer cells in blood in stain-free IFC have been achieved by digital holographic microscopy and ML [85].

This approach enables the classification of primary and metastatic colon cancer cells, along with four types of blood cells, with high precision (92.56%) at a throughput of 15 cells per second [85]. Furthermore, an image-guided cell sorting and classification system integrating microfluidics, photonics, computational microscopy, real-time image processing, and ML has been developed, offering the high throughput of FCM and the informational richness of microscopy [86]. Özel Duygan *et al.* [87] report CellCognize, an ML workflow that quickly identifies and classifies microbial cell type diversity from FCM data. After learning from multiparametric FCM data of 32 reference standards consisting of bacterial strains, yeast, and beads, CellCognize achieved an average prediction accuracy of 80% through training ANN. The system was able to resolve microbial species, identify physiological states such as growth stages, and quantify biomass production. It worked as effectively as conventional methods, such as 16S rRNA gene sequencing and 14C-labeled substrate incorporation. The pipeline was applied to both synthetic microbial communities and natural freshwater microbiota, where it accurately documented community responses to chemical amendments. Furthermore, CellCognize returned similarity scores for unknown microbiota by comparing their FCM profiles to the reference standards. This aspect shows its promise for quick, low-cost microbial diversity analysis in clinical, environmental, and industrial contexts [87]. Martis *et al.* [88] proposes a DL architecture for computerized bacterial classification that bridges the limitations of time and labor-consuming and error-prone conventional manual methods. Architecture takes advantage of transfer learning by fine-tuning existing pretrained CNN models, VGG16, MobileNet, and Inception v3, on a dataset of 12,000 images of bacteria divided into 12 classes. Preprocessing tasks like rotation and resizing standardize the images before classifying them. Among the models, VGG16 performed the highest accuracy of 99.99%, far exceeding MobileNet (73.33%) and Inception v3 (80.24%). Not only does architecture predict bacterial clusters by structure and shape but it also provides a confidence value for each prediction, providing reliability. The study determines that VGG16 performs best among models for identifying bacteria and suggests potential future system extensions for viruses and other single-celled organisms' identification [88].

8.5 Machine Learning Approaches for High-Throughput Microbial Culturing

8.5.1 ML-Driven Microbial Growth Prediction

ML-driven microbial growth prediction has emerged as a powerful approach for accelerating the development of high-performing microbial strains for the sustainable production of valuable compounds. Both constraint-based modeling (CBM) and ML techniques used independently or in combination have proven effective in analyzing and optimizing fermentation parameters [89]. These methods enable the prediction of cellular phenotypes from genotypes and uncover patterns in complex biological systems, resulting in more accurate and reliable models for fermentation processes.

In industrial contexts, ML has been successfully applied to predict biogas production performance in large-scale garage dry fermentation systems. A novel intelligent system utilizing AutoML algorithms, such as AutoGluon and H_2O, achieved high predictive accuracy for both digester and percolate tank biogas outputs. Beyond accurate forecasting, this system offered interpretable insights and early warning capabilities, underscoring the potential of ML in optimizing complex industrial fermentation workflows.

The integration of ML with HTP robotic platforms has significantly enhanced microbial culturing. An open-source robotic strain isolation platform, coupled with ML algorithms, was developed to rapidly generate microbial isolates on demand. By leveraging colony morphology and genomic data, the system maximized microbial diversity and enabled targeted selection of specific genera. When applied to human fecal samples, it successfully generated personalized gut microbiome biobanks representing over 80% of abundant taxa, highlighting its utility in systematic microbiome analysis and biobanking.

ML has also been instrumental in optimizing culture conditions in miniaturized screening platforms. The Berkeley Lights Beacon® system, an HTP, perfusion-based platform, allows real-time monitoring of strain productivity under continuous culture conditions that closely mimic industrial fed-batch fermentation [90]. This approach has facilitated the selection of high-productivity strains following whole-genome random mutagenesis.

In addition to culture optimization, ML is increasingly applied to the analysis and interpretation of experimental data. ML-enhanced spectral analysis and kinetic modeling have enabled efficient characterization of catalysts and the optimization of synthesis conditions, accelerating materials discovery in microbial culturing [91].

Innovations in culture chip technology have expanded the scope of HTP microbial culturing. A micro-engineered ceramic sheet containing up to one million growth compartments has been employed for fast viable cell counting, HTP screening, and target recovery of specific phenotypes [92]. This versatile platform enables rapid and dynamic environmental changes, facilitating the detection of specific enzymatic activities and the screening of metagenomic libraries. Together, these advances demonstrate the transformative potential of ML across the entire microbial culturing pipeline from strain selection and fermentation optimization to data analysis and biobanking.

8.5.2 AI in Microbial Cultivation Process Optimization

Advanced bioreactors and computerized culture monitoring platforms are spearheading this revolution, enabling real-time data collection and analysis for better process control [93]. Raman spectroscopy has been demonstrated to be a very useful tool for in-line monitoring of mammalian cell culture bioreactors. This nondestructive technique allows concurrent prediction of several culture parameters like glutamine, glucose, lactate, and cell density [93]. Similarly, electron-probing devices like Zn_2GeO_4:Mn@Fe^{3+} enable real-time monitoring of microbial metabolism, that is, Fe(III) respiration, and provide insight into complex metabolic activities and enhance biosynthesis [94].

RL has been highly promising in adaptive culture systems, particularly for the control of cocultures in continuous bioreactors. RL agents can maintain populations at specified levels and do better than standard controllers when infrequent sampling occurs. Astonishingly, an effective control policy can be acquired within 24 hours by running five bioreactors in parallel, which verifies the efficacy of this approach (Figure 8.4) [95]. The marriage of process analytical technology (PAT) and AI has enabled the development and use of enormous cultural datasets. Data-driven models (DDMs) can now correlate process and biological conditions with cell culture states and forecast dynamic behaviors like viable cell density, product titer, and metabolite concentrations. These models maximize computational load against accuracy and reliability, which makes it possible to make proactive decisions in bioprocess engineering [96].

A clinical illustration of AI-assisted optimization of bioprocessing for pharmaceutical microbiology is seen in the design and construction of a fully automated 24-parallel mini-bioreactors robot facility. The system, managed by a model-based adaptive input design model, supports scale-down characterization of clone libraries under conditions mimicking industrial-scale bioreactors. The cultivation strategies are optimized regularly by a macro-kinetic growth model, demonstrating the power of AI in the optimization of strain selection and process development [97].

The other breakthrough method is the HTP perfusion-based screening platform developed using the Berkeley Lights Beacon® platform. It allows real-time measurement of continuous culture strain productivity and provides phenotypes

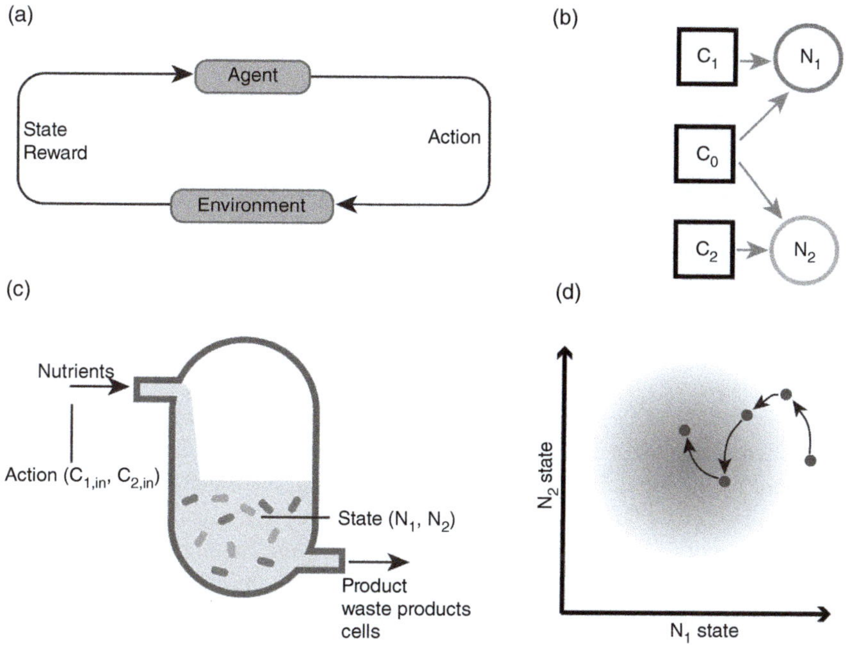

Figure 8.4 Reinforcement learning for the control of two auxotrophic species in a chemostat. (a) The basic reinforcement learning (RL) loop; the agent interacts with its environment through actions and observes the state of the environment along with a reward. The agent acts to maximize the total reward it receives (the return). (b) System of two auxotrophs dependent on two different nutrients, with competition over a common carbon source. (c) Diagram of a chemostat. The state observed by the RL agent is composed of the populations of two strains of bacteria; the actions taken by the agent control the concentration of auxotrophic nutrients flowing into the reactor. (d) Representative system trajectory. The agent's actions, taken at discrete time-points (circles), influence the state dynamics (black arrows), with the aim of fulfilling the reward condition (moving to the center of the green circle). The state is comprised of the (continuously defined) abundance of two microbial populations, N1 and N2. The agent's actions dictate the rate at which auxotrophic nutrients flow into the reactor. At each time step, the agent's reward is dependent on the distance between the current state and the target state. [95] / PLOS / CC BY 4.0.

with high correlation with behavior in industrially relevant bioreactor processes. The platform has been effectively utilized to improve high productivity strains through whole-genome random mutagenesis, producing mutants with drastically increased peak-specific productivity in bioreactors [90].

The eVOLVER platform is a scalable, HTP growth experiment DIY platform applicable for use in microbiology, molecular evolution, and systems biology. The system achieves controlled microbial cell growth conditions such that it's possible to observe subtle phenotypic differences. The eVOLVER platform has been applied to determine the adaptive niches in yeast as well as to detect temperature-sensitivity-altered strains, indicating applicability of the platform to optimizing processes [98]. Miniature bioreactors (MBRs) are also turning out to be extremely significant in the bioprocessing acceleration through enabling HTP parallel cell cultivations. MBRs find applications in media development, strain optimization, and process optimization. With automated robotic systems and other miniature process modules, MBRs can offer highly integrated, HTP cell cultivation process development solutions [99].

8.5.3 Synthetic Biology and AI-Driven Strain Engineering

Synthetic biology and AI-guided strain engineering have revolutionized the field of metabolic engineering, enabling precise genetic modifications and optimized strain design. ML plays a key role in predicting gene knockouts for enhanced growth and productivity, and AI-based programs have significantly improved CRISPR/Cas9 target prediction.

ML algorithms have been instrumental in predicting gene knockouts that can enhance microbial growth and metabolite production. ML algorithms take complex genomic and metabolic data to search for potential targets for genetic engineering. With the utilization of large datasets and complex computational models, ML can predict the effect of gene deletions or alterations on cellular metabolism and growth rates. This approach has been particularly valuable for the optimization of microbial strains for industrial applications, e.g. biofuel production or high-value chemicals. AI-powered CRISPR/Cas9 target prediction has greatly improved the efficiency and accuracy of genome editing.

Computational methods like AcRanker, an ML-based method, have been developed to predict putative anti-CRISPR proteins from just protein sequence information [100]. This has led to the discovery of new anti-CRISPRs, AcrIIA20 and AcrIIA21, which can inhibit various Cas9 proteins. These discoveries have important implications for the control of gene editing technologies and our understanding of CRISPR-Cas systems.

The CASPER (CRISPR Associated Software for Pathway Engineering and Research) tool demonstrates the ability of AI to do CRISPR target prediction. CASPER reports higher accuracy for prediction of on- and off-target activity compared to conventional methods, a 2.4% boost for on-target, and a stunning boost by 30.2% for off-target prediction [101]. This tool also enables novel applications such as multitargeting analysis and multispecies population analysis, which are critical for advanced genome editing projects and consortium-based engineering. ML has been applied in numerous aspects of strain engineering in microbial synthetic biology. For instance, in the well-known microbial cell factory *Pichia pastoris*, CRISPR-based instruments for gene editing and transcriptional regulation have been set up [102]. These tools include mechanisms for efficient gene knockout, repression, and activation. The CRISPR-ARE (Activation, Repression, and Editing) system demonstrated in *P. pastoris* allows for the simultaneous activation, repression, and editing of a gene, mirroring the versatility of CRISPR-based approaches in synthetic biology. Similarly, CRISPR/Cas9 systems have been extensively employed in *Saccharomyces cerevisiae* for various forms of genetic manipulation, including single-nucleotide resolution editing, multigene editing, transcriptional regulation, and genome-scale remodeling.

These advances have opened the door to accelerating microbial engineering, reducing labor and time requirements while maximizing our understanding of cellular genetics and physiology. The synergy between AI and synthetic biology has also led to the development of genome-scale engineering technology. Tools such as CRISPR-based control, synthetic small regulatory RNA (sRNA)-based knockdown, promoter design, and multiplex automated genome engineering (MAGE) are now very effective microbial genome-scale engineering tools. The technologies allow fine-level control of gene expression and genome editing, which makes it feasible for scientists to optimize microbial metabolism for application. Overall, the synergy of synthetic biology and AI-strain engineering opened new avenues in metabolic engineering.

8.6 Challenges and Limitations of Machine Learning in HTP Microbial Research

Several challenges and limitations still exist in its application to this area. Data quality and data standardization are among the significant challenges of HTP microbial research. The accuracy and reliability of ML models depend significantly on the quality of data input. Data collection procedures, storage, and processing strategies in microbiology can be extremely heterogeneous, adding inconsistencies and potential biases to datasets [103]. This lack of standardization can result in models that fail to generalize across laboratories or experimental systems. Furthermore, the diversity and complexity of microbial systems also make it challenging to determine shared data collection protocols, again contributing to the issue of data quality.

Bias and interpretability of ML models are another challenge. ML algorithms, particularly DL models, are "black boxes" in which it is difficult for researchers to understand the reasons behind the predictions [104]. Interpretability is an issue in microbiology because knowing the biological pathways involved in predictions is important. In addition, training dataset biases can lead to biased results, potentially reinforcing existing prejudices or overlooking important but underrepresented microbial species or phenomena [105]. To mitigate these issues, researchers are exploring XAI techniques to enhance model interpretability and transparency, as evidenced in the meibomian gland dysfunction classification study.

Computational complexity and high processing cost are significant challenges to the widespread application of ML in HTP microbiological studies. Massive genomic and metagenomic data analysis requires large computational power, which may not be accessible to every research facility [104]. The need for specialized hardware, such as high-performance computing clusters or GPU arrays, may limit accessibility and increase the cost of research. Moreover, the complexity of some ML algorithms, particularly in DL, demands significant expertise in both computer science and microbiology, leading to a shortage of skills that is not straightforward to rectify. The lack of comprehensive training datasets for some microbial species is a significant limitation to the implementation of ML on HTP microbial studies. While well-studied model organisms have abundant data, many microbial species are poorly represented in existing datasets. This lack of data can lead to poor or biased models when extrapolated to less-studied microorganisms. To overcome this shortcoming, scientists need to prioritize the generation and dissemination of data on a wider range of microbial species, as well as the development of transfer learning and domain adaptation methods that can lend knowledge from data-abundant species to make better predictions for data-scarce ones.

Ethical and regulatory concerns of AI-based microbiology are arising as ML applications in this area grow. Issues such as data confidentiality, consent to use data, and the potential for misuse of AI-derived insights within microbial research need to be carefully addressed [106]. There also has to be a set of precise rules and regulations for applying ML in microbiology, particularly where there is a direct concern with human health, e.g. the prediction of antibiotic resistance or the detection of pathogens. Having ethical standards and regulatory procedures in place that balance innovation with proper use of AI in microbiology is crucial to the long-term health and public trust of the profession. To address these problems, there are several conceivable solutions and topics for future study that can be entertained. Data quality and standardization must be addressed, prioritizing the development and implementation of standardized protocols of data acquisition, processing, and storage in microbial research [107].

8.7 Future Perspectives and Emerging Trends

The integration of ML with multi-omics technologies is revolutionizing microbial studies and their applications in numerous fields. The use of multiple technologies is resulting in a more advanced and detailed understanding of complex biological processes. In multi-omics integration, ML software is evolving essential tools to interpret and analyze the vast amount of data generated by genomics, transcriptomics, proteomics, and metabolomics studies. These approaches are particularly powerful in uncovering intricate interactions among different biological layers and identifying novel biomarkers and therapeutic targets. For instance, in the field of oncology, multi-omics technologies combined with ML enable a deeper understanding of the tumor microbiome and its role in carcinogenesis, metastasis, and treatment response. Similarly, in osteoarthritis research, multi-omics techniques are facilitating the identification of molecular signatures and cellular subpopulations, allowing for more precise diagnosis and personalized therapeutic strategies [108]. Application of ML to combine multi-omics data is not without challenges. The complexity and volume of data from heterogeneous omics platforms require sophisticated computational tools and algorithms. However, developments in ML and DL algorithms are increasingly improving our ability to process and interpret such complex datasets. With the advancement of technologies, we can expect even more accurate predictions and insights into microbial behavior and interactions. For personalized medicine applications, the integration of multi-omics data with ML is opening new avenues for targeted therapeutic interventions. This approach allows for a more precise characterization of personalized health profiles and may lead to more effective treatments with fewer side effects [109]. Pharmacometrics is also being revolutionized by these technologies, with ML techniques being integrated into model-informed drug development and discovery strategies [110]. Stepping ahead toward the future, future applications of microbial ML include possible implementation using quantum computing. Not being specifically mentioned in the provided papers, quantum computing can possibly better handle complex computations that go into the analysis of multi-omics data at a large scale using normal computers. This could lead to faster speed and accuracy in the predictions that are made in microbial research. Federated learning and decentralized AI approaches, though not directly referred to in the context presented, could be of utmost significance in microbiome studies. These approaches could facilitate collaborative research with data privacy, a critical factor in personalized medicine and clinical research. Practical applications of ML and multi-omics integration are already observed in numerous fields. In microbiology and environmental microbiology, these approaches are expanding our understanding of microbial community function and function in biogeochemical cycles [111]. In agriculture and horticulture, the integration of multi-omics information and AI is transforming plant phenotyping toward precision horticulture, predictive breeding, and improved crop management [112]. The role of XAI in microbial science, while not directly discussed in the papers presented, will become even more prominent. As ML models become more sophisticated, it will be crucial to provide transparency and interpretability of results to build confidence in AI-driven microbial science and its applications. The future of AI in microbial sciences, in general, looks bright and revolutionary. Leveraging ML with multi-omics approaches is going to speed up progress in personalized medicine, drug discovery, and environmental microbiology. As progress is made with technologies and computational power, we are likely to see better predictions, deeper insight into microbial systems, and new uses across multiple disciplines.

8.8 Conclusion

The intersection of ML with HTP microbial techniques is a paradigm shift in microbiological diagnostics and research. As traditional culture-based and molecular methods are being depleted, ML offers quick, scalable, and extremely accurate solutions for microbial identification, AMR prediction, and cultivation optimization. Given access to different data sources from metagenomic sequences and MS spectra to imaging and FCM profiles, ML models can recognize hidden patterns and make predictions with unparalleled accuracy.

This chapter illustrates how supervised, unsupervised, and RL techniques are being successfully applied across various fields of microbial science, including strain classification, growth prediction, and synthetic biology. The combination of ML with cutting-edge platforms such as MALDI-TOF MS, microfluidics, LOC devices, and CRISPR-based genome editing accelerates innovation in diagnostics, drug discovery, and personalized medicine. There are still challenges in data standardization, model interpretability, computational requirements, and regulatory approval. Resolving these issues will be crucial to making ML-based microbiology reproducible and trustworthy. Future breakthroughs, such as XAI, federated learning, and quantum computing, have the potential to continue revolutionizing the field, enabling the performance of robust, ethical, and globally collaborative microbial science. As ML expands, its role in shaping the next generation of microbiological tools will be critical to propelling progress in science and medicine.

Acknowledgments

This work was supported by the National Natural Science Foundation of China (82273890), Shenzhen Stable Support Grant (GXWD 20231130103401001), and the Shenzhen Science and Technology Program (JCYJ20240813104 817024).

References

1 Lagier, J.-C., Levasseur, A., Bilen, M. et al. (2018). Culturing the human microbiota and culturomics. *Nat Rev Microbiol* 16 (9): 540–550.

2 Shade, A., Klimowicz, A.K., Linske, M. et al. (2012). Culturing captures members of the soil rare biosphere. *Environ Microbiol* 14 (9): 2247–2252.

3 Rondon, M.R., August, P.R., Bettermann, A.D. et al. (2000). Cloning the soil metagenome: a strategy for accessing the genetic and functional diversity of uncultured microorganisms. *Appl Environ Microbiol* 66 (6): 2541–2547.

4 Rygaard, A.M., Thøgersen, M.S., Nielsen, K.F. et al. (2017). Effects of gelling agent and extracellular signaling molecules on the culturability of marine bacteria. *Appl Environ Microbiol* 83 (9): 00243–17.

5 Ravin, N.V., Skryabin, K.G., and Mardanov, A.V. (2015). Metagenomics as a tool for the investigation of uncultured microorganisms. *Russ J Genet* 51 (5): 431–439.

6 Braga, P.A.C., Gonçalves Dos Santos, V., Tata, A. et al. (2013). Bacterial identification: from the agar plate to the mass spectrometer. *RSC Adv.* 3 (4): 994–1008.

7 Fukuda, K., Saito, M., Taniguchi, H., and Ogawa, M. (2016). Molecular Approaches to Studying Microbial Communities: targeting the 16S Ribosomal RNA Gene. *J UOEH* 38 (3): 223–232.

8 Ivashko, M., Chaikovska, T., Boyko, N. et al. (2023). Microbiological diagnostics: From traditional to molecular genetic methods: a literature review. *Bulletin Of Medical And Biological Research* 5 (4): 34–41.

9 Alsulimani, A., Dar, S.A., Akhter, N. et al. (2024). The impact of artificial intelligence on microbial diagnosis. *Microorganisms* 12 (6): 1051.

10 Ercolini, D. (2013). High-throughput sequencing and metagenomics: moving forward in the culture-independent analysis of food microbial ecology. *Appl Environ Microbiol* 79 (10): 3148–3155.

11 Solieri, L., Giudici, P., and Dakal, T.C. (2012). Next-generation sequencing and its potential impact on food microbial genomics. *Ann Microbiol* 63 (1): 21–37.

12 Naresh, V. and Lee, N. (2021). A review on biosensors and recent development of nanostructured materials-enabled biosensors. *Sensors* 21 (4): 1109.

13 Walsh, A.M., Crispie, F., Claesson, M.J., and Cotter, P.D. (2017). Translating omics to food microbiology. *Annu Rev Food Sci Technol* 8 (1): 113–134.

14 De Bruyne, S., Speeckaert, M.M., Van Biesen, W., and Delanghe, J.R. (2020). Recent evolutions of machine learning applications in clinical laboratory medicine. *Crit Rev Clin Lab Sci* 58 (2): 131–152.

15 Tong, Y., Lu, W., Yu, Y., and Shen, Y. (2020). Application of machine learning in ophthalmic imaging modalities. *Eye and Vision* 7 (1): 22.

16 Budowle, B., Connell, N.D., Bielecka-Oder, A. et al. (2014). Validation of high throughput sequencing and microbial forensics applications. *Investigative Genetics* 5 (1): 1–18.

17 Drendel, G., Franks, A.E., Mathews, E.R., and Semenec, L. (2018). Microbial fuel cells, related technologies, and their applications. *Applied Sciences* 8 (12): 2384.

18 Fink, M., Cserjan-Puschmann, M., Reinisch, D., and Striedner, G. (2021). High-throughput microbioreactor provides a capable tool for early stage bioprocess development. *Scientific Reports* 11 (1): 1–10.

19 Eshaghi, S., Hedrén, M., Ignatushchenko, M. et al. (2005). An efficient strategy for high-throughput expression screening of recombinant integral membrane proteins. *Protein Science* 14 (3): 676–683.

20 Quijada, N.M., Hernández, M., and Rodríguez-Lázaro, D. (2020). High-throughput sequencing and food microbiology. *Adv Food Nutr Res* 91: 275–300.

21 Manickam, P., Mariappan, S.A., Murugesan, S.M. et al. (2022). Artificial Intelligence (AI) and Internet of Medical Things (IoMT) assisted biomedical systems for intelligent healthcare. *Biosensors* 12 (8): 562.

22 Pallen, M.J., Loman, N.J., and Penn, C.W. (2010). High-throughput sequencing and clinical microbiology: progress, opportunities and challenges. *Curr Opin Microbiol* 13 (5): 625–631.

23 Rahi, P., Prakash, O., and Shouche, Y.S. (2016). Matrix-assisted laser desorption/ionization time-of-flight mass-spectrometry (MALDI-TOF MS) based microbial identifications: Challenges and scopes for microbial ecologists. *Front Microbiol* 7: 01359. (AUG), 187691.

24 Jiang, Y., Luo, J., Huang, D. et al. (2022). Machine learning advances in microbiology: a review of methods and applications. *Front Microbiol* 13: 925454.

25 Oulas, A., Pavloudi, C., Polymenakou, P. et al. (2015). Metagenomics: tools and insights for analyzing next-generation sequencing data derived from biodiversity studies. *Bioinform Biol Insights* 9: 75–88.

26 Akinsanya, M.A., Goh, J.K., Lim, S.P., and Ting, A.S.Y. (2015). Metagenomics study of endophytic bacteria in aloe vera using next-generation technology. *Genom Data* 6: 159–163.

27 Satam, H., Joshi, K., Mangrolia, U. et al. (2023). Next-generation sequencing technology: current trends and advancements. *Biology* 12 (7): 997.

28 Barzon, L., Lavezzo, E., Militello, V. et al. (2011). Applications of next-generation sequencing technologies to diagnostic virology. *International Journal of Molecular Sciences* 12 (11): 7861–7884.

29 Zhang, L., Chen, F.X., Zeng, Z. et al. (2021). Advances in metagenomics and its application in environmental microorganisms. *Front Microbiol* 12: 766364.

30 Akaçin, İ., Ersoy, Ş., Doluca, O., and Güngörmüşler, M. (2022). Comparing the significance of the utilization of next generation and third generation sequencing technologies in microbial metagenomics. *Microbiol Res* 264: 127154.

31 Nomura, F. (2015). Proteome-based bacterial identification using matrix-assisted laser desorption ionization–time of flight mass spectrometry (MALDI-TOF MS): A revolutionary shift in clinical diagnostic microbiology. *Biochimica et Biophysica Acta (BBA) - Proteins and Proteomics* 1854 (6): 528–537.

32 Tsuchida, S., Umemura, H., Nakayama, T. et al. (2020). Current status of matrix-assisted laser desorption/ionization–time-of-flight mass spectrometry (MALDI-TOF MS) in clinical diagnostic microbiology. *Molecules* 25 (20): 4775.

33 Lartigue, M.F. (2013). Matrix-assisted laser desorption ionization time-of-flight mass spectrometry for bacterial strain characterization. *Infection, Genetics and Evolution* 13 (1): 230–235.

34 Buchan, B.W. and Ledeboer, N.A. (2013). Advances in identification of clinical yeast isolates by use of matrix-assisted laser desorption ionization-time of flight mass spectrometry. *J Clin Microbiol* 51 (5): 1359–1366.

35 Hrabák, J., Študentová, V., Walková, R. et al. (2012). Detection of NDM-1, VIM-1, KPC, OXA-48, and OXA-162 carbapenemases by matrix-assisted laser desorption ionization-time of flight mass spectrometry. *J Clin Microbiol* 50 (7): 2441–2443.

36 Du, X., Yuan, L., Gao, S. et al. (2023). Research progress on nanomaterial-based matrices for matrix-assisted laser desorption/ionization time-of-flight mass spectrometry analysis. *J Chromatogr A* 1712: 464493.

37 Tsuchida, S., Umemura, H., Nakayama, T. et al. (2020). Current status of matrix-assisted laser desorption/ionization–time-of-flight mass spectrometry (MALDI-TOF MS) in clinical diagnostic microbiology. *Molecules* 25 (20): 4775.

38 Schmidt, H. and Roahtgi, R. (2020). High-throughput flow cytometry assay to investigate TDP43 splicing function. *Bio Protoc* 10 (8): e3594.

39 Hiramatsu, K., Ideguchi, T., Yonamine, Y. et al. (2019). High-throughput label-free molecular fingerprinting flow cytometry. *Sci Adv* 5 (1): eaau0241.

40 Etcheverry, S., Faridi, A., Ramachandraiah, H. et al. (2017). High performance micro-flow cytometer based on optical fibres. *Scientific Reports* 7 (1): 1–8.

41 Piyasena, M.E. and Graves, S.W. (2014). The intersection of flow cytometry with microfluidics and microfabrication. *Lab Chip* 14 (6): 1044–1059.

42 Özyurt, C., Uludağ, İ., İnce, B., and Sezgintürk, M.K. (2023). Lab-on-a-chip systems for cancer biomarker diagnosis. *J Pharm Biomed Anal* 226: 115266.

43 Dervisevic, E., Tuck, K.L., Voelcker, N.H., and Cadarso, V.J. (2019). Recent progress in lab-on-a-chip systems for the monitoring of metabolites for mammalian and microbial cell research. *Sensors* 19 (22): 5027.

44 Kutter, J.P. (2012). Liquid phase chromatography on microchips. *J Chromatogr A* 1221: 72–82.

45 Tong, Z., Shen, C., Li, Q. et al. (2023). Combining sensors and actuators with electrowetting-on-dielectric (EWOD): advanced digital microfluidic systems for biomedical applications. *Analyst* 148 (7): 1399–1421.

46 Conde, J.P., Madaboosi, N., Soares, R.R.G. et al. (2016). Lab-on-chip systems for integrated bioanalyses. *Essays Biochem* 60 (1): 121–131.

47 Lu, Y., Wang, J., Fu, L. et al. (2023). High-throughput and separating-free phenotyping method for on-panicle rice grains based on deep learning. *Front Plant Sci* 14: 1219584.

48 Neilson, E.H., Edwards, A.M., Blomstedt, C.K. et al. (2015). Utilization of a high-throughput shoot imaging system to examine the dynamic phenotypic responses of a C4 cereal crop plant to nitrogen and water deficiency over time. *J Exp Bot* 66 (7): 1817–1832.

49 Abebe, A.M., Kim, Y., Kim, J. et al. (2023). Image-based high-throughput phenotyping in horticultural crops. *Plants* 12 (10): 2061.

50 de Groot, R., Lüthi, J., Lindsay, H. et al. (2018). Large-scale image-based profiling of single-cell phenotypes in arrayed CRISPR-Cas9 gene perturbation screens. *Mol Syst Biol* 14 (1): 8064.

51 Ivanov, S., Soliman, M., Tuomi, A. et al. (2024). Drivers of generative AI adoption in higher education through the lens of the theory of planned behaviour. *Technol Soc* 77: 102521.

52 Hamilton, A.J., Strauss, A.T., Martinez, D.A. et al. (2021). Machine learning and artificial intelligence: applications in healthcare epidemiology. *Antimicrobial Stewardship & Healthcare Epidemiology* 1 (1): 28.

53 Tack, C. (2019). Artificial intelligence and machine learningl applications in musculoskeletal physiotherapy. *Musculoskelet Sci Pract* 39: 164–169.

54 Soori, M., Arezoo, B., and Dastres, R. (2023). Machine learning and artificial intelligence in CNC machine tools, a review. *Sustainable Manufacturing and Service Economics* 2: 100009.

55 Yadav, K., Yadav, K., Bidnyk, S. et al. (2024). Artificial intelligence and machine learning in optics: tutorial. *JOSA B* 41 (8): 1739–1753.

56 Oladipupo, T. (2010) Types of Machine Learning Algorithms, institute for new technologies.

57 Dike, H.U., Zhou, Y., Deveerasetty, K.K., and Wu, Q. (2018) Unsupervised learning based on artificial neural network: a review. 322–327.

58 Tulczyjew, L., Nalepa, J., and Kawulok, M. (2021). Unsupervised feature learning using recurrent neural nets for segmenting hyperspectral images. *IEEE Geoscience and Remote Sensing Letters* 18 (12): 2142–2146.

59 Li, Y., Wang, R., and Yang, Z. (2022). Optimal scheduling of isolated microgrids using automated reinforcement learning-based multi-period forecasting. *IEEE Trans Sustain Energy* 13 (1): 159–169.

60 Bagherzadeh, J. and Asil, H. (2019). A review of various semi-supervised learning models with a deep learning and memory approach. *Iran Journal of Computer Science* 2 (2): 65–80.

61 Ayush, K., Lobell, D., Burke, M., Tanmay, K., Meng, C., Ermon, S., and Uzkent, B. (2021) Geography-aware Self-supervised learning. 10161–10170.

62 Pantanowitz, L., Pearce, T., Abukhiran, I. et al. (2025). Nongenerative artificial intelligence in medicine: advancements and applications in supervised and unsupervised machine learning. *Modern Pathology* 38 (3): 100680.

63 Al Mudawi, N. and Alazeb, A. (2022). A model for predicting cervical cancer using machine learning algorithms. *Sensors* 22 (11): 4132.

64 Al-Batah, M., Alqaraleh, M., and Salem Alzboon, M. (2025). Superior classification of Brain cancer types through machine learning techniques applied to magnetic resonance imaging. *Data and Metadata* 4: 472.

65 Alsalem, M.A., Mohammed, R., Albahri, O.S. et al. (2022). Rise of multiattribute decision-making in combating COVID-19: A systematic review of the state-of-the-art literature. *International Journal of Intelligent Systems* 37 (6): 3514–3624.

66 Peiffer-Smadja, N., Dellière, S., Rodriguez, C. et al. (2020). Machine learning in the clinical microbiology laboratory: has the time come for routine practice? *Clinical Microbiology and Infection* 26 (10): 1300–1309.

67 Yee, N., Su, C., Iwendi, C. et al. (2023). Machine learning algorithms applied to predict autism spectrum disorder based on gut microbiome composition. *Biomedicines* 11 (10): 2633.

68 Topçuoğlu, B.D., Lesniak, N.A., Ruffin, M.T. et al. (2020). A framework for effective application of machine learning to microbiome-based classification problems. *mBio* 11 (3).

69 Weis, C.V., Jutzeler, C.R., and Borgwardt, K. (2020). Machine learning for microbial identification and antimicrobial susceptibility testing on MALDI-TOF mass spectra: a systematic review. *Clinical Microbiology and Infection* 26 (10): 1310–1317.

70 Beaudry, M.S., Rhodes, O.E., Devault, A. et al. (2021). Improved microbial community characterization of 16S rRNA via metagenome hybridization capture enrichment. *Front Microbiol* 12: 644662. (e001202–20).

71 Zhou, Z., Xu, L., Zhu, L. et al. (2021). Metagenomic analysis of microbiota and antibiotic resistome in household activated carbon drinking water purifiers. *Environ Int* 148: 106394.

72 Galal, A., Talal, M., and Moustafa, A. (2022). Applications of machine learning in metabolomics: Disease modeling and classification. *Front Genet* 13: 1017340.

73 Cres, C.M., Tritt, A., Zhang, Y., and Bouchard, K.E. (2023). DL-TODA: a deep learning tool for omics data analysis. *Biomolecules* 13 (4): 585.

74 Deng, Z.-L., Mchardy, A.C., Münch, P.C., and Mreches, R. (2022). Rapid and accurate identification of ribosomal RNA sequences via deep learning. *Nucleic Acids Res* 50 (10): 60.

75 Chen, X.-F., Hou, X., Xiao, M. et al. (2021). Matrix-assisted laser desorption/ionization time of flight mass spectrometry (MALDI-TOF MS) analysis for the identification of pathogenic microorganisms: a review. *Microorganisms* 9 (7): 1536.

76 Wang, X., Su, X., Cui, X., and Ning, K. (2015, 2015). MetaBoot: a machine learning framework of taxonomical biomarker discovery for different microbial communities based on metagenomic data. *PeerJ* 7: e993.

77 Fiannaca, A., La Paglia, L., La Rosa, M. et al. (2018). Deep learning models for bacteria taxonomic classification of metagenomic data. *BMC Bioinformatics* 19 (7): 61–76.

78 Miao, J., Chen, T., Misir, M., and Lin, Y. (2024). Deep learning for predicting 16S rRNA gene copy number. *Scientific Reports* 14 (1): 1–14.

79 Dubois, D., Leyssene, D., Chacornac, J.P. et al. (2009). Identification of a variety of staphylococcus species by matrix-assisted laser desorption ionization-time of flight mass spectrometry. *J Clin Microbiol* 48 (3): 941–945.

80 Oviaño, M. and Bou, G. (2018). Matrix-assisted laser desorption ionization-time of flight mass spectrometry for the rapid detection of antimicrobial resistance mechanisms and beyond. *Clin Microbiol Rev* 32 (1): e00037–18.

81 Hamad, B.K., Noor, A.M., and Rahim, A.A. (2011). Removal of 4-chloro-2-methoxyphenol from aqueous solution by adsorption to oil palm shell activated carbon activated with K_2CO_3. *Journal of Physical Science* 22 (1): 39–55.

82 Feucherolles, M., Nennig, M., Becker, S.L. et al. (2022). Combination of MALDI-TOF mass spectrometry and machine learning for rapid antimicrobial resistance screening: the case of campylobacter spp. *Front Microbiol* 12: 804484.

83 Majchrowska, S., Drulis-Kawa, Z., Bonus, T., Roszkowiak, J., Hanas, A., Golan, T., Gula, G., Pawlowski, J., Pawlak, A., and Loch, A. (2021) AGAR a mkicrobial colony dataset for deep learning detection.

84 Ciaparrone, G., Li, X., Memmolo, P. et al. (2024). Label-free cell classification in holographic flow cytometry through an unbiased learning strategy. *Lab Chip* 24 (4): 924–932.

85 Nissim, N., Barnea, I., Shaked, N.T., and Dudaie, M. (2020). Real-time stain-free classification of cancer cells and blood cells using interferometric phase microscopy and machine learning. *Cytometry Part A* 99 (5): 511–523.

86 Gu, Y., Han, Y., Li, J. et al. (2019). Machine learning based real-time image-guided cell sorting and classification. *Cytometry Part A* 95 (5): 499–509.

87 Özel Duygan, B.D., Hadadi, N., Babu, A.F. et al. (2020). Rapid detection of microbiota cell type diversity using machine-learned classification of flow cytometry data. *Communications Biology* 3 (1): 1–13.

88 Martis, J.E., Sudeepa, K.B., Sannidhan, M.S., and Bhandary, A. (2020) A rapid automated process for organizing bacterial cluster segments using deep neural networks. *Proceedings of the 3rd International Conference on Smart Systems and Inventive Technology, ICSSIT 2020*, 963–968.

89 Khaleghi, M.K., Lewis, N.E., Shojaosadati, S.A., and Savizi, I.S.P. (2021). Synergisms of machine learning and constraint-based modeling of metabolism for analysis and optimization of fermentation parameters. *Biotechnol J* 16 (11): 2100212.

90 Rienzo, M., Jiang, H., Valley, J.K. et al. (2021). High-throughput optofluidic screening for improved microbial cell factories via real-time micron-scale productivity monitoring. *Lab Chip* 21 (15): 2901–2912.

91 Steinmann, S.N., Wang, Q., and Seh, Z.W. (2023). How machine learning can accelerate electrocatalysis discovery and optimization. *Mater Horiz* 10 (2): 393–406.

92 Ingham, C.J., Van Den Berg, A., Molenaar, D., De Vos, W.M., Sprenkels, A., Van Hylckama Vlieg, J.E.T., and Bomer, J. (2007) The micro-Petri dish, a million-well growth chip for the culture and high-throughput screening of microorganisms. *Proceedings of the National Academy of Sciences*, **104** (46), 18217–18222.

93 Abu-Absi, N.R., Hausladen, M.C., Borys, M.C. et al. (2010). Real time monitoring of multiple parameters in mammalian cell culture bioreactors using an in-line Raman spectroscopy probe. *Biotechnol Bioeng* 108 (5): 1215–1221.

94 Chen, Z., Yao, J., Ma, B. et al. (2022). A robust biocatalyst based on laccase immobilized superparamagnetic $Fe_3O_4@SiO_2–NH_2$ nanoparticles and its application for degradation of chlorophenols. *Chemosphere* 291: 132727.

95 Treloar, N.J., Ingalls, B., Fedorec, A.J.H., and Barnes, C.P. (2020). Deep reinforcement learning for the control of microbial co-cultures in bioreactors. *PLoS Comput Biol* 16 (4): e1007783.

96 Park, S., Lee, D., Park, C. et al. (2023). Data-driven prediction models for forecasting multistep ahead profiles of mammalian cell culture towardbioprocess digital twins. *Biotechnol Bioeng* 120 (9): 2494–2508.

97 Hartmann, J.F., Keppler, F., Günthel, M. et al. (2020). High spatiotemporal dynamics of methane production and emission in oxic surface water. *Environmental Science & amp Technology* 54 (3): 1451–1463.

98 Wong, B.G., Bashor, C.J., Kiriakov, S. et al. (2018). Precise, automated control of conditions for high-throughput growth of yeast and bacteria with eVOLVER. *Nat Biotechnol* 36 (7): 614–623.

99 Betts, J.I. and Baganz, F. (2006). Miniature bioreactors: current practices and future opportunities. *Microb Cell Fact* 5 (1): 21.

100 Eitzinger, S., Asif, A., Watters, K.E. et al. (2020). Machine learning predicts new anti-CRISPR proteins. *Nucleic Acids Res* 48 (9): 4698–4708.

101 Mendoza, B.J. and Trinh, C.T. (2017). Enhanced guide-RNA design and targeting analysis for precise CRISPR genome editing of single and consortia of industrially relevant and non-model organisms. *Bioinformatics* 34 (1): 16–23.

102 Liao, X., Jameel, A., Xing, X.-H. et al. (2021). A versatile toolbox for CRISPR-based genome engineering in Pichia pastoris. *Appl Microbiol Biotechnol* 105 (24): 9211–9218.

103 Bucholc, M., Shand, C., Dehsarvi, A. et al. (2023). Artificial intelligence for dementia research methods optimization. *Alzheimers Dement* 19 (12): 5934–5951.

104 Patil, D., Desai, P., Rane, N.L., and Rane, J. (2024) Machine learning and deep learning: Methods, techniques, applications, challenges, and future research opportunities, deep science.

105 Chen, C., Pan, D., Liu, G. et al. (2023). Applications of multi-omics analysis in human diseases. *MedComm (Beijing)* 4 (4): e315.

106 Husnain, A., Saeed, A., Rasool, S., and Hussain, H.K. (2023). Revolutionizing pharmaceutical research: harnessing machine learning for a paradigm shift in drug discovery. *International Journal of Multidisciplinary Sciences and Arts* 2 (2): 149–157.

107 Udegbe, F., Ekesiobi, C., Ebulue, O., and Ebulue, C. (2024). Machine learning in drug discovery: a critical review of applications and challenges. *Computer Science & amp IT Research Journal* 5 (4): 892–902.

108 Ratneswaran, A., Rockel, J.S., and Kapoor, M. (2020). Understanding osteoarthritis pathogenesis: a multiomics system-based approach. *Curr Opin Rheumatol* 32 (1): 80–91.

109 Molla, G. and Bitew, M. (2024). Revolutionizing personalized medicine: synergy with multi-omics data generation, main hurdles, and future perspectives. *Biomedicines* 12 (12): 2750.

110 Terranova, N., Venkatakrishnan, K., and Benincosa, L.J. (2021). Application of machine learning in translational medicine: current status and future opportunities. *AAPS J* 23 (4): 74.

111 Heidelberg, K.B., Gilbert, J.A., and Joint, I. (2010). Marine genomics: at the interface of marine microbial ecology and biodiscovery. *Microb Biotechnol* 3 (5): 531–543.

112 Cembrowska-Lech, D., Adamski, C., Mikiciuk, M. et al. (2023). An integrated multi-omics and artificial intelligence framework for advance plant phenotyping in horticulture. *Biology (Basel)* 12 (10): 1298.

9

Generative AI for Knowledge Mining of Synthetic Biology and Bioprocess Engineering Literature

Zhengyang Xiao and Yinjie J. Tang

Washington University in St. Louis, Department of Energy, Environmental, and Chemical Engineering, St. Louis 63130, MO, USA

9.1 Introduction

Synthetic biology (SynBio) and bioprocess engineering are historically constrained by the stochastic nature of biological systems [1, 2] and the sheer volume of multidisciplinary data [3, 4]. Over the past decade, synthetic biology has moved from a proof-of-concept field to industrial applications, producing natural products [5, 6], RNA vaccine [7], antibody drugs [8], etc. Yet, designing efficient microbial strains and optimizing bioreactor conditions remain iterative and labor-intensive processes. Researchers rely on fragmented datasets – scattered across lab notebooks, proprietary databases, and journal publications – often requiring manual curation to form actionable insights. Prior chapters often highlight the importance of curated data, but it is crucial to understand the challenges and methods involved in this curation process, which is a key focus of this chapter. GenAI and LLMs revolutionize this field by understanding relevant knowledge to perform human-like planning and tasks. LLMs have been evolving from earlier instruction-tuned model GPT-4 [9], Llama [10], and Qwen [11], to recent chain-of-thought reasoning models like OpenAI o1 and DeepSeek-R1 [12]. LLMs support applications like SynBio Q&A chatbot [13, 14], DNA sequence model [15], and autonomous protein engineering [16]. LLMs also show potential to integrate multimodal data (e.g. genetics, pathways, and bioreactor operations) into actionable insights [17].

Bioprocess development must integrate legacy biotechnologies with modern SynBio advancements – such as genetic circuits, heterologous pathways, and CRISPR-enabled strain engineering – to optimize cell performance. These multifaceted challenges highlight the need for GenAI tools to drive the next generation of biomanufacturing. The chapter begins by examining knowledge mining tools like NEKO and GraphRAG, which convert fragmented and heterogeneous literature into knowledge graphs. These tools enable researchers to pose SynBio-related questions and receive answers and SynBio information supported by cited, up-to-date evidence. Next, a protocol for using LLMs to automate the extraction of structured

Machine Learning and Big Data-enabled Biotechnology, First Edition. Edited by Hal S. Alper.

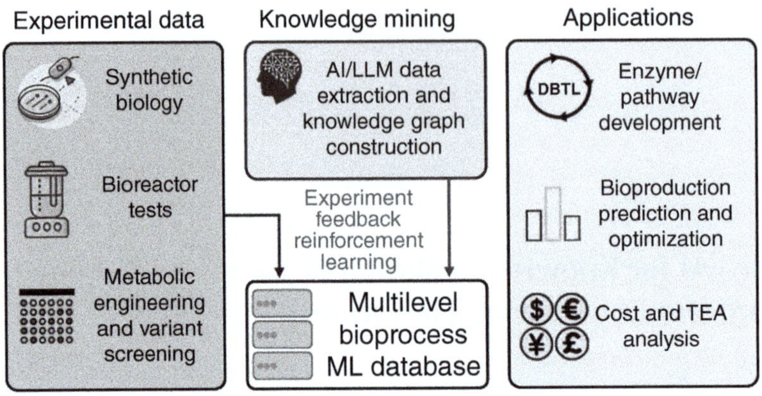

Figure 9.1 Integrating experiment data and AI-driven knowledge mining to form a multilevel bioprocess machine learning (ML) database.

bioprocess data, such as strain engineering parameters and reactor conditions, from scientific publications is introduced. This approach accelerates dataset construction for ML applications, providing valuable insights to guide future SynBio experiments and bioprocess development (Figure 9.1).

9.2 Text Mining Using Knowledge Graph Tools

LLMs excel at answering general scientific questions but are limited by their pretraining cut-off dates and cannot provide citations without RAG. Additionally, current LLMs face scaling law limitations, where exponential increases in computational resources and model parameters yield only marginal, linear improvements in performance. This reflects an inherent conflict between the vastness of available information and the finite capacity of LLM parameters. To address these challenges, protocols for augmenting LLMs with knowledge graphs are proposed, enabling more effective knowledge extraction from extensive text sources. This approach is implemented by using Ollama, an all-in-one local LLM deployment package, to run Qwen2.5-instruct-14B (Alibaba Cloud, China). Hardware constraints influence deployment choices: a 14-billion-parameter model requires approximately 12 GB of GPU/CPU memory, making cloud APIs a more viable option for users with limited computational resources. To run LLM package locally, the user could begin by installing Ollama and Python 3.11. All codes used in this chapter are available on Github https://github.com/xiao-zhengyang/Advanced-Biotechnology-Series. User gets better results with larger parameters LLM provided by cloud computing. An example code using the OpenAI API is also provided on GitHub.

9.2.1 NEKO: A Lightweight Knowledge Graph Tool

NEKO allows researchers to extract relationships between biological entities (genes, species, bioreactor parameters) from PubMed abstracts or full-text articles [18].

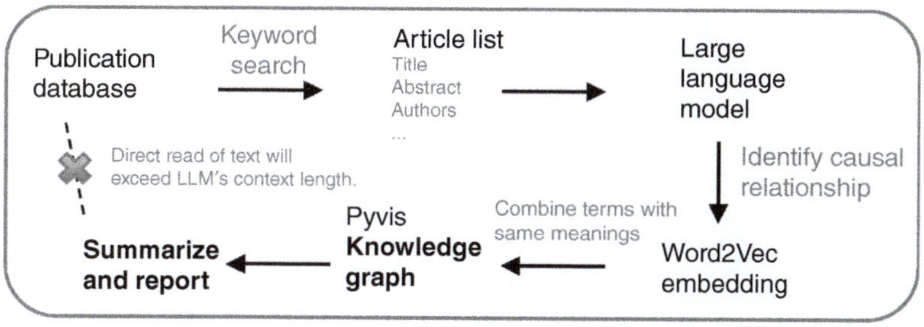

Figure 9.2 A schematic overview of the NEKO workflow for searching a publication database, identifying causal relationships, and visualizing and summarizing findings in a knowledge graph.

This system addresses two challenges: the manual effort to correlate findings from hundreds of papers, and the inability of general-purpose LLMs to provide responses with citations. Here, this section demonstrates how NEKO generates interactive knowledge graphs that map causal interactions (e.g. enzyme-substrate relationships) based on user-defined keywords (e.g. "*Yarrowia lipolytica*") (Figure 9.2). The implementation code is available on GitHub and users can run the workflow within one Jupyter notebook (https://github.com/xiao-zhengyang/Advanced-Biotechnology-Series/tree/main/NEKO).

Step 1: Literature Search and Text Collection NEKO initiates the process by retrieving relevant texts through PubMed's API. Users specify a keyword, such as "*Yarrowia lipolytica carotene production*". The `search_pubmed` function retrieves articles matching the search. The `fetch_details` function extracts article titles, journal names, and abstracts, compiling them into a structured Excel file. As in our example, the workflow fetches 500 abstracts, though this can be adjusted using the `retmax` parameter (up to ~9,000 abstracts per search). This step ensures that the input text incorporates the most recent publications, overcoming the pretraining cut-off limitations of LLMs.

Step 2: Entity and Relationship Extraction with LLM Abstracts are analyzed by the locally deployed LLM: Qwen2.5-14B via `Ollama` – to identify biological entities and their relationships. The LLM has a prompt tailored to SynBio tasks, instructing it to output entity pairs in a strict (Entity A, Entity B) format. Here, this section demonstrates an example of extracting entities from *Yarrowia* β-carotene production studies.

Step 3: Combine Entities with Similar Meanings Biological terminology varies across studies. Phrases like "C:N ratio" and "carbon-to-nitrogen ratio" refer to the same parameter. NEKO mitigates this ambiguity using an embedding to combine semantically similar terms. Entity pairs are converted into vector representations, and near-identical terms (cosine similarity >0.8) are mapped to a standardized vocabulary. For instance, "C/N ratio" and "C:N ratio" are unified under the latter term to ensure consistency. The cleaned datasets are saved for graph construction.

Step 4: Knowledge Graph, RAG, and Summarization Report This step assembles entities and relationships into a knowledge graph using `Pyvis`. Nodes

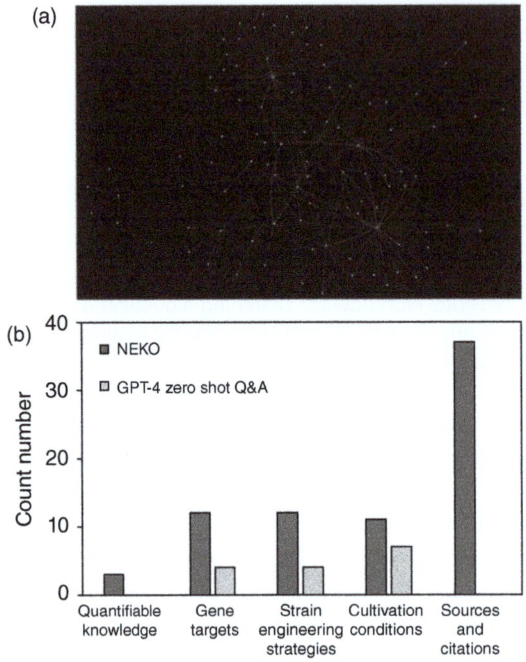

Figure 9.3 (a) A sample knowledge graph from NEKO workflow. The nodes are biological entities and the edges are labeled as citations. (b) Compared with GPT-4 zero-shot Q&A, NEKO can find β-carotene-specific titer, production genes, strain engineering methods, and cultivation conditions in the knowledge graph [18] / Elsevier / CC BY 4.0.

represent unique biological entities (e.g. genes, bioreactor parameters), while edges denote the sources (the titles of journal articles). Each relationship is traceable to its source article in the Excel file dataset. Users can query the graph to identify connections between specific nodes, for example, what are the entities related to β-carotene-producing *Yarrowia lipolytica*, and generate summaries via RAG and LLM interpretation of linked relationships (Figure 9.3a). Under best practices with a more capable 72B LLM, NEKO workflow can produce a summary report more specific and reliable than GPT-4 zero-shot question and answer (Figure 9.3b).

9.2.2 GraphRAG

GraphRAG is an advanced but computationally expensive framework to enhance LLMs in generating domain-specific outputs. In contrast to NEKO, GraphRAG integrates three distinct query modes – global, local, and DRIFT search – enabling comprehensive questioning-answering capabilities through natural language input. This section demonstrates steps for using LLM and GraphRAG to study β-carotene production in *Yarrowia lipolytica* (Figure 9.4). Codes are in GitHub repository https://github.com/xiao-zhengyang/Advanced-Biotechnology-Series/tree/main/GraphRAG.

Step 1: Configure Ollama LLM Ollama's default LLM context length (e.g. 2 k tokens) truncates long texts. To avoid this, a customized LLM with longer context is needed. Create an 8 k context length `qwen2.5:14b` LLM by running this in the terminal.

```
ollama show qwen2.5:14b --modelfile > settings.txt
```

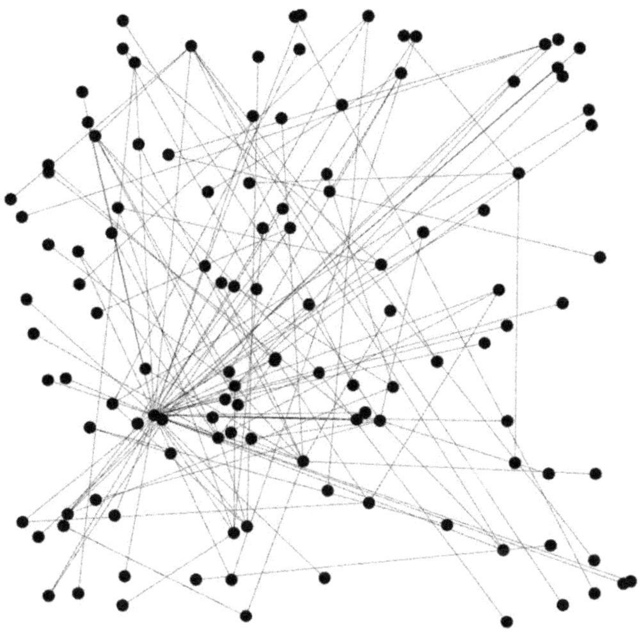

Figure 9.4 A sample knowledge graph from GraphRAG.

This command generates a `settings.txt` file in the terminal's current directory. Modify the model's `settings.txt` file by adding one line specifying the context length.

```
PARAMETER num_ctx 8196
```

Then run this command below to create a custom `qwen2.5:14b_8k model.`

```
ollama create qwen2.5:14b_8k -f settings.txt
```

Step 2: Initialize GraphRAG and Prompt Tuning The GraphRAG version used in this section was 1.2.0. The user can install it by pip.

```
pip install graphrag==1.2.0
```

The user can initialize a new GraphRAG project by typing the following in the terminal:

```
graphrag init --root. /[Yarrowia_local]
```

For illustration purposes, a preconfigured GraphRAG project, including preloaded input text, is available in the GitHub repository for user convenience. For systems without GPU, increase the `request_timeout` value to ≥`10 000` seconds in the `settings.yaml` to prevent interruptions. Prompt engineering is important, as entity-relationship extraction relies on domain-specific instructions. The user should use the autotune command to generate context-aware prompts with examples.

```
python -m graphrag prompt-tune --root. /[Yarrowia_local]
--config. / [Yarrowia_local]/settings_prompt_tune.yaml
```

Note that large-parameter LLMs (>70B) should be used for prompt generation to minimize manual corrections, as smaller LLMs may yield suboptimal prompts.

In this example, GPT-4o is used for prompt tuning. The user should substitute their own OpenAI API key in the `settings_prompt_tune.yaml` file. After automatic prompt tuning, move the generated prompts to the project's prompt folder.

Step 3: Indexing the Input Text Execute GraphRAG indexing command to initiate structured knowledge graph generation from input text. `Graphrag index --root. /[Yarrowia_local]`

Processing times vary with the user's hardware. CPU-based systems typically require more than 10 hours, while utilizing GPU acceleration reduces processing time to under 30 minutes.

Step 4: Querying and Visualization GraphRAG's three query modes enable targeted interrogation of the knowledge graph by using LLM and RAG. Global search provides high-level summaries of broad topics. Local search retrieves details. The generated knowledge graph is saved in the output folder and can be exported to visualization tools like Gephi or Cytoscape, allowing users to interactively explore its connections.

9.3 LLM-Automated Data Extraction for Machine Learning

The efficacy of ML highly depends on a large, structured database. Traditional data collection in synthetic biology relied on the manual extraction of experimental results from journal articles. This process is time-consuming, error-prone, and prone to inconsistencies due to subjective interpretations. Generative AI now offers automated extraction of data at scale. This paradigm shift not only accelerates dataset creation but also enhances reproducibility and unlocks data-driven insights. This section presents a scalable protocol for converting unstructured text into ML-ready datasets (Figure 9.5). An example code with both input text and

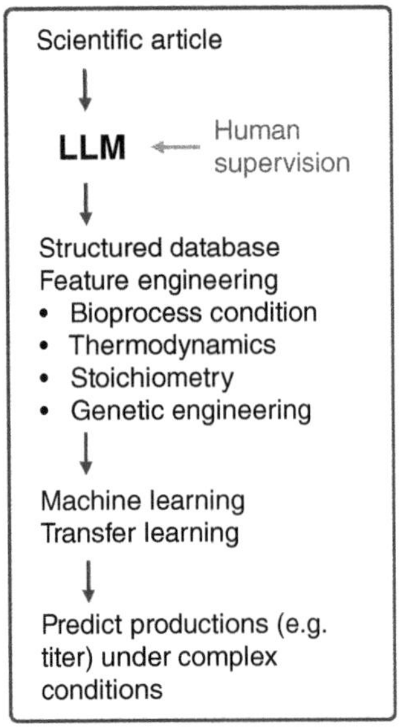

Figure 9.5 Scientific articles are converted into structured features by LLM, and then used in ML to predict production metrics under complex conditions.

output CSV is provided on GitHub https://github.com/xiao-zhengyang/Advanced-Biotechnology-Series/tree/main/Extract_data_by_LLM.

Step 1: Gathering Text and Preprocessing Peer-reviewed articles are downloaded as PDFs and converted to individual.txt files. One python code reads files and handles encoding errors common in parsed PDFs, for example, special characters or line breaks. Invalid files are flagged to prevent downstream problems.

Step 2: Structured Data Extraction via LLM The processed text files are processed by LLM in bulk (using OpenAI API). The prompt example used is:

> "*Extract a multicolumn spreadsheet from the given text | article doi | product | product concentration | Carbon source | Carbon source concentration | yield | media | time | mode of bioreactor operation | strain name | parent strain | knocked out genes | overexpressed native genes | heterologous genes| Promotor | genome integration Y/N | codon optimized | culture volume | culture vessel: Shake flask, Batch, fedbatch, chemostat, microplates | nitrogen condition: sufficient, limited |oxygen condition: sufficient, limited, anaerobic | pH | temperature | Direct evolution Y/N |When outputting the daughter strain, you need to include the expressed and knockout genes in the parent strain. Always provide full table if data is not provided, just leave as NA. Do not provide explanations and notes.*"

This prompt minimizes hallucinations by restricting responses to tabular formats and explicitly requiring null values for missing data. Here, OpenAI o1-mini is used for a balance of performance and cost.

Step 3: Automated CSV Generation and Human Validation Extracted tables are split into rows using delimiters and saved as individual.csv files. A domain expert user could review outputs to verify accuracy.

9.4 Current Limitations

While GenAI and LLMs, empowered by knowledge graph and RAG, offer significant potential for SynBio applications, their integration into real-world workflows faces several challenges. *First*, infrastructure limitations persist. The performance and accuracy of LLMs are heavily dependent on their parameter size, and models larger than 70 billion parameters typically require cloud-based deployment. This raises data security concerns for proprietary research or sensitive data, which may deter academic and industrial adoption. Transmitting sensitive information to third-party cloud services introduces risks, even with encryption. Applications governed by strict regulations (e.g. healthcare) face compliance challenges, while models trained on sensitive datasets may inadvertently memorize and expose proprietary information. Securing local deployment of LLMs requires high-end GPUs, which remain cost-prohibitive for many laboratories. Until hardware companies prioritize local AI workflows, balancing performance and privacy will continue to be a challenge. *Second*, there are scalability limits to knowledge graphs. As the input text grows beyond approximately 5,000 articles, information retrieval and query efficiency decline. Additionally, LLMs struggle to reconcile conflicting information or non-mainstream viewpoints, which can propagate errors in reasoning. *Third*, LLM-driven data extraction is constrained by predefined features, creating a "feature bias" that limits the scope of ML explanations to only those predefined inputs. Overcoming these limitations will require collaboration across the SynBio and bioprocess engineering communities to establish standardized data reporting practices that enhance ML explainability.

9.5 Conclusion

The integration of generative AI into synthetic biology and bioprocess engineering represents a new direction toward data-driven biological design and manufacturing. Tools such as NEKO and GraphRAG exemplify how knowledge mining can combine fragmented scientific literature into actionable insights, bridging gaps between experimental data, strain engineering, and bioreactor optimization. Meanwhile, LLM-driven protocols for dataset generation address the longstanding bottleneck of manual data curation, enabling predictive ML studies, self-driving labs and digital twins. Ultimately, the synergy of generative AI with synthetic biology gives a future biomanufacturing foundation model.

Acknowledgments

This study is funded by the United States National Science Foundation (NSF) award number 2225809 and DOE Energy Earthshots award number DE-SC0024702. This work was also partially supported by the NSF under Award No. 2330245, which funds the Engineering Research Center for Carbon Utilization Redesign through Biomanufacturing-Empowered Decarbonization (CURB).

References

1 Elowitz, M.B., Levine, A.J., Siggia, E.D., and Swain, P.S. (2002). Stochastic gene expression in a single cell. *Science* 297 (5584): 1183–1186.
2 Paulsson, J. (2005). Models of stochastic gene expression. *Phys Life Rev* 2 (2): 157–175.
3 Czajka, J.J., Oyetunde, T., and Tang, Y.J. (2021). Integrated knowledge mining, genome-scale modeling, and machine learning for predicting Yarrowia lipolytica bioproduction. *Metab. Eng.* 67: 227–236.
4 Winkler, J.D., Halweg-Edwards, A.L., and Gill, R.T. (2015). The LASER database: Formalizing design rules for metabolic engineering. *Metab. Eng. Commun.* 2: 30–38.
5 Ma, Y. et al. (2022). Removal of lycopene substrate inhibition enables high carotenoid productivity in Yarrowia lipolytica. *Nat. Commun.* 13 (1): 572.

6 Czajka, J.J. et al. (2018). Engineering the oleaginous yeast Yarrowia lipolytica to produce the aroma compound beta-ionone. *Microb. Cell Factories* 17 (1): 136.

7 Zhang, H. et al. (2023). Algorithm for optimized mRNA design improves stability and immunogenicity. *Nature* 621: 396–403.

8 Voronina, E.V. et al. (2016). Design of a stable cell line producing a recombinant monoclonal anti-TNFalpha antibody based on a CHO cell line. *Springerplus* 5 (1): 1584.

9 OpenAI, GPT-4 Technical report. arXiv preprint, 2023. 2303.08774.

10 Touvron, H., et al., 2023. Llama: Open and efficient foundation language models. arXiv preprint arXiv:2302.13971.

11 Bai, J., et al., 2023. Qwen technical report. arXiv preprint arXiv:2309.16609.

12 Guo, D., et al., 2025. Deepseek-r1: Incentivizing reasoning capability in llms via reinforcement learning. arXiv preprint arXiv:2501.12948.

13 Li, W. et al. (2025). Large language model for knowledge synthesis and AI-enhanced biomanufacturing. *Trends Biotechnol.* S0167-7799 (25): 00045–00049.

14 Mao, Z., et al., 2025. SynBioGPT: A retrieval-augmented large language model Platform for AI-Guided Microbial Strain Development. bioRxiv.

15 Le, N.Q.K. et al. (2021). A transformer architecture based on BERT and 2D convolutional neural network to identify DNA enhancers from sequence information. *Brief. Bioinform.* 22 (5): bbab005.

16 Singh, N. et al. (2025). A generalized platform for artificial intelligence-powered autonomous protein engineering. *bioRxiv* 16: 5648.

17 Xiao, Z. et al. (2023). Generative artificial intelligence GPT-4 accelerates knowledge mining and machine learning for synthetic biology. *ACS Synth. Biol.* 12 (10): 2973–2982.

18 Xiao, Z. et al. (2025). Network for knowledge Organization (NEKO): an AI knowledge mining workflow for synthetic biology research. *Metab. Eng.* 87: 60–67.

10

Metabolomics: Big Data Approaches

Kenya Tanaka[1,2,3], *Christopher J. Vavricka*[4], *and Tomohisa Hasunuma*[1,2,5,6]

[1]*Engineering Biology Research Center, Kobe University, 1-1 Rokkodai, Kobe 657-8501, Japan*
[2]*Graduate School of Science, Innovation and Technology, Kobe University, 1-1 Rokkodai, Kobe 657-8501, Japan*
[3]*Research Center for Solar Energy Chemistry, Graduate School of Engineering Science, Osaka University, Machikaneyama, Toyonaka, Osaka 560-8531, Japan*
[4]*Department of Biotechnology and Life Science, Graduate School of Engineering, Tokyo University of Agriculture and Technology, 2-24-16 Naka-cho, Koganei, Tokyo 184-8588, Japan*
[5]*RIKEN, Research Center for Sustainable Resource Science, 1-7-22 Suehiro, Tsurumi, Yokohama, Kanagawa 230-0045, Japan*
[6]*Department of Chemical Science and Engineering, Faculty of Engineering, Kobe University, 1-1 Rokkodai, Kobe 657-8501, Japan*

10.1 Introduction

Metabolomics is a powerful analytical approach that comprehensively analyzes small-molecule metabolites within biological systems to elucidate cellular metabolic states. Recent advancements in analytical technologies and big data processing have enabled the utilization of large-scale metabolomics datasets. However, metabolomics encompasses a diverse range of techniques, and selecting the appropriate method tailored to the specific research objective is essential for obtaining reliable data. Compared to other omics approaches, such as transcriptomics and proteomics, metabolomics deals with physicochemically diverse compounds, and metabolites generally have a much faster turnover rate than DNA or proteins. Unlike the sequence-derived function of enzymes, peptides, and DNA parts as described in prior chapter, the metabolome is more responsive and dynamic. As a result, selecting – and sometimes even developing – appropriate sample preparation and analytical methods tailored to the target metabolites is crucial [1]. Section 10.2 provides guidance on selecting suitable metabolome analytical techniques for specific research objectives.

Metabolomics has driven technological innovations across a wide range of fields, including medical science [2–7], environmental analysis [8], food evaluation [9], microbiome research [10], and synthetic biology [11, 12]. In particular, the field of synthetic biology has been rapidly advancing, allowing for the rational engineering of cells to enhance the production of target compounds. This progress has led to the development of sustainable biomanufacturing processes with reduced

Machine Learning and Big Data-enabled Biotechnology, First Edition. Edited by Hal S. Alper.

environmental impact. The analysis and applications of metabolomics data in synthetic biology will be discussed in Section 10.3. Furthermore, the use of artificial intelligence (AI)-driven metabolomics will be described in Section 10.4, followed by a discussion on future perspectives of metabolomics applications in Section 10.5. These methods of measuring metabolomes are critical for developing the models of flux analysis that are described in Chapters 11 and 12.

10.2 Methods for Metabolomics

The utilization of metabolomics can be broadly categorized into two main approaches: obtaining quantitative information and obtaining qualitative information [13]. The former involves cases where specific compounds have been predetermined for quantification, and their intracellular levels need to be measured. In such cases, standard reference compounds are analyzed in advance to confirm peak positions and establish calibration curves. The sample analysis results are then compared against these references to obtain quantitative values. This type of metabolomics analysis is referred to as targeted metabolomics. On the other hand, the latter approach is used for identifying nmetabolic differences between strains or extracting characteristic features from samples. In such cases, non-targeted analysis is employed. This method identifies compounds by mass spectrometry (MS) and database matching, eliminating the need for pre-prepared reference standards [14].

The sample preparation and analytical methods for metabolome analysis must be carefully selected based on the group of compounds being analyzed [15]. For instance, metabolic intermediates in central metabolic pathways such as glycolysis and the tricarboxylic acid cycle (TCA) cycle exhibit rapid turnover, making them highly susceptible to artificial variations caused by environmental changes during cell sampling. Conversely, metabolites with relatively slow turnover, such as lipids, are less affected by such changes. Additionally, sample preparation methods vary depending on the analytical instrument used. For example, when measuring hydrophilic compounds via GC-MS, trimethylsilylation (TMS) derivatization is often required to convert target analytes into volatile compounds [16].

10.2.1 Preparation of Samples

Metabolomics samples are generally prepared through multiple steps (Figure 10.1). The first step is quenching, where the metabolic state of the cells is rapidly halted, followed by cell collection. Next, the cells are disrupted to extract metabolites. This may be followed by protein removal or concentration through drying. The cold methanol method is the most commonly used approach for quenching [1]. In this method, cells are rapidly cooled by mixing cold methanol solution, effectively stopping metabolic reactions. The cells are then collected via low-temperature centrifugation. However, it is important to note that methanol can disrupt cell membranes, leading to leakage of intracellular metabolites before the

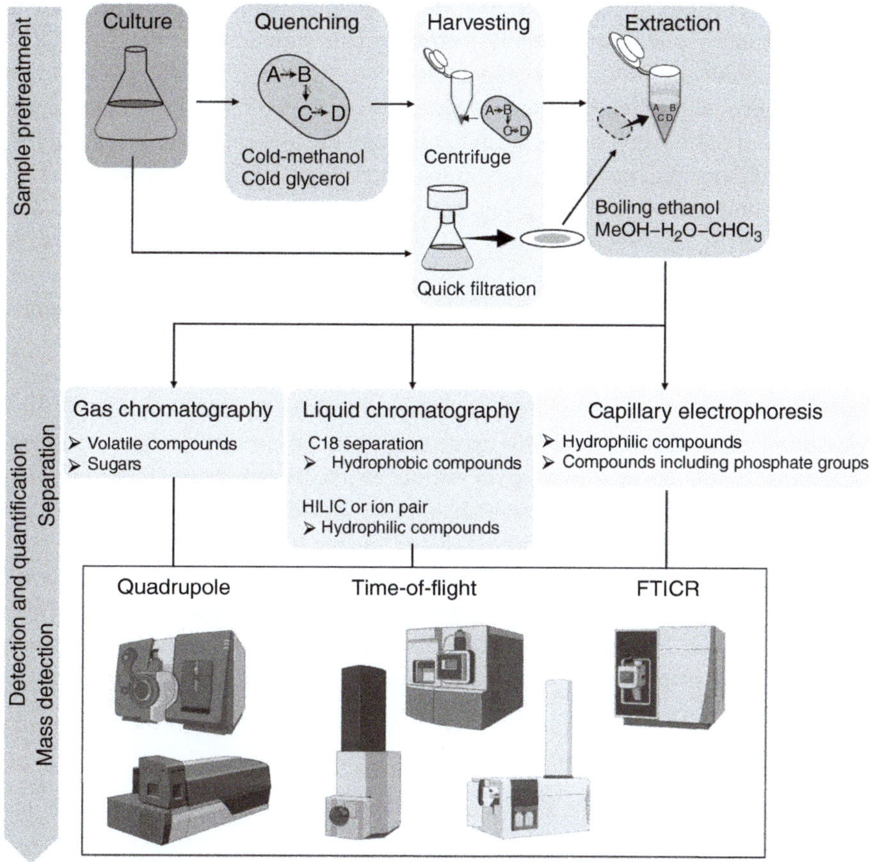

Figure 10.1 A schematic overview of a metabolomics workflow. The process begins with culturing cells, followed by quenching (e.g. cold methanol or cold glycerol) to halt metabolic activity. Harvesting is typically performed via centrifugation or quick filtration, and the extracted metabolites can be obtained using methods such as boiling ethanol or methanol–water–chloroform. Depending on the physicochemical properties of the compounds, gas chromatography (GC), liquid chromatography (LC), or capillary electrophoresis (CE) is used for separation. Finally, detection and quantification are conducted using various mass spectrometers (e.g. triple quadrupole, time-of-flight, Fourier transform ion cyclotron resonance).

extraction step. In *Chlamydomonas reinhardtii*, it has been reported that using 60% methanol containing 70 mM 2-(4-[2-hydroxyethyl]piperazin-1-yl)ethanesulfonic acid (HEPES) as the quenching solution, mixed with the sample at a 1 : 1 ratio at −40 °C, reduces metabolite leakage [17]. For cyanobacteria, enzymatic activity can be sufficiently halted using 30% methanol at −24 °C [18]. Furthermore, cold glycerol saline (GS) solution has been evaluated for intracellular metabolite recovery in various microbial models, including *Pseudomonas fluorescens*, *Streptomyces coelicolor*, and *Saccharomyces cerevisiae*. The results suggested that using GS solution led to significantly higher intracellular metabolite levels with less leakage compared to cold methanol-water solutions [19]. For plants and photosynthetic microorganisms, changes in light conditions during sampling can significantly alter metabolite intermediate concentrations. In plants, an *in situ* sampling device has been used to instantaneously switch between labeled and unlabeled $^{13}CO_2$ gas, enabling the measurement of metabolite labeling dynamics and the identification of rate-limiting steps in metabolic pathways [20]. In cyanobacteria, a phenol extraction-based rapid quenching and sampling method has been employed to preserve the *in vivo* state while measuring the metabolic state, thereby elucidating the activation dynamics of CO_2 fixation metabolism [21].

During the extraction step, intracellular metabolites are recovered via chemical lysis and/or mechanical disruption of cell walls. In microbial metabolomics, chemical lysis is the most commonly employed method. For extracting water-soluble compounds, approaches such as boiling ethanol and the buffered methanol-water-chloroform method are frequently used. However, variations in organic solvents, temperature, and pH are known to influence measurement results, and thus, employing multiple extraction methods is recommended [1]. For *Escherichia coli*, acidic acetonitrile has been suggested as the most suitable solvent for metabolite extraction [22]. For lipid extraction, traditional methods such as the Bligh and Dyer method and the Folch method, which use methanol and chloroform, have been widely used. However, alternative approaches utilizing green solvents such as ethanol and isopropanol, as well as supercritical CO_2 extraction, have been developed to reduce toxicity concerns [23].

10.2.2 Detection and Quantification

Chromatography combined with MS is widely used for metabolite analysis in metabolomics [24]. Gas chromatography (GC) and liquid chromatography (LC) are selected based on the target compounds. GC is primarily used for volatile compounds and sugars, while LC enables the analysis of organic acids, amino acids, and fatty acids. In LC, C18 columns, which utilize hydrophobic interactions, are commonly used. However, these columns are not well-suited for separating hydrophilic compounds, particularly those containing phosphate groups, such as sugar phosphates and nucleotides. The use of hydrophilic interaction LC (HILC) columns or ion-pair reagents can facilitate the analysis of hydrophilic compounds [25, 26]. Additionally, a comprehensive analytical method called unified-hydrophilic-interaction/anion-exchange LC has been developed to simultaneously analyze a wide range of compounds with diverse physicochemical properties [27]. Furthermore, capillary electrophoresis (CE) is particularly suited for separating ionic and highly polar compounds, such as phosphorylated metabolites, amino acids, and organic acids. The combination of CE with MS (CE-MS) serves as a complementary technique to GC-MS and LC-MS [28]. Recent advancements in sheathless CE-MS and improvements in sample injection techniques have enhanced sensitivity and stability [29]. These developments highlight the importance of selecting appropriate separation techniques based on the physicochemical properties of the target compounds.

The choice of MS detector connected to GC, LC, or CE also depends on the analytical objectives. Single-stage MS commonly employs quadrupole (Q) or time-of-flight (TOF-MS), where quantification is achieved by measuring molecular ion intensities. However, these methods provide limited structural information, making it challenging to distinguish between isomers or multiple molecular species. For improved structural elucidation, tandem mass spectrometry (MS/MS) techniques, such as triple quadrupole (QqQ), quadrupole-TOF (Q-TOF), and quadrupole-Fourier transform MS (Q-FTMS), are frequently used. Triple quadrupole MS (QqQ) is highly sensitive and well-suited for targeted quantification. In contrast, Q-TOF and Q-FTMS offer higher mass resolution and superior compound identification capabilities, making them ideal for non-targeted analysis. Thus, the selection of MS techniques should align with the objectives of the metabolomics study. Integrating multiple analytical techniques, including nuclear magnetic resonance (NMR), enables more comprehensive metabolite profiling and enhances the depth of metabolomic analysis [30].

10.3 Analysis and Application of Metabolomics Data for Biotechnology

Microbial production of valuable compounds using *Escherichia coli*, *Saccharomyces cerevisiae*, cyanobacteria, and other microorganisms has gained attention due to its environmentally friendly nature, operating under ambient temperature and pressure while utilizing various carbon sources such as plant biomass and CO_2 [31]. The target products range from food additives, pharmaceuticals, and pigments to commodity chemicals such as plastics and fuels. However, achieving high-yield and high-efficiency production requires microbial strain modification, screening, and optimization of cultivation conditions. These processes are typically carried out through the Design-Build-Test-Learn (DBTL) cycle, a workflow used in synthetic biology and metabolic engineering [11]. In the design phase, genome-scale metabolic models and omics data are used to design metabolic pathways, promoters, and enzymes. During the build phase, synthetic biology techniques, including DNA synthesis, CRISPR-based genome editing, and robotics, are employed to construct the desired microbial strains. In the test phase, the engineered strains are evaluated in high-throughput, not only for production levels, but also for transcriptome, proteome, and metabolome profiles. Finally, in the learn phase, the acquired data are analyzed to guide the next cycle of strain design [12]. The DBTL cycle is applied not only to strain engineering but also to microbiome engineering, where omics technologies, including metabolomics, play a crucial role in elucidating complex metabolic networks within microbiomes [32]. The following section will provide a more detailed discussion on the role of metabolomics in strain engineering.

10.3.1 Metabolomics for Identification of Pathway Bottlenecks

In metabolic pathways converting glucose or other substrates into target compounds, enzymes with low conversion rates can limit overall productivity. Enhancing the expression or activity of such enzymes can lead to increased production yields. These bottleneck reactions often result in the accumulation of their substrate intermediates. Metabolomics enables the identification of such accumulated metabolites, providing a rational approach to metabolic engineering [12, 33, 34]. This strategy is universal across various microbial strains, including *Escherichia coli*, *Saccharomyces cerevisiae*, *Corynebacterium glutamicum*, cyanobacteria, and *Streptomyces*. It is also applicable to a wide range of production targets, such as organic acids, terpenoids, aromatic compounds, and lipids, with demonstrated effectiveness (Table 10.1). For example, in an *E. coli* strain engineered for 1-propanol production, metabolomics analysis revealed the accumulation of 2-ketobutyrate, a key precursor (Figure 10.2). This suggested that the downstream enzymes, alpha-ketoisovalerate decarboxylase (KivD) or alcohol dehydrogenase (YqhD), constituted pathway bottlenecks. Enhancing YqhD activity led to a 1.38-fold increase in 1-propanol production [35]. Acetyl-coenzyme A (acetyl-CoA) is a crucial precursor for various valuable compounds, but in cyanobacteria, its accumulation is typically low, limiting the productivity of acetyl-CoA-derived chemicals. In a xylose-utilizing *Synechococcus elongatus* UTEX2973 strain, metabolomics analysis showed high accumulation of sugar phosphates in the presence of xylose, indicating insufficient metabolic flux toward acetyl-CoA. To address this, a heterologous phosphoketolase (Pkt) was introduced to convert sugar phosphates into acetyl phosphate, a direct precursor of acetyl-CoA. Additionally, to enhance the supply of fructose-6-phosphate (F6P), a phosphofructokinase (Pfk) knockout was combined with fructose-1,6-bisphosphatase (Fbp) overexpression. These modifications successfully increased acetyl-CoA accumulation, and when a 3-hydroxypropionic acid (3-HP) biosynthetic pathway was introduced, 3-HP production was enhanced 4.1-fold compared to the original strain [36].

Metabolomics not only aids in the direct identification of pathway bottlenecks but also helps discover byproduct formation and key metabolic pathways that contribute to production improvement in mutant strains. For instance, in a *Corynebacterium glutamicum* strain engineered for 3-amino-4-hydroxybenzoic acid (3,4-AHBA) production, metabolomics revealed significant accumulation of lactate. To suppress lactate formation, lactate dehydrogenase gene (*ldh*) knockout was performed, leading to a 1.33-fold increase in 3,4-AHBA production [37]. Similarly, in an *E. coli* strain engineered for succinic acid production, metabolomics identified excessive accumulation of mannitol, a byproduct. Deleting mannitol dehydrogenase gene (*mtlD*) successfully reduced mannitol synthesis and resulted in a 1.2-fold increase in succinic acid production [38]. Metabolomics-guided engineering is also effective for the production of secondary metabolites. The metabolic pathway for the opioid precursor benzylisoquinoline alkaloids contains several potential bottlenecks. A phylogenetically distinct *Bombyx mori* 3,4-dihydroxyphenylacetaldehyde synthase (DHPAAS) was identified as a bypass enzyme, enabling the direct conversion of L-3,4-dihydroxyphenylalanine (L-DOPA) to 3,4-dihydroxyphenylacetaldehyde (DHPAA), effectively skipping an enzymatic step [39]. In this study, metabolomics analysis revealed that depletion of S-adenosylmethionine (SAM) limited the methylation of downstream DHPAA-derived benzylisoquinline alkaloids to produce

Table 10.1 Metabolomics-guided strain engineering in various microorganisms.

Host organisms	Targets	Genetic modification	Fold increase	References
Artemisia annua	Artemisinin	Wild type	—	[77]
Ashbya gossypii B2	Riboflavin	Wild type	1.45	[78]
Aureobasidium pullulans CCTCC M2012223	PMA	*pyc*	1.151	[79]
Clostridium autoethanogenum DSM19630	PHB	*phaA*, *phaB*, *phaC*	12	[80]
Corynebacterium glutamicu	3-amino-4-hydroxybenzoic acid	*griH*, *griI*, Δ*ldh*	1.33	[37]
Escherichia coli	1-Propanol	Δ*ilvB*, Δ*ilvI*, Δ*leuA*, *kivd-yqhD-ilvA* (Improved *yqhD* RBSstrength), *thrAfbrBC*, *cimA2-leuBCD*	1.38	[35]
E. coli	1-Butanol	Δ*ldhA*, Δ*adhE*, Δ*frdBC*, *atoB*, *adhE2*, *crt*, *hbd*, *ter*, *fdh*	2	[81]
E. coli	Reticuline	*DHPAAS*, *6OMT*, *CNMT*, *4OMT*	—	[39]
E. coli	Taxadien-5α-ol	*dxs*, *idi*, *ispD*, *ispF*, *GGPP synthase*, *taxadiene synthase*, *taxadiene 5α hydroxylase*	15,000	[82]
E. coli	succinic acid	Δ*sdhA*Δ*ack-pta*Δ*pox*Δ*gnd*Δ*mtlD*	1.2	[38]
E. coli W3110	Methionine	Δ*metI*, Δ*metJ*, Δ*lysA*, *Trc-metH*, *Trc-cysE*, *Trc-serB*, *Trc-serC*, *metAfbr*, *yjeH*	>50	[83]
Saccharomyces cerevisiae	isoprenol	*MvaS*(A110G), *MvaE*, *ERG19*, *ERG12*, Δ*choline kinase*, *PhoA*	3.6	[84]

(Continued)

Table 10.1 (Continued)

Host organisms	Targets	Genetic modification	Fold increase	References
S. cerevisiae	Tyrosine	Δ*Aro10*, Δ*Zwf1*, *TAL*, *ARO4FBR*, *TYRC*	384	[85]
S. cerevisiae	Ergosterol	Δ*ERG13*, *dxs*, *ispC*, *ispD*, *ispE*, *ispF*, *BtispG*, *ispH*, *AtrFNR*, *BsFld*	—	[86]
S. cerevisiae Kyokai 6	SAM	Wild type	2.5	[87]
S. cerevisiae MN8140X	Ethanol	*TAL1* or *TKL1*	9.47	[88]
S. cerevisiae T73–63	β-Carotene	*crtYB*, *crtI*, *crtE*	1.39	[89]
Sarcophyton ehrenbergi	Sarcophytonolide I	Wild type	132	[90]
Streptomyces nodosus	amphotericin B	*metH*	1.264	[40]
Synechococcus elongatus	3-hydroxypropionic acid	*xylE*, *xylA*, *xylB*, *pkt*, *fbp*, *mcr*, Δ*pfk*	4.1	[36]
S. elongatus PCC 7942	Acetate	*S-CBS3-rpsD4*, *pta*	2.2	[91]
S. elongatus PCC 7942	IBA	*pyk* or *mdh*/*me*	1.6 or 1.7	[92]
S. elongatus PCC 7942	IBA	*pyk*, *pck*	>1.6	[93]
S. elongatus PCC 7942	23BD	*prk*, *rbcLXS*, *galP*, *zwf*, *gnd*, Δ*cp12*	5	[94]
S. elongatus PCC 7942	13PDO	Δ*ndhF1* or Δ*ndhD1*	1.3 or 1.2	[95]
Synechocystis sp. PCC 6803	Succinate	*ppc*	>10	[96]
Synechocystis sp. PCC 6803	PHB	*sigE*	2.3	[97]
Synechocystis sp. PCC 6803	Lactate	Δ*ackA*, *ddh*	15.6	[98]

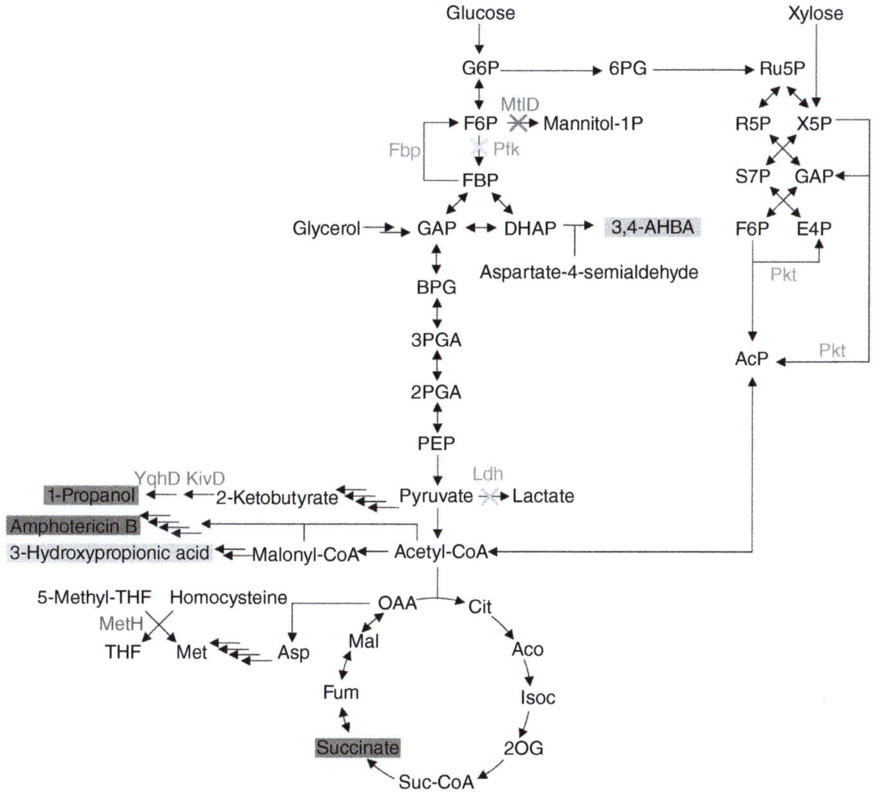

Figure 10.2 Examples of metabolomics-based pathway optimization. Through metabolomics analysis, it was discovered that overexpressing or deleting the color-coded enzyme genes corresponding to the labeled products can enhance production of those compounds. For example, increased expression of Fbp and Pkt, combined with knockout of Pfk, was shown by metabolomics to be effective for enhancing 3-hydroxypropionic acid production.

reticuline. To address this, a two-step cell production system was developed, separating SAM-dependent methylation reactions, which successfully increased the levels of downstream products, including reticuline. Additionally, in antifungal antibiotic amphotericin B (AmB) production by *Streptomyces nodosus*, a high-production mutant strain (ZJB2016050) was obtained through random mutagenesis using UV irradiation and addition of N-methyl-N-nitroso-N-nitroguanidine. To further improve productivity, metabolomics analysis was conducted on ZJB2016050 strain, revealing significant accumulation of tetrahydrofolic acid (THF) and other one-carbon metabolism-related metabolites compared to the wild-type strain. Based on this finding, folate metabolism was hypothesized as a key metabolic pathway in the mutant strain, and overexpression of methionine synthase gene (*metH*) led to a 1.26-fold increase in AmB production [40].

10.3.2 Absolute Metabolomics and Thermodynamic Analyses

The quantification accuracy of targeted metabolomics is not always precise due to factors such as compound degradation, oxidation during sample preparation, and ion suppression effects in MS [41]. To address these issues, the stable isotope dilution internal standard method has been developed [42, 43]. This method involves adding a known concentration of stable isotope-labeled internal standards (^{13}C, ^{2}H, ^{15}N-labeled compounds), which have the same chemical structure but a different mass from the target metabolites, into the sample. By comparing the signal intensity ratio of labeled (internal standard) and unlabeled (endogenous metabolite) compounds during measurement, this approach corrects for variability in ionization efficiency and metabolite loss during sample preparation, enabling more accurate absolute concentration calculations.

Furthermore, by precisely measuring intracellular metabolite concentrations, this method allows for a detailed evaluation of relative concentration ratios and energy balances within metabolic pathways. Specifically, analyzing the ATP/ADP/AMP ratio, which reflects cellular energy status, and tracking intermediate metabolite concentration changes in the TCA cycle can help identify thermodynamically constrained reactions [44, 45]. By integrating this approach with flux balance analysis (FBA) and thermodynamic flux analysis (TFA), it is possible to clearly define rate-limiting steps in metabolic pathways and optimize pathway designs for targeted compound synthesis. For example, Gibbs free energy can be calculated from measured absolute concentrations to identify reactions with high thermodynamic barriers. This information can then be incorporated into thermodynamically constrained models to predict optimal biosynthetic pathways, such as those for polyhydroxybutyrate (PHB) production [46].

10.3.3 Evaluation and Optimization of Metabolic Flux

Thus far, we have discussed methods for quantifying and evaluating metabolite pool sizes. However, analyzing metabolite concentrations alone is insufficient to assess intracellular metabolic fluxes. Metabolic flux analysis (MFA) is performed by measuring the incorporation of stable isotopes using MS. By comparing isotopic labeling patterns and positional atom data with flux distributions within a metabolic model, intracellular flux distributions can be estimated. MFA can be classified based on whether intracellular metabolite concentrations and isotopic labeling reach a steady state [47]. For instance, when cells are supplied with [$1-^{13}C$] glucose, the isotopic distribution of intracellular metabolites at steady state can be used to determine flux distributions. However, for autotrophic organisms such as cyanobacteria and plants, where CO_2 is the sole carbon source, steady-state labeling with $^{13}CO_2$ results in complete labeling of all metabolites, making this method ineffective. Instead, time-course measurements of ^{13}C incorporation are performed under conditions where metabolite concentrations are assumed to be constant. This approach enables flux distribution calculations in autotrophic systems [48]. Furthermore, in non-steady-state conditions where metabolite concentrations fluctuate dynamically, metabolic flux changes can be estimated by detailed time-series analysis of all intracellular metabolite concentrations within the metabolic model [21]. Additionally, flux variations over the entire cultivation period can be inferred from measurements of selected metabolite concentrations at different time points [49].

10.4 Artificial Intelligence (AI)-Based Metabolomics Data Processing and Analysis

AI has emerged as an indispensable tool to process and interpret metabolomics data [50, 51]. One major application is metabolite identification and annotation. Untargeted metabolomics via MS often produces many features (mass peaks) that must be assigned to actual metabolites, a process fraught with uncertainty and considered a major bottleneck in metabolomics data processing. Bayesian modeling can rigorously handle this uncertainty. For example, Integrated Probabilistic Annotation (IPA) is a Bayesian-based method developed to annotate metabolomic features by integrating various evidence sources [52]. In IPA, information such as isotope patterns, adduct formations, chromatographic retention time, and known biochemical connections is used as input to a probabilistic model. The method computes the posterior probability that a given set of spectral features corresponds to a particular metabolite. IPA not only assigns metabolite identities automatically but also quantifies the confidence in each annotation. By starting with prior probabilities (e.g. likelihood of certain metabolites based on biochemical pathways) and updating these beliefs with observed data (m/z matches, isotope ratios, etc.), the algorithm refines metabolite identities in a principled way. This provides a confidence score for each putative metabolite identification, which is lacking in traditional deterministic matching methods. IPA demonstrated that incorporating multiple evidence streams can markedly reduce false annotations and improve consistency in metabolomic analyses. In practical terms, Bayesian annotation helps researchers focus on the most probable metabolite identifications and flag ambiguous cases for further validation, thereby streamlining one of the most challenging tasks in metabolomics. Furthermore, recent advancements in machine learning-based metabolomics data processing are exemplified by the development of MS-DIAL 5 [53]. MS-DIAL 5 is an advanced multimodal MS data analysis environment that utilizes electron-activated dissociation (EAD)-based tandem MS for in-depth lipid structural elucidation. By incorporating machine learning algorithms to analyze spectral patterns and predict collision-cross section values, MS-DIAL 5 achieves a 96.4% accuracy in annotating lipid structures, including the assignment of *sn*-, OH-, and C=C positions. This platform not only facilitates detailed structural characterization of lipids but also integrates MS imaging (MSI) data to map lipid localizations. Thus, MS-DIAL 5 provides a complementary strategy to Bayesian methods like IPA by enhancing the resolution and reliability of metabolite annotations, ultimately contributing to the effective management of uncertainty and the streamlining of metabolite identification in complex biological samples.

Another key application of AI in metabolomics and synthetic biology is optimizing metabolic pathways for improved production or efficiency. For example, data-driven models can perform MFA in a novel way: instead of relying solely on predefined kinetic equations, machine learning can learn the mapping from enzyme levels and metabolite concentrations to pathway fluxes. Costello and Martin demonstrated that combining ML with abundant time-series multiomics data (proteomics and metabolomics) enabled accurate prediction of metabolic pathway dynamics, outperforming a traditional kinetic model [54]. Their ML model can integrate proteome and metabolite data to forecast how a pathway's flux changes over time or under perturbations without the need for manually deriving rate laws. AI has also been used to recommend genetic modifications for pathway optimization. By learning from prior engineering data, AI systems can suggest which genes to overexpress, knock out, or mutate to achieve a target output. A notable example is an Automated Recommendation Tool (ART) developed for multi-gene pathway engineering [55]. This tool used an ensemble of probabilistic models and Bayesian optimization to iteratively design and evaluate strain modifications. In a case study on *S. cerevisiae* tryptophan production, ART evaluated over 250 engineered strains with different promoter and gene combinations [56]. Another exciting use of AI is in discovering or engineering new pathway steps. Vavricka and coworkers sought to enhance production of benzylisoquinoline alkaloids (plant-derived compounds) in microbes. In this case, some bottleneck steps of the plant pathway are catalyzed by no known counterparts in microbes. Therefore, the team trained ML models (support vector machines, SVMs) trained with enzyme sequence data to predict "missing link" enzymes and were able to rank candidate genes from plants [57]. The candidate genes were then expressed in a microbial host and metabolomic analysis was applied to verify if the pathway was completed. Metabolomics indicated that two predicted enzymes (an aromatic acetaldehyde synthase and a phenylpyruvate decarboxylase) were functional, leading to the production of key intermediates that had been absent before. By integrating the new enzymatic steps, the engineered pathway output (a specific alkaloid) could be increased significantly. This case illustrates how AI-driven insights (predicting an enzyme from sequence space) coupled with metabolomics verification can expand biosynthetic pathways beyond their natural limits – effectively pathway engineering by AI discovery. In summary, AI methodologies empower metabolic pathway optimization by predicting pathway behavior, guiding genetic interventions, and even inventing new pathway routes, all informed and validated by comprehensive metabolomic data.

10.5 Future Direction of Metabolomics and Its Analysis

Publicly available datasets from genomics, transcriptomics, and proteomics have facilitated large-scale meta-analyses, leading to new insights. However, metabolomics data are more challenging to compare and integrate across different laboratories and experimental batches due to variations in sample preparation protocols, differences in experimental handling, and quantification inconsistencies caused by ion suppression effects. To overcome these challenges, advancements in

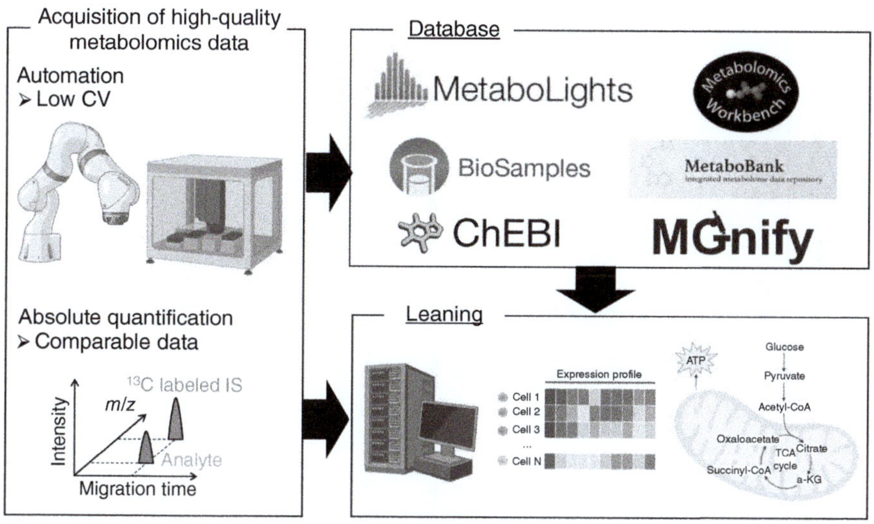

Figure 10.3 A conceptual overview of future directions in metabolomics and its analysis. Automation of sample preparation, the use of ^{13}C-labeled internal standards for improved quantification, and the expansion of public repositories are central to enhancing data quality and reproducibility. Integrating these advances with robust analytical and "learning" platforms will enable more comprehensive, cross-laboratory comparisons and large-scale meta-analyses.

automation of sample processing, adoption of absolute quantification methods, and enhancement of database platforms are crucial. For instance, studies have demonstrated that automation of sample preprocessing significantly improves reproducibility, as indicated by relative standard deviation values that ensure sufficient consistency [58]. Autonomous laboratory systems may be applicable to metabolomics-based strain engineering in the future [59]. Additionally, high-quality ^{13}C-labeled internal standards for metabolomics are being developed to enhance quantitative accuracy [60]. Further technological advancements and widespread adoption of these methods are expected to improve the overall quality of metabolomics data (Figure 10.3).

Moreover, recent innovations in MS-based fingerprinting offer promising future directions for metabolomics analysis. For example, a study on genome-scale prediction of gene ontology from MALDI-TOF mass fingerprints demonstrated that machine learning models, such as SVM and random forest algorithms, can rapidly and accurately predict gene functions from high-throughput mass spectral data [61]. This approach, applied to a comprehensive yeast knockout library, not only enabled functional annotation of genes but also highlighted the potential for optimizing bioproduction chassis selection. By integrating rapid mass spectrometric fingerprinting with machine learning algorithms, such predictive methods pave the way for more comprehensive, automated, and standardized workflows. As more high-quality metabolomics datasets are deposited into public repositories [62–77] (Table 10.2), cross-laboratory and large-scale comparative analyses will become increasingly feasible, further advancing our understanding of cellular metabolism.

Table 10.2 Representative metabolomics-related databases.

Name	Website URL	Type of data	Key features	Strengths and common use cases	Reference
MetaboLights	https://www.ebi.ac.uk/metabolights	Metabolomics studies and metadata	Supports multiple organisms and technologies; includes reference spectra.	Widely used for data sharing and reanalysis.	[62]
Metabolomics Workbench	https://www.metabolomicsworkbench.org	Metadata and raw data (MS/NMR)	Includes tools, protocols, and a large metabolite database.	Used for global dataset comparison and analysis.	[63]
MetaboBank (DDBJ)	https://www.ddbj.nig.ac.jp/metabobank	MS, NMR, imaging MS data repository	Standardized metadata format; links with global repositories.	Important Asian archive with international compatibility.	[64]
MetabolomeX-change	http://www.metabolomexchange.org	Metadata aggregator for metabolomics studies	Aggregates datasets; provides unified search across databases.	Improves discoverability of public metabolomics data.	[65]
BioSamples (EMBL-EBI)	https://www.ebi.ac.uk/biosamples	Biological sample metadata	Links and standardizes sample metadata for cross-omics use.	Enables integration across different omics studies.	[66]
ChEBI	https://www.ebi.ac.uk/chebi	Chemical entities of biological interest	Manually curated ontology of small molecules.	Used for metabolite classification and standard naming.	[67]
HMDB	https://hmdb.ca	Human metabolite reference data	Extensive data on human metabolites including spectra and pathways.	Widely used in clinical and biomedical research.	[68]
LIPID MAPS	https://www.lipidmaps.org	Lipid structure and classification data	Standardized lipid classes and curated structures.	Main resource for lipidomics analysis.	[69]

METLIN	https://metlin.scripps.edu	Spectral library of metabolites	High-resolution MS/MS spectra of known and unknowns.	Used for metabolite identification from MS/MS data.	[70]
MassBank	https://www.massbank.eu	Public MS spectra for small molecules	Curated and standardized spectral entries.	Supports metabolite identification via spectral matching.	[71]
GNPS	https://gnps.ucsd.edu	Community platform for MS/MS data	Crowdsourced spectral library and networking tools.	Used in natural product and microbiome studies.	[72]
KEGG	https://www.kegg.jp	Metabolic pathways, compounds, and reactions	Curated pathway maps linking metabolites and genes.	Used in pathway mapping and enrichment analysis.	[73]
MetaCyc	https://metacyc.org	Metabolic pathways and enzymes	Manually curated multi-organism pathway data.	Supports metabolic network modeling and annotation.	[74]
MGnify	https://www.ebi.ac.uk/metagenomics	Microbiome sequence and function data	Automated pipelines for taxonomic and functional profiling.	Links microbial data to metabolite functions.	[75]
KNApSAcK	https://www.knapsackfamily.com/KNApSAcK/	Plant species-metabolite associations	Species-specific metabolite occurrences with chemical info.	Used in plant metabolomics and phytochemistry.	[76]

References

1 Pinu, F.R., Villas-Boas, S.G., and Aggio, R. (2017). Analysis of intracellular metabolites from microorganisms: quenching and extraction protocols. *Metabolites* 7 (4): 53.

2 Khamis, M.M., Adamko, D.J., and El-Aneed, A. (2017). Mass spectrometric-based approaches in urine metabolomics and biomarker discovery. *Mass Spectrom. Rev.* 36 (2): 115–134.

3 Bujak, R., Struck-Lewicka, W., Markuszewski, M.J., and Kaliszan, R. (2015). Metabolomics for laboratory diagnostics. *J. Pharm. Biomed. Anal.* 113: 108–120.

4 Rinschen, M.M., Ivanisevic, J., Giera, M., and Siuzdak, G. (2019). Identification of bioactive metabolites using activity metabolomics. *Nat. Rev. Mol. Cell Biol.* 20 (6): 353–367.

5 Newgard, C.B. (2017). Metabolomics and metabolic diseases: where do we stand? *Cell Metab.* 25 (1): 43–56.

6 Chen, X. and Yu, D. (2019). Metabolomics study of oral cancers. *Metabolomics* 15 (2): 22.

7 Quintero, M.E., Pontes, J.G.M., and Tasic, L. (2021). Metabolomics in degenerative brain diseases. *Brain Res.* 1773: 147704.

8 Labine, L.M. and Simpson, M.J. (2020). The use of nuclear magnetic resonance (NMR) and mass spectrometry (MS)-based metabolomics in environmental exposure assessment. *Curr. Opin. Environ. Sci. Health* 15: 7–15.

9 Putri, S.P., Ikram, M.M.M., Sato, A. et al. (2022). Application of gas chromatography-mass spectrometry-based metabolomics in food science and technology. *J. Biosci. Bioeng.* 133 (5): 425–435.

10 Bauermeister, A., Mannochio-Russo, H., Costa-Lotufo, L.V. et al. (2022). Mass spectrometry-based metabolomics in microbiome investigations. *Nat. Rev. Microbiol.* 20 (3): 143–160.

11 Nielsen, J. and Keasling, J.D. (2016). Engineering cellular metabolism. *Cell* 164 (6): 1185–1197.

12 Vavricka, C.J., Hasunuma, T., and Kondo, A. (2020). Dynamic metabolomics for engineering biology: accelerating learning cycles for bioproduction. *Trends Biotechnol.* 38 (1): 68–82.

13 Koek, M.M., Jellema, R.H., van der Greef, J. et al. (2011). Quantitative metabolomics based on gas chromatography-mass spectrometry: status and perspectives. *Metabolomics* 7 (3): 307–328.

14 Cui, L., Lu, H., and Lee, Y.H. (2018). Challenges and emergent solutions for LC-MS/MS-based untargeted metabolomics in diseases. *Mass Spectrom. Rev.* 37 (6): 772–792.

15 Boness, H.V.M., de Sá, H.C., Dos Santos, E.K.P., and Canuto, G.A.B. (2023). Sample preparation in microbial metabolomics: advances and challenges. *Adv. Exp. Med. Biol.* 1439: 149–183.

16 Rontani, J.F. (2022). Use of gas chromatography-mass spectrometry techniques (GC-MS, GC-MS/MS, and GC-QTOF) for the characterization of photooxidation

and autoxidation products of lipids of autotrophic organisms in environmental samples. *Molecules* 27 (5): 1629.

17 Kapoore, R.V. and Vaidyanathan, S. (2018). Quenching for microalgal metabolomics: a case study on the unicellular eukaryotic green alga *Chlamydomonas reinhardtii*. *Metab.* 8 (4): 72.

18 Wang, B. and Young, J.D. (2022). ^{13}C-isotope-assisted assessment of metabolic quenching during sample collection from suspension cell cultures. *Anal. Chem.* 94 (22): 7787–7794.

19 Villas-Bôas, S.G. and Bruheim, P. (2007). Cold glycerol-saline: the promising quenching solution for accurate intracellular metabolite analysis of microbial cells. *Anal. Biochem.* 370 (1): 87–97.

20 Hasunuma, T., Harada, K., Miyazawa, S. et al. (2010). Metabolic turnover analysis by a combination of *in vivo* ^{13}C-labelling from $^{13}CO^2$ and metabolic profiling with CE-MS/MS reveals rate-limiting steps of the C^3 photosynthetic pathway in *Nicotiana tabacum* leaves. *J. Exp. Bot.* 61 (4): 1041–1051.

21 Tanaka, K., Shirai, T., Vavricka, C.J. et al. (2023). Dark accumulation of downstream glycolytic intermediates initiates robust photosynthesis in cyanobacteria. *Plant Physiol.* 191 (4): 2400–2413.

22 Rabinowitz, J.D. and Kimball, E. (2007). Acidic acetonitrile for cellular metabolome extraction from *Escherichia coli*. *Anal. Chem.* 79 (16): 6167–6173.

23 Saini, R.K., Prasad, P., Shang, X., and Keum, Y.S. (2021). Advances in lipid extraction methods—a review. *Int. J. Mol. Sci.* 22 (24): 13643.

24 Fraga-Corral, M., Carpena, M., Garcia-Oliveira, P. et al. (2022). Analytical metabolomics and applications in health, environmental, and food science. *Crit. Rev. Anal. Chem.* 52 (4): 712–734.

25 Enmark, M., Harun, S., Samuelsson, J. et al. (2021). Selectivity limits of and opportunities for ion pair chromatographic separation of oligonucleotides. *J. Chromatogr. A* 1651: 462269.

26 Bartlett, M.G. (2024). Current state of hydrophilic interaction liquid chromatography of oligonucleotides. *J. Chromatogr. A* 1736: 465378.

27 Nakatani, K., Izumi, Y., Takahashi, M., and Bamba, T. (2022). Unified-hydrophilic-interaction/anion-exchange liquid chromatography-mass spectrometry (Unified-HILIC/AEX/MS): a single-run method for comprehensive and simultaneous analysis of the polar metabolome. *Anal. Chem.* 94 (48): 16877–16886.

28 Monton, M.R. and Soga, T. (2007). Metabolome analysis by capillary electrophoresis-mass spectrometry. *J. Chromatogr. A* 1168 (1–2): 237–246. discussion 236.

29 Zhang, W. and Ramautar, R. (2021). CE-MS for metabolomics: developments and applications in the period 2018–2020. *Electrophoresis* 42 (4): 381–401.

30 Jeppesen, M.J. and Powers, R. (2023). Multiplatform untargeted metabolomics. *Magn. Reson. Chem.* 61 (12): 628–653.

31 Lee, S.Y., Kim, H.U., Chae, T.U. et al. (2019). A comprehensive metabolic map for production of bio-based chemicals. *Nat. Catal.* 2: 18–33.

32 Lawson, C.E., Harcombe, W.R., Hatzenpichler, R. et al. (2019). Common principles and best practices for engineering microbiomes. *Nat. Rev. Microbiol.* 17 (12): 725–741.

33 Kato, Y., Inabe, K., Hidese, R. et al. (2022). Metabolomics-based engineering for biofuel and bio-based chemical production in microalgae and cyanobacteria: a review. *Bioresour. Technol.* 344 (Pt A): 126196.

34 Tanaka, K., Bamba, T., Kondo, A., and Hasunuma, T. (2024). Metabolomics-based development of bioproduction processes toward industrial-scale production. *Curr. Opin. Biotechnol.* 85: 103057.

35 Ohtake, T., Kawase, N., Pontrelli, S. et al. (2022). Metabolomics-driven identification of the rate-limiting steps in 1-propanol production. *Front. Microbiol.* 13: 871624.

36 Yao, J., Wang, J., Ju, Y. et al. (2022). Engineering a xylose-utilizing *Synechococcus elongatus* UTEX 2973 chassis for 3-hydroxypropionic acid biosynthesis under photomixotrophic conditions. *ACS Synth. Biol.* 11: 678–688.

37 Kawaguchi, H., Hasunuma, T., Ohnishi, Y. et al. (2021). Enhanced production of γ-amino acid 3-amino-4-hydroxybenzoic acid by recombinant *Corynebacterium glutamicum* under oxygen limitation. *Microb. Cell Factories* 20: 228.

38 Valle, A., Soto, Z., Muhamadali, H. et al. (2022). Metabolomics for the design of new metabolic engineering strategies for improving aerobic succinic acid production in *Escherichia coli*. *Metabolomics* 18: 56.

39 Vavricka, C.J., Yoshida, T., Kuriya, Y. et al. (2019). Mechanism-based tuning of insect 3,4-dihydroxyphenylacetaldehyde synthase for synthetic bioproduction of benzylisoquinoline alkaloids. *Nat. Commun.* 10 (1): 2015.

40 Zhang, B., Chen, Y., Jiang, S.X. et al. (2021). Comparative metabolomics analysis of amphotericin B high-yield mechanism for metabolic engineering. *Microb. Cell Factories* 20: 66.

41 Annesley, T.M. (2003). Ion suppression in mass spectrometry. *Clin. Chem.* 49 (7): 1041–1044.

42 Wu, L., Mashego, M.R., van Dam, J.C. et al. (2005). Quantitative analysis of the microbial metabolome by isotope dilution mass spectrometry using uniformly ^{13}C-labeled cell extracts as internal standards. *Anal. Biochem.* 336 (2): 164–171.

43 Bennett, B.D., Kimball, E.H., Gao, M. et al. (2009). Absolute metabolite concentrations and implied enzyme active site occupancy in *Escherichia coli*. *Nat. Chem. Biol.* 5 (8): 593–599.

44 Nishino, S., Okahashi, N., Matsuda, F., and Shimizu, H. (2015). Absolute quantitation of glycolytic intermediates reveals thermodynamic shifts in *Saccharomyces cerevisiae* strains lacking *PFK1* or *ZWF1* genes. *J. Biosci. Bioeng.* 120 (3): 280–286.

45 Park, J.O., Rubin, S.A., Xu, Y.F. et al. (2016). Metabolite concentrations, fluxes, and free energies imply efficient enzyme usage. *Nat. Chem. Biol.* 12 (7): 482–489.

46 Janasch, M., Crang, N., Asplund-Samuelsson, J. et al. (2022). Thermodynamic limitations of PHB production from formate and fructose in *Cupriavidus necator*. *Metab. Eng.* 73: 256–269.

47 Antoniewicz, M.R. (2021). A guide to metabolic flux analysis in metabolic engineering: methods, tools, and applications. *Metab. Eng.* 63: 2–12.

48 Cheah, Y.E. and Young, J.D. (2018). Isotopically nonstationary metabolic flux analysis (INST-MFA): putting theory into practice. *Curr. Opin. Biotechnol.* 54: 80–87.

49 Leighty, R.W. and Antoniewicz, M.R. (2011). Dynamic metabolic flux analysis (DMFA): a framework for determining fluxes at metabolic non-steady state. *Metab. Eng.* 13 (6): 745–755.

50 Lawson, C.E., Martí, J.M., Radivojevic, T. et al. (2021). Machine learning for metabolic engineering: a review. *Metab. Eng.* 63: 34–60.

51 Chi, J., Shu, J., Li, M. et al. (2024). Artificial intelligence in metabolomics: a current review. *Trends Anal. Chem.* 178: 117852.

52 Del Carratore, F., Schmidt, K., Vinaixa, M. et al. (2019). Integrated probabilistic annotation: a Bayesian-based annotation method for metabolomic profiles integrating biochemical connections, isotope patterns, and adduct relationships. *Anal. Chem.* 91 (20): 12799–12807.

53 Takeda, H., Matsuzawa, Y., Takeuchi, M. et al. (2024). MS-DIAL 5 multimodal mass spectrometry data mining unveils lipidome complexities. *Nat. Commun.* 15 (1): 9903.

54 Costello, Z. and Martin, H.G. (2018). A machine learning approach to predict metabolic pathway dynamics from time-series multiomics data. *NPJ Syst. Biol. Appl.* 4: 19.

55 Radivojević, T., Costello, Z., Workman, K., and Garcia Martin, H. (2020). A machine learning automated recommendation tool for synthetic biology. *Nat. Commun.* 11 (1): 4879.

56 Zhang, J., Petersen, S.D., Radivojevic, T. et al. (2020). Combining mechanistic and machine learning models for predictive engineering and optimization of tryptophan metabolism. *Nat. Commun.* 11 (1): 4880.

57 Vavricka, C.J., Takahashi, S., Watanabe, N. et al. (2022). Machine learning discovery of missing links that mediate alternative branches to plant alkaloids. *Nat. Commun.* 13 (1): 1405.

58 Malinowska, J.M., Palosaari, T., Sund, J. et al. (2022). Automated sample preparation and data collection workflow for high-throughput *in vitro* metabolomics. *Metabolites* 12 (1): 52.

59 Fushimi, K., Nakai, Y., Nishi, A. et al. (2025). Development of the autonomous lab system to support biotechnology research. *Sci. Rep.* 15 (1): 6648.

60 Soma, Y., Takahashi, M., Fujiwara, Y. et al. (2022). Quantitative metabolomics for dynamic metabolic engineering using stable isotope-labeled internal standards mixture (SILIS). *J. Biosci. Bioeng.* 133 (1): 46–55.

61 Vavricka, C.J., Mochizuki, M., Yuzawa, S., Murata, M., Yoshida, T., Watanabe, N., Nakatsui, M., Ishii, J., Hara, K., Alper, H.S., Hasunuma, T., Kondo, A., and Araki, M. (2024). Genome-scale prediction of gene ontology from mass fingerprints reveals new metabolic gene functions. *bioRxiv*, 2024.11.27.625594. https://doi.org/10.1101/2024.11.27.625594.

62 Yurekten, O., Payne, T., Tejera, N. et al. (2024). MetaboLights: open data repository for metabolomics. *Nucleic Acids Res.* 52 (D1): D640–D646.

63 Sud, M., Fahy, E., Cotter, D. et al. (2016). Metabolomics Workbench: an international repository for metabolomics data and metadata, metabolite standards, protocols, tutorials and training, and analysis tools. *Nucleic Acids Res.* 44 (D1): D463–D470.

64 Ara, T., Kodama, Y., Tokimatsu, T. et al. (2024). DDBJ update in 2023: the MetaboBank for metabolomics data and associated metadata. *Nucleic Acids Res.* 52 (D1): D67–D71.

65 Salek, R.M., Neumann, S., Schober, D. et al. (2015). Coordination of Standards in Metabolomics (COSMOS): facilitating integrated metabolomics data access. *Metabolomics* 11 (6): 1587–1597.

66 Courtot, M., Gupta, D., Liyanage, I. et al. (2022). BioSamples database: FAIRer samples metadata to accelerate research data management. *Nucleic Acids Res.* 50 (D1): D1500–D1507.

67 Hastings, J., Owen, G., Dekker, A. et al. (2016). ChEBI in 2016: improved services and an expanding collection of metabolites. *Nucleic Acids Res.* 44 (D1): D1214–D1219.

68 Wishart, D.S., Guo, A., Oler, E. et al. (2022). HMDB 5.0: the human metabolome database for 2022. *Nucleic Acids Res.* 50 (D1): D622–D631.

69 Conroy, M.J., Andrews, R.M., Andrews, S. et al. (2024). LIPID MAPS: update to databases and tools for the lipidomics community. *Nucleic Acids Res.* 52 (D1): D1677–D1682.

70 Guijas, C., Montenegro-Burke, J.R., Domingo-Almenara, X. et al. (2018). METLIN: a technology platform for identifying knowns and unknowns. *Anal. Chem.* 90 (5): 3156–3164.

71 Horai, H., Arita, M., Kanaya, S. et al. (2010). MassBank: a public repository for sharing mass spectral data for life sciences. *J. Mass Spectrom.* 45 (7): 703–714.

72 Wang, M., Carver, J.J., Phelan, V.V. et al. (2016). Sharing and community curation of mass spectrometry data with global natural products social molecular networking. *Nat. Biotechnol.* 34 (8): 828–837.

73 Kanehisa, M., Furumichi, M., Sato, Y. et al. (2023). KEGG for taxonomy-based analysis of pathways and genomes. *Nucleic Acids Res.* 51 (D1): D587–D592.

74 Caspi, R., Billington, R., Fulcher, C.A. et al. (2018). The MetaCyc database of metabolic pathways and enzymes. *Nucleic Acids Res.* 46 (D1): D633–D639.

75 Richardson, L., Allen, B., Baldi, G. et al. (2023). MGnify: the microbiome sequence data analysis resource in 2023. *Nucleic Acids Res.* 51 (D1): D753–D759.

76 Afendi, F.M., Okada, T., Yamazaki, M. et al. (2012). KNApSAcK family databases: integrated metabolite–plant species databases for multifaceted plant research. *Plant Cell Physiol.* 53 (2): e1.

77 Ma, D.M., Wang, Z., Wang, L. et al. (2015). A genome-wide scenario of terpene pathways in self-pollinated *Artemisia annua*. *Mol. Plant* 8 (11): 1580–1598.

78 Schwechheimer, S.K. (2018). Improved riboflavin production with *Ashbya gossypii* from vegetable oil based on ^{13}C metabolic network analysis with combined labeling analysis by GC/MS, LC/MS, 1D, and 2D NMR. *Metab. Eng.* 47: 357–373.

79 Feng, J., Yang, J., Yang, W. et al. (2018). Metabolome- and genome-scale model analyses for engineering of *Aureobasidium pullulans* to enhance polymalic acid and malic acid production from sugarcane molasses. *Biotechnol. Biofuels* 11: 94.

80 de Souza Pinto Lemgruber, R., Valgepea, K., Tappel, R. et al. (2019). Systems-level engineering and characterisation of *Clostridium autoethanogenum* through heterologous production of poly-3-hydroxybutyrate (PHB). *Metab. Eng.* 53: 14–23.

81 Pontrelli, S., Fricke, R.C.B., Sakurai, S.S.M. et al. (2018). Directed strain evolution restructures metabolism for 1-butanol production in minimal media. *Metab. Eng.* 49: 153–163.

82 Ajikumar, P.K., Xiao, W.H., Tyo, K.E. et al. (2010). Isoprenoid pathway optimization for Taxol precursor overproduction in *Escherichia coli*. *Science* 330 (6000): 70–74.

83 Huang, J.F., Shen, Z.Y., Mao, Q.L. et al. (2018). Systematic analysis of bottlenecks in a multibranched and multilevel regulated pathway: the molecular fundamentals of *L*-methionine biosynthesis in *Escherichia coli*. *ACS Synth. Biol.* 7 (11): 2577–2589.

84 Kim, J., Baidoo, E.E.K., Amer, B. et al. (2021). Engineering *Saccharomyces cerevisiae* for isoprenol production. *Metab. Eng.* 64: 154–166.

85 Gold, N.D., Gowen, C.M., Lussier, F.X. et al. (2015). Metabolic engineering of a tyrosine-overproducing yeast platform using targeted metabolomics. *Microb. Cell Factories* 14: 73.

86 Kirby, J., Dietzel, K.L., Wichmann, G. et al. (2016). Engineering a functional 1-deoxy-*D*-xylulose 5-phosphate (DXP) pathway in *Saccharomyces cerevisiae*. *Metab. Eng.* 38: 494–503.

87 Hayakawa, K., Matsuda, F., and Shimizu, H. (2016). Metabolome analysis of *Saccharomyces cerevisiae* and optimization of culture medium for *S*-adenosyl-*L*-methionine production. *AMB Express* 6 (1): 38.

88 Hasunuma, T., Sanda, T., Yamada, R. et al. (2011). Metabolic pathway engineering based on metabolomics confers acetic and formic acid tolerance to a recombinant xylose-fermenting strain of *Saccharomyces cerevisiae*. *Microb. Cell Factories* 10 (1): 2.

89 Bu, X., Sun, L., Shang, F., and Yan, G. (2017). Comparative metabolomics profiling of engineered *Saccharomyces cerevisiae* leads to a strategy for improving β-carotene production by acetate supplementation. *PLoS One* 12 (11): e0188385.

90 Farag, M.A., Al-Mahdy, D.A., Meyer, A. et al. (2017). Metabolomics reveals biotic and abiotic elicitor effects on the soft coral *Sarcophyton ehrenbergi* terpenoid content. *Sci. Rep.* 7 (1): 648.

91 Sawa, N., Tatsuke, T., Ogawa, A. et al. (2019). Modification of carbon metabolism in *Synechococcus elongatus* PCC 7942 by cyanophage-derived sigma factors for bioproduction improvement. *J. Biosci. Bioeng.* 127 (2): 256–264.

92 Jazmin, L.J., Xu, Y., Cheah, Y.E. et al. (2017). Isotopically nonstationary ^{13}C flux analysis of cyanobacterial isobutyraldehyde production. *Metab. Eng.* 42: 9–18.

93 Cheah, Y.E., Xu, Y., Sacco, S.A. et al. (2020). Systematic identification and elimination of flux bottlenecks in the aldehyde production pathway of *Synechococcus elongatus* PCC 7942. *Metab. Eng.* 60: 56–65.

94 Kanno, M., Carroll, A.L., and Atsumi, S. (2017). Global metabolic rewiring for improved CO^2 fixation and chemical production in cyanobacteria. *Nat. Commun.* 8: 14724.

95 Hirokawa, Y., Matsuo, S., Hamada, H. et al. (2017). Metabolic engineering of *Synechococcus elongatus* PCC 7942 for improvement of 1,3-propanediol and glycerol production based on *in silico* simulation of metabolic flux distribution. *Microb. Cell Factories* 16 (1): 212.

96 Hasunuma, T., Matsuda, M., Kato, Y. et al. (2018). Temperature-enhanced succinate production concurrent with increased central metabolism turnover in the cyanobacterium *Synechocystis* sp. PCC 6803. *Metab. Eng.* 48: 109–120.

97 Osanai, T., Numata, K., Oikawa, A. et al. (2013). Increased bioplastic production with an RNA polymerase sigma factor SigE during nitrogen starvation in *Synechocystis* sp. PCC 6803. *DNA Res.* 20 (6): 525–535.

98 Hidese, R., Matsuda, M., Osanai, T. et al. (2020). Malic enzyme facilitates *D*-lactate production through increased pyruvate supply during anoxic dark fermentation in *Synechocystis* sp. PCC 6803. *ACS Synth. Biol.* 9 (2): 260–268.

11

Strain Engineering, Flux Design, and Metabolic Production Using Big Data: Ongoing Advances and Opportunities

Rafael S. Costa[1,2,3] *and Rui Henriques*[4]

[1] *UCIBIO – Applied Molecular Biosciences Unit, Department of Chemistry, School of Science and Technology, Universidade NOVA de Lisboa, 2829-516 Caparica, Portugal*
[2] *Associate Laboratory i4HB – Institute for Health and Bioeconomy, School of Science and Technology, Universidade NOVA de Lisboa, 2829-516 Caparica, Portugal*
[3] *LAQV-REQUIMTE, DQ, NOVA School of Science and Technology, Universidade NOVA de Lisboa, Caparica, 2829-516, Portugal*
[4] *INESCID and IST, Universidade de Lisboa, Av. Rovisco Pais, 1, Lisbon 1049-001, Portugal*

11.1 Introduction

The biotech industry is being transformed by the Big Data advent, made possible by the unprecedented capacity to massively profile and process data, resulting in an explosion of usable information on individual biosystems. Representative sources of biotech Big Data include (bio)process data and various types of omics data (e.g. genomics, transcriptomics, proteomics, metabolomics) that are systematically stored and managed at several publicly available repositories for a wide range of healthcare, agriculture, industrial, and environmental endpoints [1–4]. Although the production of data in the most diverse biotechnology areas is increasing at a high rate, nevertheless, computational techniques are necessary to process such data and subsequently tap into their full potential. The power of machines to transform and learn from biodata is reshaping the current outlook [5]. However, the structural complexity, heterogeneity, high-dimensionality, and increasingly massive volumes of bioprocess and omics data, coupled with the context-sensitive and stochastic nature of living systems, challenge the capacity to reach the desired phenotypes.

Complementarily, recent advancements in molecular engineering have revolutionized strain engineering and flux design, enabling the redirection of metabolic flux toward desired products [5, 6] (for further details about metabolic flux analysis, see Chapter 12). The metabolic engineering field relies on genetic modifications (activation, underexpression, knockout, and/or modulation of regulatory elements) to reprogram native metabolic pathways for producing a wide variety of valuable compounds. This process has been further streamlined by the Design-Build-Test-Learn (DBTL) cycle framework for strain

Machine Learning and Big Data-enabled Biotechnology, First Edition. Edited by Hal S. Alper.

engineering, leveraging on high-throughput profiling to accelerate and optimize the development of strains [7]. In this pipeline, the Design component outlines the strategic approach, while the Build step involves the techniques used to introduce sequence diversity. The Test phase includes the workflows for phenotyping, and the Learn component involves the application of computational tools to analyze the data collected during the testing. While computational tools are integrated throughout the cycle (from statistical analyses to complex simulations) to guide the process (analyze experimental results and identify new modification targets), the Learn stage remains an area with significant room for improvement [8].

A combination of multi-experiment data and data-driven analytical approaches (machine learning (ML) techniques) offers a compelling solution to address the introduced challenges while complying with the DBTL cycle. Currently, many biomanufacturing companies are investing in the digitalization of biotechnological systems and processes, recurring on various computational modeling formalisms, including mechanistic constraint-based modeling (CBM), ML, and hybrid modeling approaches. CBM approaches are based on fundamental principles such as mass and energy conservation, knowledge of the stoichiometry of metabolic networks that describe the range of steady-state flux distributions, using, for instance, the popular flux balance analysis (FBA) approach [9], and require relatively less experimental data. On the other hand, ML explores the capability of computers to perform tasks without explicit programming and no necessity to integrate background knowledge on metabolic processes. To carry out these tasks, ML algorithms can identify patterns, derive predictions, and continuously improve their performance through adaptive mechanisms [10, 11]. Moreover, the adaptive mechanisms present in ML allow computers to learn from datasets. These learning mechanisms, including ongoing deep learning (DL) breakthroughs, enable computers to progressively improve performance in diverse biotechnological applications, such as enzyme optimization, discover relevant genes in biosynthetic pathways, and propose new designs for the next DBTL cycle [12–14]. In this regard, sharing large-scale, high-quality data from bioreactor experiments and monitored organisms is crucial for learning purposes. ML techniques able to represent, integrate, and process high-dimensional data are an invaluable asset in any future biotechnology area to achieve a deep understanding of biosystems [15]. In parallel to the emergence of pure ML models, integrative approaches that combine mechanistic CBM with ML (e.g. hybrid ML/FBA models) are gaining traction in computational strain design [16, 17].

This chapter is structured as follows. The first part reviews critical ML-based approaches for the descriptive and predictive modeling of Big Data in the biotech domain. An overview of hybrid ML/CBM approaches is also provided. The second part of the chapter presents several successful case studies and associated methodologies in the context of metabolic engineering research for the principled illustration of the introduced methods. Lastly, open challenges and future perspectives are outlined.

11.2 Big Data in Biotechnology: Prerequisites for ML-Based Approaches

The design of experimental studies in the biotechnological domain should abide by fundamental principles for the optimal application of subsequent learning tasks, including devising standardized protocols across experiments to ensure comparability between samples and reproducibility, recording all relevant endogenous and exogenous features that can impact the study (e.g. bioreactor cultivation conditions, strain backgrounds), guaranteeing a good coverage of diverse experimental parameterizations (e.g. unbiased sampling techniques for gene knockouts, media composition, environmental conditions) to promote the learning capacity, selecting control groups whenever applicable, isolating causes of noise and artifacts, calibrating instruments regularly and validating measurements to ensure data quality and accuracy, collecting replicates to assess/reduce variability and improve reliability, ensuring sufficient sample size to enable robust statistical analysis and generalization capacity, collecting data across various scales (e.g. multiple omics layers) for holistic learning, and monitoring longitudinal experiments at fine and consistent sampling rates. Complementarily to design considerations, the following sections devise subsequent data preprocessing principles to handle the multimodal, high-dimensional, longitudinal, and noisy nature of bioprocess and omics data.

11.2.1 Data Multimodality and Heterogeneity

In metabolic engineering experiments, data can be collected from a wide variety of biological and physicochemical modalities. First, multiple modalities can be identified at each omic layer, including genomics (genotyping, methylation), transcriptomics (RNAseq, chromatin accessibility), proteomics (abundance, post-translational modifications), and metabolomics (see Chapter 10). Complementarily, molecular interactions (interactomics) between metabolites and other molecules, such as proteins, whether binding or regulatory, can be profiled. Second, metabolic flux analysis (fluxomics) allows the estimation of intracellular fluxes. Third, accompanying biophysical and biochemical measurements are generally relevant to assessing enzyme activity, kinetics, binding affinities, and reaction energetics. Fourth, imaging data, including fluorescence, electron, or confocal microscopy for structural and functional analysis, as well as cellular imaging, may also be available for specific ends. Fifth, phenotype profiling (phenomics) can encompass growth rate measurements (e.g. biomass accumulation), metabolite yield and productivity (production rates of desired compounds), stress response, and morphological analyses (e.g. characterization of cell size, shape, and structure). Finally, and most central, physiological measurements of environmental parameters, nutrient availability (e.g. concentrations of substrates, cofactors) and byproduct accumulation (e.g. toxic or inhibitory compounds) are generally critical.

Complementarity, data heterogeneity can arise due to variations in the monitoring protocols, instrumentation, environmental setups (e.g. temperature, pH, media composition, oxygen availability, and other conditions), (epi)genotypic and phenotypic diversity of strains and organisms, data annotation process sampling rates on longitudinal studies, and feature coverage (e.g. specific omics and fluxes only profiled on a subset of all samples leading to blockwise missing data). Such sources of heterogeneity – whether caused by collection methods, populations, or environments – can introduce several types of biases that hinder the subsequent learning goals, including sampling biases toward specific conditions that lead to generalization difficulties, measurement biases associated with systematic differences that can be assigned to the protocol and not the measured phenomenon, confounding biases due to spurious associations between variables or temporal patterns, and batch effects capturing systematic nonbiological forms of variation [18]. Differences in the extent of recorded features or denormalized protocol variations can lead to feature spaces where large segments of missing entries will occur in blocks, generally claiming for dedicated handling principles, ranging from manifold alignment and domain-aware imputation to multi-source ensemble learning (models learned from subsamples and subspaces without block-wise missings) [19].

To guarantee the capacity to learn from multimodal and heterogeneous sources of data, several principles should be considered (e.g. [20]), including data integration principles to consolidate datasets from different modalities using mapping tables and ontologies for cross-referencing or more advanced principles (e.g. kernel-based fusion, hierarchical integration); standardization principles to handle heterogeneity such as scaling methods, batch effect correction (e.g. ComBat, RUV), and quantile normalization; adding protocol variations as covariates for ensuring sensitivity to protocol-specific biases; integrative learning principles such as graph-based approaches (e.g. heterogeneous networks) matrix factorization techniques (e.g. NMF) to find common patterns; quality assessment principles to quantify signal-to-noise ratios and subsequently apply filtering criteria to retain high-quality data; sampling principles to guarantee an uniform penetration of heterogeneity sources across training, validation, and testing observations; and harmonization principles such as cross-dataset alignment using canonical correlation analysis (CCA), Procrustes analysis, and embedding techniques; representation learning and multi-view learning principles can be used to learn from experimental data combining two or more distinct data structures (e.g. time series data, imaging data).

11.2.2 Stratification and Data Transformations

Complementarily, data sampling challenges pertaining to strata representativity, as well as the presence of categorical targets exhibiting class imbalance and numerical targets displaying skewed distributions, are common and stem from distinct underlying causes, including biases of experimental setups toward certain outcomes (e.g. neglection of certain conditions or emphasis on specific outputs of interest), experimental design choices, annotation inconsistencies, as well as natural biological constraints that make specific outcomes typically rare (e.g. successful strain designs, high-yield production), a common case with targeted modifications where successful interventions are relatively few compared to the vast number of possible perturbations.

Strategies to mitigate these challenges include resampling to reduce imbalances and skewness, incorporating domain-specific knowledge to transform and engineer input and/or output features with more adequate statistical distributions, optimizing the subsequent learning approaches (e.g. adjust decision thresholds, penalize misprediction of minority classes and numeric ranges via dedicated losses) and opt for their pretraining on related data, and acquiring new experimental data from underrepresented strata. Complementarily, data augmentation strategies recurring to variational autoencoders, metabolic models, perturbations to the existing data samples, or generative adversarial networks can be applied to generate synthetic data to handle limits pertaining to data availability, sampling biases, and imbalance of targets. Finally, pairwise learning principles under contrastive loss functions [21] can be used to counteract generalization difficulties.

Given the nature of (bio)process and omics data, alternative data transformations are further recommended to aid descriptive and predictive learning tasks, including normalization of molecular variables to avoid biases from varying scales (e.g. moving from absolute countings toward transactions per million to account for sampling differences, scaling features to a shared distribution for distance-dependent ML approaches), log transformation to address distribution skewness (e.g. gene expression or concentration measurements), dimensionality reduction (DR) approaches to reduce complexity and noise (e.g. PCA, tSNE, UMAP, feature selection methods), encoding of categorical variables (e.g. embedding of experimental conditions, numerical encodings of ordinal variables, one-hot-encoding of nominal variables), discretization of features, particularly targets, for simplifying complex relationships and enhancing interpretability, correcting batch effects [18], and time series or signal transformations for frequency domain representations (e.g. spectral analysis of electrical measurements) and noise reduction (e.g. piecewise aggregation, moving averages, wavelet transforms, filtering).

11.2.3 Handling of Missing Values and Outliers

In the context of metabolic engineering, *missing values* can arise from a variety of causes, including misaligned longitudinal sampling, measurement errors, abnormal conditions, instrumentation sensitivity limits (e.g. mass spectrometry or chromatography may not cover all metabolites in the sample, technical variability of single-cell RNA sequencing). Missing data can be preserved to be handled at the subsequent learning step or imputed using domain-specific knowledge, mechanistic models, centrality statistics, interpolation or regression in time series, predictive methods (e.g. nearest neighbors), and ensemble views (e.g. multiple imputation) [22, 23]. Less commonly, missing values can be removed by dropping the corresponding records or variables, depending on criteria such as data size and missing rates.

Similarly, the presence of *outliers* – observations that significantly deviate from the expected data patterns (beyond natural biological variability) – can arise from a variety of factors, including experimental errors related with instrumentation (fluctuations in sensitivity or signal), operation (fluctuations in expected experimental conditions), sampling biases (e.g. parts of a culture experiencing localized abnormal growth or metabolic activity), and contamination (e.g. by substances or microbes); variations in strain or genotypes; and unexpected metabolic instability due to unaccounted conditions, and experimental design choices (e.g. stress tests such as oxidative stress, nutrient limitations, extreme pH changes or other perturbations). The detection of outliers should, whenever possible, be guided by domain knowledge and follow contextual stances since the experimental context can create meaningful deviations. The detection can be aided by statistical methods to measure deviations from model expectations (ranging from univariate z-score/IQR views toward multivariate views) and nonparametric methods such as clustering approaches [24]. Outlier observations, when meaningful or informative, can be preserved to reshape the learning of descriptors and predictors. To reduce their impact on the learning stage, outliers can be further treated using capping (Winsorization) and multivariate corrections.

11.2.4 Longitudinal Studies

Metabolic engineering experiments involve the profiling and/or optimization of dynamic systems. The temporal dimension is generally present, whether monitoring strain adaptation, fed-batch fermentation (e.g. cell growth, metabolite production, substrate consumption), regulatory responses, bioprocesses, cellular adaptation and recovery from stressors, changes in pathway activation and metabolite accumulation after genetic or environmental perturbations, among others. These experiments generate time-varying data in the form of multivariate time series data when the systems are measured at approximately regular periods or timestamped multivariate event data when measurements occur at an irregular pace or after specific phenomena or perturbations. In this context, multiple principles should be considered to aid descriptive and predictive tasks (e.g. [18, 25]), including the alignment of data points to a common timeline; the application of interpolation or resampling techniques (e.g. spline, Gaussian processes) to standardize time intervals; the normalization of measurements per time point or over the entire time course to prevent cumulative views on metabolic products from creating biases; the extraction of statistical features such as trends, peaks, spectral/frequency features, rates of change, and oscillations over the entire time course or along sliding windows; the creation of lagged variables to capture dependencies along time; the application of dynamic mode decomposition or recurrent autoencoders for temporal compression; the application of smoothing filters to reduce experimental noise and power transformations to stabilize variance; the selection of dedicated ML methods to learn from multivariate time series and event data; the partitioning of data into training, validation, and test sets ensuring temporal coherence (e.g. walk-forward or rolling-window schema).

11.3 Types of ML-Based Approaches for the Design of Microbial Cell Factories

With Big Data being generated at an unprecedented rate, ML-based approaches are becoming a prevalent option for their integrative analysis and interpretation. This section introduces essential background on the application and mechanisms of data-driven predictive and descriptive modeling approaches, with a focus on ML-based techniques, including pure ML and hybrid ML able to incorporate constraint-based methods. Figure 11.1 illustrates the major ML-based approaches described in this section.

11.3.1 Machine Learning (ML) Models

ML provides a powerful framework for developing computational algorithms that learn from experimental data. By leveraging a diverse range of algorithms, ML enables the automatic construction of data-driven models for descriptive, predictive and prescriptive ends. Descriptors can be used to understand bioprocess dynamics for knowledge acquisition, predictors can estimate latent features of interest or iteratively recommend the next steps in lab experiments, and prescriptors

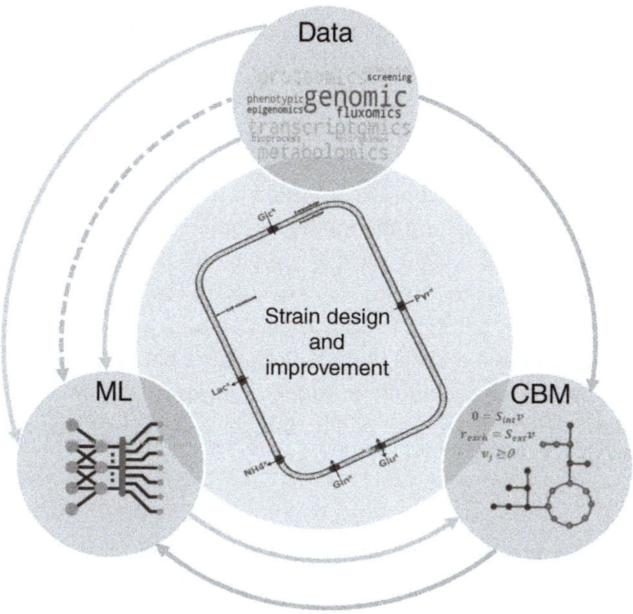

Figure 11.1 ML-based pipelines for cell factory design and improvement using big biological data. Pure ML uses data (particularly omics and exogenous features) for predictive and knowledge acquisition ends (blue arrow). In multi-omics data analysis, ML enables a deeper understanding of metabolic systems by incorporating condition-specific fluxomic results (orange arrows). ML can also use single/multi-omics data as input to identify key features, which can then be integrated with CBM to predict fluxes and improve metabolic models, aiding the analysis of metabolic fluxes (purple arrows).

can leverage longitudinal bioreactor studies to explore the suitability of unseen conditions for ultimately enhancing various process outcomes [26, 27]. Numerous ML algorithms have been developed and are readily available via open-source Python packages (e.g. [28]). Generally, ML models can be categorized into two main classes: supervised and unsupervised learning methods (reinforcement learning (RL) will not be addressed due to the limited examples in the context of engineering microbial cell factories). The selection of the most suitable ML method depends on the desired outputs. Here we provide an overview of ML methods commonly employed in the field of industrial biotechnology, highlighting their applications.

11.3.2 Supervised ML

Supervised learning methods [29], including regression (for continuous outcomes) and classification (for categorical outcomes), single-output and multiple-output predictors, aim to establish a relationship between input features (denoted as $\mathbf{x}$) and corresponding outcomes ($\mathbf{y}$) using training data, $\{(\mathbf{x}_1, \mathbf{y}_1), \dots, (\mathbf{x}_n, \mathbf{y}_n)\}$. This relation is mathematically expressed as:

$$\hat{\mathbf{y}} = f(\mathbf{x}|\theta), \tag{11.1}$$

where θ represents the model parameters. In parametric ML approaches, the objective is to optimize θ by minimizing a loss function that measures the error between the model's predictions and the true values, $\mathscr{L}(\mathbf{y}, \hat{\mathbf{y}})$. The loss function can be assessed in training data or, more rigorously, a set-aside validation (out-of-sample) dataset to ensure that the predictor is able to generalize toward unseen observations, thereby mitigating overfitting risks.

After the pretreatment steps described in Section 11.2, the essential stages involved in a typical supervised learning workflow consists of problem scoping, feature engineering (select and create key variables), and the model selection, hyperparameterization, training, and evaluation (elaborating on this topic is not the scope of the chapter; we redirect interested readers to [30] for more details on these stages).

Selected supervised ML methods applied to various biotechnological tasks are briefly introduced. Some of the targeted tasks include optimizing pathways, classifying cellular phenotypes, predicting metabolic production, optimizing metabolic fluxes, and designing genetic modifications, all of which involve extracting insights from complex biological data.

11.3.2.1 Neural Networks

Inspired by biological neural systems, neural networks (NNs) consist of interconnected processing units, known as perceptrons or nodes, generally organized into layers [31]. Each node receives input from connected nodes, typically from previous layers, processes this input, and generates an output that influences nodes in subsequent layers. The network, f, receives an external signal, $\mathbf{x}$, at the input layer and iteratively transforms it at each intermediate hidden layer, $\mathbf{x}^{[p+1]} = \varphi^{[p]}(\mathbf{x}^{[p]})$, until it returns the system's outputs, $\hat{\mathbf{y}}$, at the final output layer. Different NN architectures with particular topologies and node activation functions have been proposed for different applications [32].

Recent advances in DL, the learning of NNs with a high number of parameters (θ), have renewed interest in their application across various domains, including, in the context of metabolic engineering, enzyme and pathway design [33], and pathway optimizing [34, 35]. DL have also proven useful for productivity optimization in bioprocesses using historical data [36]. From an ML stance, advances in DL have been mostly driven by expressive layering (e.g. recurrent, convolutional, contrastive, residual, attention-based), alternative learning schema (e.g. self-supervision, transfer learning), novel architectures (e.g. transformers, generative adversarial networks), and data augmentation strategies [37].

11.3.2.2 Decision Trees and Ensembles

Decision trees (DTs), as originally described by Quinlan [38], are simplistic nonparametric supervised functions of input features. The function, $f(\mathbf{x})$, is a composition of rules according to a tree structure that is learnt from training data, where each node's rule defines a feature test according to its predictive information. Each data observation falls within a decision path from root to a leaf based on the tested features ($\mathbf{x}$) along the path, and is annotated with the decision outcomes – e.g. continuous and/or categorical values – observed at the corresponding leaf ($\hat{\mathbf{y}}$). The depth of DTs is controlled during the training process to mitigate both underfitting and overfitting risks associated with shallow and deep trees, respectively. DTs are widely used due to their simplicity, ease of implementation, and high interpretability.

In strain engineering and bioprocesses, they have been applied to predict which protein expression levels result in high production yield or to identify input process conditions of various fermentation runs that maximize the production efficiency [39].

The Random Forest (RF) algorithm is an ensemble of DTs based on two major procedures: bootstrapped data sampling, along with feature subspace selection, to learn multiple independent trees (i.e. bagging) with reduced correlation. The ensemble of $k \in \mathbb{N}^+$ trees, $\{f_1(\mathbf{x}), \ldots, f_k(\mathbf{x})\}$, is then used to produce corresponding output estimates, $\{\hat{\mathbf{y}}_1, \ldots, \hat{\mathbf{y}}_k\}$, which are subsequently used to produce the final estimate $\hat{\mathbf{y}}$ through uniform weighting schema, preventing both underfitting and overfitting risks [40].

Alternative successful tree ensembles, such as extreme gradient boosting (XGBoost) [41], are able to learn trees sequentially, where each tree corrects errors made by the previous ones, and optimally assigns efficacy scores for nonuniform weighting schema.

Tree ensembles are versatile, being capable of handling arbitrarily high dimensionality, missing data, numerical and/or categorical outcomes, and feature importance analysis. This flexibility makes them suitable for a wide range of biotechnological applications. For instance, they have been used for pathway optimization [42], differential metabolic flux analysis from varying physical traits [43], and predicting recombinant protein concentrations based on online process parameters [44].

11.3.2.3 Alternative Predictive Approaches

Bayesian approaches for ML use data to update beliefs, where beliefs are given by parameters that describe the distribution of inputs and their associative dependency with targets. Parameters are generally estimated from frequentist views or prior knowledge if available, making Bayesian predictors eligible candidates for hybrid ML in the biotech domain. Predictions for new inputs are inferred from maximum likelihood estimates using the existing beliefs, $p(\mathbf{y}|\mathbf{x}, \theta) \propto p(\mathbf{x}|\mathbf{y}, \theta)p(\mathbf{y}|\theta)$. Joint distributions can be factorized by identifying a network of conditional variable dependencies for greater capacity to learn from high-dimensional inputs. Ultimately, naïve Bayesian approaches place an independence assumption between all input variables, offering a simple baseline reference of performance for more sophisticated ML predictors.

In contrast with Bayesian approaches, similarity, a central concept for pattern-centric learning, is at the center of analogizer approaches. In its simplest form, the nearest-neighbor algorithm predicts the targets for a new observation by weighting the targets recovered by the most similar observations in the training data, $\hat{\mathbf{y}} = f(\mathbf{x}) = \tau(\{w_i \mathbf{y}_i\}_{\mathbf{x}_i \in N(\mathbf{x})})$, where N is the neighborhood of $\mathbf{x}$, w_i is the similarity degree between $\mathbf{x}$ and $\mathbf{x}_i$, and τ is a metric of centrality (weighted mode, mean, or median).

Support vector machines (SVMs), a different class of analogizer approaches, rely on kernels to map the original space into a new space where critical observations (support vectors) can be identified to learn (one or more) hyperplanes that define the classification boundaries or regression surface (herein SVMs are commonly referred to as support vector regressors). Despite the linearity of SVMs, the kernel trick generally promotes their capacity to handle high-dimensional data [45].

SVMs have been shown to be useful in addressing cell factory design challenges. Examples include multigene pathway optimization problems to maximize production titers [35] and automated enzyme search to predict protein–chemical interactions (i.e. whether the enzyme catalyzes a reaction) using both protein sequence and chemical information [46].

11.3.2.4 Selected Case Studies

Two application examples are now introduced to show how supervised ML can be used, highlighting important advances in biotechnological projects: (1) prediction of metabolic fluxes [47] and (2) maximization of 1-dodecanol production [48]. These examples were chosen as they closely align with the routine tasks of a systems metabolic engineer.

The first example explores the use of ML models to predict metabolic fluxes [47]. While constraint-based models (CBMs) are valuable for modeling metabolic networks, they require extensive prior knowledge, tuning objective functions, and rely heavily on predefined assumptions, which limit their applicability under diverse conditions (e.g. less-studied perturbations, infrequent exogenous conditions). Gonçalves *et al.* [47] highlight the growing potential of ML approaches, particularly in leveraging large-scale omics data, such as proteomics and transcriptomics, for metabolic flux prediction. The authors benchmarked state-of-the-art ML models, including RFs, SVMs, DTs, and NNs, against the parsimonious version of FBA (pFBA) [49], using omics data as inputs. The ML models were subjected to nested cross-validation for decoupling hyperparameterization from evaluation. The analysis revealed that ML models, particularly RFs, are competitive in predicting both intracellular and extracellular fluxes compared to pFBA, achieving statistical significance superiority for specific conditions. These findings were consistent when using transcriptomic and proteomic layers individually and jointly. The results suggest that ML approaches can effectively overcome traditional methods in scenarios where these methods struggle. To evaluate generalization, the ML models were also tested on an independent dataset. While RFs demonstrated superior performance in certain cases, their generalization was nevertheless limited, given the limited number of training examples and distinct experimental conditions.

The second example, conducted by Opgenorth and colleagues [48], explored the optimization of 1-dodecanol production in *Escherichia coli* using a combination of ML and two iterative DBTL cycles. This systematic framework represents a significant shift from traditional trial-and-error approaches in metabolic engineering. The study focused on leveraging proteomics data to guide strain development, ultimately improving production efficiency. In the first DBTL cycle, the authors examined 36 engineered *E. coli* strains, systematically modulating ribosome-binding sites (RBSs) and acyl-ACP/acyl-CoA reductases within a single operon. Measured variables included dodecanol titers and pathway protein levels. The results revealed that *Maqu_2507* reductase consistently outperformed other reductases for dodecanol production, whereas *Maqu_2220* yielded the highest total fatty alcohol production. This initial cycle generated a valuable dataset for training ML models, including RFs and NNs, to predict optimal protein profiles for further strain improvement. The second DBTL cycle implemented the predictions generated by the ML models from the first cycle. Fourteen new strains were designed, with the ML models guiding the adjustments of protein expression levels. The results revealed a 21% increase in dodecanol titer, reaching 0.83 g/L, which is significantly higher than previously reported values for batch cultures. However, certain challenges, such as inaccuracies in the predicted strength of binding sites and unexpected operon-wide effects, limited the extent of improvements. The significance of this work lies in showcasing how ML can effectively complement systematic frameworks like DBTL to advance metabolic engineering, while underscoring the potential of iterative cycles to achieve improvements in production over time. The study further emphasizes the importance of balancing exploration and exploitation within the DBTL framework. While initial cycles should focus on exploring diverse design spaces, subsequent iterations should leverage the accumulated knowledge for a more targeted optimization. This balanced approach is crucial for maximizing the efficacy of ML algorithms and refining strain engineering processes. In conclusion, this study provides a compelling example of how integrating ML with high-throughput DBTL cycles can advance metabolic engineering. While challenges remain, such as refining predictive tools and addressing data limitations, the results highlight the presence of a promising way toward more efficient production of bio-based products.

11.3.3 Unsupervised ML

Unsupervised learning constitutes a crucial aspect of ML for acquiring knowledge on the behavior of biosystems, particularly considering the significant effort often required for their annotation [50]. The main goal is to identify patterns in complex, multivariate input data by either grouping similar instances, learning representations, retrieving local patterns, or modeling dynamics. Common examples of unsupervised ML techniques in biotechnology include clustering [51], representation learning [52], biclustering [53], parametric modeling [54], and DR [55].

11.3.3.1 Clustering

Clustering aims at grouping data observations in clusters to uncover relevant structures. Common approaches including hierarchical clustering, which builds a nested hierarchy of clusters using agglomerative (bottom-up) or divisive (top-down) methods; density-based clustering, which groups points based on dense regions; and model-based clustering, which assumes data is generated from a mixture of probabilistic distributions, like Gaussian Mixture Models (GMMs), to learn soft structures, i.e. structures where observations are assigned a membership probability to each cluster. Partition-based clustering, where K-means is the paradigmatic example, assigns data into a parameterizable number of K clusters by minimizing intra-cluster variance,

$$\mathbf{c}_i = \text{argmin}_j\, d(\mathbf{x}_i, \overline{\mathbf{c}}_j)^2 \tag{11.2}$$

i.e. cluster assignment based on the distance (d) between an observation ($\mathbf{x}_i$) and the nearest cluster centroid ($\overline{\mathbf{c}}_j$). Distances can be customized depending on the variable types (e.g. Gower distance for mixed variables) and data structure (e.g. DTW for time series data). The algorithm iteratively updates the cluster centroids,

$$\overline{\mathbf{c}}_j = \frac{1}{|\mathbf{c}_j|} \sum_{i=1}^{|\mathbf{c}_j|} \mathbf{x}_i \tag{11.3}$$

and reassigns observations until it meets a stopping criterion – no further changes occur in data point assignments or when the total sum of distances within clusters is minimized.

Clustering has been used for different ends [56, 57]: categorize microbial strains based on their metabolic profiles, aiding in the selection of best candidates for bio-production; differentiate enzymatic reactions, aiding in the identification of novel biosynthetic pathways; and grouping metabolites from mass spectrometry data, distinguishing between core metabolic pathways and potential byproducts. Clustering methods, however, have some limitations. For instance, they generally group molecular entities based on all available features, which may hide relationships that exist only within specific subsets of features. Furthermore, each entity is restricted to a single group in hard clustering structures, even though they can participate in multiple groups depending on their biological functions. These shortcomings can be addressed by biclustering, making it valuable for identifying patterns of co-regulation [58].

11.3.3.2 Biclustering

Biclustering, originated in the context of gene expression data analysis [59], is a unique category of clustering techniques that can group observations and variables simultaneously, hence often positioned as a pattern discovery method. The problem, also referred to as subspace clustering, aims at identifying a set of biclusters, such that each bicluster B_k satisfies specific criteria of interest. To guide the formation of these biclusters, homogeneity criteria are commonly employed, typically implemented through the use of a merit function [60]. A given bicluster $B = (I, J)$, where I and J are, respectively, the subset of observations and variables, is generally characterized by its pattern, φ_J, i.e. the set of expected values on the given variables. Different types of homogeneity can be pursued, including:

- constant overall if for all $\mathbf{x}_i \in I$ and $y_j \in J$, $a_{ij} = \mu + \eta_{ij}$, where μ is the typical value and η_{ij} is the observed noise;
- constant on variables, i.e. pattern on observations, if $a_{ij} = \mu_j + \eta_{ij}$, where μ_j represents the expected value in variable y_j;
- additive (multiplicative) if for all $i \in I$ and $j \in J$, $a_{ij} = \mu_j + \gamma_i + \eta_{ij}$ $(a_{ij} = \mu_j \times \gamma_i + \eta_{ij})$ where μ_j is the expected value in variable y_j, and γ_i the adjustment for observation $\mathbf{x}_i$;
- order-preserving on variables if there is a permutation of J variables, π_J, under which the sequence of values in every observation in I is strictly increasing.

In addition to homogeneity, biclustering searches can incorporate statistical significance criteria [61] to ensure that a bicluster's occurrence deviates from what is expected under a null model, and dissimilarity criteria [62] to prevent redundancies among biclusters.

In biotechnology, biclustering has numerous applications, including the identification of co-regulated metabolic pathways and the discovery of drugs with similar activity patterns [63]. Biclustering can provide valuable insights on putative regulatory modules (e.g. coordinated sets of genes and metabolites) or correlated metabolic fluxes and enzymatic activities under specific conditions [64, 65], as well as understand how microbes or engineered strains adapt to different environments by clustering metabolic responses under stress conditions [66].

11.3.3.3 Representation Learning and Dimensionality Reduction

DR focuses on decreasing the number of features in a dataset while retaining most of its relevant information or structural patterns [67]. DR techniques are generally categorized into two main approaches: feature selection and feature extraction/transformation. Feature selection involves identifying a subset of features based on specific criteria, such as minimizing information loss or maximizing predictive power, via filter techniques, embedded methods, and wrapper models [68]. In contrast, feature extraction generates new features by transforming or combining existing ones, often through nonlinear transformations. One of the most widely used DR techniques in biotechnology is Principal Component Analysis (PCA) due to its inherent simplicity. PCA reduces the dimensionality of a dataset by capturing the maximum variance with the fewest possible dimensions, effectively summarizing the data while preserving its essential patterns [69]. The resulting k dimensions (components) are a linear projection of original dimensions given by the eigenvectors of the covariance matrix from centered data, with the explained variance computed according to the corresponding eigenvalues, $\sum_{j=1}^{k} \lambda_j^2 / \sum_{j=1}^{n} \lambda_j^2$, where $k \leq m$.

DR techniques like t-distributed Stochastic Neighbor Embedding (t-SNE) and Uniform Manifold Approx. Projection (UMAP) are also particularly suitable to visualize correlations and select variables in highly complex multidimensional data [70–72].

They have been consistently applied to characterize cellular types and states, including discrimination of stressed cells [73] and healthy cells from diseased ones [74]. Recently, they have also been applied for classifying cells as high producers or low producers [75].

Representation learning offers an alternative to classic DR methods by leveraging DL techniques such as autoencoders and variational inference for feature extraction [76]. It is increasingly applied in metabolic engineering to acquire numeric representations (embeddings) of complex data, such as high-dimensional, multilayered, temporal, and/or imaging data. Representation learning has been applied to learn embeddings from metabolic reactions to predict novel biosynthetic pathways, protein sequences to predict enzyme activities (novel biocatalysts), multi-omics to optimize microbial strains, or context-aware genotypic data to predict strain phenotypes [77].

11.3.4 Hybrid ML and Constraint-Based Models

A promising avenue for analyzing high-dimensional biological data involves the integration of mechanistic modeling methods [78, 79], such as the widely utilized FBA approach (for a comprehensive review of CBM, see [80] and [81]) with ML stances. Hybrid ML/CBM approaches combine state-of-the-art ML techniques (either supervised or unsupervised learning paradigms) with mechanistic CBM methods (here referred to as FBA approach) for exploring synergies and opportunities of both formalisms [27]. This integration allows ML to leverage mechanistic knowledge, ultimately translating into an improved predictive capacity and interpretability, as reviewed recently [16, 82].

Various data- and knowledge-driven pipelines can be established depending on the available data types and specific research goals. Currently, the different workflows/approaches for integrating constraint-based models and ML using omics data can be classified into two main groups: i) CBM-FBA as input for ML, in which condition-specific CBMs are generated using multiple datasets from high-throughput technologies to obtain metabolic fluxes prediction and from which ML techniques can infer important patterns, and ii) ML as input for CBM-FBA, in which ML methods are directly applied to single/multi-omics datasets for feature selection that are then used as input constraints to build/improve CBM-FBA models. In this section, we highlight some of these integrative studies, providing an overview of the steps involved in each workflow.

11.3.4.1 CBM-FBA as Input for ML

When available, experimental data can be combined with FBA-derived fluxes to create multi-omic features, enabling the prediction of targets of interest [83, 84]. The fluxes generated by metabolic models can then feed the ML algorithms for further analysis. Plaimas *et al.* [85] supplied ML with descriptors derived from FBA, biochemical knowledge and transcriptomic data to predict essential enzymes/reactions within a metabolic network. By employing an SVM algorithm, this system demonstrated a significant improvement in the prediction accuracy compared to FBA approaches alone. A different hybrid pipeline integrating FBA and ML was described by Vijayakumar *et al.* [86, 87]. This novel approach involves constructing condition-specific metabolic models by incorporating transcriptomic data as constraints that are then simulated using a multi-objective FBA model. The ML model is then trained on transcriptomic data and the predicted fluxomics profile, enabling the identification of key genes/reactions under different conditions. Similarly, another approach is to incorporate input fluxes predicted from pFBA combined with omics data to improve the predictive power of microbial growth rates [88].

A recent innovation is the development of a hybrid approach that embeds FBA directly into ML (artificial NNs) [89]. Unlike traditional FBA, which requires extensive experimental data to define medium uptake fluxes, the proposed artificial metabolic network (AMN) incorporates a neural preprocessing layer to capture relationships between medium composition and uptake fluxes, facilitating accurate phenotype predictions without additional experimental measurements. By constraining ML with CBM, this hybrid approach achieved higher predictive accuracy for metabolic fluxes and growth rates while reducing data requirements. The AMN framework also shows potential for broader applications, including designing gene knockout strategies, thus representing a significant advancement in strain engineering. Alternatively, unsupervised modeling strategies, including PCA, have also been applied with FBA-derived results [90, 91].

Numerous computational methodologies exist for data integration to develop context- and condition-specific metabolic models [92]. An interesting direction is to combine condition-specific metabolic modeling with Bayesian ML to infer the metabolic pathway's responsiveness to various environmental conditions [93]. Furthermore, alternative ML approaches (bridging supervised and unsupervised algorithms) can also be integrated with FBA results to estimate the main metabolic features of CBM [94].

While FBA-based approaches have successfully been used in analyzing multi-omics data to address specific limitations inherent to each data type, they require prior knowledge of the appropriate objective function (which is often unknown) of an organism [95]. Additionally, these approaches necessitate several preprocessing steps to effectively integrate CBM with multi-modal data, which can be a complex and challenging process [96].

11.3.4.2 ML as Input for CBM-FBA

An alternative approach involves incorporating input features from ML into FBA-based models for improving their predictive power or aiding model creation.

Instead of solely relying on direct flux estimates generated by FBA, ML can be employed to analyze multi-omics data and generate data-driven flux boundaries. These can then be passed as additional constraints within the FBA model, effectively narrowing the solution space and improving the accuracy of the predictions.

A representative example integrates ML with constraint programming to predict metabolic fluxes [97]. This approach has demonstrated the power of ML to generate data-derived fluxes, which can then be incorporated as input constraints within the FBA framework, thereby guiding predictions. By integrating quadratic programming, the approach ensures that the predicted flux profiles satisfy stoichiometric constraints, enhancing the biological relevance of the results. Additionally, ML methods can also be used to estimate enzyme catalytic turnover numbers (k_{cat} in vitro or $k_{app,max}$) at the genome-scale to parameterize FBA models, leading to more accurate predictions [98].

Also, efforts on unsupervised approaches to obtain additional constraints for the FBA formulation have shown promising results [99]. An interesting strategy is the learning of representations to characterize strain variations and identify metabolic states, which can be further utilized by CBM [100]. Similarly, an unsteady-state FBA framework has been successfully applied by incorporating PCA of the time-dependent metabolomic data, which helps identify the linear metabolic states to predict metabolic fluxes under dynamic conditions [101]. Beyond dimensional reduction techniques, clustering methods can also be applied in combination with FBA to group flux distributions (e.g. [102]).

Figure 11.1 synthesizes the major pipelines for integrating ML and CBM. The pipeline selection depends essentially on the desired output and the type of data available. Despite their successful application, these pipelines remain challenged by the high dimensionality, quality, and heterogeneity of multi-omics data, often showing significant forms of imbalance and skewness [103, 104]. To overcome this, more research is needed to further assess new ways to integrate complex data with these computational approaches.

11.4 ML-Based Approaches in Microbial Cell Factory Design: Case Studies

The rapid progress of ML, coupled with the increasingly wide data collection efforts, jointly offers a flexible and efficient toolkit for overcoming challenges in biotechnology and microbial biomanufacturing. This section illustrates how various computational approaches described in Section 11.3 can leverage Big Data to address various challenges in cell factory development using reference case studies. Particularly, some examples are highlighted to show how integrating ML with multiple recorded experiments can promote a variety of applications ranging from strain engineering and flux design to metabolic production.

11.4.1 ML-Based Approaches in Strain Engineering and Flux Design

Strain engineering aims to identify genetic strategies to optimize microbial performance. Traditionally, this was achieved through random alterations of genes, such as knockouts, changes promoting overexpression, or the introduction of heterologous pathways. A key focus of modern strain engineering is the iterative DBTL cycle [105], where large datasets are acquired to evaluate strain performance and orient improvements. In this regard, ML can play a critical role in extracting meaningful insights from Big Data, thereby guiding successive rounds of strain engineering and pathway design [106]. To accelerate the adoption of ML-driven strain development, significant efforts have been made to integrate these technologies into the DBTL framework, leading to increased efficiency and precision in bioproduction. As a first example, an integrated DBTL pipeline for strain optimization of fine chemicals was proposed by Carbonell *et al.* [107]. The workflow integrates computational pathway design, robotic-assisted DNA assembly, high-throughput screening, and ML-based learning cycles to efficiently prototype metabolic pathways. The Design stage employs bioinformatics tools for enzyme selection and pathway construction, followed by Build automation through robotic DNA assembly. The Test phase uses high-throughput metabolite screening, feeding results into the Learn phase, where ML models refine pathway performance predictions. By leveraging intelligent sampling and statistical learning, this approach significantly reduced experimental workload while accelerating strain development, achieving a 500-fold improvement in the production of (2S)-pinocembrin. Later, HamediRad and coworkers [108] demonstrated the power of ML-driven optimization within the DBTL cycle by developing an automated algorithm-driven platform called BioAutomata. BioAutomata employs Bayesian optimization to efficiently explore the vast design space of possible gene expression levels. This Bayesian approach effectively balances exploration of the design space with exploitation of promising regions, enabling efficient identification of optimal gene expression levels while minimizing the number of experiments required. By optimizing the lycopene biosynthetic pathway, their system evaluated less than 1% of possible pathway variants yet outperformed random screening by 77%. The combination of predictive modeling and high-throughput automation enabled rapid discovery of optimal pathway configurations, reducing both experimental cost and design complexity. Another interesting example is the Automated Recommendation Tool (ART) [109], an ML-based platform that integrates probabilistic modeling and a Bayesian ensemble ML approach to systematically recommend strain designs in the DBTL cycle. Using protein expression levels or promoter combinations as input, ART has been showed to be capable of predicting production levels of target molecules. A significant strength of ART is its capacity to quantify uncertainty in its predictions, allowing researchers to identify and explore under-characterized regions of the design space. A different ML-assisted workflow (called MiYA) designed for the combinatorial optimization of heterologous metabolic pathways was demonstrated successfully by Zhou and collaborators [110]. This innovative system combines artificial NNs with the YeastFab Assembly strategy to enhance the DBTL cycle. To demonstrate its effectiveness, MiYA was applied to optimize two biosynthetic pathways in *Saccharomyces cerevisiae* to improve violacein production. These studies highlight the significant potential of integrating ML algorithms into the DBTL cycle to accelerate strain development efforts.

More recently, deep NNs have demonstrated significant promise in rational strain engineering, including the discovery of pathways via retrosynthesis (i.e. identifying a series of biochemical reactions from substrates to desired molecules). Biswas *et al.* [111], introduced a data-efficient DL model for low-N protein engineering, utilizing the UniRep framework [33] trained on labeled mutants for extrapolating functional predictions of 10 million sequences via *in silico* directed evolution. Their approach was tested on two distinct proteins (avGFP and TEM-1 β-lactamase), where top AI-designed candidates exhibited up to 500% improvement in function from a single assay cycle. By leveraging global protein sequence knowledge and fine-tuning representations on local mutational landscapes, their method demonstrated superior generalization compared to classical protein engineering strategies. In another study, Segler and coworkers [112] employed a Monte Carlo Tree Search (MCTS)-based retrosynthesis system that integrates deep NNs and symbolic AI to predict synthetic routes for small organic molecules. Their system incorporates three NNs: an expansion policy network to guide the search, a filter network to assess reaction feasibility, and a rollout network to estimate the best retrosynthetic moves. Trained on 12.4 million reactions from the Reaxys database (www.reaxys.com), the model significantly outperformed traditional rule-based retrosynthesis planning, solving almost twice as many synthesis problems 30 times faster than prior methods.

Various other strategies that allow precise control of gene expression and pathway optimization can also be enhanced using ML. A typical approach involves fine-tuning gene expression by modifying promoter and RBS elements. In this regard, ML has been used to guide the design of promoters and RBS sequences in a data-driven manner, a common strategy for regulating gene expression in rational pathway engineering. For instance, Meng *et al.* [113] demonstrated the power of NNs in quantitatively designing regulatory elements critical for controlling gene expression, such as promoters and RBSs. In this study, an NN was trained from a library of 100 mutated Trc promoters and RBS sequences, capturing a wide range of expression strengths. The model achieved a high correlation coefficient for both training and test datasets, indicating its strong predictive accuracy. Recently, ML has also been utilized to optimize multigene pathways for chemical production. For example, Jervis and coworkers [35] implemented an ML-guided approach (using SVM and feedforward NNs) to optimize the translational control of biochemical pathways in *E. coli*. This methodology was demonstrated on the mevalonate pathway, a nine-step metabolic route leading to the production of monoterpenoids. By analyzing a 5184-member RBS library, their model accurately predicted high-producing pathway variants while screening only 3% of the library. Applied to a recombinant monoterpenoid biosynthetic pathway, this strategy increased production titers by over 60%.

These two last examples demonstrate the potential of systemically integrating high-throughput strain construction, testing, and ML to optimize multistep pathways for improving titers and yield. However, these strategies depend on the creation of extensive mutant strain libraries, restricting the applicability of ML to microorganisms with well-developed genetic engineering tools and protocols. One possible alternative approach is to feed an NN trained on available mRNA expression data to a multi-objective optimizer, such as the non-dominated sorting genetic algorithm-II (NSGA-II). This method was utilized by Kang and colleagues [114] to identify the optimal strain design that was showed to be capable of improving the lycopene titer, reducing the necessity for a large library. While less explored in the context of strain engineering, RL holds significant promise for optimizing strain performance. As a proof of concept, Sabzevari *et al.* [115] introduced a multi-agent RL approach, validated on a combinatorial strain library of *Saccharomyces cerevisiae*, focused on enhancing L-tryptophan biosynthesis. Their results highlight the potential of RL-based strain optimization, reducing reliance on trial-and-error methods.

Combining ML and CBM approaches using large-scale data has also emerged in strain engineering and flux design (i.e. selection of gene targets and their subsequent optimization). For instance, a comprehensive framework for multi-omics integration and predictive modeling in *E. coli* was introduced by Kim and coworkers [84]. They developed a multi-omics compendium integrating 4389 genome-wide profiles across 649 conditions, including transcriptomics, proteomics, metabolomics, and fluxomic data. Using this resource, they constructed a multi-scale predictive model – integrating recurrent NNs, LASSO regression, and constraint-enriched FBA – trained on omics data and interaction networks to predict cellular states under unexplored conditions, outperforming previous models. Another interesting example of integrating FBA with data-driven approaches involves a hybrid framework that combines FBA with Bayesian factor modeling to predict metabolic pathway activation under various environmental conditions [93]. This approach uses multi-omic data, such as gene expression profiles, to augment an FBA model, enabling the integration of fluxomic and transcriptomic layers. By linking gene expression levels to metabolic flux constraints, the method captures cellular responses to dynamic environmental stimuli. The study focused on *E. coli* as a model organism and demonstrated the framework's capability to infer pathway activation profiles across time and under varying conditions, such as stress responses. This work highlights the ability of hybrid frameworks to elucidate genotype and phenotype relationships, providing valuable insights into bacterial metabolic responses for industrial strain optimization.

ML-based approaches have also been used to acquire fluxomic information from enzyme or gene expression data [116]. For example, Oulia and coauthors [117] applied DL techniques to predict metabolic fluxes based on enzyme levels in the glycolysis pathway of *Entamoeba histolytica*. In their study, they demonstrated how a deep NN model, trained on a large dataset of 68950 instances generated through synthetic data augmentation, outperforms classical ML. Similarly, Costello *et al.* [118] introduced an ML approach to predict metabolic pathway dynamics from time-series multi-omics data. Using metabolomics and proteomics, they demonstrated that supervised ML models (including RF and SVM) outperform traditional kinetic models in capturing metabolic flux dynamics. The model was validated on two biofuel-producing pathways, limonene and isopentenol, showing that predictions improved significantly with increased training data.

Biclustering has also been applied in strain engineering and flux design with relevant results. Alzahrani *et al.* [119] applied a graph-based biclustering algorithm to analyze growth phenotype datasets and identify co-fit genes that exhibit similar essentiality or dispensability under different stress conditions. By representing growth phenotype data as a two-dimensional matrix, they mapped the problem into the search for subspace patterns with constant values on columns (i.e. genes with similar growth patterns). More recently, Karim *et al.* [120] applied biclustering to analyze species-volatile organic compound (VOC) relationships by constructing bipartite networks linking species to their emitted VOCs. Biclustering was used to categorize species based on shared VOC emissions, uncovering evolutionary adaptations and enabling the engineering of VOC biosynthesis pathways.

11.4.2 Application of ML-Based Approaches for Metabolic Production

Metabolic production involves the use of microbes to synthesize valuable compounds, such as pharmaceuticals, biofuels, and industrial biochemicals from renewable bioresources. Following the design and construction of biosynthetic pathways within a microbial host, the efforts focus on enhancing the metabolic production of the target compound. In addition, scaling up the bioprocesses in microbial cell factories is essential for successful commercialization. While conventional approaches depend on trial and error, generally inefficient and time-consuming, ML-based approaches accelerate this process by analyzing large datasets to predict metabolic production and subsequently optimize fermentation conditions. Several studies in literature report advances from the application of ML-based approaches to enhance metabolic production [39, 44]. For instance, Colletti *et al.* [121] and Varman *et al.* [122] developed pure ML models to a priori estimate the chemical productivity of engineered *E. coli* and *S. cerevisiae* strains, given a set of inputs such as the number of biosynthesis steps, cultivation conditions, and genetic modifications. These regression models accurately predicted the product yields, providing important guidelines in fermentation processes. Another study examined the enhanced xylitol production by *Debaryomyces nepalensis* in a bioreactor through NN regressors coupled with a GA [123]. Using a dataset containing fermentation variables such as temperature, pH, oxygen transfer rate, biomass, and glycerol concentrations from previous time points, NN models were trained to predict the biomass growth and xylitol yield. By coupling these predictors with GA, they optimized fermentation conditions, demonstrating that ANN-GA outperformed regression-GA, improving xylitol production from 59.4 g/L to 65.73 g/L.

Other approaches have also shown reliable results in improving metabolic production of valuable compounds using large-scale data by combining metabolic modeling with ML. For example, Zhang *et al.* [124] integrated CBM (FBA) and ML models to optimize tryptophan metabolism in yeast. By employing CBM, they identified key metabolic engineering targets and constructed a combinatorial metabolic pathway library of potential strain designs controlled by various promoter combinations. To facilitate high-throughput characterization, they developed a biosensor screening system, generating a fluorescence-based dataset of over 124000 experimental time-series data points from more than 500 different strain designs. The dataset, consolidating genotype data, growth profiles, and the biosensor output, was then used to train ML algorithms (EVOLVE [125] and ART [109]), which successfully predicted optimized metabolic pathways. The acquired results showed that this hybrid approach achieved up to 74% higher titer of tryptophan compared to the best initial design. In another study [126], a hybrid ML framework was integrated with genome-scale metabolic models to assess microbial factory performance. Using *E. coli* as an example, the authors manually curated a dataset of approximately 1,200 engineered strains from around 100 research articles, including key genetic modifications, bioprocess conditions and production metrics for over 20 different compounds, including succinate, L-lactate, and isobutanol. Then, by augmenting this dataset with FBA simulations, they generated additional metabolic features. Employing ensemble ML techniques, they demonstrated a high cross-validation accuracy in predicting strain performance based on titer and yield. Similarly, Czajka and colleagues [127] introduced an integrated framework combining knowledge mining, CBM, and an ML ensemble model using gene expression data and environmental conditions to predict bioproduction yields in *Yarrowia lipolytica*, spanning various products like organic acids, terpenoids, and lipids. While the framework showed promising results, its success depends on the availability of broad-scope data sources.

Recently, in 2020, a novel hybrid ML/CBM approach for controlling nutrient availability in a fed-batch culture has been developed [128]. Schinn and collaborators developed a hybrid computational approach integrating CBM with ML to predict the amino acid profile from bioprocess measurements in CHO cell cultures. By utilizing bioprocess data from 10 CHO clones with varying growth and productivity profiles, they trained a metabolic model to estimate amino acid uptake rates based on observed cell proliferation and recombinant protein production. Then, by applying statistical regression models, they corrected systematic biases in metabolic model predictions and implemented an exponential decay function to extend forecasts beyond steady-state assumptions. Their approach correctly forecasted the amino acid consumption dynamics over an 8-day culture period with an accuracy of up to 86%, enabling anticipatory control of nutrient feeding in bioreactors.

In the context of multi-omics integration, Zampieri *et al.* [129] employed a poly-omics ML approach to predict lactate production in CHO cells, integrating gene expression data with metabolic flux simulations. By reconstructing condition-specific genome-scale metabolic models based on gene expression from 127 CHO cultures, they estimated metabolic reaction rates using FBA. Then, by applying ML methods such as PCA and elastic net regression, they refined metabolic predictions and trained generalized linear models to predict lactate yields under different bioprocess conditions. In a second study, a CBM of *Pseudomonas putida* was developed, incorporating gene expression data and FBA to predict rhamnolipid biosynthesis

[90]. Statistical and ML techniques were then applied to FBA-generated flux distributions to identify biomarkers and metabolic pathways for increasing rhamnolipids production.

Unsupervised approaches have also been used to improve the production of desired chemicals in microbial systems. For example, Morris *et al.* [130] proposed an unsupervised big data analytics pipeline focused on multivariate associations for identifying metabolic inhibitors and promoters to optimize the production process of monoclonal antibodies (mAbs) in CHO cell cultures. By applying the analytical pipeline to metabolomics data under varying fed-batch bioreactor conditions, the authors identified key metabolites influencing production titers. This analysis enabled the prioritization of the top metabolites, then used for metabolic pathway enrichment studies to better understand their roles in productivity. To validate these findings, the authors integrated transcriptomics data with metabolomics results, correlating gene expression changes with the production of specific metabolites. They identified amino acids that either inhibited or enhanced metabolite production and proposed modifications to feed media based on these insights. The optimized feed media resulted in an 8.2% improvement in mAb production, underscoring the practical value of this integrated analytical approach. A PCA-based framework was also applied to proteomics data to identify specific enzymes that limit the production of target molecules [131]. The analysis guided the manipulation of gene expression levels of the mevalonate pathway enzymes in *E. coli* and improved the production of limonene up to 40%.

11.5 Conclusions and Perspectives

This chapter covered notable advances in the application of ML-based approaches to address ongoing challenges in metabolic engineering to show how Big Data is revolutionizing industrial biotechnology. To this end, we first established essential grounds on descriptive and predictive modeling techniques in this domain, along with critical principles for experimental study design and data preprocessing. Particular attention is placed on examining the synergy between ML and existing approaches, including CBM and DBTL cycle framework, to fully tap into the transformative potential of hybrid approaches in metabolic engineering.

Real-world case studies in strain engineering, flux design, and metabolic production were explored to form a comprehensive picture of how ML-based approaches are driving innovations, paying particular attention to the quantification of their impact when possible.

Despite the opportunities of both supervised and unsupervised ML stances, challenges pertaining to the full exploration of multimodal data sources, the longitudinal nature of most experiments, emerging profiling technologies (e.g. single-cell multi-omics, fine-grained molecular binding activity), and ongoing DL advances, remain open. In addition, the design of the "best" ML-based approach is not trivial and is critically dependent on the specificities of the target problem and available data. In this context, broader efforts toward systematic comparisons and the automation of ML-based pipelines are expected [132]. In addition, as ML algorithms may show limits of extrapolating beyond the given data training ranges, particular care to the generalization capacity of the models (e.g. uncertainty-bounded estimates) is key to guarantee their suitability for unseen combinations of endogenous-exogenous factors, a critical need when optimizing metabolic systems.

Perspectives

A greater penetration of ***Deep Learning*** approaches in the metabolic engineering domain is expected in the upcoming years [133]. Deep generative models (e.g. GANs, VAEs) can be used to guide novel strain designs with improved metabolite production. Attention-based models yield the potential to capture complex relationships between metabolic inputs and fluxes. Sequence-to-function models can leverage transformer architectures for designing synthetic regulatory elements. Transformers like AlphaFold and ProtGPT2 are already assisting enzyme design for enhanced metabolic pathways. Finally, the capacity of DL architectures to capture the inherent dynamic nature of bioprocesses using, for instance, recurrent layering, as well as complex multilevel relationships using graph NNs, makes them appropriate candidates to directly learn from raw multivariate time series data produced by longitudinal studies without laborious feature engineering needs.

Transfer Learning can leverage knowledge from related tasks or organisms for improved predictive and descriptive capacity (e.g. [134]). Paradigmatic examples include transferring learned representations from well-characterized organisms (e.g. *E. coli* and *S. cerevisiae*) to predict metabolic functions in less-studied organisms, pretraining models on simulated or experimental pathway data to predict pathway performance across different conditions or strains, or fine-tuning models trained on fluxomics data from one condition (e.g. aerobic growth) to predict fluxes under different culture conditions or bioprocess setups.

Complementarily, ***Multitask Learning*** (MTL) can be applied to simultaneously solve multiple related prediction tasks by sharing knowledge across tasks (e.g. [135]), thereby leveraging synergies to improve model generalization and performance. MTL can be applied to predict multiple metabolic traits (e.g. yield, productivity, robustness) during strain design, predict molecular properties (e.g. binding affinity, toxicity, stability, solubility, activity) from shared molecular representations, design genetic circuits with optimal performance across diverse conditions or hosts, predict diverse phenotypes from omics data, or leverage different microbial strains or conditions to generalize metabolic engineering strategies.

We further expect the community to translate the ongoing potentials of ***Self-Supervision*** into the metabolic engineering domain by devising artificial tasks that do not require manual labeling for pretraining models (e.g. [136]), which can then be fine-tuned for specific downstream tasks. In this context, fluxomics can be reconstructed from incomplete or noisy data by predicting masked flux values to learn the underlying network structures. Similarly, shared embeddings of unlabeled multi-omics data can be learnt by masking whole layers, phenotypes, and molecular markers, as well as latent representations of bioprocess data by predicting masked growth rates, metabolite abundance or yields under varying conditions.

Finally, the emergence of ML approaches with guarantees of ***trust*** (e.g. uncertainty quantification), ***explainability***, and ***traceability*** [137] can be further leveraged by metabolic engineering tasks for a diversity of ends, including avoiding unintended consequences like toxic by-product formation or biosafety hazards, acquiring new insights into metabolism and pathway interactions, enhancing transparency for regulatory compliance, and strengthening agency in collaborative innovation and industrial applications.

References

1 Apweiler, R., Martin, M.J., and al. e. (2013). Update on activities at the universal protein resource (UniProt) in 2013. *Nucleic Acids Res.* 41 (D1): D43–D47.

2 Benson, D.A., Karsch-Mizrachi, I., Clark, K. et al. (2012). GenBank. *Nucleic Acids Res.* 40 (Database issue): D48–D53.

3 Kim, S., Chen, J., Cheng, T. et al. (2025). PubChem 2025 update. *Nucleic Acids Res.* 53 (D1): D1516–D1525.

4 Morell, W.C., Birkel, G.W., Forrer, M. et al. (2017). The experiment data depot: a web-based software tool for biological experimental data storage, sharing, and visualization. *ACS Synth. Biol.* 6 (12): 2248–2259.

5 Antoniewicz, M.R. (2021). A guide to metabolic flux analysis in metabolic engineering: methods, tools and applications. *Metab. Eng.* 63: 2–12.

6 Zhang, J., Chen, Y., Fu, L. et al. (2021). Accelerating strain engineering in biofuel research via build and test automation of synthetic biology. *Curr. Opin. Biotechnol.* 67: 88–98.

7 Esvelt, K.M. and Wang, H.H. (2013). Genome-scale engineering for systems and synthetic biology. *Mol. Syst. Biol.* 9: 641.

8 Petzold, C.J., Chan, L.J., Nhan, M., and Adams, P.D. (2015). Analytics for metabolic engineering. *Front. Bioeng. Biotechnol.* 3: 135.

9 Orth, J.D., Thiele, I., and Palsson, B.O. (2010). What is flux balance analysis? *Nat. Biotechnol.* 28 (3): 245–248.

10 Khaleghi, M.K., Savizi, I.S.P., Lewis, N.E., and Shojaosadati, S.A. (2021). Synergisms of machine learning and constraint-based modeling of metabolism for analysis and optimization of fermentation parameters. *Biotechnol. J.* 16 (11): e2100212.

11 Helm, J.M., Swiergosz, A.M., Haeberle, H.S. et al. (2020). Machine learning and artificial intelligence: definitions, applications, and future directions. *Curr. Rev. Musculoskelet. Med.* 13 (1): 69–76.

12 Liao, X., Ma, H., and Tang, Y.J. (2022). Artificial intelligence: a solution to involution of design-build-test-learn cycle. *Curr. Opin. Biotechnol.* 75: 102712.

13 Kim, G.B., Kim, W.J., Kim, H.U., and Lee, S.Y. (2020). Machine learning applications in systems metabolic engineering. *Curr. Opin. Biotechnol.* 64: 1–9.

14 Srinivasan, P. and Smolke, C.D. (2021). Engineering cellular metabolite transport for biosynthesis of computationally predicted tropane alkaloid derivatives in yeast. *Proc. Natl. Acad. Sci. USA* 118 (25).

15 Volk, M.J., Lourentzou, I., Mishra, S. et al. (2020). Biosystems design by machine learning. *ACS Synth. Biol.* 9 (7): 1514–1533.

16 Rana, P., Berry, C., Ghosh, P., and Fong, S.S. (2020). Recent advances on constraint-based models by integrating machine learning. *Curr. Opin. Biotechnol.* 64: 85–91.

17 Sahu, A., Blatke, M.A., Szymanski, J.J., and Topfer, N. (2021). Advances in flux balance analysis by integrating machine learning and mechanism-based models. *Comput. Struct. Biotechnol. J.* 19: 4626–4640.

18 Yu, Y., Zhang, N., Mai, Y. et al. (2023). Correcting batch effects in large-scale multiomics studies using a reference-material-based ratio method. *Genome Biol.* 24 (1): 201.

19 Xiang, S., Yuan, L., Fan, W. et al. (2014). Bi-level multi-source learning for heterogeneous block-wise missing data. *Neuroimage* 102 (Pt 1): 192–206.

20 Cavill, R., Jennen, D., Kleinjans, J., and Briede, J.J. (2016). Transcriptomic and metabolomic data integration. *Brief. Bioinform.* 17 (5): 891–901.

21 Zeng, Z.S., Guo, J., Jin, J., and Luo, X.Z. (2025). CLAIRE: a contrastive learning-based predictor for EC number of chemical reactions. *J. Chem.* 17 (1).

22 Emmanuel, T., Maupong, T., Mpoeleng, D. et al. (2021). A survey on missing data in machine learning. *J. Big Data* 8 (1): 140.

23 Armitage, E.G., Godzien, J., Alonso-Herranz, V. et al. (2015). Missing value imputation strategies for metabolomics data. *Electrophoresis* 36 (24): 3050–3060.

24 Aggarwal, C. (2017). *An Introduction to Outlier Analysis*. Springer.

25 Graw, S., Chappell, K., Washam, C.L. et al. (2021). Multi-omics data integration considerations and study design for biological systems and disease. *Mol Omics.* 17 (2): 170–185.

26 Oyetunde, T., Bao, F.S., Chen, J.W. et al. (2018). Leveraging knowledge engineering and machine learning for microbial bio-manufacturing. *Biotechnol. Adv.* 36 (4): 1308–1315.

27 Jordan, M.I. and Mitchell, T.M. (2015). Machine learning: trends, perspectives, and prospects. *Science* 349 (6245): 255–260.

28 Pedregosa, F., Varoquaux, G., Gramfort, A. et al. (2011). Scikit-learn: machine learning in python. *J. Mach. Learn. Res.* 12: 2825–2830.

29 Singh, A., Thakur, N., and Sharma, A. (2016). A review of supervised machine learning algorithms. *IEEE* 1310–1315.

30 Kampezidou, S.I., Ray, A.T., Bhat, A.P. et al. (2024). Fundamental components and principles of Supervised machine learning workflows with numerical and categorical data. *Eng Basel.* 5 (1): 384–416.

31 Probert, M. (2021). Machine learning with neural networks: an introduction for scientists and engineers. *Contemp. Phys.* 62 (4): 236–237.

32 Krogh, A. (2008). What are artificial neural networks? *Nat. Biotechnol.* 26 (2): 195–197.

33 Alley, E.C., Khimulya, G., Biswas, S. et al. (2019). Unified rational protein engineering with sequence-based deep representation learning. *Nat. Methods* 16 (12): 1315–1322.

34 Kotopka, B.J. and Smolke, C.D. (2020). Model-driven generation of artificial yeast promoters. *Nat. Commun.* 11 (1): 2113.

35 Jervis, A.J., Carbonell, P., Vinaixa, M. et al. (2019). Machine learning of designed translational control allows predictive pathway optimization in Escherichia coli. *ACS Synth. Biol.* 8 (1): 127–136.

36 Treloar, N.J., Fedorec, A.J.H., Ingalls, B., and Barnes, C.P. (2020). Deep reinforcement learning for the control of microbial co-cultures in bioreactors. *PLoS Comput. Biol.* 16 (4): e1007783.

37 Larochelle, H., Bengio, Y., Louradour, J., and Lamblin, P. (2009). Exploring strategies for training deep neural networks. *J. Mach. Learn. Res.* 10: 1–40.

38 Quinlan, J.R. (1986). Induction of decision trees. *Mach. Learn.* 1 (1): 81–106.

39 Coleman, M.C., Buck, K.K., and Block, D.E. (2003). An integrated approach to optimization of Escherichia coli fermentations using historical data. *Biotechnol. Bioeng.* 84 (3): 274–285.

40 Becker, T., Rousseau, A.J., Geubbelmans, M. et al. (2023). Decision trees and random forests. *Am. J. Orthod. Dentofacial Orthop.* 164 (6): 894–897.

41 Wiens, M., Verone-Boyle, A., Henscheid, N. et al. (2025). A tutorial and use case example of the eXtreme gradient boosting (XGBoost) artificial intelligence algorithm for drug development applications. *Cts-Clin Transl Sci.* 18 (3).

42 Groher, A.C., Jager, S., Schneider, C. et al. (2019). Tuning the performance of synthetic riboswitches using machine learning. *ACS Synth. Biol.* 8 (1): 34–44.

43 Matthews, M.L., Wang, J.P., Sederoff, R. et al. (2021). A multiscale model of lignin biosynthesis for predicting bioenergy traits in Populus trichocarpa. *Comput. Struct. Biotechnol. J.* 19: 168–182.

44 Melcher, M., Scharl, T., Spangl, B. et al. (2015). The potential of random forest and neural networks for biomass and recombinant protein modeling in Escherichia coli fed-batch fermentations. *Biotechnol. J.* 10 (11): 1770–1782.

45 Saraswat P. Supervised Machine learning algorithm: a review of classification techniques. *International Conference on Intelligent Emerging Methods of Artificial Intelligence & Cloud Computing 2022*. p. 477–82.

46 Faulon, J.L., Misra, M., Martin, S. et al. (2008). Genome scale enzyme-metabolite and drug-target interaction predictions using the signature molecular descriptor. *Bioinformatics* 24 (2): 225–233.

47 Goncalves, D.M., Henriques, R., and Costa, R.S. (2023). Predicting metabolic fluxes from omics data via machine learning: moving from knowledge-driven towards data-driven approaches. *Comput. Struct. Biotechnol. J.* 21: 4960–4973.

48 Opgenorth, P., Costello, Z., Okada, T. et al. (2019). Lessons from two design-build-test-learn cycles of Dodecanol production in Escherichia coli aided by machine learning. *ACS Synth. Biol.* 8 (6): 1337–1351.

49 Lewis, N.E., Hixson, K.K., Conrad, T.M. et al. (2010). Omic data from evolved E-coli are consistent with computed optimal growth from genome-scale models. *Mol. Syst. Biol.* 6: 390.

50 Larrañaga, P., Calvo, B., Santana, R. et al. (2006). Machine learning in bioinformatics. *Brief. Bioinform.* 7 (1): 86–112.

51 Yu, H., Wen, G.Q., Gan, J.Z. et al. (2020). Self-paced learning for K-means clustering algorithm. *Pattern Recogn. Lett.* 132: 69–75.

52 Bengio, Y., Courville, A., and Vincent, P. (2013). Representation learning: a review and new perspectives. *IEEE Trans. Pattern Anal. Mach. Intell.* 35 (8): 1798–1828.

53 Alexandre, L., Costa, R.S., and Henriques, R. (2021). DI2: prior-free and multi-item discretization of biological data and its applications. *BMC Bioinformatics.* 22 (1): 426.

54 Jia, G., Stephanopoulos, G., and Gunawan, R. (2012). Incremental parameter estimation of kinetic metabolic network models. *BMC Syst. Biol.* 6: 142.

55 Bartenhagen, C., Klein, H.U., Ruckert, C. et al. (2010). Comparative study of unsupervised dimension reduction techniques for the visualization of microarray gene expression data. Bmc. *Bioinformatics* 11: 567.

56 Nutzmann, H.W. and Osbourn, A. (2014). Gene clustering in plant specialized metabolism. *Curr. Opin. Biotechnol.* 26: 91–99.

57 Chavali, A.K. and Rhee, S.Y. (2018). Bioinformatics tools for the identification of gene clusters that biosynthesize specialized metabolites. *Brief. Bioinform.* 19 (5): 1022–1034.

58 Xie, J., Ma, A., Fennell, A. et al. (2019). It is time to apply biclustering: a comprehensive review of biclustering applications in biological and biomedical data. *Brief. Bioinform.* 20 (4): 1449–1464.

59 Cheng, Y. and Church, G.M. (2000). Biclustering of expression data. *Proc Int Conf Intell Syst Mol Biol.* 8: 93–103.

60 Madeira, S.C. and Oliveira, A.L. (2004). Biclustering algorithms for biological data analysis: a survey. *IEEE/ACM Trans. Comput. Biol. Bioinform.* 1 (1): 24–45.

61 Henriques, R. and Madeira, S.C. (2018). BSig: evaluating the statistical significance of biclustering solutions. *Data Min. Knowl. Disc.* 32 (1): 124–161.

62 Henriques, R., Ferreira, F.L., and Madeira, S.C. (2017). BicPAMS: software for biological data analysis with pattern-based biclustering. Bmc. *Bioinformatics* 18: 82.

63 Castanho, E.N., Aidos, H., and Madeira, S.C. (2024). Biclustering data analysis: a comprehensive survey. *Brief. Bioinform.* 25 (4): bbae342.

64 Lakizadeh, A. and Jalili, S. (2016). BiCAMWI: a genetic-based Biclustering algorithm for detecting dynamic protein complexes. *PLoS One* 11 (7): e0159923.

65 Reiss, D.J., Baliga, N.S., and Bonneau, R. (2006). Integrated biclustering of heterogeneous genome-wide datasets for the inference of global regulatory networks. *BMC Bioinformatics.* 7: 280.

66 Kacmarczyk, T., Waltman, P., Bate, A. et al. (2011). Comparative microbial modules resource: generation and visualization of multi-species biclusters. *PLoS Comput. Biol.* 7 (12): e1002228.

67 Chhikara, P., Jain, N., Tekchandani, R., and Kumar, N. (2022). Data dimensionality reduction techniques for industry 4.0: research results, challenges, and future research directions. *Software Pract Exper.* 52 (3): 658–688.

68 Chandrashekar, G. and Sahin, F. (2014). A survey on feature selection methods. *Comput. Electr. Eng.* 40 (1): 16–28.

69 Lever, J., Krzywinski, M., and Atman, N. (2017). POINTS OF SIGNIFICANCE principal component analysis. *Nat. Methods* 14 (7): 641–642.

70 van der Maaten, L. and Hinton, G. (2008). Visualizing Data using t-SNE. *J. Mach. Learn. Res.* 9: 2579–2605.

71 Armstrong, G., Martino, C., Rahman, G. et al. (2021). Uniform manifold approximation and projection (UMAP) reveals composite patterns and resolves visualization Artifacts in microbiome data. *mSystems* 6 (5): e0069121.

72 Dai, D., Horvath, N., and Varner, J. (2018). Dynamic sequence specific constraint-based Modeling of cell-free protein synthesis. *PRO* 6 (8): 132.

73 Shaik, R. and Ramakrishna, W. (2014). Machine learning approaches distinguish multiple stress conditions using stress-responsive genes and identify candidate genes for broad resistance in rice. *Plant Physiol.* 164 (1): 481–495.

74 Velev, J., LeBien, J., and Roche-Lima, A. (2023). Unsupervised machine learning method for indirect estimation of reference intervals for chronic kidney disease in the Puerto Rican population. *Sci. Rep.* 13 (1): 17198.

75 Alexandre, L., Costa, R.S., and Henriques, R. (2022). DISA tool: discriminative and informative subspace assessment with categorical and numerical outcomes. *PLoS One* 17 (10): e02766253.

76 Selmy, H.A., Mohamed, H.K., and Medhat, W. (2024). Big data analytics deep learning techniques and applications: a survey. *Inf. Syst.* 120: 102318.

77 Jang, W.D., Kim, G.B., Kim, Y., and Lee, S.Y. (2022). Applications of artificial intelligence to enzyme and pathway design for metabolic engineering. *Curr. Opin. Biotechnol.* 73: 101–107.

78 Zampieri, G., Vijayakumar, S., Yaneske, E., and Angione, C. (2019). Machine and deep learning meet genome-scale metabolic modeling. *PLoS Comput. Biol.* 15 (7): e1007084.

79 Pinto, J.P., Ramos, J.R.C., Costa, R.S., and Oliveira, R. (2023). A general hybrid Modeling framework for systems biology applications: combining mechanistic knowledge with deep neural networks under the SBML standard AI. 4 (1): 303–318.

80 Ramon, C., Gollub, M.G., and Stelling, J. (2018). Integrating -omics data into genome-scale metabolic network models: principles and challenges. *Syst. Biol.* 62 (4): 563–574.

81 Thiele, I. and Palsson, B.O. (2010). A protocol for generating a high-quality genome-scale metabolic reconstruction. *Nat. Protoc.* 5 (1): 93–121.

82 Antonakoudis, A., Barbosa, R., Kotidis, P., and Kontoravdi, C. (2020). The era of big data: genome-scale modelling meets machine learning. *Comput. Struct. Biotechnol. J.* 18: 3287–3300.

83 Kavvas, E.S., Yang, L., Monk, J.M. et al. (2020). A biochemically-interpretable machine learning classifier for microbial GWAS. *Nat. Commun.* 11 (1): 2580.

84 Kim, M., Rai, N., Zorraquino, V., and Tagkopoulos, I. (2016). Multi-omics integration accurately predicts cellular state in unexplored conditions for Escherichia coli. *Nat. Commun.* 7: 13090.

85 Plaimas, K., Mallm, J.P., Oswald, M. et al. (2008). Machine learning based analyses on metabolic networks supports high-throughput knockout screens. *BMC Syst. Biol.* 2: 67.

86 Vijayakumar, S. and Angione, C. (2021). Protocol for hybrid flux balance, statistical, and machine learning analysis of multi-omic data from the cyanobacterium Synechococcus sp. PCC 7002. *STAR Protoc.* 2 (4): 100837.

87 Vijayakumar, S., Rahman, P., and Angione, C. (2020). A hybrid flux balance analysis and machine learning pipeline elucidates metabolic adaptation in cyanobacteria. *iScience* 23 (12): 101818.

88 Culley, C., Vijayakumar, S., Zampieri, G., and Angione, C. (2020). A mechanism-aware and multiomic machine-learning pipeline characterizes yeast cell growth. *Proc. Natl. Acad. Sci. USA* 117 (31): 18869–18879.

89 Faure, L., Mollet, B., Liebermeister, W., and Faulon, J.L. (2023). A neural-mechanistic hybrid approach improving the predictive power of genome-scale metabolic models. *Nat. Commun.* 14 (1): 4669.

90 Occhipinti, A., Eyassu, F., Rahman, T.J. et al. (2018). In silico engineering of pseudomonas metabolism reveals new biomarkers for increased biosurfactant production. *PeerJ.* 6: e6046.

91 Angione, C. and Lio, P. (2015). Predictive analytics of environmental adaptability in multi-omic network models. *Sci. Rep.* 5: 15147.

92 Jensen, K., Gudmundsson, S., and Herrgård, M.J. (2018). Enhancing metabolic models with genome-scale experimental data. In: *Systems Biology* (ed. N. Rajewsky, S. Jurga, and J. Barciszewski), 337–250. Springer.

93 Angione, C., Pratanwanich, N., and Lio, P. (2015). A hybrid of metabolic flux analysis and Bayesian factor Modeling for Multiomic temporal pathway activation. *ACS Synth. Biol.* 4 (8): 880–889.

94 Jalili, M., Scharm, M., Wolkenhauer, O. et al. (2021). Exploring the metabolic heterogeneity of cancers: a benchmark study of context-specific models. *J Pers Med.* 11 (6).

95 Wintermute, E.H., Lieberman, T.D., and Silver, P.A. (2013). An objective function exploiting suboptimal solutions in metabolic networks. *BMC Syst. Biol.* 7: 98.

96 Machado, D. and Herrgard, M. (2014). Systematic evaluation of methods for integration of transcriptomic data into constraint-based models of metabolism. *PLoS Comput. Biol.* 10 (4): e1003580.

97 Wu, S.G., Wang, Y., Jiang, W. et al. (2016). Rapid prediction of bacterial heterotrophic Fluxomics using machine learning and constraint programming. *PLoS Comput. Biol.* 12 (4): e1004838.

98 Heckmann, D., Lloyd, C.J., Mih, N. et al. (2018). Machine learning applied to enzyme turnover numbers reveals protein structural correlates and improves metabolic models. Nature. *Communications* 9 (1): 5252.

99 Sanchez, B.J., Li, F., Kerkhoven, E.J., and Nielsen, J. (2019). SLIMEr: probing flexibility of lipid metabolism in yeast with an improved constraint-based modeling framework. *BMC Syst. Biol.* 13 (1): 4.

100 Brunk, E., George, K., Keasling, J. et al. (2016). Characterizing strain variation in engineered E. Coli using a multi-omics based workflow. *Abstr. Pap. Am. Chem. Soc.* 2 (5): 335–346.

101 Bordbar, A., Yurkovich, J.T., Paglia, G. et al. (2017). Elucidating dynamic metabolic physiology through network integration of quantitative time-course metabolomics. *Sci Rep-Uk.* 7: 46249.

102 Patané, A., Jansen, G., Conca, P. et al. (2019). Multi-objective optimization of genome-scale metabolic models: the case of ethanol production. *Ann. Oper. Res.* 276 (1–2): 211–227.

103 Ching, T., Himmelstein, D.S., Beaulieu-Jones, B.K. et al. (2018). Opportunities and obstacles for deep learning in biology and medicine. *J. R. Soc. Interface* 15 (141).

104 Xu, C.M. and Jackson, S.A. (2019). Machine learning and complex biological data. *Genome Biol.* 20 (1): 76.

105 Cummins, B., Vrana, J., Moseley, R.C. et al. (2023). Robustness and reproducibility of simple and complex synthetic logic circuit designs using a DBTL loop. *Synth Biol (Oxf).* 8 (1): ysad005.

106 Moreno-Paz, S., van der Hoek, R., Eliana, E. et al. (2024). Machine learning-guided optimization of p-Coumaric acid production in yeast. *ACS Synth. Biol.* 13 (4): 1312–1322.

107 Carbonell, P., Jervis, A.J., Robinson, C.J. et al. (2018). An automated design-build-test-learn pipeline for enhanced microbial production of fine chemicals. *Commun Biol.* 1: 66.

108 HamediRad, M., Chao, R., Weisberg, S. et al. (2019). Towards a fully automated algorithm driven platform for biosystems design. *Nat. Commun.* 10 (1): 5150.

109 Radivojevic, T., Costello, Z., Workman, K., and Garcia, M.H. (2020). A machine learning automated recommendation tool for synthetic biology. *Nat. Commun.* 11 (1): 4879.

110 Zhou, Y., Li, G., Dong, J. et al. (2018). MiYA, an efficient machine-learning workflow in conjunction with the YeastFab assembly strategy for combinatorial optimization of heterologous metabolic pathways in Saccharomyces cerevisiae. *Metab. Eng.* 47: 294–302.

111 Biswas, S., Khimulya, G., Alley, E.C. et al. (2021). Low-N protein engineering with data-efficient deep learning. *Nat. Methods* 18 (4): 389–396.

112 Segler, M.H.S., Preuss, M., and Waller, M.P. (2018). Planning chemical syntheses with deep neural networks and symbolic AI. *Nature* 555 (7698): 604–610.

113 Meng, H., Wang, J., Xiong, Z. et al. (2013). Quantitative design of regulatory elements based on high-precision strength prediction using artificial neural network. *PLoS One* 8 (4): e60288.

114 Kang, C.K., Shin, J., Cha, Y. et al. (2023). Machine learning-guided prediction of potential engineering targets for microbial production of lycopene. *Bioresour. Technol.* 369: 128455.

115 Sabzevari, M., Szedmak, S., Penttila, M. et al. (2022). Strain design optimization using reinforcement learning. *PLoS Comput. Biol.* 18 (6): e1010177.

116 Ajjolli Nagaraja, A., Fontaine, N., Delsaut, M. et al. (2019). Flux prediction using artificial neural network (ANN) for the upper part of glycolysis. *PLoS One* 14 (5): e0216178.

117 Oulia, F., Charton, P., Lo-Thong-Viramoutou, O. et al. (2024). Metabolic fluxes using deep learning based on enzyme variations: application to glycolysis in Entamoeba histolytica. *Int. J. Mol. Sci.* 25 (24).

118 Costello, Z. and Martin, H.G. (2018). A machine learning approach to predict metabolic pathway dynamics from time-series multiomics data. *NPJ Syst Biol Appl.* 4: 19.

119 Alzahrani, M., Kuwahara, H., Wang, W., and Gao, X. (2017). Gracob: a novel graph-based constant-column biclustering method for mining growth phenotype data. *Bioinformatics* 33 (16): 2523–2531.

120 Karim, M.B., Huang, M., Ono, N. et al. (2020). BiClusO: a novel Biclustering approach and its application to species-VOC relational data. *IEEE/ACM Trans. Comput. Biol. Bioinform.* 17 (6): 1955–1965.

121 Colletti, P.F., Goyal, Y., Varman, A.M. et al. (2011). Evaluating factors that influence microbial synthesis yields by linear regression with numerical and ordinal variables. *Biotechnol. Bioeng.* 108 (4): 893–901.

122 Varman, A.M., Xiao, Y., Leonard, E., and Tang, Y.J. (2011). Statistics-based model for prediction of chemical biosynthesis yield from Saccharomyces cerevisiae. *Microb. Cell Factories* 10: 45.

123 Pappu, S.M.J. (2017). Gummadi SN. Artificial neural network and regression coupled genetic algorithm to optimize parameters for enhanced xylitol production by in bioreactor. *Biochem. Eng. J.* 120: 136–145.

124 Zhang, J., Petersen, S.D., Radivojevic, T. et al. (2020). Combining mechanistic and machine learning models for predictive engineering and optimization of tryptophan metabolism. *Nat. Commun.* 11 (1): 4880.

125 TeselaGen Biotechnology. Evolve. https://teselagen.com.

126 Oyetunde, T., Liu, D., Martin, H.G., and Tang, Y.J. (2019). Machine learning framework for assessment of microbial factory performance. *PLoS One* 14 (1): e0210558.

127 Czajka, J.J., Oyetunde, T., and Tang, Y.J. (2021). Integrated knowledge mining, genome-scale modeling, and machine learning for predicting Yarrowia lipolytica bioproduction. *Metab. Eng.* 67: 227–236.

128 Schinn, S.M., Morrison, C., Wei, W. et al. (2021). A genome-scale metabolic network model and machine learning predict amino acid concentrations in Chinese hamster ovary cell cultures. *Biotechnol. Bioeng.* 118 (5): 2118–2123.

129 Zampieri G, Coggins M, Valle G, Angione C. A poly-omics machine learning method to predict metabolite production in CHO cells. *2nd International Electronic Conference on Metabolomics2017*. p. 9.

130 Morris, C., Polanco, A., Yongky, A. et al. (2020). Bigdata analytics identifies metabolic inhibitors and promoters for productivity improvement and optimization of monoclonal antibody (mAb) production process. *Bioresour. Bioprocess.* 7 (1).

131 Alonso-Gutierrez, J., Kim, E.M., Batth, T.S. et al. (2015). Principal component analysis of proteomics (PCAP) as a tool to direct metabolic engineering. *Metab. Eng.* 28: 123–133.

132 Karmaker, S.K., Hassan, M.M., Smith, M.J. et al. (2021). AutoML to date and beyond: challenges and opportunities. *ACM Comput. Surv.* 54 (8).

133 Lui X. Deep learning for metabolic pathway design, *Metabolic Engineering* 80, 130-141 2024. doi: https://doi.org/10.1016/j.ymben.2023.09.012.

134 Rogers, A.W., Vega-Ramon, F., Yan, J. et al. (2022). A transfer learning approach for predictive modeling of bioprocesses using small data. *Biotechnol. Bioeng.* 119 (2): 411–422.

135 Lui, R., Guan, D., and Matthews, S. (2023). Mechanistic task groupings enhance multitask deep learning of strain-specific Ames mutagenicity. *Chem. Res. Toxicol.* 36 (8): 1248–1254.

136 Kirchoff, K.E. (2024). *Learning Improved Representations through Informed Self-Supervision: Applications to Drug Discovery*. University of North Carolina at Chapel Hill.

137 Wheeler, N.E. (2025). Responsible AI in biotechnology: balancing discovery, innovation and biosecurity risks. *Front. Bioeng. Biotechnol.* 13: 1537471.

12

Next-Generation Metabolic Flux Analysis Using Machine Learning

Ahmed Almunaifi[1], *Richard C. Law*[1], *Samantha O'Keeffe*[1], *Kartikeya Pande*[1], *Tongjun Xiang*[1], *Onyedika Ukwueze*[2], *Aranaa Odai-Okley*[2], *Pin-Kuang Lai*[2], *and Junyoung O. Park*[1]

[1] *University of California, Department of Chemical and Biomolecular Engineering, 420 Westwood Plaza, Boelter Hall 5531, Los Angeles, CA, USA*
[2] *Stevens Institute of Technology, Department of Chemical Engineering and Materials Science, 1 Castle Point Terrace, Hoboken, NJ, USA*

12.1 Introduction

For much of the 19th and 20th centuries, biochemistry and computing have taken independent paths. Following the crowning achievements of solving the double helix structure of DNA and building the first modern computer in the mid-20th century, the two fields grew exponentially and started influencing each other (Figure 12.1). Neurons and their network formation in the human brain inspired mathematicians and computer scientists to develop artificial neural networks (ANNs). Automation and information handling in computers inspired biologists and biochemists to develop next-generation sequencing (NGS) and bioinformatics. As the two fields become more intertwined, innovative solutions to global challenges in health and energy are arising. One of the outcomes is the development of advanced metabolic profiling frameworks that propel biotechnology and medicine. These approaches build on the efforts of metabolic models (described in Chapters 1 and 2), metabolomics (described in Chapter 10), and flux models (described in Chapters 9 and 11). Here we focus our attention on the dynamics of metabolism and explore the role of machine learning (ML) in elucidating them.

12.2 Dynamic Nature of Metabolism

Organisms consist of several complex networks of biochemical reactions (see Chapters 1 and 2). Metabolism is one of them that is fast-acting and responsible for generating cellular energy currency and biochemical building blocks in changing environments. In other words, while the structures of metabolic networks are invariable, they are dynamic and constantly regulated to maintain homeostasis.

Machine Learning and Big Data-enabled Biotechnology, First Edition. Edited by Hal S. Alper.

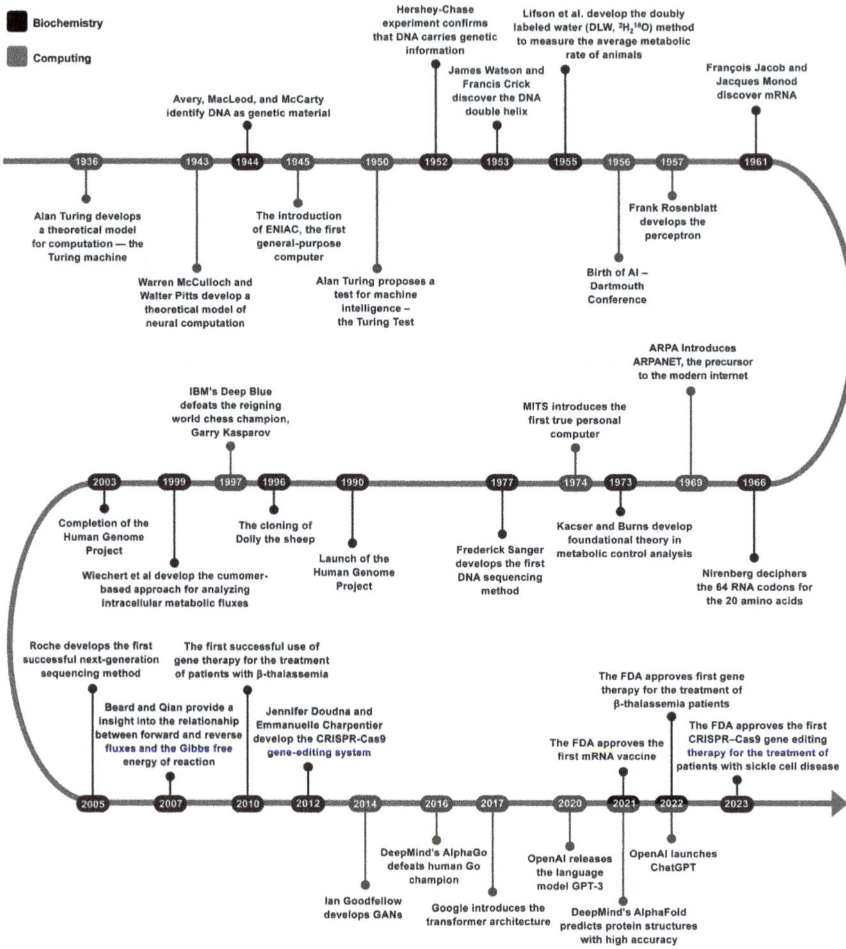

Figure 12.1 Events timeline. A timeline of significant events in biochemistry and computing that shaped quantitative and computational biology.

Metabolic networks consist mostly of enzyme-catalyzed reactions and some spontaneous reactions. In both reaction types, increasing substrate concentrations leads to higher reaction rates. In the former, however, a finite number of enzyme active sites leads to saturation behavior of reaction kinetics, which is known as Michaelis-Menten kinetics. Enzyme activity, or the rate of enzyme-catalyzed reaction, is also influenced by the concentrations of products and effectors, which can exert positive or negative regulation. Thus, each metabolite wears many hats, acting as a substrate and a product as well as an activator and an inhibitor to various enzymes. Putting together enzymes and metabolites engenders a metabolic network and an emergent property, autoregulation, which guides transient metabolic states to stable steady states.

Let us look at an example of a reaction in glycolysis that is catalyzed by phosphofructokinase (PFK).

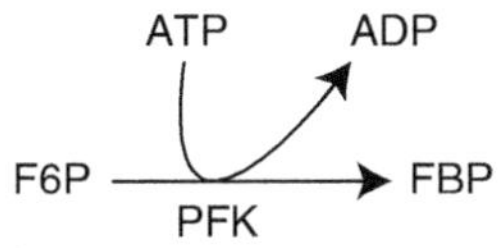

PFK catalyzes the phosphorylation of fructose-6-phosphate (F6P) to fructose-1,6-bisphosphate (FBP) by coupling the conversion of adenosine triphosphate (ATP) to adenosine diphosphate (ADP). The reaction rate, v_{PFK}, may be expressed as a reversible Michaelis-Menten-like equation.

$$v_{PFK} = v_{max}^{+} \frac{\frac{[F6P]}{k_{F6P}} \frac{[ATP]}{k_{ATP}}}{\left(1 + \frac{[F6P]}{k_{F6P}} + \frac{[FBP]}{k_{FBP}}\right)\left(1 + \frac{[ATP]}{k_{ATP}} + \frac{[ADP]}{k_{ADP}}\right)} \left(1 - \frac{Q}{K_{eq}}\right) f_{regulation} \tag{12.1}$$

where $K_{eq} = e^{-\Delta G^\circ/RT}$, $Q = \frac{[FBP][ADP]}{[F6P][ATP]}$, $f_{regulation} = \begin{cases} \frac{[A]/k_A}{1+[A]/k_A}, \textit{for allosteric activator A} \\ \frac{1}{1+[I]/k_I}, \textit{for allosteric inhibitor I} \end{cases}$

v_{max}^{+} = the maximum forward rate of the reaction,

[X] = Concentration of the metabolite X, where X is a substrate or a product (F6P, ATP, FBP, ADP),

[A] = Concentration of the allosteric activator A,

[I] = Concentration of the allosteric inhibitor I,

k_X = Dissociation constant for metabolite X, where X is a substrate or a product (F6P, ATP, FBP, ADP),

k_A = Activator dissociation constant,

k_I = Inhibitor dissociation constant,

Q = Reaction quotient defined as the product of the concentrations of products divided by the product of the concentrations of substrates, and

K_{eq} = Equilibrium constant of the reaction.

In the PFK reaction, ATP is a substrate, but it also acts as an inhibitor. Therefore, the following factor should be applied to the enzyme kinetic equation:

$$f_{regulation} = \frac{1}{1 + [ATP]/k_{ATP}}$$

Determination of reaction rates requires the concentrations of substrates, products, effectors, and enzymes as well as kinetic, thermodynamic, and regulatory information. In practice, determination of fluxes across metabolic networks is challenging due to the number of concentration measurements and kinetic parameters that are necessary to compute fluxes.

Mathematical models of metabolism that detail reaction kinetics and thermodynamics confer a quantitative knowledge of time-dependent metabolic activities. However, the kinetic and thermodynamic parameters constituting mathematical models (Michaelis constants k_m, catalytic constants k_{cat}, equilibrium constants K_{eq}, etc.) are seldom known for exact physiological conditions of interest (e.g. pH, temperature, and ionic strength). The use of mathematical optimization, ensemble modeling, and deep learning obviates the need for measuring the full set of model parameters. The concentrations of enzymes and metabolites, with the help of the measured and predicted parameters, determine reaction rates, which in turn constitute ordinary differential equations (ODEs) describing the time evolution of metabolite levels through mass balance.

An example of a global optimization technique used to infer kinetic parameters is a genetic algorithm (GA) [1]. The method generates multiple starting points for local optimization, crosses, mutates, and selects improved solutions iteratively. GA has been used to determine the parameters of large kinetic models starting from a set of "guessed" values as well as time course experimental data of metabolite concentrations. The input values are used to parameterize kinetic models and simulate the concentrations of metabolites, which are then compared to experimental measurements. The results are a set of model parameters that best represent observation [1, 2]. GA approaches have been used to study allosteric regulation of enzymes in modeling a metabolic futile cycle, glycolysis, and the pentose phosphate pathway (PPP) in *E. coli*, and fed-batch fermentation processes, to name a few [3–5].

Ensemble modeling is another approach that circumvents the need for complete information on enzyme kinetics of reactions in a metabolic network [6]. In this method, an "ensemble" of kinetic models is created to reach the same steady state in terms of metabolite concentrations and flux distribution. The kinetic models are primed to a steady state. The steady-state flux can either be measured from experiments (often nutrient uptake and product secretion rates) or obtained from a flux balance of each metabolite. Thermodynamic constraints can be incorporated to parameterize the ensemble as well as the inclusion of enzyme regulation information if available. Ensemble kinetic modeling has been used in several studies, including its application in central carbon metabolism of *E. coli* and cyanobacteria, and the trehalose pathway in *Saccharomyces cerevisiae* [6–8].

Neural networks are emerging alternatives in kinetic modeling. RENAISSANCE (reconstruction of dynamic models through stratified sampling using ANNs and concepts of evolution strategies) is an ML framework that can parameterize kinetic models with physiological relevance without the need for training data. RENAISSANCE captures the nonlinear behavior of parameterized kinetic models through feedforward neural networks (FFNNs) and optimizes them with natural evolution strategies (NES). The ML framework has been used to model *Escherichia coli* metabolism in agreement with experimentally observed steady states and dynamic responses [9]. Another example of an ML framework that has been used for dynamic modeling is REKINDLE (reconstruction of kinetic models using deep learning). REKINDLE is an unsupervised deep learning-based method that utilizes generative adversarial networks (GANs) [10]. REKINDLE trains the GANs with data generated from existing kinetic modeling frameworks,

Table 12.1 Overview of computational frameworks and resources that are available for quantitative analysis of dynamic metabolism.

Platform	Application	Reference
RENAISSANCE	Efficient parameterization of large-scale kinetic models by integrating diverse omics data.	[9]
REKINDLE	Efficient generation of kinetic models of metabolism.	[10]
ML-Flux	Determination of metabolic fluxes by deep learning of steady-state isotope labeling patterns.	[11]
COBRA Toolbox	Constraint-based modeling that includes flux balance analysis (FBA) and flux variability analysis (FVA).	[12]
MetaboAnalyst	Interpretation of metabolomics data and functional analysis on a pathway level.	[13]
METRAN	Determination of metabolic fluxes from steady-state isotope labeling patterns.	[14]
INCA	Determination of steady-state metabolic fluxes from steady and unsteady isotope labeling patterns.	[15]

such as Monte Carlo sampling. These computational techniques have facilitated kinetic modeling despite the limited availability of kinetic parameters (Table 12.1).

12.3 Flux Balance Analysis and Metabolic Flux Analysis

The combination of biochemical measurements and computational tools has enabled kinetic modeling of metabolism. Let us consider a scenario in which we have incomplete knowledge of enzyme regulation. The knowledge gap would lead to incorrect kinetic equations for reaction steps with missing regulation. Would a kinetic model built upon the incorrect equations accurately depict biochemical reality and physiology? Likely not, but perhaps under limited circumstances. How can we quantify metabolic activities and elucidate the dynamics of metabolism under any circumstances without knowing the full details of reaction kinetics and regulation?

Flux balance analysis (FBA) and metabolic flux analysis (MFA) come to our rescue by providing a systems-level overview of metabolic pathway utilization. Both analyses are computational approaches that yield flux distributions throughout a metabolic network without reference to underlying catalytic and regulatory mechanisms. In mammalian systems, including human cells and organoids as well as research animals, flux analysis reveals anomalies and dysregulation in metabolism that lead to or accompany diseases. In model systems such as *E. coli* or *S. cerevisiae*, flux analysis reveals bottlenecks in metabolic pathways for bioproduct synthesis and suggests metabolic engineering strategies. In non-model organisms, whose metabolism is even less understood, FBA and MFA enable the mapping of metabolic activities and provide the basis for deeper mechanistic studies.

(a) Reaction Network

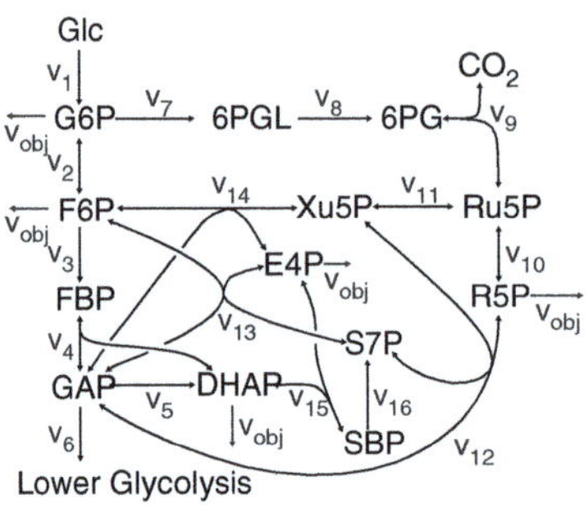

(c) Stoichiometric Matrix S

	v_1	v_2	v_3	v_4	v_5	v_6	v_7	v_8	v_9	v_{10}	v_{11}	v_{12}	v_{13}	v_{14}	v_{15}	v_{16}	v_{obj}
G6P	1	−1	0	0	0	0	-1	0	0	0	0	0	0	0	0	0	−0.205
F6P	0	1	−1	0	0	0	0	0	0	0	0	0	1	1	0	0	−0.0709
FBP	0	0	1	−1	0	0	0	0	0	0	0	0	0	0	0	0	0
DHAP	0	0	0	1	−1	0	0	0	0	0	0	0	0	0	−1	0	−0.129
GAP	0	0	0	1	1	−1	0	0	0	0	0	1	−1	1	0	0	0
6PGL	0	0	0	0	0	0	1	−1	0	0	0	0	0	0	0	0	0
6PG	0	0	0	0	0	0	0	1	−1	0	0	0	0	0	0	0	0
Ru5P	0	0	0	0	0	0	0	0	1	−1	−1	0	0	0	0	0	0
R5P	0	0	0	0	0	0	0	0	0	1	0	−1	0	0	0	0	−0.8977
Xu5P	0	0	0	0	0	0	0	0	0	0	1	−1	0	−1	0	0	0
S7P	0	0	0	0	0	0	0	0	0	0	0	1	−1	0	0	1	0
E4P	0	0	0	0	0	0	0	0	0	0	0	0	1	−1	−1	0	−0.361
SBP	0	0	0	0	0	0	0	0	0	0	0	0	0	0	1	−1	0

(b) Steady-State Mass Balance

$$\frac{dG6P}{dt} = v_1 - v_2 - v_7 - 0.205 v_{obj} = 0$$

$$\frac{dF6P}{dt} = v_2 - v_3 + v_{13} + v_{14} - 0.0709\, v_{obj} = 0$$

$$\frac{dFBP}{dt} = v_3 - v_4 = 0$$

$$\frac{dDHAP}{dt} = v_4 - v_5 - v_{15} - 0.129\, v_{obj} = 0$$

$$\frac{dGAP}{dt} = v_4 + v_5 - v_6 + v_{12} - v_{13} + v_{14} = 0$$

$$\frac{d6PGL}{dt} = v_7 - v_8 = 0$$

$$\frac{d6PG}{dt} = v_8 - v_9 = 0$$

$$\frac{dRu5P}{dt} = v_9 - v_{10} - v_{11} = 0$$

$$\frac{dR5P}{dt} = v_{10} - v_{12} - 0.8997\, v_{obj} = 0$$

$$\frac{dXu5P}{dt} = v_{11} - v_{12} - v_{14} = 0$$

$$\frac{dS7P}{dt} = v_{12} - v_{13} + v_{16} = 0$$

$$\frac{dE4P}{dt} = v_{13} - v_{14} - v_{15} - 0.361\, v_{obj} = 0$$

$$\frac{dSBP}{dt} = v_{15} - v_{16} = 0$$

Figure 12.2 Overview of creating a stoichiometric matrix S for a network of reactions. (a) A metabolic network that includes upper glycolysis and the pentose phosphate pathway (PPP) with an objective function that represents biomass production. (b) The steady-state mass balance equations for the reaction network. (c) The stoichiometric matrix S corresponding to the reaction network.

How do FBA and MFA compute fluxes and how are they different? Unlike the parameterized kinetic modeling approach, FBA and MFA rely solely on steady-state mass balances on individual metabolites in the metabolic network. In FBA, steady-state mass balances are expressed as a system of linear equations with reaction fluxes as variables. The linear system can be expressed in the matrix form $S\boldsymbol{v}=\boldsymbol{0}$, where S is the stoichiometric matrix and $\boldsymbol{v}$ is the vector of steady-state fluxes that satisfy mass balances for all metabolites (Figure 12.2). The solutions to this linear system lie in the null space of S, and a specific vector of fluxes for all reactions in the metabolic network are chosen by linear optimization of a physiologically relevant objective function (e.g. maximizing ATP production, maximizing growth rate, maximizing product synthesis, etc.) within linear constraints defined based on thermodynamics and experimental measurements. Due to its computational efficiency, FBA is widely employed with genome-scale models containing hundreds to thousands of metabolic reactions.

While flux solutions from MFA must also lie in the null space of S, MFA requires an experimental input to discern the "best" fluxes that correspond to experimentally measured isotopic labeling patterns imprinted on metabolites when cells metabolize nutrients that contain heavy isotopes [16]. The isotopes

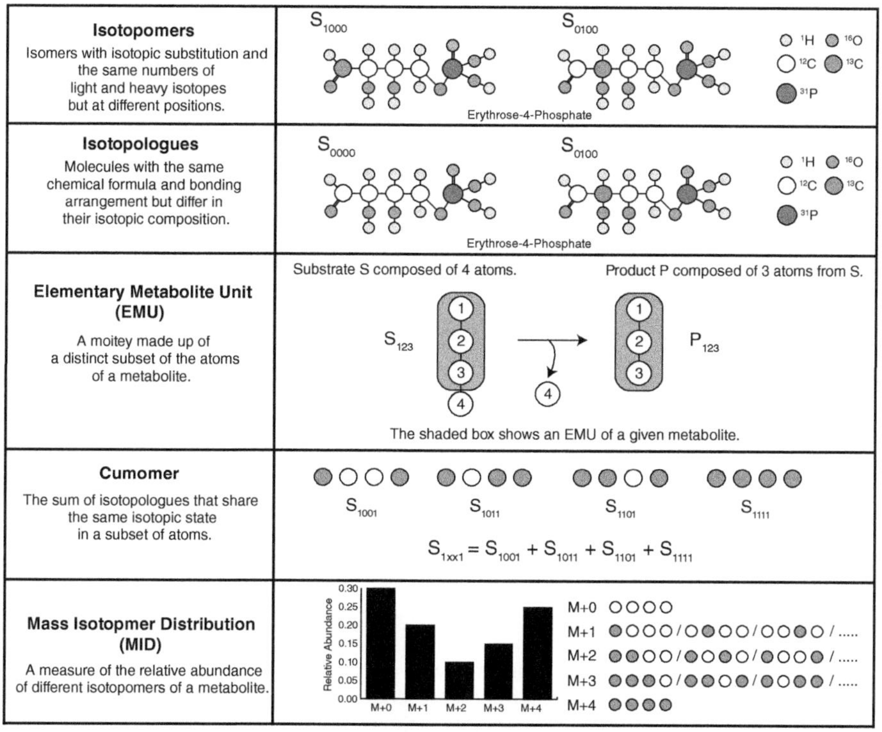

Figure 12.3 Defining isotopomer, isotopologue, cumomer, elementary metabolite unit, and mass isotopomer distribution. Various representations of isotopically labeled erythrose-4-phosphate are demonstrated.

can be traced across reaction networks using mass spectrometry (MS) and nuclear magnetic resonance (NMR) spectroscopy. MFA solves the nonlinear least squares problem of fitting measured isotope labeling patterns, which adds computational costs with increasing numbers of reactions in the metabolic network. Thus, MFA most often employs pathway-scale metabolic models containing a dozen to a few dozen reactions. The cumomer [17] and the elementary metabolite unit (EMU) [18] frameworks simplify the mass balance equations by reducing the number of equations that need to be solved. A cumomer represents cumulated isotopologues that share the same isotopic state in a subset of atoms in a compound (Figure 12.3). An EMU is a subset of atoms in a compound. These frameworks facilitate the simulation of isotopic labeling patterns on metabolites without using the complete set of isotopologues. Software applications, such as INCA, Metran, 13CFLUX2, mfapy, and freeflux, integrate isotopomer measurements from MS and NMR for MFA (Table 12.1) [14, 15, 19–21].

Carbon-13 (^{13}C) is the most frequently used isotope for flux determination. Carbon makes up the backbones of metabolites and biomass; thus, studying ^{13}C labeling patterns unravels the different metabolic pathways utilized to create them [22]. MFA uses mathematical optimization to obtain fluxes that simulate labeling patterns that are most consistent with experimentally measured isotope labeling patterns [23].

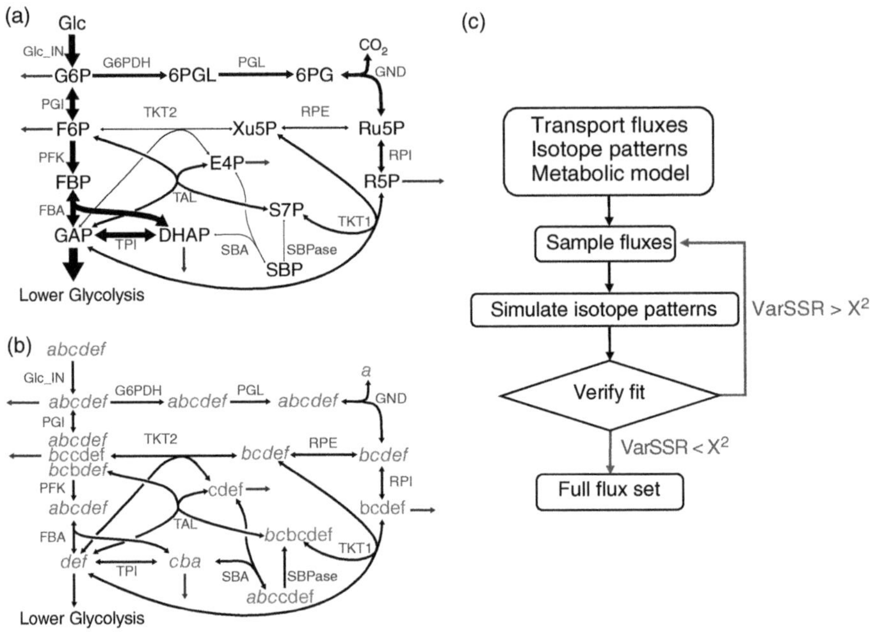

Figure 12.4 Overview of metabolic flux analysis. (a) Upper glycolysis and the PPP as an example of metabolic network. Each arrow represents a reaction, the arrow sizes represent the flux through the reaction. (b) Atom mapping showing the carbons are shuffled in metabolites as they go through metabolic reactions. (c) Metabolic flux analysis workflow.

Simulated isotope patterns are compared to measured isotope patterns by the variance-weighted (σ^2) sum of squared residuals (VarSSR) of mass isotopomer distributions (MIDs) of individual metabolites (Figure 12.4).

$$VarSSR = \sum \frac{(MID_{measured} - MID_{simulated})^2}{\sigma^2_{measured\ MIDs}} \tag{12.2}$$

Since atom mapping differs between pathways, pathways of interest dictate which tracer(s) should be used for flux determination. Thus, the accuracy of fluxes determined by ^{13}C MFA depends on the composition and the positional labeling of tracers [24].

12.4 Incorporating Machine Learning into Metabolic Flux Analysis

FBA identifies the limits of a metabolic network and predicts overall phenotypes such as growth rate, gene essentiality, and feasibility of bioproduct synthesis. Compared to FBA, MFA predicts fluxes more accurately by incorporating isotope tracing information; however, MFA comes with a steeper learning curve. Expertise in atom mapping, isotope tracing, metabolic model construction, and mathematical optimization is required for implementing MFA.

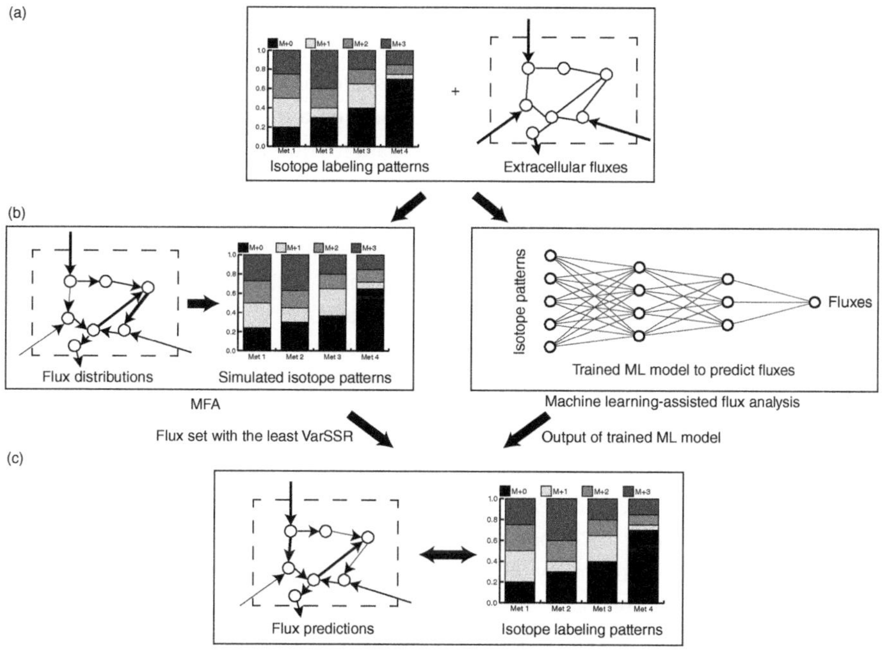

Figure 12.5 Machine learning to streamline flux quantification: (a) Experimental measurements of isotope labeling patterns and measured transport fluxes (nutrient uptake and product secretion rates) are the ingredients for determining fluxes. (b) MFA simulates isotope labeling patterns until they match experimental observation satisfactorily. Machine learning (ML) model captures the relationship between isotope labeling patterns and fluxes. (c) Both approaches aim to determine fluxes that are encoded in isotope labeling patterns.

The complexity of MFA has motivated the exploration of ML techniques to streamline MFA (Figure 12.5). A variety of ML algorithms exist today with their unique strengths and weaknesses. These algorithms can be broadly categorized into supervised and unsupervised learning methods. Supervised learning involves training models on labeled data, where the input features and desired outputs are known, while unsupervised learning aims to discover patterns and relationships in unlabeled data. The performance and applicability of ML models derived depend on the quality and quantity of the datasets used during the training stage.

12.4.1 Challenges in Applying ML to MFA

Before describing the different ML tools available for use in MFA, we discuss some challenges in applying ML to MFA. These challenges are rooted in data availability and the computational power needed to train ML models.

12.4.1.1 A Lack of Isotope Labeling Patterns and Flux Data for Training Machine Learning Models

Training datasets for flux analysis may be sourced from prior MFA results [25–27], including isotope labeling patterns [28] from MS or NMR [29]. While these methods provide detailed insights into metabolic fluxes, the availability of comprehensive and high-quality flux data is limited. The increasing scale and complexity of metabolic models present further challenges. For instance, the *E. coli* iAF1260 model [30], a widely studied genome-scale metabolic network, consists of 2,077 internal reactions, 304 exchange reactions, and a biomass objective function. Generating sufficient experimental and MFA data to train ML models for such large systems is laborious and costly.

Experimental results from different sources and different measurement techniques also pose undesirable variability, which may lead to discrepancies that can propagate through the training data [31]. Missing data, corruption, and errors impact the ML models' training and prediction accuracy. Addressing these problems through data cleaning is essential to ensure effective learning [32]. Consistent annotation of data, whether human- or machine-generated, is also critical.

12.4.1.2 Variable-Size Input of Isotope Labeling Patterns for Flux Prediction

Fluxes may be obtained through the use of different isotope tracers and the measurement of different metabolites. For example, for glucose, which has six carbons, there are 62 positionally ^{13}C-labeled glucose tracers (2^6=64 less two trivial glucose isotopologues that are uniformly ^{13}C-labeled and uniformly unlabeled). Depending on the organism of interest and available instrumentation, measurable metabolites would also vary. Therefore, the size and type of input isotope labeling patterns to ML models have large variability.

12.4.1.3 Incorporation of Disparate Data Types

Multi-omics data (e.g. transcriptomics and proteomics) and thermodynamic constraints constitute additional information about metabolic fluxes. However, since these data and constraints are in different formats, integration of heterogeneous datasets for training ML models and predicting fluxes poses challenges. Furthermore, the relationship between other omics data and fluxes is not fully established except in the simplest cases (e.g. gene knockout or silencing resulting in zero or negligible flux through the reaction that the corresponding enzyme catalyzes), adding to the challenge of incorporating datasets beyond isotope labeling patterns for flux prediction.

12.4.1.4 Initial Computational Cost

Substantial computing power is needed for gathering, generating, and cleaning up training data as well as for training, validating, and testing ML models. Furthermore, if the structure of metabolic networks or atom mapping across metabolic networks is unknown, constructing a metabolic model suitable for flux analysis would be computationally and experimentally costly. On the upside, once ML models are trained, using them should be less computationally intensive than running mathematical optimization to obtain fluxes.

12.4.2 Available Computational Tools

ML models are based on algorithms that learn patterns from training data to make predictions or decisions. We demonstrate computational tools and examples of ML algorithms that have been used and may be adapted for various tasks in flux determination.

12.4.2.1 Sampling Metabolic Fluxes for Training ML Models

Since an ML model is only as good as its training data, sampling fluxes that are feasible and physiologically relevant is an important task. One of the most common ways to sample fluxes is Monte Carlo (MC) sampling, a randomized algorithm that explores constrained flux spaces. When the flux space is large (genome-scale models), MC sampling, such as the hit-and-run (HR) algorithm, is efficient. Improved HR algorithms, such as the artificial centering hit and run (ACHR) and the coordinate hit and run with rounding (CHRR), have been developed to satisfy narrow solution spaces [33]. COnstraints-Based Reconstruction and Analysis (COBRA) is a constraint-based FBA toolbox available in MATLAB and Python that facilitates imposing physicochemical constraints on a metabolic network to obtain steady-state fluxes [12]. COBRA utilizes the MC-HR techniques mentioned above to estimate flux distributions in metabolic networks. ANNs and GANs are ML frameworks in which MC sampling algorithms can be integrated to generate flux training datasets.

12.4.2.2 Atom Mapping Throughout Metabolic Networks

Determining the fate of atoms through a metabolic network forms the basis of isotope tracing. Atom mapping across individual reactions from the substrate sides to the respective product sides provides information about the relationship between flux distributions and isotope labeling patterns. Atom mapping models may be developed by integrating the knowledge of reaction mechanisms of metabolic enzymes in literature, but it becomes tedious as the scope of the metabolic model increases and with unknown reaction mechanisms. Therefore, computational methods have played a complementary role in mapping atom transitions [34]. Potential ML tools for atom mapping include graph convolutional network (GCN) and transformer architectures. GCNs are a specialized type of neural network designed to operate on graph-structured data. GCNs have been employed for studying cell-wise metabolic flux using single-cell RNA sequencing data [35] and predicting gene essentiality in *E. coli* [36]. Transformer encoder models are an unsupervised ML approach commonly used for text analysis. They have been used for mapping chemical reactions exclusively from reaction SMILES as training data [37]. The transformer architecture has also been used to develop an ML language model called BERT (bidirectional encoder representations from transformer) that has been widely used in synthetic chemistry for predicting chemical reactions and retrosynthetic pathways [22–24].

12.4.2.3 Selection of Information-Rich Isotope Tracers

Information-rich isotope tracers should provide a quantitative knowledge of fluxes through metabolic pathways of interest. Software packages such as INCA have been developed to aid in the selection of information-rich isotope tracers. INCA allows users to simulate isotope labeling patterns from metabolic fluxes [38]. Supervised learning ML models called support vector machines (SVMs) are used for classification and regression tasks and have been used to analyze gene expression, protein abundance data, and metabolomics data [39]. The ability of SVMs to separate and categorize data makes them a promising computational tool in the selection of isotope tracers. Another example of a supervised ML model is random forests (RFs). RFs are ensemble learning methods that combine multiple decision trees and aggregate them to enhance prediction accuracy and robustness. RF models have been employed in drug discovery for the identification of bioactive molecules, in studying cancer metabolic pathways, and risk prediction of heart failure patients [40–42]. RFs may also be used to identify the utility of isotope tracers for flux prediction.

12.4.2.4 Other Helpful ML Tools

Reinforcement learning is a paradigm in which an agent learns to interact with an environment and maximize cumulative rewards through trial and error [43]. Reinforcement learning can be used to optimize dynamic pathways, which involve the temporal regulation of enzyme expression, and control gene expression in response to environmental cues or cellular states, allowing for precise regulation of metabolic processes. Transfer learning is a technique that leverages knowledge acquired from one domain to enhance performance in another related domain [44]. Transfer learning has shown promise in reconstructing gene regulatory networks by transferring knowledge from well-studied organisms to less-characterized ones [45], thereby accelerating network modeling in data-scarce contexts. Unsupervised ML tools that can be used for different tasks in MFA are clustering and association algorithms. Clustering algorithms analyze unlabeled raw data, find patterns and then group the data into clusters based on similarities [46]. K-means clustering algorithms are capable of handling large datasets and have been used to aid in the mapping of metabolite distributions in human tonsil tissue using unlabeled data [47]. Association algorithms are another way of finding patterns in datasets and correlations between variables [48]. These algorithms can be harnessed to discover new patterns in complex systems such as metabolic networks and isotope tracing therein.

12.4.3 A General Workflow for Machine Learning-Based Metabolic Flux Analysis

ML-based flux determination consists of four steps, which can find approximate counterparts in MFA. It begins with metabolic model construction, followed by training data generation, ML model training, and testing (Figure 12.6). To train and test ML models for flux quantitation, paired data consisting of isotope patterns and fluxes are needed. The differences in ML-based and optimization-based strategies have implications for the execution of flux computation and the analysis of the resulting fluxes. Nonetheless, both flux determination approaches rely on the same fundamental relationship between fluxes and isotope labeling patterns, and the ingredients needed for flux determination remain the same.

Ingredients for flux quantitation are:

1) Metabolic network
2) Isotope labeling patterns
3) (optional) Rates of nutrient uptake and product secretion

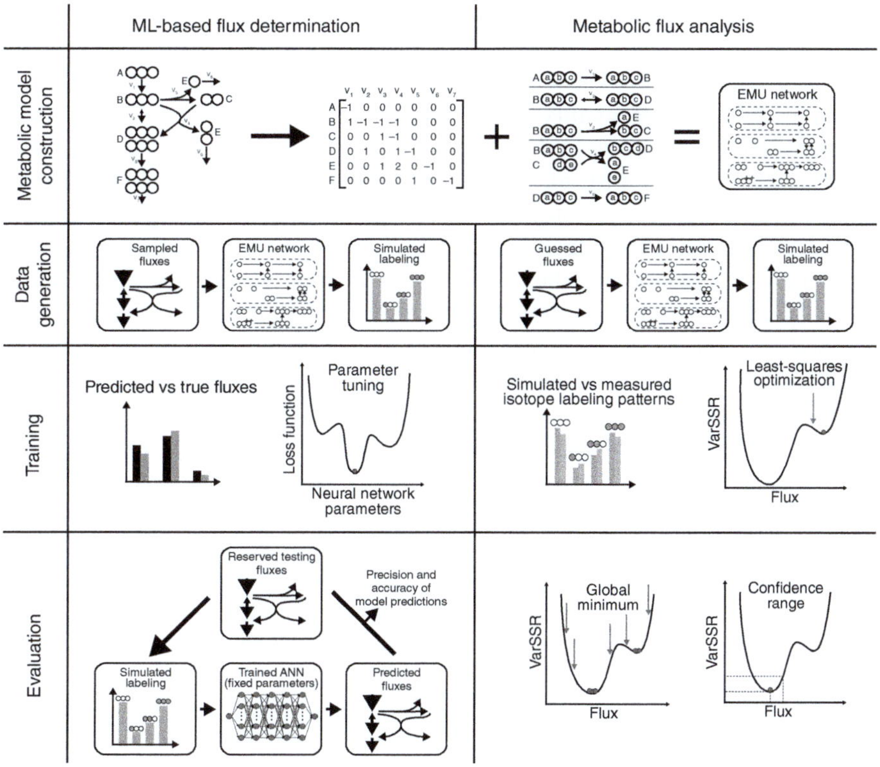

Figure 12.6 Steps of ML-based flux analysis. Four steps in ML-based MFA construction are described alongside an analogous process in traditional MFA. Both approaches apply the same fundamental frameworks for metabolic model construction. However, simulations for data generation occur prior to training for ML-based approach, whereas simulations for optimization happen during the runtime of flux determination. Similarly, training in ML-based MFA frontloads computational demand, whereas the computational demand of optimization-based MFA is the highest during runtime. Lastly, ML-based MFA globally evaluates the accuracy and precision of flux prediction using reserved testing data, while optimization-based MFA individually evaluates accuracy and precision via additional rounds of simulation and optimization.

12.4.3.1 Metabolic Model Construction

As a first of flux determination, a metabolic model that describes the relationship between metabolic fluxes and isotope labeling patterns must be constructed. A metabolic model consists of (1) a stoichiometric matrix and (2) atom mapping. A stoichiometric matrix (S) numerically represents a metabolic model based on the stoichiometric coefficients of substrates and products for individual reactions (Figure 12.2). Each row of S corresponds to a metabolite, and each column corresponds to a reaction in the model. Metabolic models are available from various databases such as BiGG [49]. Atom mapping describes how the atoms are reshuffled from substrates to products through biochemical reactions (Figure 12.4). Atom mapping information is available from databases such as MetaCyc [50].

12.4.3.2 Acquiring Training Data

ML models require a set of true input-output pairs for training. In the context of flux determination, input consists of measured isotope labeling patterns (Figure 12.3) and, if available, transport fluxes, and output is the complete set of intracellular and transport fluxes. However, there is often no measurement-based "ground truth" for intracellular fluxes. Instead, training data may be generated by simulating isotope labeling patterns (e.g. isotopologues and MIDs) using physiologically relevant fluxes. An important consideration for data generation by simulation is the sampling of fluxes. Constraining the flux space based on physiology, thermodynamics, and/or heuristics makes sampling more effective. To minimize bias in flux prediction, sampling should evenly cover the constrained flux space [15, 51]. Once the flux space and sampling algorithms are chosen, the size of training dataset should be determined to balance between computational efficiency (which may mean insufficient coverage of the flux space) and accuracy (which may require exhaustive flux sampling). Once a set of fluxes has been chosen for labeling simulation, data generation for ML-based MFA is analogous to the step in MFA where isotope labeling patterns are simulated in the search process of optimization. The key difference is that while optimization-based MFA requires iterative simulations for each flux prediction, ML-based MFA frontloads all simulations and learns the relationship between isotope labeling patterns and fluxes for all future flux predictions.

12.4.3.3 Training Machine Learning Models

The paired data, consisting of isotope labeling patterns and metabolic fluxes, are split into three bins for training, validating, and testing ML models. Training involves ML model (algorithm) selection and parameter tuning. The performances of ML models are evaluated on the accuracy and precision of their flux predictions. A loss function that quantifies the discrepancy between the true and predicted fluxes is used to guide the training and evaluation of ML models.

$$Loss = f(v_{\text{true}}, v_{\text{predicted}}) \tag{12.3}$$

The loss function in ML model training and evaluation contrasts with the objective function in optimization-based MFA, which quantifies the discrepancy between the measured and simulated labeling patterns.

$$Loss = f(MID_{\text{measured}}, MID_{\text{simulated}}) \tag{12.4}$$

The different loss functions mean that optimization-based MFA aims for the accurate simulation of isotope labeling patterns, whereas ML prediction aims for accurate flux prediction. While it is expected that the two approaches would yield similar fluxes from the same isotope labeling pattern input, the results may not necessarily be similar because the same or similar isotope patterns may arise from disparate flux sets.

12.4.3.4 Evaluation of Machine Learning Models

The trained ML models may be evaluated for their accuracy and precision of flux prediction using the testing data that ML models have not seen during the training stage. Comparison between the predicted and true fluxes from the reserved testing data quantifies the bias and width of the error distribution, which sets the basis of accuracy and precision of ML flux predictions. The evaluation process in ML-based MFA contrasts with that of optimization-based MFA, which involves multiple rounds of optimization starting from different initial flux sets to map local and global minima and confidence interval determination by how sensitive VarSSR is to the fluctuation of fluxes at the identified minimum VarSSR. In evaluating ML models, goodness-of-fit analyses (e.g. coefficient of determination, reduced X^2 test, etc.) should be performed.

12.4.3.5 Execution: Computing Metabolic Fluxes Directly from Isotope Labeling Patterns

Major advantages of ML-based MFA are ease of use, minimal computational cost, and accessibility. Unlike optimization-based MFA, which requires iterative simulations of isotope labeling patterns, ML models require a single round of computation. Furthermore, while the present optimization-based MFA software is a powerful tool for interpreting isotope labeling patterns and computing metabolic fluxes, expertise in metabolic model construction and programming is required to obtain reliable flux output. The execution of ML-based MFA requires one simple step of inputting experimental measurements of isotope labeling patterns. Given these considerations, prebuilt metabolic models and pretrained ML models streamline flux determination and open up the utility of MFA to the broader research community.

12.4.4 Improved Speed and Accuracy of ML-Based MFA

Various ML architectures and overall software designs may be implemented to improve MFA. Recent work has demonstrated that ML models predict fluxes accurately and rapidly from isotope labeling patterns [11]. ML-Flux, a two-stage ML framework, employs deep-learning-based imputation to handle variable measurements of isotope labeling patterns and successive neural networks to convert the full isotope patterns into metabolic fluxes (Figure 12.7 and Table 12.1).

To understand the ML-Flux framework, it is helpful to begin by describing the second-stage ML model, which is a fully connected feedforward ANN that takes in isotope labeling patterns of all metabolites within a metabolic model and outputs linearly independent free fluxes. Free fluxes are expanded using linear algebra

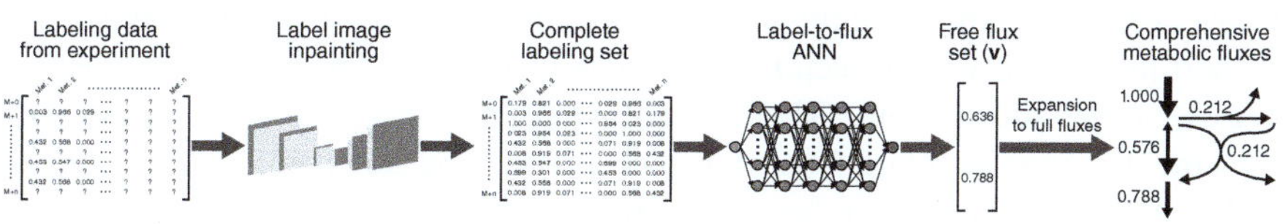

Figure 12.7 A sample workflow for ML-based MFA. Experimental labeling data, organized into a stacked array of MIDs for all measured metabolites, are input into a PCNN model that imputes the isotope labeling patterns of unmeasured metabolites. The complete isotope pattern information is input into an ANN that predicts free fluxes. Using linear algebra, free fluxes reveal a comprehensive metabolic flux map.

and knowledge of the network solution space to produce the full flux distribution through the metabolic model. This approach ensures that no redundant flux predictions are made and that all metabolites are mass balanced.

One issue with using the ANN model trained with complete isotope labeling datasets is that the requirement for flux prediction is a comprehensive measurement of isotope labeling patterns of all metabolites, which is technically challenging due to low metabolite concentrations and stability. To bridge this gap between experimental measurement and rigid ANN input, ML-Flux employs another deep learning model that uses a partial convolutional neural network (PCNN). PCNNs have been used for image inpainting, in which damaged parts of an image are restored. Since isotope labeling fractions take values between 0 and 1, analogous to grayscale pixels in black and white images, ML-Flux uses PCNN-based inpainting to fill in the isotope labeling pattern matrix that may have missing elements to metabolite measurement challenges.

The two-stage ML platform for flux prediction demonstrates how deep learning can streamline flux quantitation from experimentally measured isotope labeling patterns. The results of ML-Flux show promise as flux predictions were as accurate, if not more accurate, compared with optimization-based MFA software. Some other unmistakable advantages of ML-Flux are its speed and accessibility as flux prediction is done in a single step within a few seconds. Development of advanced ML architectures and their application to MFA would further increase the accuracy and speed of flux predictions.

12.4.5 Toward Dynamic Flux and Isotope Labeling Analysis

Dynamic metabolic flux analysis (DMFA) tracks metabolic states that change over time, whereas classical MFA maps metabolic steady states. In batch and fed-batch cultures, cellular metabolism does not stay constant throughout the whole culture process. Thus, DMFA is needed to extend flux analysis to more real-world applications.

DMFA assumes that metabolism is at pseudo-steady state; the process dynamics occurs at a time scale that is slower than the time scales of intracellular metabolic changes. For example, consider a fed-batch reactor in which *E. coli* is cultured to produce acetate. The extracellular environment changes as glucose is consumed and acetate is produced. Intracellular metabolism responds and adjusts to these changes.

Two common techniques used in studying the dynamics of metabolic systems are the average flux method and the piecewise flux dynamics method. The average flux method provides a way to acquire snapshots of metabolism broken into distinct phases of metabolic state. The average transport fluxes are calculated and implemented into standard MFA for each phase. The downside of this method is that fluxes are not time-resolved. To determine time-resolved fluxes, data smoothing techniques such as spline fitting, moving average, and polynomial fitting followed by applying MFA consecutively over narrow time intervals are used. Alternatively, using piecewise linear flux function over time segments, the piecewise flux dynamics method enables the direct fitting of the time series data to determine fluxes [52].

Another approach to conducting DMFA is a hybrid of dynamic modeling and MFA. The hybrid method is a quasi-dynamic analysis that integrates a series of steady-state MFA models with a system of differential equations representing the dynamics of biochemical networks. The MFA equations cover the reactions with unknown kinetic properties, whilst the ODEs include the reactions with known parameters [53]. DMFA becomes computationally demanding with increasing time resolution and kinetic details. The use of neural network layers can augment kinetic and regulatory insights, which impart mechanistic insights useful for drug discovery and treating diseases. ML frameworks have been developed to predict metabolic pathway dynamics better than traditional kinetic models through abundant proteomic and metabolomic datasets [54]. This shows the potential of data-driven models for predicting pathway dynamics, though the abundance of data remains a challenge.

A class of deep learning architecture, Physics-Informed Neural Networks (PINNs), are neural networks that are constrained by physical laws. PINNs constrained with known kinetics can solve ODEs faster than traditional solvers and thus can be used in solving DMFA problems. PINNs enabled studying the dynamics of wound healing assays, considering both biological and reaction-diffusion laws [41]. Another ML approach that can potentially be applied to DMFA is long short-term memory (LSTM). LSTM models are a type of recurrent neural network (RNN) designed to learn long-term dependencies over sequential data. LSTM models have been shown to predict real-time data about process variables using multivariate input for process monitoring and control [55]. Therefore, LSTM may be used for predicting dynamic fluxes from the time evolution of isotope labeling patterns.

12.5 Future Outlook

MFA has provided quantitative and actionable insights into biotechnology and medicine [56]. Researchers are tackling increasingly complex biological systems such as microbial communities [57] and genome-scale metabolic networks [58, 59]. The increased complexity of metabolic networks demands a new paradigm of MFA. Furthermore, experimental limitations on metabolite measurement by MS or NMR pose challenges for resolving metabolic fluxes in large models.

To further improve flux quantitation in complex systems, the development of ML algorithms designed specifically for solving biological problems, including metabolic fluxes, rather than borrowing algorithms designed for different purposes, is essential. Since black box algorithms hamper the interpretability of the relationship between isotope labeling patterns and fluxes, explainable artificial intelligence (XAI) approaches are desirable so that we do not blindly rely on ML predictions [60]. By uncovering patterns and relationships within complex datasets, the application of ML to MFA imparts intuition for the inner workings of metabolism. In conjunction with the new knowledge, the development of sophisticated ML models would fully realize the next generation of MFA and its full potential in determining fluxes in metabolically heterogeneous populations of cells, such as microbial communities and multicellular organisms.

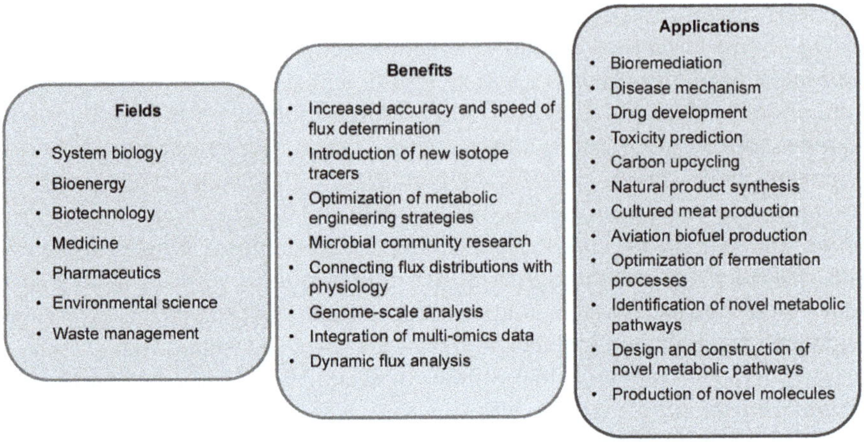

Figure 12.8 Future outlook. Widespread flux determination will provide quantitative insights that benefit diverse application areas of biological sciences.

On the experimental front, use of novel isotope tracers and comprehensive measurement of metabolites would spur flux analysis beyond central carbon metabolism. Nitrogen tracing (e.g. using ^{15}N-labeled glutamine) has already been used for mapping amino acid and nucleotide metabolism in cancer research, neuroscience and immunology, and gut microbiota [61, 62]. Once ^{13}C and ^{15}N tracing is conquered, with advances in metabolite detection and quantification, other isotope tracers such as ^{2}H, ^{18}O, and ^{34}S will broaden the horizon of MFA. We are cautiously optimistic that artificial intelligence will complement advances in biochemistry and analytical chemistry to become a transformative tool for improving quantitative metabolic analysis, imparting new knowledge, and advancing biotechnology and medicine (Figure 12.8).

References

1 Tohsato, Y., Ikuta, K., Shionoya, A. et al. (2013). Parameter optimization and sensitivity analysis for large kinetic models using a real-coded genetic algorithm. *Gene* 518 (1): 84–90.

2 Ono I, Kita H, Kobayashi S. (2003). A real-coded genetic algorithm using the unimodal normal distribution crossover. [cited 2025 Feb 25]; 213–37. https://link.springer.com/chapter/10.1007/978-3-642-18965-4_8.

3 Gilman, A. and Ross, J. (1995 [cited 2025 Mar 1]). Genetic-algorithm selection of a regulatory structure that directs flux in a simple metabolic model. *Biophys. J.* 69 (4): 1321–1333. https://www.cell.com/action/showFullText?pii=S0006349595799994.

4 Kikuchi, S., Tominaga, D., Arita, M. et al. (2003 [cited 2025 Mar 1]). Dynamic modeling of genetic networks using genetic algorithm and S-system. *Bioinformatics* 19 (5): 643–650. https://dx.doi.org/10.1093/bioinformatics/btg027.

5 Chen, L.Z., Nguang, S.K., Chen, X.D., and Li, X.M. (2004). Modelling and optimization of fed-batch fermentation processes using dynamic neural networks and genetic algorithms. *Biochem. Eng. J.* 22 (1): 51–61.

6 Tran, L.M., Rizk, M.L., and Liao, J.C. (2008). Ensemble modeling of metabolic networks. *Biophys. J.* 95 (12): 5606–5617.

7 Nishiguchi, H., Hiasa, N., Uebayashi, K. et al. (2019). Transomics data-driven, ensemble kinetic modeling for system-level understanding and engineering of the cyanobacteria central metabolism. *Metab. Eng.* 1 (52): 273–283.

8 Jia, G., Stephanopoulos, G., and Gunawan, R. (2012 [cited 2025 Mar 1]). Ensemble kinetic modeling of metabolic networks from dynamic metabolic profiles. *Metabolites* 2 (4): 891–912. https://pubmed.ncbi.nlm.nih.gov/24957767/.

9 Choudhury, S., Narayanan, B., Moret, M. et al. (2024, 2024 [cited 2025 Feb 5]). Generative machine learning produces kinetic models that accurately characterize intracellular metabolic states. *Nat. Catal.* 7 (10): 1086–1098. https://www.nature.com/articles/s41929-024-01220-6.

10 Choudhury, S., Moret, M., Salvy, P. et al. (2022 [cited 2025 Feb 5]). Reconstructing kinetic models for dynamical studies of metabolism using generative adversarial networks. *Nat. Mach. Intell.* 4 (8): 710–719. Available from: https://www.nature.com/articles/s42256-022-00519-y.

11 Law, R.C., Pande, K., O'Keeffe, S. et al. (2025 2023.11.06.565907). Accurate and rapid determination of metabolic flux by deep learning of isotope patterns. *bioRxiv* Available from: http://biorxiv.org/content/early/2025/04/15/2023.11.06.565907.abstract.

12 Schellenberger, J., Que, R., Fleming, R.M.T. et al. (2011 [cited 2025 Mar 15]). Quantitative prediction of cellular metabolism with constraint-based models: the COBRA Toolbox v2.0. *Nat. Protoc.* 6 (9): 1290–1307. https://www.nature.com/articles/nprot.2011.308.

13 Xia, J., Psychogios, N., Young, N., and Wishart, D.S. (2009 [cited 2025 Apr 30]). MetaboAnalyst: A web server for metabolomic data analysis and interpretation. *Nucleic Acids Res.* 37 (SUPPL. 2): https://pubmed.ncbi.nlm.nih.gov/19429898/.

14 MIT Technology Licensing Office, METRAN - software for 13C-metabolic flux analysis | [cited 2025 Apr 2]. https://tlo.mit.edu/industry-entrepreneurs/available-technologies/metran-software-13c-metabolic-flux-analysis.

15 Rahim, M., Ragavan, M., Deja, S. et al. (2022 [cited 2025 Mar 18]). INCA 2.0: a tool for integrated, dynamic modeling of NMR- and MS-based isotopomer measurements and rigorous metabolic flux analysis. *Metab. Eng.* 69: 275–285. https://pubmed.ncbi.nlm.nih.gov/34965470/.

16 Mutlib, A.E. (2008 [cited 2025 Mar 23]). Application of stable isotope-labeled compounds in metabolism and in metabolism-mediated toxicity studies. *Chem. Res. Toxicol.* 21 (9): 1672–1689. https://pubs.acs.org/doi/full/10.1021/tx800139z.

17 Wiechert, W., Mö, M., Isermann, N. et al. (1999 [cited 2025 Apr 2];). Bidirectional reaction steps in metabolic networks: III. explicit solution and analysis of isotopomer labeling systems. & sons. *Inc. Biotechnol. Bioeng.* 66: 69–85. https://analyticalsciencejournals.onlinelibrary.wiley.com/doi/10.1002/.

18 Antoniewicz, M.R., Kelleher, J.K., and Stephanopoulos, G. (2007). Elementary metabolite units (EMU): a novel framework for modeling isotopic distributions. *Metab. Eng.* 9 (1): 68–86. Available from: https://doi.org/10.1016/j.ymben.2006.09.001.

19 Weitzel, M., Nöh, K., Dalman, T. et al. (2013). 13CFLUX2 - High-performance software suite for 13C-metabolic flux analysis. *Bioinformatics* 29 (1): 143–145.

20 Matsuda, F., Maeda, K., Taniguchi, T. et al. (2021). mfapy: An open-source python package for 13C-based metabolic flux analysis. *Metab. Eng. Commun.* 1 (13): e00177.

21 Wu, C., Guarnieri, M., and Xiong, W. (2023 [cited 2025 Apr 2]). Freeflux: A python package for time-efficient isotopically nonstationary metabolic flux analysis. *ACS Synth. Biol.* 12 (9): 2707–2714. https://pubs.acs.org/doi/full/10.1021/acssynbio.3c00265.

22 Allen, D.K. and Young, J.D. (2020 [cited 2025 Mar 9]). Tracing metabolic flux through time and space with isotope labeling experiments. *Curr. Opin. Biotechnol.* 64: 92–100. https://pubmed.ncbi.nlm.nih.gov/31864070/.

23 Alseekh, S., Aharoni, A., Brotman, Y. et al. (2021 [cited 2025 Mar 9]). Mass spectrometry-based metabolomics: a guide for annotation, quantification and best reporting practices. *Nat. Methods* 18 (7): 747–756. https://www.nature.com/articles/s41592-021-01197-1.

24 Buescher, J.M., Antoniewicz, M.R., Boros, L.G. et al. (2015). A roadmap for interpreting 13C metabolite labeling patterns from cells. *Curr. Opin. Biotechnol.* 1 (34): 189–201.

25 Yizhak, K., Gaude, E., Le Dévédec, S. et al. (2014). Phenotype-based cell-specific metabolic modeling reveals metabolic liabilities of cancer. *Elife* 21: 3.

26 Folger, O., Jerby, L., Frezza, C. et al. (2011). Predicting selective drug targets in cancer through metabolic networks. *Mol. Syst. Biol.* 7 (1).

27 Agren, R., Mardinoglu, A., Asplund, A. et al. (2014). Identification of anticancer drugs for hepatocellular carcinoma through personalized genome-scale metabolic modeling. *Mol. Syst. Biol.* 10 (3).

28 Jang, C., Chen, L., and Rabinowitz, J.D. (2018). Metabolomics and Isotope Tracing. *Cell* 173 (4): 822–837.

29 Beckonert, O., Keun, H.C., Ebbels, T.M.D. et al. (2007). Metabolic profiling, metabolomic and metabonomic procedures for NMR spectroscopy of urine, plasma, serum and tissue extracts. *Nat. Protoc.* 2 (11): 2692–2703.

30 Feist, A.M., Henry, C.S., Reed, J.L. et al. (2007). A genome-scale metabolic reconstruction for *Escherichia coli* K-12 MG1655 that accounts for 1260 ORFs and thermodynamic information. *Mol. Syst. Biol.* 3 (1).

31 Bourguignon, P.Y., Samal, A., Képès, F. et al. (2010). Challenges in experimental data integration within genome-scale metabolic models. *Algorithms Mol. Biol.* 5 (1): 20.

32 Han, Y. and Styczynski, M.P. (2024). Assessing structural uncertainty of biochemical regulatory networks in metabolic pathways under varying data quality. *npj Syst. Biol. Appl.* 10 (1): 94.

33 Fallahi, S., Skaug, H.J., and Alendal, G. (2020 [cited 2025 Mar 15]). A comparison of Monte Carlo sampling methods for metabolic network models. *PLoS One* 15 (7): e0235393. https://pmc.ncbi.nlm.nih.gov/articles/PMC7329079/.

34 Latendresse, M., Malerich, J.P., Travers, M., and Karp, P.D. (2012). Accurate Atom-mapping computation for biochemical reactions. *JCIM* 52 (11): 2970–2982. <Go to ISI>://WOS:000311461400017.

35 Alghamdi, N., Chang, W., Dang, P. et al. (2021 [cited 2025 Mar 15]). A graph neural network model to estimate cell-wise metabolic flux using single-cell RNA-seq data. *Genome Re* 31 (10): 1867–1884. https://pmc.ncbi.nlm.nih.gov/articles/PMC8494226/.

36 Hasibi, R., Michoel, T., and Oyarzún, D.A. (2024 [cited 2025 Mar 15]). Integration of graph neural networks and genome-scale metabolic models for predicting gene essentiality. *npj Syst. Biol. Appl.* 10 (1): 1–10. https://www.nature.com/articles/s41540-024-00348-2.

37 Schwaller, P., Hoover, B., Reymond, J.L. et al. (2021 [cited 2025 Mar 16]). Extraction of organic chemistry grammar from unsupervised learning of chemical reactions. *Sci. Adv.* 7 (15): https://www.science.org/doi/10.1126/sciadv.abe4166.

38 Young, J.D. (2014 [cited 2025 Mar 16]). INCA: a computational platform for isotopically non-stationary metabolic flux analysis. *Bioinformatics* 30 (9): 1333. https://pmc.ncbi.nlm.nih.gov/articles/PMC3998137/.

39 Mahadevan, S., Shah, S.L., Marrie, T.J., and Slupsky, C.M. (2008 [cited 2025 Mar 16]). Analysis of metabolomic data using support vector machines. *Anal. Chem.* 80 (19): 7562–7570. Available from: https://pubs.acs.org/doi/full/10.1021/ac800954c.

40 Cano, G., Garcia-Rodriguez, J., Garcia-Garcia, A. et al. (2017). Automatic selection of molecular descriptors using random forest: application to drug discovery. *Expert Syst. Appl.* 15 (72): 151–159.

41 Pang, H., Lin, A., Holford, M. et al. (2006 [cited 2025 Mar 16]). Pathway analysis using random forests classification and regression. *Bioinformatics* 22 (16): 2028–2036. https://dx.doi.org/10.1093/bioinformatics/btl344.

42 Chen, H., Jiang, R., Huang, W. et al. (2022 [cited 2025 Mar 16]). Identification of energy metabolism-related biomarkers for risk prediction of heart failure patients using random forest algorithm. *Front. Cardiovasc. Med.* 9: 993142. http://www.gsea-msigdb.org/gsea/msigdb/.

43 Pecioski, D., Gavriloski, V., Domazetovska, S., and Ignjatovska, A. (2023). An overview of reinforcement learning techniques. In: *12th Mediterranean Conference on Embedded Computing*. MECO.

44 Hosna, A., Merry, E., Gyalmo, J. et al. (2022 [cited 2025 Mar 17]). Transfer learning: a friendly introduction. *J. Big Data* 9 (1, 1): –19. Available from: https://journalofbigdata.springeropen.com/articles/10.1186/s40537-022-00652-w.

45 Moore, B.M., Wang, P., Fan, P. et al. (2020 [cited 2025 Mar 17]). Within- and cross-species predictions of plant specialized metabolism genes using transfer learning. in silico plants. 2 (1): 1–19. Available from: https://dx.doi.org/10.1093/insilicoplants/diaa005.

46 Rodriguez, M.Z., Comin, C.H., Casanova, D. et al. (2019 [cited 2025 Mar 17]). Clustering algorithms: a comparative approach. *PLoS One* 14 (1): e0210236. https://pmc.ncbi.nlm.nih.gov/articles/PMC6333366/.

47 Ganesh, S., Hu, T., Woods, E. et al. (2021 [cited 2025 Mar 26]). Spatially resolved 3D metabolomic profiling in tissues. *Sci. Adv.* 7 (5): 957–984. https://www.science.org/doi/10.1126/sciadv.abd0957.

48 Agrawal, R., Imieliński, T., and Swami, A. (1993 [cited 2025 Mar 17]). Mining association rules between sets of items in large databases. *ACM SIGMOD Record* 22 (2): 207–216. https://www.researchgate.net/publication/200043124_Mining_Association_Rules_Between_Sets_of_Items_in_Large_Databases_SIGMOD_Conference.

49 King, Z.A., Lu, J., Dräger, A. et al. (2016). BiGG models: a platform for integrating, standardizing and sharing genome-scale models. *Nucleic Acids Res.* 44 (D1): D515–D522.

50 MetaCyc: Metabolic Pathways From all Domains of Life [cited 2025 Mar 17]. Available from: https://metacyc.org/

51 Matsuda, F., Maeda, K., Taniguchi, T. et al. (2021). mfapy: an open-source python package for 13C-based metabolic flux analysis. *Metab. Eng. Commun.* 1 (13): e00177.

52 Antoniewicz, M.R. (2013). Dynamic metabolic flux analysis — tools for probing transient states of metabolic networks. *Curr. Opin. Biotechnol.* 24 (6): 973–978.

53 Yugi, K., Nakayama, Y., Kinoshita, A., and Tomita, M. (2005 [cited 2025 Mar 19]). Hybrid dynamic/static method for large-scale simulation of metabolism. *Theor. Biol. Med. Model.* 2 (1): 1–11. https://tbiomed.biomedcentral.com/articles/10.1186/1742-4682-2-42.

54 Costello, Z. and Martin, H.G. (2018 [cited 2025 Mar 18]). A machine learning approach to predict metabolic pathway dynamics from time-series multiomics data. *npj Syst. Biol. Appl.* 4 (1): 1–14. https://www.nature.com/articles/s41540-018-0054-3.

55 Williams, T., Kalinka, K., Sanches, R. et al. (2023 [cited 2025 Mar 18]). Machine learning and metabolic modelling assisted implementation of a novel process analytical technology in cell and gene therapy manufacturing. *Sci. Rep.* 2023 13:1 13 (1): 1–19. Available from: https://www.nature.com/articles/s41598-023-27998-2.

56 Law, R.C., Lakhani, A., O'Keeffe, S. et al. (2022 [cited 2025 Mar 18]). Integrative metabolic flux analysis reveals an indispensable dimension of phenotypes. *Curr. Opin. Biotechnol.* 75: 102701. https://pubmed.ncbi.nlm.nih.gov/35278746/.

57 Antoniewicz, M.R. (2020). A guide to deciphering microbial interactions and metabolic fluxes in microbiome communities. *Curr. Opin. Biotechnol.* 1 (64): 230–237.

58 Neal, M., Brakewood, W., Betenbaugh, M., and Zengler, K. (2024 [cited 2025 Mar 18]). Pan-genome-scale metabolic modeling of Bacillus subtilis reveals functionally distinct groups. Faust K, editor. *mSystems* 9 (11): Available from: https://pubmed.ncbi.nlm.nih.gov/39365060/.

59 Correia, K. and Mahadevan, R. (2020 [cited 2025 Mar 18]). Pan-genome-scale network reconstruction: harnessing phylogenomics increases the quantity and quality of metabolic models. *Biotechnol. J.* 15 (10): 1900519. Available from: https://onlinelibrary.wiley.com/doi/full/10.1002/biot.201900519.

60 Ali, S., Abuhmed, T., El-Sappagh, S. et al. (2023). Explainable Artificial Intelligence (XAI): what we know and what is left to attain trustworthy Artificial Intelligence. *Inf. Fusion* 1 (99): 101805.

61 Reisz, J.A. and A, D.'.A. (2017 [cited 2025 Mar 8]). Measurement of metabolic fluxes using stable isotope tracers in whole animals and human patients. *Curr. Opin. Clin. Nutr. Metab. Care* 20 (5): 366. https://pmc.ncbi.nlm.nih.gov/articles/PMC5794022/.

62 AJL, C. (2001 [cited 2025 Mar 8]). Role of glutamine in cerebral nitrogen metabolism and ammonia neurotoxicity. *Ment. Retard. Dev. Disabil. Res. Rev.* 7 (4): 280–286. https://pubmed.ncbi.nlm.nih.gov/11754523/.

13

Streamlining the Design-Build-Test-Learn Process in Automated Biofoundries

Enrico Orsi[1], *Nicolás Gurdo*[1,2], *and Pablo I. Nikel*[1]

[1] *Technical University of Denmark, The Novo Nordisk Foundation Center for Biosustainability, Søltofts Plads 1, Kongens Lyngby, Denmark*
[2] *Microbial Cultivation Process Development Department, Novo Nordisk A/S, Novo Alle 1, Bagsværd, Denmark*

13.1 Introduction

In the era of synthetic biology, the design-build-test-learn cycle (DBTLc) has become a formalized pipeline for standardized, iterative optimization of biological parts [1–4]. Each module in the cycle serves a specific function and fulfills distinct objectives, each presenting its own set of challenges. Notably, the standardization inherent in the DBTLc framework can significantly enhance the throughput of iterations, enabling faster and more efficient advancements in synthetic biology [3].

As we are witnessing the pervasion of big data, artificial intelligence (AI), and machine learning (ML) technologies in all areas of biology [5–8], the DBTLc demands high-throughput approaches to generate and analyze vast amounts of data efficiently. In principle, the entire workflow in industrial biotechnology – from DNA synthesis to bioprocess optimization with engineered strains – can now be studied while generating (and analyzing) high-throughput data at every step [9]. Each phase of this pipeline is supported by iterative rounds of DBTLc, allowing researchers to refine designs systematically based on experimental results.

Successfully executing these high-throughput iterative cycles within biofoundries represents one of the grand challenges of this decade. To address this aspect, a global alliance of biofoundries has been established as a collaborative network [10]. This alliance aims to support the sustainable implementation of biofoundry infrastructures, advancing practices that contribute to a sustainable bioeconomy [11, 12].

This chapter provides an overview of how automated biofoundries accommodate high-throughput DBTLc and their applications in industrial biotechnology. It begins with an overview of the classic DBTLc framework, accompanied by examples of successful applications that highlight its transformative potential. The discussion then transitions to adaptations required for implementing DBTLc pipelines in automated biofoundry setups, emphasizing how automation enhances efficiency, scalability, and standardization in high-throughput data generation and collection.

Machine Learning and Big Data-enabled Biotechnology, First Edition. Edited by Hal S. Alper.

Subsequently, we explore the global development of biofoundries, offering insights into their characteristics and contributions to synthetic biology. Then, we address current challenges that hinder biofoundries from reaching their full potential. These include technological bottlenecks, integration issues, and the need for further standardization and collaboration. Finally, we conclude the chapter with further perspectives and trends that are likely to streamline the workflow of automated biofoundries. By examining these aspects comprehensively, this chapter aims to shed light on how biofoundries can be optimized to drive innovation in synthetic biology while supporting sustainable practices in the global bioeconomy.

13.2 The Design-Build-Test-Learn Cycle (DBTLc)

The DBTLc paradigm for standardization and iteration in synthetic biology (Figure 13.1) is a concept that has its roots in (electronic) engineering principles [13]. By default, the DBTLc can be applied to the optimization of any biological part of interest, e.g. DNA assembly [14–18], protein engineering [19], metabolic rewiring [20–22], or even a whole bioprocess [23]. However, as detailed in subsequent sections of Chapter 13, a significant body of literature describes the use of the DBTLc for microbial strain engineering [13]. In this setup, microorganisms are studied and improved as cell factories for performing bioconversions of industrial relevance. For the sake of simplicity, we will adhere to this scope for describing the DBTLc modules in the next section.

13.2.1 The DBTLc Components

Within the context of engineering microorganisms, the DBTLc phases are structured in four segments:

Design. This phase has the scope of determining the strategy for the whole iteration. It includes choices of target modification to implement as well as the design of the biological parts to synthesize and introduce into the host for implementing such modifications. It could include, e.g. choice of heterologous genes to express, or the *in silico* design of plasmids for gene deletions. In some cases, use of genome-scale metabolic models, such as those described in Chapters 1, 2, and 9, can guide the design phase by identifying target genes to delete or overexpress [24]. Some public databases are also available for guiding the design phase. The Kyoto Encyclopedia of Genes and Genomes (KEGG) is one of the pioneering databases designed for the systematic analysis of gene functions, cellular processes, chemical compounds, and enzymes [25]. Similarly, the BRaunschweig ENzyme DAtabase (BRENDA) serves as a comprehensive resource for enzyme-ligand interactions, enabling the exploration of functional and molecular parameters of enzymatic reactions [26]. MetaCyc provides a curated catalog of metabolic reactions and enzymes across various microorganisms [27]. Software tools developed for extracting data from these databases and identifying viable design strategies have facilitated the rational engineering of metabolic networks. Notable advancements in this domain

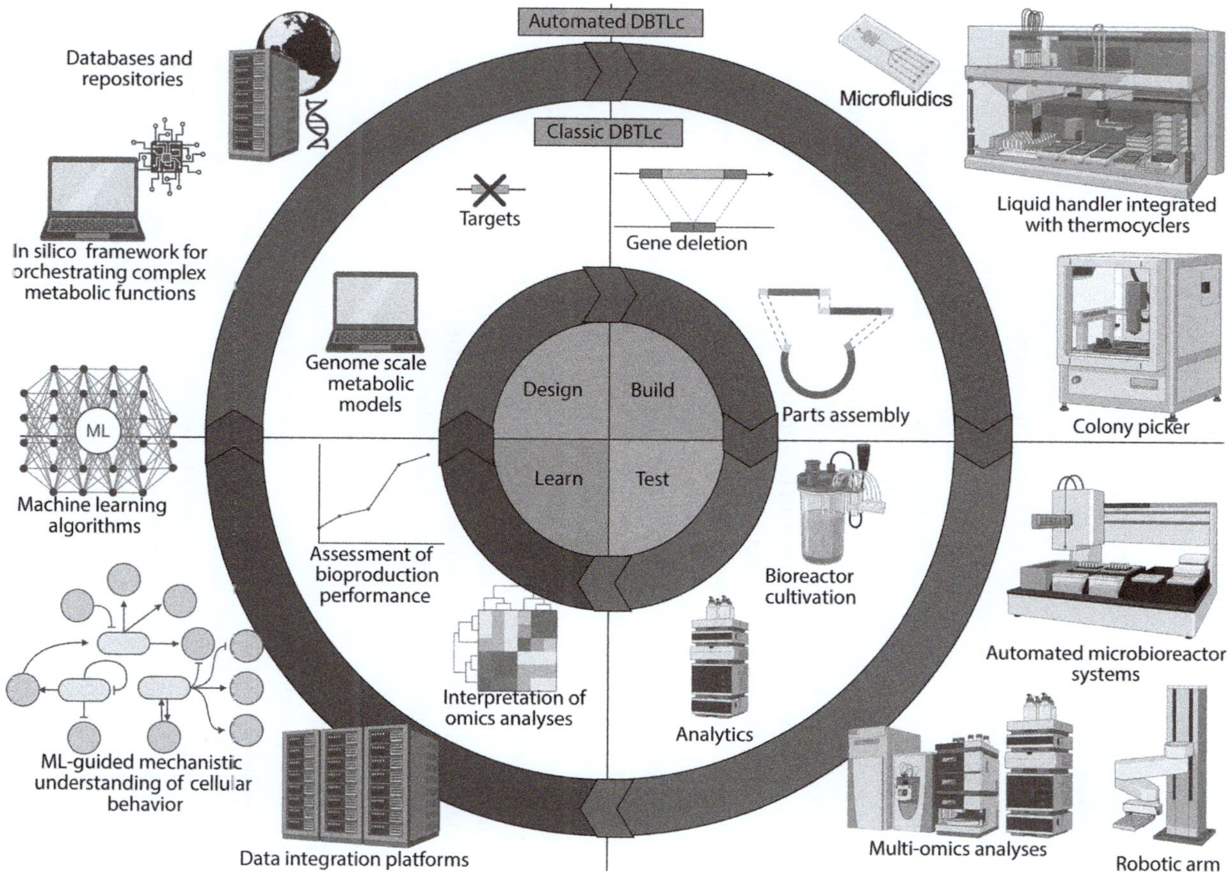

Figure 13.1 Overview of the design-build-test-learn cycle (DBTLc) in its classic form (inner circle) and in a high-throughput, automated setup (outer circle). We outline the most common techniques for iterative strain engineering in the classic DBTLc; the automated DBTLc indicates the infrastructures that allow a higher throughput in the cycle. Created with BioRender.com.

include Optknock [28], Fastknock [29], Retropath2.0 [30], SelenzymeRF [31], and EnzymeMiner [32]. Once the components for a given metabolic pathway have been selected, the corresponding DNA sequences must be designed and synthesized. This process leverages established frameworks such as BioBricks [33, 34], the SEVA plasmid system [35–37], and computational tools (e.g. RBS Calculator v2.1 [38, 39]), which enable automated, physics-based modeling of gene expression, an aspect covered in Chapter 3.

Build. The aim of this step is to create and assemble the biological parts required for the iteration. In strain engineering, this involves assembling the biological components to be tested, such as plasmids for heterologous gene expression or DNA cassettes for gene inactivation. Routine techniques include, for example, Gibson assembly [40], a one-step isothermal reaction that joins overlapping DNA fragments using exonuclease, polymerase, and ligase; Golden Gate assembly [41], a restriction enzyme-based method that utilizes type IIS restriction enzymes for seamless cloning; USER cloning [42], which uses uracil-containing primers to amplify PCR products, removes uracil residues, and assembles fragments through complementary overhangs; and traditional restriction-ligation cloning. This step also includes transforming these constructs into the final host. Recent advances in genome editing techniques, particularly the CRISPR/Cas toolkit [43–45], have significantly enhanced this process to enable the introduction of diverse types of mutations, including in non-model microorganisms [46–48]. When the goal of the DBTLc is strain characterization, this step can, in principle, be omitted.

Test. Here the scope is to measure the effect of the modifications implemented in the previous phases. It often involves cultivation assays followed by the quantification of a phenotype (e.g. a product molecule), typically through the calculation of titers (g/L), rates (g/L/h), and yields (g/g) [49]. This phase heavily relies on analytical techniques and is consequently considered the technical bottleneck of the DBTLc cycle. When performing multilevel strain characterization, such as through omics analyses [50], the throughput of these techniques is high, generating large amounts of data that require complex interpretation [51]. Genomics provides the foundational blueprint of an organism by analyzing DNA sequences to identify genetic variations and modifications in engineered constructs. Transcriptomics examines gene expression patterns, e.g. through RNA sequencing (RNA-seq), to understand regulatory responses and optimize modifications [52]. Proteomics investigates protein expression, modifications, and interactions, offering insights into the functional outcomes of genetic and transcriptional changes. Metabolomics provides detailed insights into metabolic states by profiling intracellular and extracellular metabolites, enabling a comprehensive understanding of biochemical pathway activities and cellular physiology [53–55].

Learn. Finally, the learning phase evaluates the iteration's success and determines whether a new design phase is needed. This step is often less structured, relying on *ad hoc* observations and literature data [56]. However, with the increasing amount of data generated per experiment, there is a growing need for a more structured, data-driven approach to this phase. In recent years, the application of ML in biotechnological processes has expanded significantly, driven by the emergence of next-generation algorithms such as deep learning – a subset of neural network methodologies that employs hierarchical architectures – and the availability of user-friendly development frameworks [57, 58]. As discussed in detail in the next section, these approaches enable data interpretation in the context of high-throughput data generation of a biofoundry.

13.2.2 Application of the DBTLc for Strain Engineering

Some examples of DBTLc use have become widely recognized and are now considered the gold standard in the field. One of the earliest DBTLc iterations in metabolic engineering was the decade-long effort to engineer *Escherichia coli* and *Saccharomyces cerevisiae* for the production of the antimalarial drug artemisinin. Beginning in 2003 with the heterologous expression of the eukaryotic mevalonate pathway in *E. coli*, approximately 25 μg/L of amorphadiene – the first committed precursor to artemisinin [59] – were produced. The authors reported improved titers from 25 to 25 g/L, a remarkable million-fold increase, ultimately leading to the one-step chemical conversion of artemisinic acid to artemisinin by 2013 [60]. The rationale behind switching from *E. coli* to *S. cerevisiae* as the production host, along with the full roadmap of this achievement, is extensively reviewed in a study summarizing the entire scientific journey [61].

Another notable DBTLc success story from the early 2010s is the engineering of *E. coli* for the production of 1,4-butanediol, a non-natural chemical building block used in plastics, polyesters, and spandex fibers. This effort, guided by genome-scale modeling in its design phase to inform rational engineering modifications in the build phase, achieved titers of up to 18 g/L [62]. Other DBTLc-driven advancements in biomanufacturing include strain engineering efforts that have achieved titers exceeding 100 g/L for various commodity chemicals, e.g. succinic acid [63], lactic acid [64], L-lysine [65], and glutaric acid [66]. Additionally, DBTLc approaches have been instrumental in optimizing both prokaryotic and eukaryotic strains as cell factories, expanding the number of domesticated strains available for biomanufacturing [67–75].

The growing success of DBTLc campaigns has driven the development of dedicated infrastructure to support these workflows. As we will explore in the next section, biofoundries represent a significant leap forward in standardizing and automating strain engineering. By integrating computation-driven design with robot-assisted workflows, biofoundries enhance throughput and accelerate strain optimization.

13.2.3 Description of Biofoundries and Their Operating Parts

A biofoundry is an advanced laboratory that leverages automation, robotics, AI, and high-throughput technologies to execute and optimize the DBTL cycle (Figure 13.1). These infrastructures are designed to standardize workflows, addressing challenges such as time-intensive manual procedures, variability in results and the increasing complexity of biological systems. The core components of a biofoundry include the following modules:

Design. These systems are blueprinted using computational tools to design DNA sequences, metabolic pathways and synthetic circuits [76, 77]. Advanced software, powered by ML algorithms, predicts the performance of engineered systems, facilitating a more targeted approach to bioengineering. As discussed above, several updated and refined databases provide extensive repositories containing gene functions, cellular processes, chemical compounds, and enzymes from several microorganisms, serving as an initial step for pathway design. These databases undergo continuous updates and refinements, establishing them as primary resources for selecting enzymatic activities and pathways in metabolic engineering and pathway design. The individual components are carefully selected and integrated into an *in silico* framework, collectively orchestrating complex functions that culminate in a blueprint prototype to be assembled in subsequent steps.

Build. In this step, the focus is on the physical assembly and synthesis of biological constructs through advance laboratory automation steps, including robotic liquid handlers for DNA assembly, high-throughput cloning platforms, automated colony pickers, and microfluidic devices for precise manipulation of biological samples [78]. In recent years, specialized toolboxes have been developed to incorporate novel functions into the microbial host of choice. Genome-wide modifications integrate rapid multipart DNA assembly methods with advanced genome engineering techniques [79–84]. As discussed above, a significant advancement in molecular biology has been the discovery of CRISPR and their associated Cas proteins, which have been adapted for gene and genome editing applications [85–87]. To enhance the scope of these technologies, multiplex-editing strategies have been developed, enabling the simultaneous modification of multiple *loci* within the eukaryotic and prokaryotic genomes [88]. When automated, these technologies enhance both the speed and extent of achievable mutations. Additionally, when optimizing DNA sequences at the single base-pair level, employing diverse libraries expands the range of potential solutions, particularly in the context of directed evolution efforts [89–91]. In certain cases, *in vivo* hypermutators can be leveraged to facilitate and refine these evolutionary processes [92, 93].

Test. Automated high-throughput screening (HTS) technologies are essential for evaluating the performance of engineered biological constructs, enabling rapid and efficient assessment of various biological and biochemical properties [94–96]. Automated bioreactors are essential for HTS of engineered biological constructs, providing precisely controlled environments for cell culture while continuously monitoring and adjusting critical parameters such as pH, oxygen levels, nutrient supply, and biomass production [97]. These systems enable real-time data collection, improving reproducibility and scalability in bioprocess development [98, 99]. In parallel, omics technologies play a crucial role in testing and characterizing engineered constructs [100]. The integration of automated cultivations with omics-based testing generates comprehensive, high-resolution datasets, facilitating advanced computational analysis, predictive modeling and iterative design improvements in synthetic biology and biotechnology.

Learn. SynBio relies on both mechanistic and ML models to extract insights from omics data, ultimately refining strain engineering strategies in the DBTLc [101–104].

Integration of advanced ML algorithms with automation enables the efficient utilization of large datasets, facilitating the prediction of novel interventions for process optimization [105]. In practice, mechanistic models represent molecular components and their interactions within a system, which provide enhanced interpretability, transparency, and explainability. These models help researchers understand the underlying mechanisms governing cellular behavior and guide hypothesis-driven engineering. In contrast, ML techniques, which leverage statistical and computational approaches, are particularly effective at identifying complex, nonobvious patterns in high-dimensional omics datasets [106–108]. By distinguishing key features that differentiate strains and experimental conditions, ML models can refine mechanistic models by highlighting missing or inaccurate components and interactions. This integration improves predictive accuracy and aids in designing more informative experiments based on data-driven model topology and simulation outputs. While ML models may lack full interpretability and transparency, they play a crucial role in extracting meaningful insights from large-scale biological datasets, ultimately expanding our understanding of cellular systems and driving the next iteration of strain optimization [109–111].

Finally, biofoundries include data integration platforms that rely on centralized databases and data management tools to ensure seamless integration and retrieval of information across the DBTLc. These platforms serve as the backbone of biofoundry operations by standardizing data formats, enabling interoperability between various software and laboratory automation systems and facilitating real-time data sharing [110].

13.2.4 Laboratory Workflows with Potential of Automation in Biofoundries

As described above, the DBTLc can be applied to several biological scopes. Biofoundries enable the automation of a wide range of workflows in biology, significantly reducing costs and accelerating timelines. A brief overview of the different workflows with potential for automation is presented in Table 13.1. Some examples of DBTLc that can be performed in biofoundries include but are not limited to:

DNA Synthesis and Assembly. The ability to synthesize large quantities of DNA and generate combinatorial libraries is fundamental in synthetic biology [112, 113]. Automated platforms in biofoundries can perform error-free synthesis and assembly of DNA constructs, creating libraries with thousands of variants for the subsequent steps [114]. Once synthesized, DNA fragments can be assembled into larger constructs. Several assembly methods have been developed, including both Gibson Assembly and Golden Gate Assembly. Both techniques have several advantages, being widely used in synthetic biology due to their efficiency and scarless integration of DNA fragments. For example, these techniques can be automated on liquid-handling robotic platforms. Researchers have automated a DNA assembly protocol using the Type-IIS restriction-based Modular Cloning (MoClo) technique on liquid-handling robots by optimizing pipetting parameters for different liquid viscosities (e.g. enzymes, DNA, water, and buffers) [115]. In addition, they

Table 13.1 Overview of biological workflows which can be automated in a biofoundry.

Workflow	What is Automated	How	Examples	References
DNA synthesis/ Library construction	DNA design, synthesis, assembly, and error correction	Robotic systems perform oligonucleotide synthesis, error-checking, and assembly	CRISPR library construction; metabolic pathway engineering	[114–116]
Protein engineering	Directed evolution workflows, mutagenesis, and screening	Automated mutagenesis and high-throughput screening	Enzyme optimization for biofuels	[92, 123]
Cell factories	Engineering and testing cellular production systems	High-throughput genetic transformation and phenotype screening	Production of insulin, ethanol, or bioplastics	[125–127]
Prototyping	Testing several design iterations	High-throughput platforms evaluate prototypes under controlled conditions	Synthetic circuits; biosensors	[126, 132]
Process optimization	Scaling and tuning up production processes	Automated fermentation systems and process analytical technologies	Industrial-scale enzyme production	[128, 130, 131, 133]

developed a web-based tool (mocloassembly.com) that automates script generation and library construction from GenBank files. The tool includes optimized liquid classes, labware settings, and deck layouts, with DNA parts available via Addgene and Boston University's BDC ICE Registry. This workflow enables high-throughput, error-reduced DNA assembly, with sequencing data showing ~95% accuracy and requiring only 4% of the manual effort [115]. Lastly, the improved version of MoClo, known as the RoboMoClo framework, employs a hybrid approach that integrates both hierarchical and iterative gene assembly strategies while utilizing fewer destination vectors compared to conventional MoClo systems [116].

Library Construction and Cloning. Combinatorial DNA libraries are crucial for exploring genetic diversity and optimizing biological functions. Libraries can be generated through site-directed mutagenesis, recombination-based methods or random mutagenesis. Key approaches include: (1) oligonucleotide-based mutagenesis, which introduces specific mutations at defined locations within a gene [117]; (2) error-prone PCR, which incorporates random mutations during PCR amplification to create genetic diversity [118]; (3) *in vivo* hypermutators to increase the mutational accumulation rate in the biological system [119]; (4) DNA shuffling, which involves recombination of different DNA fragments to generate novel sequences [120]; and (5) CRISPR-based library construction, relying on utilizing CRISPR-Cas9 to generate genome-wide knockout or activation libraries [121]. These libraries can contain thousands to millions of variants, enabling directed evolution and functional screening for optimized enzymes, regulatory elements, or metabolic pathways [122].

Protein Engineering. Biofoundries also facilitate the development of novel proteins with tailored properties by integrating automated technologies that streamline protein engineering workflows [92]. Automated directed evolution workflows, encompassing mutagenesis and HTS, enable researchers to identify optimized proteins for industrial, pharmaceutical, and agricultural applications. Directed evolution serves as a powerful technique in protein engineering, emulating natural selection to evolve proteins toward predefined functional objectives. This iterative process involves generating a diverse library of protein variants through random mutagenesis, followed by selection or screening to identify variants with enhanced or novel functionalities [123]. The automation of this workflow within biofoundries expedites the evolution process, allowing for the rapid identification of optimized proteins for various applications. HTS plays a crucial role in evaluating large libraries of protein variants to identify those with desirable characteristics [96]. Automated HTS platforms facilitate the rapid assessment of activity, stability, and specificity in thousands to millions of variants, significantly reducing the time and labor required compared to traditional manual methods. Furthermore, the integration of HTS with automated directed evolution, coupled with ML algorithms, enhances the efficiency of protein engineering by enabling predictive modeling of beneficial mutations. These computational approaches refine mutant library design and facilitate the identification of enzyme variants with improved catalytic properties, thereby advancing the capabilities of biofoundries in protein optimization [124].

Strain Engineering for Manufacturing. Engineering microbial or mammalian cells to function as "factories" for the production of valuable compounds, such as biofuels, pharmaceuticals, and specialty chemicals, is a major focus of biofoundries. Automation enables rapid prototyping and iterative optimization of these cellular systems. High-throughput prototyping entails evaluating multiple design iterations to identify the most effective solutions. Automated workflows in biofoundries enable the simultaneous processing of dozens or even hundreds of prototypes, significantly enhancing the efficiency of the DBTLc. This approach is particularly valuable for optimizing genetic circuits, metabolic pathways, and enzyme functionality [125–127]. For example, an automatic platform has been implemented for the biosynthesis of the flavonoid (2*S*)-pinocembrin in *E. coli* to demonstrate effectiveness of automated, iterative DBTL cycles. Through two DBTL iterations, the production pathway was enhanced by 500-fold, achieving competitive titers of up to 88 mg/L. Furthermore, the application of this pipeline to optimize an alkaloid biosynthetic pathway highlights its potential for accelerating microbial strain engineering and optimizing the production of diverse chemical compounds.

Process Optimization and Scale-up. Once a design has been validated, biofoundries play a crucial role in optimizing production processes for industrial scalability. Automated systems can fine-tune parameters such as fermentation conditions, media compositions, and downstream processing steps, ensuring maximum yield and cost-efficiency while maintaining product quality [128–131].

13.3 Geographical Distribution of Biofoundries Around the Globe

The implementation of successful workflows in biofoundries comes with its own challenges. As this new type of infrastructure is rapidly being adopted worldwide, the Global Biofoundry Alliance (GBA) was established in 2018 to accelerate the successful implementation of biofoundries through collaboration and knowledge sharing [10]. Initially comprising 15 noncommercial biofoundries from four continents, the GBA consortium has grown to 35 members as of 2025 (https://www.biofoundries.org/members), demonstrating the success of this initiative. Yet, the four continents represented remain North America, Europe, Asia, and Oceania. Despite being intrinsically similar in their equipment and organization, some of these biofoundries have specialized in specific objectives, leading to a certain degree of scope differentiation (Table 13.2).

The high capital costs of building a biofoundry make it challenging for emerging economies to develop such infrastructure. Establishing a biofoundry is complex and expensive, requiring specialized equipment and expert personnel, with operating costs far exceeding those of standard laboratories [11]. However, automation can significantly enhance speed, reproducibility, and reliability, positioning biofoundries as key hubs for synthetic biology programs and drivers of economic development. Moreover, as computational and robotic technologies continue to advance, their costs are expected to decrease, making biofoundries more accessible over time. Therefore, we hope that emerging economies will strategically invest in national biofoundries as part of their development agenda, anticipating future reductions in technological barriers. Finally, the current focus is on enhancing collaboration, optimizing workflows, and ensuring the reproducibility and reuse of efforts within the field. Biofoundries are already playing a crucial role in supporting the growth of startups and small-to-medium-sized enterprises (SMEs), while facilitating cooperation among private companies, academic institutions, and policymakers. This collaborative environment is driving the establishment of reliable, cost-effective standards for synthetic biology that are gaining broad acceptance among users. Additionally, local governments and regional funding bodies are actively supporting platform technologies within biofoundry hubs, helping reduce the risks associated with private investment in new ventures, while also fostering a thriving ecosystem of companies advancing synthetic biology technologies [134].

Table 13.2 Examples of noncommercial biofoundries in the Global Biofoundry Alliance with specialized scopes beyond general DBTLc for strain engineering.

Biofoundry and location	Scope	Website
Agile Biofoundry, Emeryville, USA	Sustainable industrial production of renewable fuels and chemicals	https://agilebiofoundry.org/
Australian Institute for Bioengineering and Nanotechnology Biofoundry, Brisbane, Australia	Integrating capabilities in synthetic biology for advanced biomanufacturing, including biofuels, chemicals, biologics, and novel bioinspired devices	https://aibn.uq.edu.au/welcome-biofoundry
Biofactorial, Vancouver, Canada	Providing access to unique metagenomic and chemical compound libraries for developing screening paradigms and discovering biocatalysts or drug candidates	https://biofactorial.microbiology.ubc.ca/
CompuGene, TU Darmstadt, Darmstadt, Germany	Enabling the realization of complex genetic circuits in biological systems	https://www.compugene.tu-darmstadt.de/compugene/index.de.jsp
DTU Biosustain Biofoundry, Lyngby, Denmark	Helping bridge the technological "valley of death" and advancing selected research projects in developing cell factories for chemical or protein production	https://www.biosustain.dtu.dk/technologies/biofoundry
Edinburgh Genome Foundry, Edinburgh, United Kingdom	Modular assembly of DNA constructs using a highly automated robotic platform	https://biology.ed.ac.uk/research/facilities/edinburgh-genome-foundry
GeneMill, Liverpool, United Kingdom	Expression of custom protein at competitive rates	https://www.liverpool.ac.uk/research/facilities/genemill/
Lara, Laboratory Automation Robotic Assistant Biochemistry, Greifswald, Germany	Fully automatized high-throughput protein screening	https://lara.uni-greifswald.de/
London Biofoundry, Imperial College, London, UK	Traditional DBTLc, refocused during SARS-CoV-2 pandemic to develop alternative patient sample testing pipelines	https://www.londonbiofoundry.org/
Living Measurement Systems Biofoundry (NIST), Maryland, United States of America	Enabling the engineering of cells for reliable and safe use in dynamic and unpredictable environments for applications, e.g. living therapeutics, environmental sensing, and structured materials fabrication	https://www.nist.gov/programs-projects/nist-living-measurement-systems-foundry
SYNBIOCHEM, Manchester, United Kingdom	Synthetic biology of fine and specialty chemicals production	https://synbiochem.co.uk/

13.4 Challenges for Implementing Fully Automated Biofoundries

The long-term vision for biofoundries is to develop fully automated bioengineering pipelines that can execute the DBTL cycles seamlessly. However, several practical limitations currently hinder the realization of this goal. These challenges can be grouped into distinct categories based on their underlying causes (Figure 13.2).

One of the most significant hurdles in biofoundry automation is the lack of standardization across different platforms, equipment, and protocols. Unlike traditional engineering disciplines, biology inherently exhibits high variability due to differences in host organisms, genetic constructs, and environmental conditions. Most laboratory instruments still operate using proprietary software and non-standardized data formats, complicating integration efforts. Ensuring compatibility between robotic liquid handlers, analytical instruments, and computational design tools remains a substantial challenge. Without universally accepted standards, reproducibility is hindered, and cross-institutional collaboration becomes more difficult. Addressing these limitations is essential to unlocking the full potential of biofoundries.

Another major challenge lies in the ability of biofoundries to support scaling-up efforts. While small-scale automated workflows have demonstrated success in high-throughput strain engineering, scaling up strain performance to industrial levels remains complex. Many automated workflows require customization for specific applications, making it difficult to establish universally applicable automation pipelines. Additionally, large-scale fermentations present different gradients in nutrient availability as well as gas-liquid transfer dynamics, exposing strains to environments that differ significantly from benchtop reactors. The transition from benchtop automation to large-scale biofoundries necessitates significant investment in infrastructure and process refinement. Furthermore, biofoundries should be flexible enough to accommodate a wide range of biological systems, from microbial strains to mammalian cells. Recent efforts in this direction have been directed to implement "versatile" biofoundries, which can be adapted depending on specific requirements (e.g. anoxic conditions). However, this adaptability is hindered by the limitations of current robotic systems, which are often designed for specific workflows and cannot easily transition between different experimental setups. Overcoming these scalability and flexibility constraints requires innovations in modular automation systems that can be readily reconfigured based on experimental demands.

Biological systems are inherently complex, exhibiting dynamic behaviors, stochasticity as well as intricate regulatory networks. This complexity challenges the predictability and control required for biofoundry automation. Unlike mechanical or electronic systems, biological parts interact in nonlinear ways, making it difficult to design predictable, scalable workflows. Engineering biology at scale necessitates robust modeling and simulation tools capable of accurately capturing cellular behaviors under diverse conditions. Current computational models often oversimplify biological interactions, limiting their predictive power. Furthermore, emergent properties such as genetic instability, metabolic burden,

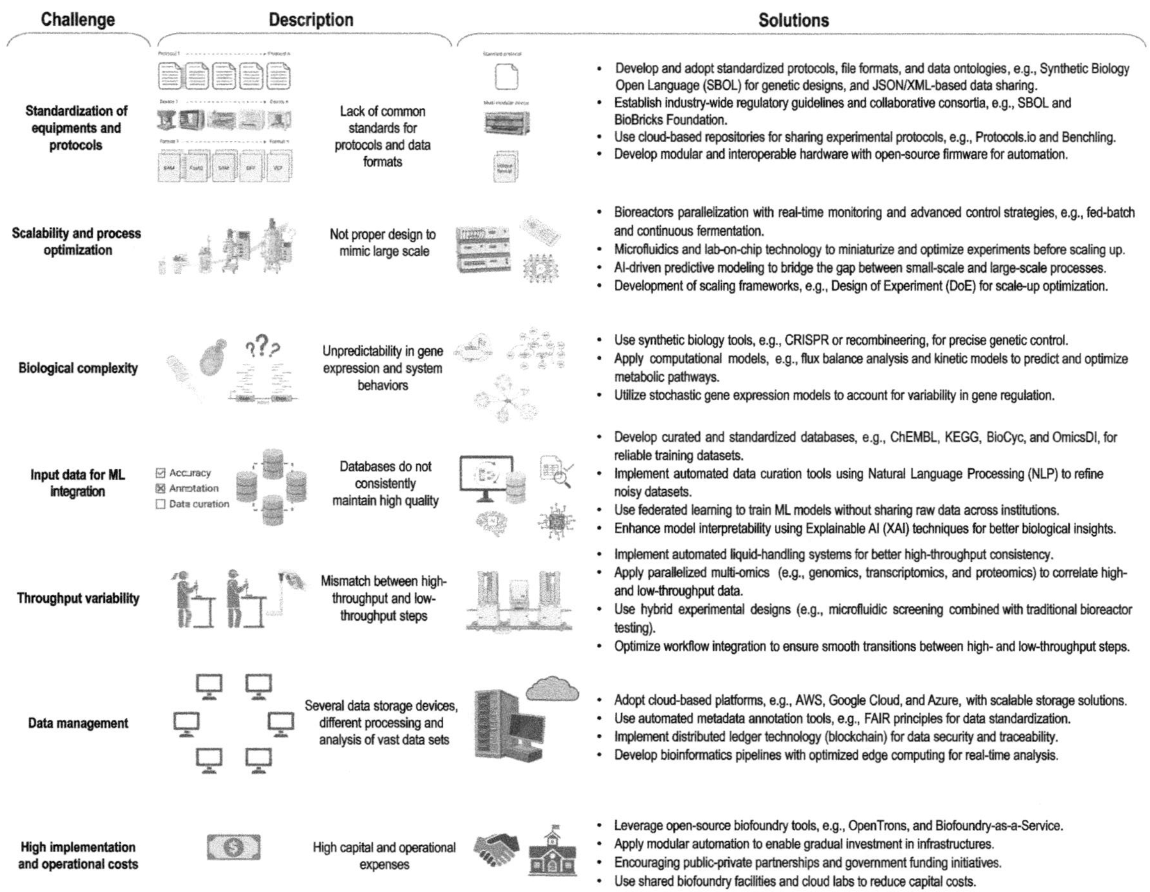

Figure 13.2 Key challenges for implementing a fully automated biofoundry. Created with BioRender.com.

and unintended interactions can introduce unexpected variability, complicating system optimization. Advancements in computational biology and HTS methods will be essential to overcoming these challenges.

AI and ML are poised to revolutionize biofoundries by optimizing experimental design, predicting biological behaviors and accelerating data analysis. However, the effective integration of AI-driven decision-making into fully automated workflows remains a formidable challenge. One key limitation is the availability of high-quality, structured datasets required for training AI models. Biological data often exhibits significant noise, variability, and inconsistencies, making it difficult to derive reliable predictive models. Furthermore, implementing AI in biofoundries requires real-time data acquisition, processing, and feedback loops. Developing robust AI frameworks that can dynamically adjust experimental parameters based on real-time results is an ongoing challenge. Additionally, ensuring that AI-driven decisions align with biological constraints and experimental feasibility is critical for successful implementation.

Another major bottleneck is the throughput variability across the DBTL cycle steps. Differences in the efficiency and duration of each phase can create imbalances, leading to bottlenecks and delays in workflow completion. For instance, while design and build phases may proceed rapidly with automated DNA synthesis and assembly, testing and learning steps often require complex biological assays that take days or weeks to yield meaningful data. Variability in cell growth rates, gene expression levels, and assay reproducibility can further complicate standardization across different DBTL iterations. Addressing these disparities requires improved automation of phenotypic characterization and data integration to ensure smooth progression through all phases.

The massive datasets generated by automated biofoundries present significant challenges in data management, storage, and analysis. High-throughput experiments produce vast amounts of sequencing data, metabolomic profiles, and imaging datasets that require efficient processing and interpretation. Data heterogeneity further complicates integration efforts, as different experiments generate diverse data types that must be harmonized for meaningful insights. Standardized ontologies, metadata frameworks, and cloud-based storage solutions are essential for managing large-scale biological datasets. Additionally, developing AI-driven data analytics pipelines will be key to extracting actionable insights, optimizing workflows, and improving system predictability. Also, high capital costs make it challenging for emerging economies to develop biofoundry infrastructures. Developing and maintaining automated biofoundries require substantial financial investment. High costs associated with robotic instrumentation, protocol implementation, AI-driven software development, and continuous system upgrades pose significant barriers to widespread adoption. While large-scale research institutions and companies can afford these investments, smaller laboratories and startups struggle to access automated biofoundry resources. Infrastructure constraints, including space requirements, power consumption, and maintenance demands, further complicate implementation. Traditional laboratories cannot support large-scale automation, necessitating extensive retrofitting or construction of dedicated facilities.

Reducing automation costs and developing compact, modular biofoundries is key to enabling broader accessibility.

The automation of synthetic biology workflows raises concerns regarding biosecurity and ethical implications. Fully automated biofoundries possess the capability to rapidly design and construct genetically modified organisms (GMOs), which could potentially be misused. Establishing biosecurity measures, e.g. access control, screening mechanisms, and regulatory compliance, is essential to mitigate potential risks. Ethical considerations also arise in the context of intellectual property and equitable access to biofoundry technologies. The increasing privatization of synthetic biology tools and automation technologies may create barriers for smaller research institutions and developing nations. Addressing these concerns requires international regulatory frameworks that promote responsible innovation and ensure equitable distribution of biofoundry capabilities.

13.5 Perspectives and Outlook

We have seen how biofoundries can increase the throughput of the DBTLc and assist in iterating through the four phases using automated setups. The challenges mentioned in the previous section can serve as a foundation for developing trajectories that the field may adopt to maximize the impact of these infrastructures, which is the focus of this section.

Regarding the issue of standardization, it is desirable to establish standardized procedures and data management strategies as early as possible within the biofoundry framework – ideally through a pilot campaign before the first official one begins. Standardization should involve all team members across the DBTLc to ensure that generated data is correctly stored and promptly processed by the team responsible for the next phase. Implementing this strategy will help maintain high standards and ensure compliance throughout the entire cycle. Additionally, if the biofoundry's purpose is to serve customers, it should ensure that generated and transferred data is discussed and clearly understood by the customer. The further development of universally accepted standards will enhance reproducibility, streamline automation, and facilitate seamless collaboration between different biofoundries. We believe that associations such as the GBA might favor while developing technical roadmaps for implementing these universally accepted standards.

Optimizing and streamlining procedures within a biofoundry will inevitably emphasize certain types of processes or applications. This trend is already partially evident in Table 13.2, where some biofoundries specialize in specific DBTL cycles. We believe that a well-functioning biofoundry should have a clear mission and focus on refining procedures that align with its objectives – rather than serving as a generalized high-throughput platform for a wide range of scopes and applications. That said, biofoundries are also highly flexible in adapting to new workflows. This adaptability was demonstrated by the temporary repurposing of London's biofoundry for SARS-CoV-2 test screening, showcasing how these facilities can rapidly pivot to address urgent scientific and societal needs [135].

One example of such applications could be biofoundries that specialize in improving the scalability of microbial cultures for large-scale cultivation. There is an inherent limitation in simulating such cultivations in small-scale setups as a biofoundry would offer. However, adopting batteries of miniaturized stirred bioreactors for scale-down experimentation can offer a solution for simulating large-scale fermentation [136], including the investigation of multiple parameters that can hinder the transition from bench-top to cubic-meter fermenters.

When dealing with the unpredictability of biological systems, we believe that by focusing on a specific purpose for the biofoundry (e.g. mammalian cells), it will be possible to learn how the biological samples running through the DBTLc behave. This approach will allow us to understand which variables most affect standardization, thereby reducing variability. In the case of using different species of microorganisms, one potential solution to reduce cross-contamination is to introduce kill-switch mechanisms, e.g. temperature-sensitive mechanisms and CRISPR-based strategies [44], in the strains. This would help prevent their persistence in machinery when those systems are used for other species.

ML can aid in managing the unpredictability of biological systems by employing neural networks to predict biological features. For instance, in protein engineering evolution, models are now capable of creating artificial proteins with desired functions by training the network against millions of sequences available in databases [137]. Moreover, AI is expected to contribute to coordinating dry lab and wet lab activities across the different phases of the DBTLc. Although we are not there yet, we are getting closer and closer [3].

When dealing with different throughput sizes across the DBTL modules, one option could be to simplify (as much as possible) the amount of data to process during the data-heavy test phase. For example, using growth-coupled designs and associated screening campaigns for modules linked to growth fitness can be a useful ally for testing metabolic activities [77, 138–142] or evolving enzymatic functions of interest [92, 143–146]. Additionally, biosensors that provide fluorescence signals in the presence of a molecule could become useful tools to reduce the burden of the test phase [83, 147–149].

Establishing biofoundries requires high investments in capital and operational costs. To promote the emergence of biofoundries beyond Europe, North America, Asia, and Oceania, early-stage biofoundries could benefit from adopting cheaper and simpler apparatuses for their operational units. For example, small-scale (<20 mL volume) mini-bioreactors that are open source allow customers to interact and program them as needed [150]. We believe that this type of solution offers a good trade-off between throughput, controlled conditions, and lower capital expenses.

We foresee that the implementation of one or more of these perspectives will likely be adopted when developing new biofoundries or improving pipelines in existing ones. These approaches will enable biofoundries to optimize their workflows, reduce costs, and enhance scalability, ultimately fostering the growth and success of synthetic biology across various industries.

Acknowledgments

The authors would like to acknowledge the work of many researchers in the field of automation who have made authoritative contributions to the field, the work of whom could not always be cited due to space constraints. This chapter includes key literature published up to March 2025. The Nikel Lab gratefully acknowledges financial support from The Novo Nordisk Foundation through grants NNF20CC0035580, LiFe (NNF18OC0034818), TARGET (NNF21OC0067996), FM·*Pseudomonas* (NNF24OC0091501), and NovoF (NNF23OC0083631), and the European Union's Horizon 2020 Research and Innovation Programme under grant agreements No. 814418 (SinFonia) and No. 101082049 (TOLERATE).

Ethics Declarations

Competing interests—The authors declare no competing interests.

References

1 Liu, Y. and Nielsen, J. (2019). Recent trends in metabolic engineering of microbial chemical factories. *Curr. Opin. Biotechnol.* 60: 188–197.

2 Petzold, C.J., Chan, L.J.G., Nhan, M., and Adams, P.D. (2015). Analytics for metabolic engineering. *Front. Bioeng. Biotechnol.* 3: 135.

3 Gurdo, N., Volke, D.C., McCloskey, D., and Nikel, P.I. (2023). Automating the design-build-test-learn cycle towards next-generation bacterial cell factories. *New Biotechnol.* 74: 1–15.

4 Gurdo, N., Volke, D.C., and Nikel, P.I. (2022). Merging automation and fundamental discovery into the design–build–test–learn cycle of nontraditional microbes. *Trends Biotechnol.* 40: 1148–1159.

5 Tarca, A.L., Carey, V.J., Chen, X.W. et al. (2007). Machine learning and its applications to biology. *PLoS Comput. Biol.* 3: e116.

6 Greener, J.G., Kandathil, S.M., Moffat, L., and Jones, D.T. (2022). A guide to machine learning for biologists. *Nat. Rev. Mol. Cell Biol.* 23: 40–55.

7 Zhou, J. and Huang, M. (2024). Navigating the landscape of enzyme design: from molecular simulations to machine learning. *Chem. Soc. Rev.* 53: 8202–8239.

8 Li, W., Mao, Z., Xiao, Z. et al. (2025). Large language model for knowledge synthesis and AI-enhanced biomanufacturing. *Trends Biotechnol.* 43: 1864–1875.

9 Liao, X., Ma, H., and Tang, Y.J. (2022). Artificial intelligence: a solution to involution of design-build-test-learn cycle. *Curr. Opin. Biotechnol.* 75: 102712.

10 Hillson, N., Caddick, M., Cai, Y. et al. (2019). Building a global alliance of biofoundries. *Nat. Commun.* 10: 2040.

11 Hołówko, M.B., Frow, E.K., Reid, J.C. et al. (2021). Building a biofoundry. *Synth. Biol.* 6: ysaa026.

12 Timmis, K., Karahan, Z.C., Ramos, J.L. et al. (2025). Microbes saving lives and reducing suffering. *Microb. Biotechnol.* 18: e70068.

13 Nielsen, J. and Keasling, J.D. (2016). Engineering cellular metabolism. *Cell* 164: 1185–1197.

14 Young, R., Haines, M., Storch, M., and Freemont, P.S. (2021). Combinatorial metabolic pathway assembly approaches and toolkits for modular assembly. *Metab. Eng.* 63: 81–101.

15 Casini, A., Storch, M., Baldwin, G.S., and Ellis, T. (2015). Bricks and blueprints: methods and standards for DNA assembly. *Nat. Rev. Mol. Cell Biol.* 16: 568–576.

16 Kozaeva, E., Nielsen, Z.S., Nieto-Domínguez, M., and Nikel, P.I. (2024). The pAblo·pCasso self-curing vector toolset for unconstrained cytidine and adenine base-editing in Gram-negative bacteria. *Nucleic Acids Res.* 52: e19.

17 Volke, D.C., Martino, R.A., Kozaeva, E. et al. (2022). Modular (de)construction of complex bacterial phenotypes by CRISPR/nCas9-assisted, multiplex cytidine base-editing. *Nat. Commun.* 13: 3026.

18 Wirth, N.T., Kozaeva, E., and Nikel, P.I. (2020). Accelerated genome engineering of *Pseudomonas putida* by I-*Sce*I—mediated recombination and CRISPR-Cas9 counterselection. *Microb. Biotechnol.* 13: 233–249.

19 Zeymer, C. and Hilvert, D. (2018). Directed evolution of protein catalysts. *Annu. Rev. Biochem.* 87: 131–157.

20 Wirth, N.T., Gurdo, N., Krink, N. et al. (2022). A synthetic C2 auxotroph of *Pseudomonas putida* for evolutionary engineering of alternative sugar catabolic routes. *Metab. Eng.* 74: 83–97.

21 Wirth, N.T. and Nikel, P.I. (2021). Combinatorial pathway balancing provides biosynthetic access to 2-fluoro-*cis,cis*-muconate in engineered *Pseudomonas putida*. *Chem Catal.* 1: 1234–1259.

22 Kozaeva, E., Volkova, S., Matos, M.R.A. et al. (2021). Model-guided dynamic control of essential metabolic nodes boosts acetyl-coenzyme A–dependent bioproduction in rewired *Pseudomonas putida*. *Metab. Eng.* 67: 373–386.

23 Neubauer, P., Cruz, N., Glauche, F. et al. (2013). Consistent development of bioprocesses from microliter cultures to the industrial scale. *Eng. Life Sci.* 13: 224–238.

24 Gong, Z., Chen, J., Jiao, X. et al. (2024). Genome-scale metabolic network models for industrial microorganisms metabolic engineering: current advances and future prospects. *Biotechnol. Adv.* 72: 108319.

25 Kanehisa, M., Furumichi, M., Sato, Y. et al. (2022). KEGG for taxonomy-based analysis of pathways and genomes. *Nucleic Acids Res.* 51: D587–D592.

26 Schomburg, I., Chang, A., and Schomburg, D. (2002). BRENDA, enzyme data and metabolic information. *Nucleic Acids Res.* 30: 47–49.

27 Karp, P.D., Riley, M., Paley, S.M., and Pellegrini-Toole, A. (2002). The MetaCyc Database. *Nucleic Acids Res.* 30: 59–61.

28 Burgard, A.P., Pharkya, P., and Maranas, C.D. (2003). OptKnock: a bilevel programming framework for identifying gene knockout strategies for microbial strain optimization. *Biotechnol. Bioeng.* 84: 647–657.

29 Hassani, L., Moosavi, M.R., Setoodeh, P., and Zare, H. (2024). FastKnock: an efficient next-generation approach to identify all knockout strategies for strain optimization. *Microb. Cell Factories* 23: 37.

30 Delépine, B., Duigou, T., Carbonell, P., and Faulon, J.L. (2018). RetroPath2.0: a retrosynthesis workflow for metabolic engineers. *Metab. Eng.* 45: 158–170.

31 Stoney, R.A., Hanko, E.K.R., Carbonell, P., and Breitling, R. (2023). SelenzymeRF: updated enzyme suggestion software for unbalanced biochemical reactions. *Comput. Struct. Biotechnol. J.* 21: 5868–5876.

32 Hon, J., Borko, S., Stourac, J. et al. (2020). *EnzymeMiner*: Automated mining of soluble enzymes with diverse structures, catalytic properties and stabilities. *Nucleic Acids Res.* 48: W104–W109.

33 Anderson, J.C., Dueber, J.E., Leguia, M. et al. (2010). *BglBricks*: a flexible standard for biological part assembly. *J. Biol. Eng.* 4: 1.

34 Ellis, T., Adie, T., and Baldwin, G.S. (2011). DNA assembly for synthetic biology: from parts to pathways and beyond. *Integr. Biol.* 3: 109–118.

35 Martínez-García, E., Fraile, S., Algar, E. et al. (2023). SEVA 4.0: an update of the standard European vector architecture database for advanced analysis and programming of bacterial phenotypes. *Nucleic Acids Res.* 51: D1558–D1567.

36 Silva-Rocha, R., Martínez-García, E., Calles, B. et al. (2013). The standard European vector architecture (SEVA): a coherent platform for the analysis and deployment of complex prokaryotic phenotypes. *Nucleic Acids Res.* 41: D666–D675.

37 Volke, D.C., Calero, P., and Nikel, P.I. (2020). *Pseudomonas putida*. *Trends Microbiol.* 28: 512–513.

38 Espah Borujeni, A. and Salis, H.M. (2016). Translation initiation is controlled by RNA folding kinetics *via* a ribosome drafting mechanism. *J. Am. Chem. Soc.* 138: 7016–7023.

39 Reis, A.C. and Salis, H.M. (2020). An automated model test system for systematic development and improvement of gene expression models. *ACS Synth. Biol.* 9: 3145–3156.

40 Gibson, D.G., Young, L., Chuang, R.Y. et al. (2009). Enzymatic assembly of DNA molecules up to several hundred kilobases. *Nat. Methods* 6: 343–345.

41 Engler, C., Kandzia, R., and Marillonnet, S. (2008). A one pot, one step, precision cloning method with high throughput capability. *PLoS One* 3: e3647.

42 Geu-Flores, F., Nour-Eldin, H.H., Nielsen, M.T., and Halkier, B.A. (2007). *USER fusion*: a rapid and efficient method for simultaneous fusion and cloning of multiple PCR products. *Nucleic Acids Res.* 35: e55.

43 Donohoue, P.D., Barrangou, R., and May, A.P. (2018). Advances in industrial biotechnology using CRISPR-Cas systems. *Trends Biotechnol.* 36: 134–146.

44 Krink, N., Nikel, P.I., and Beisel, C.L. (2024). A Hitchhiker's guide to CRISPR editing tools in bacteria. *EMBO Rep.* 25: 1694–1699.

45 de Lorenzo, V., Pérez-Pantoja, D., and Nikel, P.I. (2024). *Pseudomonas putida* KT2440: the long journey of a soil-dweller to become a synthetic biology *chassis*. *J. Bacteriol.* 206: e00136–e00124.

46 Volke, D.C., Orsi, E., and Nikel, P.I. (2023). Emergent CRISPR-Cas-based technologies for engineering non-model bacteria. *Curr. Opin. Microbiol.* 75: 102353.

47 Lu, L., Shen, X., Sun, X. et al. (2022). CRISPR-based metabolic engineering in non-model microorganisms. *Curr. Opin. Biotechnol.* 75: 102698.

48 Favoino, G., Krink, N., Schwanemann, T. et al. (2024). Enhanced biosynthesis of poly(3-hydroxybutyrate) in engineered strains of *Pseudomonas putida via* increased malonyl-CoA availability. *Microb. Biotechnol.* 17: e70044.

49 Konzock, O. and Nielsen, J. (2024). TRYing to evaluate production costs in microbial biotechnology. *Trends Biotechnol.* 42: 1339–1347.

50 Amer, B. and Baidoo, E.E.K. (2021). Omics-driven biotechnology for industrial applications. *Front. Bioeng. Biotechnol.* 9: 613307.

51 Davy, A.M., Kildegaard, H.F., and Andersen, M.R. (2017). Cell factory engineering. *Cell Syst.* 4: 262–275.

52 Turlin, J., Puiggené, Ò., Donati, S. et al. (2023). Core and auxiliary functions of one-carbon metabolism in *Pseudomonas putida* exposed by a systems-level analysis of transcriptional and physiological responses. *mSystems.* 8: e00004–e00023.

53 Becker, J. and Wittmann, C. (2018). From systems biology to metabolically engineered cells—an omics perspective on the development of industrial microbes. *Curr. Opin. Microbiol.* 45: 180–188.

54 Volke, D.C., Gurdo, N., Milanesi, R., and Nikel, P.I. (2023). Time-resolved, deuterium-based fluxomics uncovers the hierarchy and dynamics of sugar processing by *Pseudomonas putida*. *Metab. Eng.* 79: 159–172.

55 Danchin, A. and Nikel, P.I. (2019). Why nature chose potassium. *J. Mol. Evol.* 87: 271–288.

56 Nielsen, A.A., Der, B.S., Shin, J. et al. (2016). Genetic circuit design automation. *Science* 352: aac7341.

57 Jin, S., Zeng, X., Xia, F. et al. (2021). Application of deep learning methods in biological networks. *Brief. Bioinform.* 22: 1902–1917.

58 Lawson, C.E., Martí, J.M., Radivojević, T. et al. (2021). Machine learning for metabolic engineering: a review. *Metab. Eng.* 63: 34–60.

59 Martin, V.J., Pitera, D.J., Withers, S.T. et al. (2003). Engineering a mevalonate pathway in *Escherichia coli* for production of terpenoids. *Nat. Biotechnol.* 21: 796–802.

60 Paddon, C.J., Westfall, P.J., Pitera, D.J. et al. (2013). High-level semi-synthetic production of the potent antimalarial artemisinin. *Nature* 496: 528–532.

61 Paddon, C.J. and Keasling, J.D. (2014). Semi-synthetic artemisinin: a model for the use of synthetic biology in pharmaceutical development. *Nat. Rev. Microbiol.* 12: 355–367.

62 Yim, H., Haselbeck, R., Niu, W. et al. (2011). Metabolic engineering of *Escherichia coli* for direct production of 1,4-butanediol. *Nat. Chem. Biol.* 7: 445–452.

63 Ahn, J.H., Seo, H., Park, W. et al. (2020). Enhanced succinic acid production by *Mannheimia* employing optimal malate dehydrogenase. *Nat. Commun.* 11: 1970.

64 Weusthuis, R.A., Mars, A.E., Springer, J. et al. (2017). *Monascus ruber* as cell factory for lactic acid production at low pH. *Metab. Eng.* 42: 66–73.

65 Becker, J., Zelder, O., Häfner, S. et al. (2011). From zero to hero—design-based systems metabolic engineering of *Corynebacterium glutamicum* for L-lysine production. *Metab. Eng.* 13: 159–168.

66 Han, T., Kim, G.B., and Lee, S.Y. (2020). Glutaric acid production by systems metabolic engineering of an L-lysine-overproducing *Corynebacterium glutamicum*. *Proc. Natl. Acad. Sci. USA* 117: 30328–30334.

67 Whitford, C.M., Cruz-Morales, P., Keasling, J.D., and Weber, T. (2021). The Design-Build-Test-Learn cycle for metabolic engineering of *Streptomycetes*. *Essays Biochem.* 65: 261–275.

68 Orsi, E., Beekwilder, J., Eggink, G. et al. (2021). The transition of *Rhodobacter sphaeroides* into a microbial cell factory. *Biotechnol. Bioeng.* 118: 531–541.

69 Weldon, M. and Euler, C. (2025). Physiology-informed use of *Cupriavidus necator* in biomanufacturing: a review of advances and challenges. *Microb. Cell Factories* 24: 30.

70 Zha, J., Zhao, Z., Xiao, Z. et al. (2023). Biosystem design of *Corynebacterium glutamicum* for bioproduction. *Curr. Opin. Biotechnol.* 79: 102870.

71 Herrmann, L.W., Letti, L.A.J., Penha, R.O. et al. (2024). *Bacillus* genus industrial applications and innovation: first steps towards a circular bioeconomy. *Biotechnol. Adv.* 70: 108300.

72 Charubin, K., Bennett, R.K., Fast, A.G., and Papoutsakis, E.T. (2018). Engineering *Clostridium* organisms as microbial cell-factories: challenges and opportunities. *Metab. Eng.* 50: 173–191.

73 Ye, J.W., Lin, Y.N., Yi, X.Q. et al. (2023). Synthetic biology of extremophiles: a new wave of biomanufacturing. *Trends Biotechnol.* 41: 342–357.

74 Cairns, T.C., Nai, C., and Meyer, V. (2018). How a fungus shapes biotechnology: 100 years of *Aspergillus niger* research. *Fungal Biol. Biotechnol.* 5: 13.

75 Calero, P. and Nikel, P.I. (2019). Chasing bacterial *chassis* for metabolic engineering: a perspective review from classical to non-traditional microorganisms. *Microb. Biotechnol.* 12: 98–124.

76 Medema, M.H., van Raaphorst, R., Takano, E., and Breitling, R. (2012). Computational tools for the synthetic design of biochemical pathways. *Nat. Rev. Microbiol.* 10: 191–202.

77 Orsi, E., Schulz-Mirbach, H., Cotton, C.A.R. et al. (2025). Computation-aided designs enable developing auxotrophic metabolic sensors for wide-range glyoxylate and glycolate detection. *Nat. Commun.* 16: 2168.

78 Torres-Acosta, M.A., Lye, G.J., and Dikicioglu, D. (2022). Automated liquid-handling operations for robust, resilient, and efficient bio-based laboratory practices. *Biochem. Eng. J.* 188: 108713.

79 Volke, D.C., Friis, L., Wirth, N.T. et al. (2020). Synthetic control of plasmid replication enables target- and self-curing of vectors and expedites genome engineering of *Pseudomonas putida*. *Metab. Eng. Commun.* 10: e00126.

80 Volke, D.C., Turlin, J., Mol, V., and Nikel, P.I. (2020). Physical decoupling of XylS/*Pm* regulatory elements and conditional proteolysis enable precise control of gene expression in *Pseudomonas putida*. *Microb. Biotechnol.* 13: 222–232.

81 Federici, F., Luppino, F., Aguilar-Vilar, C. et al. (2025). CIFR (*Clone-Integrate-Flip-out-Repeat*): a toolset for iterative genome and pathway engineering of Gram-negative bacteria. *Metab. Eng.* 88: 180–195.

82 Fernández-Cabezón, L., Cros, A., and Nikel, P.I. (2021). Spatiotemporal manipulation of the mismatch repair system of *Pseudomonas putida* accelerates phenotype emergence. *ACS Synth. Biol.* 10: 1214–1226.

83 Hernández-Sancho, J.M., Boudigou, A., Alván-Vargas, M.V.G. et al. (2024). A versatile microbial platform as a tunable whole-cell chemical sensor. *Nat. Commun.* 15: 8316.

84 Wirth, N.T., Rohr, K., Danchin, A., and Nikel, P.I. (2023). Recursive genome engineering decodes the evolutionary origin of an essential thymidylate kinase activity in *Pseudomonas putida* KT2440. *MBio* 14: e01081–e01023.

85 Spangler, J.R., Leski, T.A., Schultzhaus, Z. et al. (2022). Large scale screening of CRISPR guide RNAs using an optimized high throughput robotics system. *Sci. Rep.* 12: 13953.

86 Iwai, K., Wehrs, M., Garber, M. et al. (2022). Scalable and automated CRISPR-based strain engineering using droplet microfluidics. *Microsyst. Nanoeng.* 8: 31.

87 Batianis, C., Kozaeva, E., Damalas, S.G. et al. (2020). An expanded CRISPRi toolbox for tunable control of gene expression in *Pseudomonas putida*. *Microb. Biotechnol.* 13: 368–385.

88 Wannier, T.M., Ciaccia, P.N., Ellington, A.D. et al. (2021). Recombineering and MAGE. *Nat. Rev. Methods Primers.* 1: 7.

89 Wang, Y., Xue, P., Cao, M. et al. (2021). Directed evolution: methodologies and applications. *Chem. Rev.* 121: 12384–12444.

90 Fernández-Cabezón, L., Cros, A., and Nikel, P.I. (2019). Evolutionary approaches for engineering industrially-relevant phenotypes in bacterial cell factories. *Biotechnol. J.* 14: 1800439.

91 Turlin, J., Dronsella, B., De Maria, A. et al. (2022). Integrated rational and evolutionary engineering of genome-reduced *Pseudomonas putida* strains promotes synthetic formate assimilation. *Metab. Eng.* 74: 191–205.

92 Orsi, E., Schada von Borzyskowski, L., Noack, S. et al. (2024). Automated *in vivo* enzyme engineering accelerates biocatalyst optimization. *Nat. Commun.* 15: 3447.

93 Molina, R.S., Rix, G., Mengiste, A.A. et al. (2022). *In vivo* hypermutation and continuous evolution. *Nat. Rev. Methods Primers.* 2: 36.

94 Sarnaik, A., Liu, A., Nielsen, D., and Varman, A.M. (2020). High-throughput screening for efficient microbial biotechnology. *Curr. Opin. Biotechnol.* 64: 141–150.

95 Sánchez-Pascuala, A., de Lorenzo, V., and Nikel, P.I. (2017). Refactoring the Embden-Meyerhof-Parnas pathway as a whole of portable *GlucoBricks* for implantation of glycolytic modules in Gram-negative bacteria. *ACS Synth. Biol.* 6: 793–805.

96 Bozkurt, E.U., Ørsted, E.C., Volke, D.C., and Nikel, P.I. (2025). *Accelerating Enzyme Discovery and Engineering with High-throughput Screening*. Royal Society of Chemistry.

97 Hemmerich, J., Noack, S., Wiechert, W., and Oldiges, M. (2018). Microbioreactor systems for accelerated bioprocess development. *Biotechnol. J.* 13: e1700141.

98 Junker, B.H. and Wang, H.Y. (2006). Bioprocess monitoring and computer control: key roots of the current PAT initiative. *Biotechnol. Bioeng.* 95: 226–261.

99 Wirth, N.T., Funk, J., Donati, S., and Nikel, P.I. (2023). *QurvE*: user-friendly software for the analysis of biological growth and fluorescence data. *Nat. Protoc.* 18: 2401–2403.

100 Schaub, J., Clemens, C., Kaufmann, H., and Schulz, T.W. (2012). Advancing biopharmaceutical process development by system-level data analysis and integration of omics data. *Adv. Biochem. Eng. Biotechnol.* 127: 133–163.

101 Helleckes, L.M., Hemmerich, J., Wiechert, W. et al. (2023). Machine learning in bioprocess development: from promise to practice. *Trends Biotechnol.* 41: 817–835.

102 Patra, P., Disha, B.R., Kundu, P. et al. (2023). Recent advances in machine learning applications in metabolic engineering. *Biotechnol. Adv.* 62: 108069.

103 Calero, P., Volke, D.C., Lowe, P.T. et al. (2020). A fluoride-responsive genetic circuit enables *in vivo* biofluorination in engineered *Pseudomonas putida*. *Nat. Commun.* 11: 5045.

104 Pardo, I., Bednar, D., Calero, P. et al. (2022). A nonconventional Archaeal fluorinase identified by *in silico* mining for enhanced fluorine biocatalysis. *ACS Catal.* 12: 6570–6577.

105 Carbonell, P., Radivojevic, T., and García Martín, H. (2019). Opportunities at the intersection of synthetic biology, machine learning, and automation. *ACS Synth. Biol.* 8: 1474–1477.

106 Pedersen, H.K., Forslund, S.K., Gudmundsdottir, V. et al. (2018). A computational framework to integrate high-throughput '-omics' datasets for the identification of potential mechanistic links. *Nat. Protoc.* 13: 2781–2800.

107 Haas, R. and Nikel, P.I. (2023). Challenges and opportunities in bringing nonbiological atoms to life with synthetic metabolism. *Trends Biotechnol.* 41: 27–45.

108 Arce-Rodríguez, A., Volke, D.C., Bense, S. et al. (2019). Non-invasive, ratiometric determination of intracellular pH in *Pseudomonas* species using a novel genetically encoded indicator. *Microb. Biotechnol.* 12: 799–813.

109 Volk, M.J., Lourentzou, I., Mishra, S. et al. (2020). Biosystems design by machine learning. *ACS Synth. Biol.* 9: 1514–1533.

110 Carbonell, P., Le Feuvre, R., Takano, E., and Scrutton, N.S. (2020). *In silico* design and automated learning to boost next-generation smart biomanufacturing. *Synth. Biol.* 5: ysaa020.

111 Cros, A., Alfaro-Espinoza, G., de Maria, A. et al. (2022). Synthetic metabolism for biohalogenation. *Curr. Opin. Biotechnol.* 74: 180–193.

112 Hoose, A., Vellacott, R., Storch, M. et al. (2023). DNA synthesis technologies to close the gene writing gap. *Nat. Rev. Chem.* 7: 144–161.

113 Hughes, R.A. and Ellington, A.D. (2017). Synthetic DNA synthesis and assembly: putting the synthetic in synthetic biology. *Cold Spring Harb. Perspect. Biol.* 9: a023812.

114 Hess, J.F., Kohl, T.A., Kotrová, M. et al. (2020). Library preparation for next generation sequencing: a review of automation strategies. *Biotechnol. Adv.* 41: 107537.

115 Ortiz, L., Paván, M., McCarthy, L. et al. (2017). Automated robotic liquid handling assembly of modular DNA devices. *J. Vis. Exp.* 130: e54703.

116 Kang, D.H., Ko, S.C., Heo, Y.B. et al. (2022). RoboMoClo: a robotics-assisted modular cloning framework for multiple gene assembly in biofoundry. *ACS Synth. Biol.* 11: 1336–1348.

117 Aparicio, T., Jensen, S.I., Nielsen, A.T. et al. (2016). The Ssr protein (*T1E_1405*) from *Pseudomonas putida* DOT-T1E enables oligonucleotide-based recombineering in platform strain *P. putida* EM42. *Biotechnol. J.* 11: 1309–1319.

118 Gloeckner, C., Kranaster, R., and Marx, A. (2010). Directed evolution of DNA polymerases: construction and screening of DNA polymerase mutant libraries. *Curr. Protoc. Chem. Biol.* 2: 89–109.

119 Rix, G. and Liu, C.C. (2021). Systems for *in vivo* hypermutation: a quest for scale and depth in directed evolution. *Curr. Opin. Chem. Biol.* 64: 20–26.

120 Coco, W.M., Levinson, W.E., Crist, M.J. et al. (2001). DNA shuffling method for generating highly recombined genes and evolved enzymes. *Nat. Biotechnol.* 19: 354–359.

121 Cong, L., Ran, F.A., Cox, D. et al. (2013). Multiplex genome engineering using CRISPR/Cas systems. *Science* 339: 819–823.

122 Zhou, Y., Zhu, S., Cai, C. et al. (2014). High-throughput screening of a CRISPR/Cas9 library for functional genomics in human cells. *Nature* 509: 487–491.

123 Zhong, Z., Wong, B.G., Ravikumar, A. et al. (2020). Automated continuous evolution of proteins *in vivo*. *ACS Synth. Biol.* 9: 1270–1276.

124 Wu, Z., Kan, S.B.J., Lewis, R.D. et al. (2019). Machine learning-assisted directed protein evolution with combinatorial libraries. *Proc. Natl. Acad. Sci. USA* 116: 8852–8858.

125 Rosch, T.M., Tenhaef, J., Stoltmann, T. et al. (2024). *AutoBioTech*—a versatile biofoundry for automated strain engineering. *ACS Synth. Biol.* 13: 2227–2237.

126 Robinson, C.J., Carbonell, P., Jervis, A.J. et al. (2020). Rapid prototyping of microbial production strains for the biomanufacture of potential materials monomers. *Metab. Eng.* 60: 168–182.

127 Zhang, J., Chen, Y., Fu, L. et al. (2021). Accelerating strain engineering in biofuel research via build and test automation of synthetic biology. *Curr. Opin. Biotechnol.* 67: 88–98.

128 Krausch, N., Kim, J.W., Lucia, S. et al. (2022). Optimal operation of parallel mini-bioreactors in bioprocess development using multi-stage MPC. In: *Computer Aided Chemical Engineering* (ed. L. Montastruc and S. Negny), 1069–1074. Elsevier.

129 Kaspersetz, L., Schröder-Kleeberg, F., Mione, F.M. et al. (2024). Automation of experimental workflows for high throughput robotic cultivations. In: *Computer Aided Chemical Engineering* (ed. F. Manenti and G.V. Reklaitis), 2971–2976. Elsevier.

130 Kim, J.W., Krausch, N., Aizpuru, J. et al. (2023). Model predictive control and moving horizon estimation for adaptive optimal bolus feeding in high-throughput cultivation of *E. coli*. *Comput. Chem. Eng.* 172: 108158.

131 Kaspersetz, L., Waldburger, S., Schermeyer, M.T. et al. (2022). Automated bioprocess feedback operation in a high-throughput facility *via* the integration of a mobile robotic lab assistant. *Front. Chem. Eng.* 4: 812140.

132 Carbonell, P., Jervis, A.J., Robinson, C.J. et al. (2018). An automated Design-Build-Test-Learn pipeline for enhanced microbial production of fine chemicals. *Commun. Biol.* 1: 66.

133 Coley, C.W., Thomas, D.A., Lummiss, J.A.M. et al. (2019). A robotic platform for flow synthesis of organic compounds informed by AI planning. *Science* 365: eaax1566.

134 Farzaneh, T. and Freemont, P.S. (2021). Biofoundries are a nucleating hub for industrial translation. *Synth. Biol.* 6: ysab013.

135 Crone, M.A., Priestman, M., Ciechonska, M. et al. (2020). A role for biofoundries in rapid development and validation of automated SARS-CoV-2 clinical diagnostics. *Nat. Commun.* 11: 4464.

136 Tajsoleiman, T., Mears, L., Krühne, U. et al. (2019). An industrial perspective on scale-down challenges using miniaturized bioreactors. *Trends Biotechnol.* 37: 697–706.

137 Madani, A., Krause, B., Greene, E.R. et al. (2023). Large language models generate functional protein sequences across diverse families. *Nat. Biotechnol.* 41: 1099–1106.

138 Orsi, E., Claassens, N.J., Nikel, P.I., and Lindner, S.N. (2021). Growth-coupled selection of synthetic modules to accelerate cell factory development. *Nat. Commun.* 12: 5295.

139 Wenk, S., Schann, K., He, H. et al. (2020). An "energy-auxotroph" *Escherichia coli* provides an *in vivo* platform for assessing NADH regeneration systems. *Biotechnol. Bioeng.* 117: 3422–3434.

140 Aslan, S., Noor, E., Benito Vaquerizo, S. et al. (2020). Design and engineering of *E. coli* metabolic sensor strains with a wide sensitivity range for glycerate. *Metab. Eng.* 57: 96–109.

141 Satanowski, A., Dronsella, B., Noor, E. et al. (2020). Awakening a latent carbon fixation cycle in *Escherichia coli*. *Nat. Commun.* 11: 5812.

142 Wu, T., Gómez-Coronado, P.A., Kubis, A. et al. (2023). Engineering a synthetic energy-efficient formaldehyde assimilation cycle in *Escherichia coli*. *Nat. Commun.* 14: 8490.

143 Li, Z., Deng, Y., and Yang, G.Y. (2023). Growth-coupled high throughput selection for directed enzyme evolution. *Biotechnol. Adv.* 68: 108238.

144 Chen, J., Wang, Y., Zheng, P., and Sun, J. (2022). Engineering synthetic auxotrophs for growth-coupled directed protein evolution. *Trends Biotechnol.* 40: 773–776.

145 Nielsen, J.R., Weusthuis, R.A., and Huang, W.E. (2023). Growth-coupled enzyme engineering through manipulation of redox cofactor regeneration. *Biotechnol. Adv.* 63: 108102.

146 He, H., Gómez-Coronado, P.A., Zarzycki, J. et al. (2024). Adaptive laboratory evolution recruits the promiscuity of succinate semialdehyde dehydrogenase to repair different metabolic deficiencies. *Nat. Commun.* 15: 8898.

147 De Paepe, B., Peters, G., Coussement, P. et al. (2017). Tailor-made transcriptional biosensors for optimizing microbial cell factories. *J. Ind. Microbiol. Biotechnol.* 44: 623–645.

148 Chacón, M. and Dixon, N. (2024). Genetically encoded biosensors for the circular plastics bioeconomy. *Metab. Eng. Commun.* 19: e00255.

149 Bowman, E.K., Wagner, J.M., Yuan, S.F. et al. (2021). Sorting for secreted molecule production using a biosensor-in-microdroplet approach. *Proc. Natl. Acad. Sci.* 118: e2106818118.

150 Steel, H., Habgood, R., Kelly, C.L., and Papachristodoulou, A. (2020). *In situ* characterisation and manipulation of biological systems with Chi. *Bio. PLoS Biol.* 18: e3000794.

14

Machine Learning-Enhanced Hybrid Modeling for Phenotype Prediction and Bioreactor Optimization

Oliver Pennington[1], Yirong Chen[2], Youping Xie[2], and Dongda Zhang[1]

[1] *The University of Manchester, Department of Chemical Engineering, Oxford Road, Manchester, UK*
[2] *Fuzhou University, College of Biological Science and Engineering, North Wulongjiang Avenue, Fuzhou, China*

14.1 Bioprocess Modeling and Optimization

14.1.1 Challenges in Bioprocess Modeling

The fourth industrial revolution, often referred to as "Industry 4.0," requires the integration of smart technologies into manufacturing and industrial processes, playing a key role in the improvement of bioprocess efficiency, sustainability, and profitability [1, 2]. Due to the desire to further explore the potential of bioprocess automation, digitalization is currently at the forefront of bioprocess development [3]. Digitalization itself relies on the development of digital twins, a concept first proposed in 2002, in which bidirectional information transfer between the physical and digital domains occurs automatically. An effective digital twin facilitates real-time decision-making to maximize productivity, efficiency, and sustainability [4]. Modeling plays a key role in the development of a digital twin by simulating the physical domain in real-time, as well as potentially being used to predict unobservable states using soft sensors [5–7]. The main challenge associated with the development of a bioprocess digital twin is having a sufficiently accurate and validated mathematical model of the desired bioprocess to make rapid predictions and process adjustments [6], which is why official records show the first implementation of a digital twin within an industrial-scale bioprocess to be as recent as 2019 [8].

Developing models for fermentation bioprocesses is immensely appealing due to their various potential applications in frameworks such as process control and optimization, design of experiments (DoEs), and robust Quality by Design (QbD) for the ever-expanding production of biofuels, biochemicals, and biopharmaceuticals. In an ideal world, a fully descriptive mechanistic model could be constructed and applied to a bioprocess. However, designing such models and identifying their parameters to an adequate degree is highly challenging in bioprocessing; a fully descriptive bioprocess model would need to include various considerations such as self-inhibitory and self-catalytic feedback interactions, nonhomogeneous cell

Machine Learning and Big Data-enabled Biotechnology, First Edition. Edited by Hal S. Alper.

populations and phases, pH and temperature dependence, and even mass transfer limitations. As a result, simple models, such as first-order kinetics (which are commonly seen in chemical reaction systems [9]), are generally only applicable in narrow operating spaces. Therefore, when constructing a predictive bioprocess dynamic model, a balance must be found between the incorporation of sufficient complexity and the identifiability of the model parameters.

Bioprocess models found in literature can generally be categorized somewhere on the spectrum between machine learning ("data-driven" or "black box") and mechanistic ("kinetic" or "white box") [10]. Due to the aforementioned challenges associated with constructing a detailed mechanistic model, similar to those described in Chapters 1–2, 9, and 11–12, a sensible approach can be to build a simplified mechanistic model based on fundamental assumptions, often referred to as "first principles" [11]. Such models contain physically interpretable parameters for the system at hand, providing a degree of process insight. However, when assumptions are made for model derivation, the model itself will only be valid within the operating space where the assumptions hold true, meaning that its generalizability is lost. An example of a relatively simple kinetic model is the Monod model, which can fit some batch processes well; however, it ignores the potential inhibitory effects of high cell density (HCD) and high substrate concentrations [12]. Improving generalizability can be achieved by constructing a more detailed mechanistic model by introducing further process understanding; however, this will require more data, especially if additional state and control variables are introduced [10]. Moreover, adding further detail to a kinetic model without adequate hypothesis testing introduces inductive bias, which can actually reduce predictive capabilities of the model if the additional "knowledge" is not reflective of the system. In addition, increasing the intricacy of a kinetic model can increase the number of model parameters, thus reducing parameter identifiability and increasing model uncertainty [13].

On the other end of the spectrum, machine learning models have no prerequisite assumptions and can have strong generalization capabilities. For example, artificial neural networks (ANNs) are known to be able to approximate any function [14]. Employing pure machine learning models may sound appealing; however, large quantities of data are required to sufficiently capture the potentially nonlinear relationships, such as the relationships between substrates and cell growth, that are known to exist in biosystems. In addition, experiments are expensive to run, with some states being challenging to measure, leading to bioprocess data typically being multisource, multi-scale, noisy, and heterogeneous (a mixture of online and offline, and binary and continuous) [15]. Therefore, it is uncommon to have access to an adequate availability of data to train particularly nonlinear machine learning models for biosystems while avoiding overfitting. Moreover, machine learning models provide limited insight since the model structure itself is not physically interpretable, which is reflected by the poor extrapolative capabilities of machine learning models [10].

Irrespective of the model built, challenges such as batch-to-batch variation can also lead to issues with model training and validation. In addition, models can fall short during upscaling due to various phenomena, including differing flow regimes and more prevalent localized nutrient and cell population heterogeneity. As a result, upscaling often observes different process dynamics that result in lower yields and productivity or even total failure of the culture. Therefore, it is important to develop bioprocess models with the strongest possible extrapolative capabilities and generalizability to maximize both the accuracy of predictions and the ability to incorporate new information. As such, navigating the hurdles of developing a state-of-the-art bioprocess model is no simple task.

In this chapter, we will explore the use of machine learning to enhance our understanding of bioprocesses for modeling and optimization of fermentation processes, while making the most of the limited available experimental data and fundamental process knowledge.

14.1.2 Machine Learning for Bioprocess Modeling

Machine learning is a powerful tool for tackling a lack of process knowledge and high system complexity, which is often the case when assessing biosystems. The ability to handle intricate and poorly understood systems has encouraged the application of machine learning to a multitude of bioprocesses using a variety of techniques. In this section, we will explore a brief overview of different machine learning applications, both recent and not, in biosystems.

14.1.2.1 Principal Component Analysis and Partial Least Squares

Principal component analysis (PCA) is a linear dimensionality reduction technique that generates latent variables, called principal components, to explain the most variance in the original variables as possible [16]. PCA has been applied to several biochemical applications, including the identification of patterns between operating factors for biogas and biofuel production [17, 18], cell phenotype discrimination [19], process abnormality detection [20], and determining statistical similarities and differences between product distributions [21]. Due to the innate complexity of biosystems, dimensionality reduction can be particularly useful for problem simplification; a study combining PCA with an ANN for the productive capacity of *Eucalyptus* forest sites in Brazil showed that using the latent variables improved the ANN predictive capabilities [22]. Naturally, the biggest limitation of applying PCA to model bioprocesses is the inability to capture nonlinear relationships that can be observed within some biosystems due to the linear nature of PCA.

Partial Least Squares (PLS) is similar in many ways to PCA; however, instead of only using latent variables, it directly relates inputs to outputs in a linear fashion. For Multiple Input Multiple Output (MIMO) models, PLS can be directly used; however, in some cases, multiple models are used in parallel for each individual output in a Multiple Input Single Output (MISO) format [23]. In terms of biosystem applications, PLS has seen multiple applications, including classification and authentication of operating conditions in a yeast fermentation bioreactor [24], bioprocess monitoring [25], bioreactor state estimation and control [26, 27], mapping amino acid sequences with protein functions [28], and identifying flux relationships [29]. Similar to PCA, PLS is limited to applications where nonlinear relationships do not need to be fully captured.

14.1.2.2 Artificial Neural Networks

ANNs can capture complex nonlinear relationships and generally come in the form of an MIMO or MISO model. The inspiration for ANNs came from mimicking the human brain, where a network of neurons can be "activated" to pass on information [14]. Due to the ability of an ANN to capture nonlinear relationships, ANNs have seen a variety of applications in biosystems. For example, previous studies have utilized the potential of ANNs for predicting methane content and biogas production [30], predicting fermentation dynamics online [31], recognizing appropriate kinetic model structures for batch processing [32], optimizing the production of algal biofuel [33], predicting protein structures [34], and classifying the three-dimensional (3D) structures of enzymes [35]. ANNs are a powerful tool; however, the most prominent limitation in the application of ANNs to biosystems is the large amount of data that is required to capture complex nonlinear relationships. Such quantities of data are, at best, time-consuming or expensive to obtain and can be, at worst, infeasible to practically obtain.

However, in certain cases, the fundamental ANN structure can be expanded to enhance its ability to capture different relationships. For example, deep neural networks (DNNs) are a subcategory of ANNs that possess multiple hidden layers of nodes, allowing the approximation of more complex relationships. Successful incorporation of DNNs to biosystems in literature includes the construction of a hybrid model using a DNN for industry-scale fermentation [36], as well as the detection of "optimum tea fermentation," where the tea has fermented enough to reach a desired quality but is yet to degrade due to over-fermentation [37]. However, since DNNs have multiple layers, they tend to have a higher number of parameters, which means they need even more training data than a shallow ANN to avoid overfitting. Recurrent neural networks (RNNs) are another type of ANN that have been applied to biochemical processes. What makes them "recurrent" compared to a feed-forward neural network is the addition of an internal cycle; possessing this internal memory facilitates the learning of sequential data, such as time-series [38]. The capability of RNNs to identify sequential relationships has been harnessed in various biosystem applications, such as statistically summarizing amino acid sequences [39], and recognizing promoter sequences in genetic engineering [40]. In addition, an RNN demonstrated improved fitting and predictive capabilities when compared to a feed-forward neural network in monitoring simultaneous saccharification and fermentation of ethanol [41]. As with most ANNs, RNNs are not always applicable due to a lack of data availability for biosystems.

14.1.2.3 Gaussian Processes

Gaussian processes (GPs) return a probabilistic function, instead of a scalar, for each output variable from a given multivariate input. If no noise is assumed, a GP will fit exactly to the dataset [42]. GPs are particularly useful for uncertainty analysis as they can directly estimate the mean and variance of the output, unlike other machine learning methods. Several applications of GPs to biosystems have been presented in literature, including dynamic modeling and optimization of microalgal lutein production under uncertainty [43], ensuring the meeting of probabilistic constraints in bioreactor control [44], and estimating the cell density distribution in δ-endotoxins batch production from *Bacillus thuringiensis* [45]. A notable limitation of GPs is the high computational expense of calculating the inverse of the posterior, which scales poorly with higher-dimensional problems [46]. However, studies are being conducted on approximating symmetric positive definite matrix inversions under reduced computational cost, which extends the applicability of GPs [47].

14.1.2.4 Ensemble Learning

Ensemble learning (EL) combines several inducers (machine learning models that relate an input to an output) to generate a single output based on the outputs of each individual inducer. EL structures usually involve either a series structure, such as gradient boosting (GB), where the first inducer provides an initial prediction that is corrected by subsequent inducers, or a parallel structure, such as Random Forest (RF), where inducers in parallel have their predictions averaged or weighted to give a single output [48]. With RFs being the more widely applied EL approach to biosystem modeling [23], application examples include the optimization of fermentation for Rifamycin B production [49], building soft-sensors for industrial-scale fed-batch fermentation [50], and estimating the cetane number of fatty acid methyl esters (FAMEs) for biodiesel synthesis [51]. Although less utilized, GB has still been applied to cases such as liquid–liquid bioseparations [52] and predicting biohydrogen production [53]. Despite being a powerful modeling tool, EL can suffer from overfitting due to the large number of parameters involved and can be computationally expensive to run if the system has an excessive number of inducers.

14.1.2.5 Reinforcement Learning

Reinforcement Learning (RL) is a trial-and-error approach to learning optimal input sequences through the provision of rewards and penalties [54]. RL has proven to be highly effective in the gaming world, with chess engines such as the free-to-access "Stockfish Engine" outperforming the best human players in the world in a fraction of the thinking time [55]. With the ability to identify optimal input sequences, there exists a broad range of potential applications to biosystems, with previous studies exploring topics such as identifying an optimal control strategy for fed-batch biopharmaceutical processes [56], bioprocess kinetics identification [57, 58], and policy optimization subject to chance constraints for a stochastic dynamic bioreactor system [59]. Despite having the potential to be utilized in model-free applications [23], RL can also be limited by large data requirements; a trial-and-error approach can require a large number of experiments that, in certain cases, are too expensive and time-consuming to conduct.

14.1.2.6 Future Prospects for Machine Learning

As the push toward digitalization continues, new applications of machine learning are continually being explored. For example, studies have been conducted on automatic model generation through symbolic regression, which allows the automated generation of interpretable mathematical models for chemical and biochemical systems [60–62]. Automatic model generation has the potential to save large amounts of time by accelerating iterative model design, particularly when coupled with DoE for optimal gain of knowledge [62].

Alternatively, the concept of hybrid modeling, which will be explored in more detail in Section 14.1.3, has seen a recent surge in applications to bioprocesses [63, 64], despite hybrid modeling being first proposed as early as 1992 [8]. Section 14.1.3 will explore the flexibility and applicability of hybrid modeling to infer why this methodology is increasing in popularity over 30 years after its conception.

14.1.3 Introduction to Hybrid Modeling

14.1.3.1 Literature Review

The term "hybrid modeling" falls under the umbrella term "grey box" modeling as it lies on the spectrum between pure mechanistic "white box" and empirical "black box" modeling, as depicted in Figure 14.1. But why build a hybrid model? Combining the two approaches to make a hybrid model allows improvement upon the accuracy of a mechanistic model by integrating a machine learning component that avoids the introduction of extra inductive bias from additional mechanistic terms developed using further assumptions. In addition, less data is required to train a hybrid model than a pure machine learning model because the mechanistic backbone of the model can capture some nonlinear relationships within the system, therefore leaving fewer nonlinear relationships that need to be approximated by the machine learning model.

There are two common structures for hybrid models: parallel and series. In a series hybrid model structure, either the mechanistic or machine learning component feeds into the other, which subsequently yields the output, as shown in Figure 14.2a, meaning the mechanistic and machine learning models do not act independently, though the first model may be independent from the second. Alternatively, in a parallel hybrid model structure, the mechanistic and machine learning components lie independently of one another; their respective outputs combine to give a single output, as shown in Figure 14.2b. The decision on whether a series or parallel structure should be used depends on the problem at hand. Table 14.1 highlights various examples of the applications of hybrid models in biosystems seen in literature, as well as the respective structures of the hybrid models built. The most common motivation for constructing a hybrid model

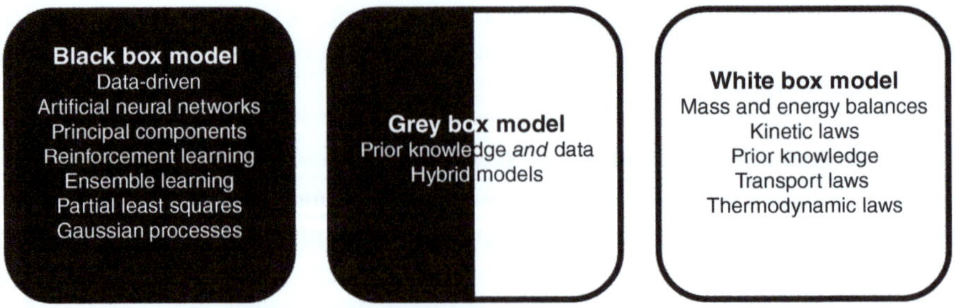

Figure 14.1 Grey box modeling: the concept.

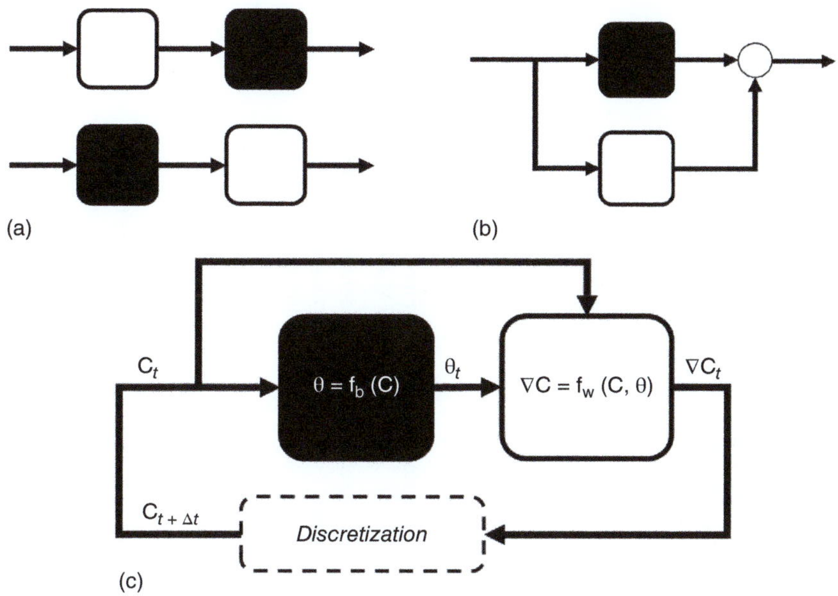

Figure 14.2 Illustrations of hybrid model structures for sequential series structures (a), where the white and black box come one-after-the-other in any order, parallel structure (b), where the white and black box models operate independently and combine to yield a single solution, and a dynamic hybrid structure (c), where a black box model updates time-varying parameters as a function of state variables to be used within a dynamic mechanistic white box model.

appears to be improving model predictive capabilities for subsequent monitoring, optimization, and control applications. As can be seen in Table 14.1, a common hybrid model structure is that where a data-driven model feeds reaction rates (or time-varying kinetic parameters), in series, to a form of mechanistic model. However, it should be noted that such a hybrid model configuration is not a generic series or parallel structure since, for dynamic systems, the hybrid model updates the state variables, which will in turn update the data-driven model and subsequently the mechanistic component; it is more of a cyclical series structure as illustrated in Figure 14.2c.

14.1.3.2 Hybrid Modeling for Biosystem Optimization

As mentioned in Section 14.1.3, improving model predictive capabilities using hybrid modeling is particularly beneficial for optimization applications. For example, the machine learning component can be easily retrained upon running a new experiment. A potential approach for training and validating a hybrid model is to run an experiment from DoE, whether it be a predicted optimum or an area of poor knowledge, and use the new data to update the machine learning component; repeating this process iteratively can be an efficient way to identify a suitable "optimal" solution while simultaneously generating a well-trained model.

Table 14.1 Hybrid modeling of biosystems from literature.

Simulated states	Mechanistic component	Machine learning component	Hybrid coupling	Application	Motivation	Source
C_X, C_S, C_P	Monod structure KM accounting for light intensity	2nd order polynomial	Parallel	Lab-scale lutein production from microalgae	Predictive simulation and optimization	[65]
C_X, $C_{S,1}$, $C_{S,2}$, C_I, C_P	KM from previous study	Deep NN	State fed to DNN and KM: DNN provides parameter for KM	Industrial scale fermentation	Improving upon KM accuracy and robustness	[36]
C_X, $C_{S,1}$, $C_{S,2}$, C_P	Intracellular KM from previous study	Three parallel PLS models	Parallel	Lignocellulosic ethanol fermentation	Real-time state estimation	[66]
C_X, $C_{S,1}$, O_2, C_P, T	Comprehensive KM from previous study	ANN	Parallel	Yeast fermentation	Improving upon KM accuracy	[67]
C_X, $C_{S,1}$, C_P	Monod structure KM	RL	Series: RL takes potential model structures to determine a correct KM structure and time-varying parameter trajectories	In-silico case studies	Identifying correct model structures and time-varying parameter values	[58]
C_X, $C_{S,1}$, C_P	Basic first-principles KM	ANN	Series: ANN gives growth rate and time-varying kinetic parameter for KM to use	Sodium gluconate fermentation from *Aspergillus niger*	Overcome missing mechanism knowledge	[68]

$C_{X,T}$, $C_{X,P}$, $C_{X,P'}$, $C_{S,1}$	KMs of varying detail	ANN	Series: ANN feeds time-varying parameter values to KM	y-linolenic acid (GLA) fermentation from *Cunninghamella achinulata* X-15 batch experiment	Investigating effects of hybrid model greyness	[69]
C_X, $C_{S,1}$, C_P	Mass balance with basic fist principles KM	ANN	Series: ANN feeds reaction rates to first-principles model	*E. coli* fed-batch fermentation (synthetic data)	Increasing hybrid model predictive capabilities through bootstrapping	[70]
$C_{X,T}$, $C_{X,V}$, $C_{S,1}$, $C_{S,2}$, C_P, $C_{BP,1}$, $C_{BP,2}$	Mass balance with basic first-principles KM	ANN	Series: ANN feeds reaction rates to first-principles model	Fed-batch cultivation of Chinese Hamster Ovary (CHO) DP-12 cells for antibody production	Comparing mechanistic and hybrid models	[71]
$C_{X,T}$, $C_{X,V}$, $C_{S,1}$, $C_{S,2}$, C_P, $C_{BP,1}$, $C_{BP,2}$	Monod structure KM	ANN	Series: ANN feeds reaction rates to first-principles model	Batch production of monoclonal antibodies	Demonstrating a data augmentation methodology to address data-shortage challenges	[72]
C_X, $C_{S,1}$, $C_{S,2}$, C_P	Basic first-principles KM	Gaussian process (GP)	Series: GP feeds reaction rates to first-principles model	*Xantho-phyllomyces dendrorhous* fermentation for astaxanthin production	Exploring the use of mixed-sugar fermentation and the suitability of kinetic and hybrid models	[73]

(Continued)

Table 14.1 (Continued)

Simulated states	Mechanistic component	Machine learning component	Hybrid coupling	Application	Motivation	Source
C_X, $C_{S,1}$, $C_{S,2}$	Mass balance and basic first-principles KM	Various: ANN, extra tree regressor, light gradient boosting machine, and extreme gradient boosting	Series: ML component estimates $C_{S,1}$ state to feed into mechanistic model to estimate C_X and $C_{S,2}$	Multiscale fungal fermentation of *Aspergillus niger*	Developing a soft-sensor for C_X, $C_{S,1}$, and $C_{S,2}$	[74]
9 chemical states	Anaerobic digestion model 1 (ADM1)	Various: support vector machine, ANN, random forest	Series: ML model provides kinetic parameters	Anaerobic digestion under various feedstocks	Optimization of continuous anaerobic digestion process	[75]
$C_{X,T}$, $C_{X,V}$, $C_{S,1}$, $C_{S,2}$, C_P, $C_{BP,1}$, $C_{BP,2}$	Monod structure KM	Physics-informed neural network	Model itself is an ANN, but the training incorporates data and dynamics of the KM	Fed-batch cultivation of Chinese Hamster Ovary (CHO) cells	Developing a robust model for real-time process monitoring and control	[76]

Note:
$C_{S,i}$ = Substrate *i* concentration, C_X = Cell density, $C_{X,T}$ = Total cell density, $C_{X,V}$ = Viable cell density, $C_{X,P}$ = Product-containing cell density, $C_{X,P'}$ = Product-free cell density, C_P = Product concentration, C_I = Inhibitor concentration, O_2 = Oxygen concentration, $C_{BP,i}$ = By-product *i* concentration, KM = Kinetic Model, NN = Neural Network, ANN = Artificial Neural Network, PLS = Partial Least Squares, ML = Machine Learning, RL = Reinforcement Learning.

Table 14.2 Optimization of biosystems using hybrid modeling from literature.

Objective	Degrees of freedom	Application	Source
Geometric mean of productivity and efficiency	$C_{S,0}$, $C_{X,0}$, T, t_f, N	Ethanol production from cashew apple juice using *Saccharomyces cerevisiae* CCA008 (with the modified gen FLO5α)	[77]
Final product titre	$T(t)$, $pH(t)$, $F_1(t)$ (glycerol), $F_2(t)$ (methanol)	Single-chain antibody fragment (scFv) production from *Pichia pastoris* GS115	[78]
Maximise secretion of product	$F_1(t)$	Fed-batch heterologous protein production from *Saccharomyces cerevisiae* SEY2102 yeast	[79]
Maximise batch productivity	t_f, t_{ind}, C_{ind} at t_{ind}	*Escherichia coli* expressing the β-galactosidase enzyme	[79]
Maximise profitability	$F_1(t)$ (glucose), $F_2(t)$ (inducer)	Production of β-galactosidase by recombinant *E. coli* D1210 containing the plasmid pSDD8	[79]
Maximise total mass of active product obtained	$F_1(t)$ (glucose), $F_2(t)$ (glutamine)	Fed-batch BHK-21 cultures expressing the human fusion glycoprotein IgG1–IL2	[80]

Note:
C_S,0 = Initial substrate concentration, C_X,0 = Initial cell density, T = Temperature, $T(t)$ = Dynamic temperature profile, t_f = Final time, t_{ind} = Induction time, N = Rotor speed, $pH(t)$ = Dynamic pH profile, $F_i(t)$ = Dynamic feed i profile, $Cind$ = Inducer concentration, KM = Kinetic Model, ANN = Artificial Neural Network.

Some examples of biosystem optimization studies using hybrid modeling are presented in Table 14.2. As one may expect, Table 14.2 highlights that objectives are often based on maximizing growth, yield, product titer, productivity, or profitability of the biosystem. Table 14.2 also shows that optimization from hybrid modeling is applicable over a variety of domains, whether it be mammalian cell lines, microalgal cultures, or fungi.

Overall, ANNs and hybrid modeling were first proposed in 1943 and 1992, respectively [8]. However, despite being a modeling technique that is now over 30 years old, hybrid modeling is still proving to be a capable approach for advancing biosystem development at an increasing rate. The relevance of hybrid modeling is thanks to a multitude of factors, including improved computational power, as well as access to new tools and techniques for data collection, such as free-floating wireless sensors, soft-sensors, and chemometrics [81]. Since hybrid modeling possesses a machine learning component, new data does not need to be incorporated directly into the mechanistic model if process knowledge is limited, thus keeping hybrid modeling relevant even as techniques for data collection advance.

14.2 Methodology for Hybrid Modeling

In this section, we explore the construction of a hybrid model for making predictions of a fermentation process. We start with the development of a preliminary mechanistic model before employing a machine learning component to improve model accuracy. Hereafter, uncertainty analysis is incorporated, and process optimization for a fed-batch fermentation system is illustrated. As a running example to show how the described methods are implemented, we imagine a batch process with homogeneous cell culture with a single substrate and a single product.

14.2.1 Mechanistic Model Construction

Preliminary mechanistic modeling aims to incorporate mechanistic information into the mathematical model. Implementing any mechanistic model structures introduces some inductive bias as all mechanistic models rely upon at least one assumption. As a result, it is important to stick to first principles and not venture into overly elaborate mechanistic structures, as they will introduce inductive bias and hinder the accuracy of a hybrid model. An example of an appropriate mechanistic relationship would be the assumption that each cell has the same specific uptake rate of substrate q_S, the same specific production rate of product q_P, and the same specific growth rate μ, irrespective of cell density, resulting in the system of ODEs shown in Eqs. 14.1a–14.1c:

$$\frac{dC_X(t)}{dt} = \mu \cdot C_X(t), \tag{14.1a}$$

$$\frac{dC_S(t)}{dt} = -q_S \cdot C_X(t), \tag{14.1b}$$

$$\frac{dC_P(t)}{dt} = q_P \cdot C_X(t). \tag{14.1c}$$

where C_X, C_S, and C_P represent cell density, substrate concentration, and product concentration, respectively.

In this instance, we know that a model is needed to describe specific rates because, otherwise, the model structure implies that the cell culture will infinitely grow and excrete product while continuing to consume substrate even after substrate depletion, phenomena which we know to be incorrect. Although this may be sufficient mechanistic information for cases with more data, it can be beneficial to include further assumptions to prevent phenomena we know to be unrealistic. Since introducing further assumptions introduces inductive bias, the assumptions should be simple in nature. For example, introducing Monod kinetics for growth and substrate ensures that non-negative substrate concentrations will be maintained while allowing kinetic relationships between zero-order and first-order with respect to the substrate concentration [82]. In order to prevent infinite growth of cells, we can also assume a first-order decay rate with respect to cell density. The resulting mechanistic model is described in Eqs. 14.2a–14.2c:

$$\frac{dC_X(t)}{dt} = \mu_{\text{max}} \cdot \frac{C_S(t)}{C_S(t) + K_S} \cdot C_X(t) - \mu_{\text{d}} \cdot C_X(t), \tag{14.2a}$$

$$\frac{dC_S(t)}{dt} = -v_{\max_S} \cdot \frac{C_S(t)}{C_S(t) + K_S} \cdot C_X(t), \tag{14.2b}$$

$$\frac{dC_P(t)}{dt} = q_P \cdot C_X(t), \tag{14.2c}$$

where $\mu_{\max}$ is the maximum specific growth rate, μ_d is the specific death rate, $v_{\max_S}$ is the maximum specific substrate consumption rate, and K_S is the saturation constant (the substrate concentration at which half the maximum specific consumption rate is achieved [82]). It should be noted that although the specific death rate and specific product production rate are likely to be some function of state variables (since products can be self-inhibitory or lead to cell death), we may not have enough knowledge of the system to make valid assumptions. Therefore, it is better to leave unknown relationships to the machine learning component that commences construction in Section 14.2.2.

The next step for the mechanistic component construction is parameter estimation of the mechanistic model, which involves solving the optimization framework described in Framework 14.3:

$$\min_{\beta} \frac{1}{2} \sum_i \sum_t \left(\frac{C_i(t, \beta) - C_{i,t}^{\text{meas}}}{\sigma_{i,t}^{\text{meas}}} \right)^2, \tag{14.3a}$$

$$\text{s.t.}: \quad \boldsymbol{\beta} \geq \mathbf{0}, \qquad \boldsymbol{\beta} = [\mu_{\max}, \mu_{\text{d}}, v_{\max_S}, K_S, q_P]^T, \tag{14.3b}$$

where the objective function in Eq. 14.3a minimizes the difference between the simulated state variables $C_i(t, \beta)$ and their corresponding experimental measurements $C_{i,t}^{meas}$and measurement standard deviation $\alpha_{i,t}^{meas}$for state i at time point t up until the final measurement time t_f.

Optimization Framework 14.3 can be carried out using a variety of tools. Gradient descent algorithms such as interior point optimization (IPOPT) are appropriate and can be freely accessed using the *fmincon* function in MATLAB or the Pyomo and SciPy packages in Python [83–85]. Particle Swarm Optimization (PSO) can also be used to generate good initial guesses in cases where gradient descent algorithms converge to poor solutions as a result of sensitivity to the initial guess. However, its gradient-free nature prevents it from guaranteeing convergence to any local optima if used on its own.

Uncertainty analysis of the mechanistic model should also be conducted to assess the model's fitting performance. This can be addressed by calculating the uncertainty of each parameter from the Hessian of the objective function in Eq. 14.3a (referred to here as Z for simplicity) with respect to the estimated parameters β^*. The Hessian matrix of the parameter estimation objective function with respect to the estimated parameters $\boldsymbol{H}(Z, \beta^*)$ is inverted to generate the corresponding covariance matrix $\boldsymbol{Cov}(Z, \beta^*)$ of the model, as in Eq. 14.4:

$$\boldsymbol{Cov}(Z, \boldsymbol{\beta}^*) = \boldsymbol{H}(Z, \boldsymbol{\beta}^*)^{-1}, \tag{14.4}$$

where the diagonal of the covariance matrix represents the individual parameter variances [86]. Monte Carlo sampling can be used to generate parameter confidence intervals (in this chapter, we recommend the use of 95% confidence intervals, which corresponds to two standard deviations from the mean) and subsequently model trajectory confidence intervals. Confidence intervals are important for gauging the predictive capabilities of the model; a model may appear to capture data well using the nominal parameter values, but if confidence intervals are large, it will be less likely to make predictions with high precision. It should be noted that, at this point, if the mechanistic model predicts well within the desired operating range with adequately sized confidence intervals, the model is unlikely to require the further addition of a machine learning component. However, if predictive capabilities are poor and/or confidence intervals are inappropriately large, introducing a machine learning component could be greatly beneficial, the first step of which is described in Section 14.2.2.

14.2.2 Time-Varying Parameter Estimation

A mechanistic modeling does not always yield the predetermined desired predictive capabilities with sufficiently low uncertainty, making them inadequate for their applications. Under the assumption that further alterations to the mechanistic model cannot *confidently* be made, the introduction of a machine learning component could be advantageous as it can increase model accuracy while maintaining good generalizability.

Since we have a mechanistic backbone, it can be assumed that poor model performance is due to missing information from the model structure, meaning some of the constant parameters within the model are not constant parameters but time-varying functions of state variables and operating conditions. Henceforth, we will refer to such hidden functions embedded within parameters as *time-varying parameters*. The most appropriate parameters to dedicate to be time-varying are parameters that have the highest uncertainty, which are identified using the covariance matrix calculated in Eq. 14.4; parameters with the highest standard deviation to nominal value ratio are deemed "most uncertain." Normalizing standard deviation with respect to the parameter nominal value avoids potential bias introduced from the units chosen to describe the parameter. Sensitivity analysis must subsequently be conducted to ensure the chosen parameter has an appropriate effect on the mechanistic model trajectories; if not, the parameter should be removed altogether as it has a negligible impact on the model's accuracy.

Once the time-varying parameters have been identified, the end goal is to simulate them as a function of state variables and operating conditions using a machine learning model. Training the machine learning model is conducted in two stages: (1) identifying appropriate time-varying parameter profiles over the entire time course, and (2) training a machine learning model to estimate time-varying parameter profiles using corresponding state variable profiles and process conditions. Assuming that the most uncertain parameters in Eqs. 14.2a–14.2c are found to be μ_d, K_S, and q_P, identifying their respective time-varying profiles is achieved with a simple adjustment to Eqs. 14.2a–14.2c and parameter estimation Framework 14.3.

The resulting parameter estimation framework is presented in Framework 14.5:

$$\min_{\theta_{\mu_d},\theta_{K_S},\theta_{q_P}} \frac{1}{2}\sum_i\sum_t\left(\frac{C_i(t)-C_{i,t}^{\text{meas}}}{\sigma_{i,t}^{\text{meas}}}\right)^2 + \sum_k \lambda_{\theta_k}\sum_{t_\theta}(\theta_k(t_\theta)-\theta_k(t_\theta+\Delta t_\theta))^2, \tag{14.5a}$$

$$\text{s.t. :}\ \frac{dC_X(t)}{dt} = \mu_{\max}\cdot\frac{C_S(t)}{C_S(t)+\theta_{K_S}(t)}\cdot C_X(t) - \theta_{\mu_d}(t)\cdot C_X(t), \tag{14.5b}$$

$$\frac{dC_S(t)}{dt} = -v_{\max_S}\cdot\frac{C_S(t)}{C_S(t)+\theta_{K_S}(t)}\cdot C_X(t), \tag{14.5c}$$

$$\frac{dC_P(t)}{dt} = \theta_{q_P}(t)\cdot C_X(t), \tag{14.5d}$$

where λ_{θ_k} is a weight for the regularization term that penalizes dramatic deviations time-varying parameter k, and $\theta_k(t_\theta)$ is the time-varying parameter of the nominal parameter k from time t_θ to time $t_\theta + \Delta t_\theta$. We recommend that each time-varying parameter discretization step Δt_θ to be no less than the time difference between consecutive measurements in order to minimize overfitting. Increasing the value of λ_{θ_k} results in stronger penalization of time-varying parameter deviations, thus reducing the overfitting of noisy data. However, this will also reduce the fitting accuracy as the data-fitting term in Eq. 14.5a is prioritized less.

Once time-varying parameter profiles have been identified, the simulation of state variables should have increased in accuracy. Therefore, the time-varying parameter profiles can be assigned to corresponding state variables and necessary operating conditions for training of the machine learning component, which is discussed in Section 14.2.3.

14.2.3 Machine Learning Model Construction

Regardless of the machine learning component chosen, the overall training aim is the same: generate a data-driven model that takes state variables and key process operating conditions as inputs to predict corresponding time-varying parameter values as outputs. For the simple batch example, we have multiple inputs, $C_X(t)$, $C_S(t)$, and $C_P(t)$, and multiple outputs $\theta\mu_d(t)$, $\theta_{K_S}(t)$, and $\theta q_P(t)$, at each time-point t from solving Framework 5. The inputs and outputs can then be related in the form of an MIMO machine learning model that is not directly dependent on time. In order to train the machine learning component, a loss function that minimizes the difference between the data-driven models predicted time-varying parameters and Section 14.2.2 estimated time-varying parameters is required. A general form of the loss function is shown in Eq. 14.6:

$$\min_{\boldsymbol{w},\boldsymbol{b}} \frac{1}{n}\sum_k\sum_{t_\theta}|\theta_k(\boldsymbol{C}(t_\theta),\boldsymbol{u}(t_\theta),\boldsymbol{w},\boldsymbol{b}) - \theta_k(t_\theta)|, \tag{14.6}$$

which represents a Mean Absolute Error (MAE) loss function (often referred to as L1 Loss) [87, 88] where $u(t_\theta)$ represents the control action at time t_θ (for example, an adjustment of flowrate), w and b are the weights and biases of the machine learning component, C is the vector of state variables $[C_X, C_S.C_P]$, and n is the number of datapoints used in training. A similar framework uses the Mean Squared Error (MSE) loss function (often referred to as L2 loss) [88], as shown in Eq. 14.7:

$$\min_{\boldsymbol{w},\boldsymbol{b}} \frac{1}{n} \sum_{k} \sum_{t_\theta} (\theta_k(\boldsymbol{C}(t_\theta), \boldsymbol{u}(t_\theta), \boldsymbol{w}, \boldsymbol{b}) - \theta_k(t_\theta))^2. \tag{14.7}$$

Once training of the machine learning component is complete, the final hybrid model, with fixed weights w and biases b, takes the form shown in Eqs. 14.8a–14.8c:

$$\frac{dC_X(t)}{dt} = \mu_{\max} \cdot \frac{C_S(t)}{C_S(t) + \theta_{K_S}(\boldsymbol{C}(t), \boldsymbol{u}(t))} \cdot C_X(t) - \theta_{\mu_d}(\boldsymbol{C}(t), \boldsymbol{u}(t)) \cdot C_X(t), \tag{14.8a}$$

$$\frac{dC_S(t)}{dt} = -v_{\max_S} \cdot \frac{C_S(t)}{C_S(t) + \theta_{K_S}(\boldsymbol{C}(t), \boldsymbol{u}(t))} \cdot C_X(t), \tag{14.8b}$$

$$\frac{dC_P(t)}{dt} = \theta_{q_P}(\boldsymbol{C}(t), \boldsymbol{u}(t)) \cdot C_X(t). \tag{14.8c}$$

Section 14.2.3 will consistently refer to the machine learning model as a single MIMO model. However, it should be noted that using several MISO models (one for each time-varying parameter) is theoretically sound, but it may require a larger number of parameters. In this section, we will use an ANN as the machine learning model as an example. As highlighted in Section 14.1.3, ANNs are commonly implemented within hybrid models as the machine learning component due to their ability to capture highly nonlinear relationships. Smaller networks (networks with fewer hidden layers and fewer nodes within each layer) are preferred where possible, as they contain fewer weights and biases, thus reducing overfitting of the training data. However, the size of the network will vary case-by-case and a variety of different architectures should be trialed to gauge the best performance on both training and validation.

In order to mitigate overfitting, the dataset should be split into three independent datasets; a training dataset that is used to train the ANN; a separate dataset for validation of the ANN that is not used during training, where hyperparameter retuning can take place, and a third dataset for testing the final model [89, 90]. The partitioning of a dataset into training, validation, and testing subsets depends on the quantity and diversity of data available; however, multiple permutations of data partitioning should be tested in order to assess the sensitivity of training to the data partitioning. Once the dataset has been partitioned, each variable should be individually standardized or normalized. In this chapter, we follow the commonly used standardization approach where each variable is linearly transformed to have a mean of zero and a standard deviation of unity [91], which has been known to improve the training speed and accuracy of neural networks [92, 93].

Once the ANNs of varying structures are set up and the data has been appropriately partitioned, training can commence. There are several choices to be made when training an ANN that change case-by-case, examples of which include incorporating regularization terms, the number of epochs, batch size, and finally, the optimization algorithm used to train the ANN.

Regularization terms can be added to the training objective function to reduce or remove unnecessary parameters and avoid overfitting [94]. Several variations of regularization terms exist, including L1 and L2 regularization that penalize the absolute ANN parameter values and squared ANN parameter values, respectively [95]. For example, introducing the L1 regularization term to the MAE loss function results in the objective function illustrated in Eq. 14.9:

$$\min_{\boldsymbol{w},\boldsymbol{b}} \frac{1}{n} \sum_k \sum_t |\theta_k(\boldsymbol{C}(t), \boldsymbol{u}(t), \boldsymbol{w}, \boldsymbol{b}) - \theta_k(t)| + \lambda_{\mathrm{L1}} \sum_q |w_q|, \tag{14.9}$$

where λ_{L1} is the penalty weight that can be increased to penalize weight magnitude more at the cost of ANN fitting. Similarly, adding an L2 regularization term to the MAE loss function yields the objective function shown in Eq. 14.10:

$$\min_{\boldsymbol{w},\boldsymbol{b}} \frac{1}{n} \sum_k \sum_t |\theta_k(\boldsymbol{C}(t), \boldsymbol{u}(t), \boldsymbol{w}, \boldsymbol{b}) - \theta_k(t)| + \lambda_{\mathrm{L2}} \sum_q w_q^2, \tag{14.10}$$

Epoch number and batch size refer to the way in which data is fed to the training algorithm. Batch size dictates how many datapoints per batch to process before updating the ANN parameters [96]. Batch size can be reduced from the size of the full training dataset in cases where the training dataset is too large to process in a single iteration [97]. Epochs are simply how many times the entire dataset is seen by the optimization algorithm [96]. So, increasing the number of epochs can improve the fitting of the ANN to the training data at the cost of longer computation time and an increased risk of overfitting. Early stopping is a form of regularization that assesses the trained ANN performance on validation data and terminates training if a stopping criterion is met [98], thus avoiding overfitting from using an excessive number of epochs.

Training an ANN generally involves back-propagation [99], which uses the chain rule to calculate the gradient of the loss function with respect to the ANN weights and biases. Several tools exist for ANN training, one of which is the Python package PyTorch [100], in which different loss functions, regularization terms (and weights), batch sizes, and epochs can be chosen.

14.2.4 Hybrid Model Uncertainty Estimation

Quantifying uncertainty is an important step in process model development, offering valuable insight into prediction confidence by assessing its expected precision. This effectively means that uncertainty analysis can be viewed as a form of model validation [101], despite often being overlooked as such, as excessive model uncertainty on a validation dataset can be a sign of overfitting of the training dataset [102]. Although some level of uncertainty is inevitable, especially in biosystems, minimizing model uncertainty supports the development of optimal control strategies, such as strategies used in QbD methodologies.

In the context of hybrid modeling where the most uncertain and influential parameters have been regressed by the machine learning model, it can subsequently be assumed that hybrid model uncertainty will mostly stem from the data-driven model. One such methodology for uncertainty estimation is bootstrapping, a method widely applied in machine learning for quantifying uncertainty and improvising model stability and robustness [103, 104]. Bootstrapping involves using different random samples of the training dataset to train multiple independent machine learning models. Each model is then combined with the mechanistic component to make its own hybrid model. Having several hybrid models instead of one means multiple state variable trajectories are generated; sampling the different trajectories facilitates the calculation of confidence intervals that quantify the model uncertainty.

Data sampling for bootstrapping can be completely random, or they can follow a strategy; a previous study conducted bootstrapping such that all data over predetermined periods of time within the experimental time horizon were eliminated for the training of each model [105]. Simulating scenarios where the system experiences missing information over a given duration helps assess how well a model generalizes to periods of incomplete data, an example of which is when sensors in an industrial bioprocess fail and require a fixed time horizon to be repaired. It should be noted that using GPs to simulate the machine learning component, instead of an ANN, can be beneficial for uncertainty analysis as they automatically generate the confidence interval of time-varying parameters and, therefore, the overall hybrid model. However, GPs are not always simple to implement due to their lack of an explicit model structure and are not suitable for systems with high-dimensional inputs, as discussed in Section 14.1.2.3.

14.2.5 Considerations for Hybrid Model Construction

There are multiple options when constructing a hybrid model. This section provides a brief overview of some key considerations that can be made during the construction of a hybrid model, particularly when using an ANN as the machine learning component.

14.2.5.1 Hybrid Model Greyness

Hybrid model "greyness" refers to the ratio of white box to black box modeling within the hybrid model. No metric exists for the quantification of "greyness;" however, we know when "greyness" is being increased or decreased relative to an initial model design. For example, implementing more mechanistic knowledge would be moving toward the "white" end of the "greyness" scale. Although the hybrid model "greyness" will vary case-by-case, a study investigating the effects of differing hybrid model "greyness" found that incorporating more specific mechanistic knowledge can increase the risk of incorporating incorrect inductive bias into the model [69], meaning mechanistic knowledge should only be incorporated in high confidence.

14.2.5.2 Discrepancy Hybrid Modeling

Hybrid modeling does not always require a machine learning model for the simulation of time-varying parameters; discrepancy hybrid models take the output of the mechanistic component and correct it with a data-driven component, similar to GB EL. For instance, when applied to the mechanistic model described in Eqs. 14.2a–14.2c, a discrepancy hybrid model would take the form shown in Eqs. 14.11a–14.11c:

$$\frac{dC_X(t)}{dt} = \mu_{\max} \cdot \frac{C_S(t)}{C_S(t) + K_S} \cdot C_X(t) - \mu_{\mathrm{d}} \cdot C_X(t) + \varepsilon_X(\boldsymbol{C}(t), \boldsymbol{u}(t), \boldsymbol{w}, \boldsymbol{b}), \tag{14.11a}$$

$$\frac{dC_S(t)}{dt} = -v_{\max_S} \cdot \frac{C_S(t)}{C_S(t) + K_S} \cdot C_X(t) + \varepsilon_S(\boldsymbol{C}(t), \boldsymbol{u}(t), \boldsymbol{w}, \boldsymbol{b}), \tag{14.11b}$$

$$\frac{dC_P(t)}{dt} = q_P \cdot C_X(t) + \varepsilon_P(\boldsymbol{C}(t), \boldsymbol{u}(t), \boldsymbol{w}, \boldsymbol{b}), \tag{14.11c}$$

where $\varepsilon_i(C(t), u(t), w, b)$ is the discrepancy term for state i that is itself a function of state variables and operating conditions in the form of a machine learning model with weights w and biases b. Comparing discrepancy hybrid models to hybrid models with time-varying parameters was investigated in a previous study for a chemical system, which found the discrepancy model to perform worse than other hybrid model structures for a dynamic system [106, 107].

14.2.5.3 Machine Learning Component

As discussed in Section 14.1.2, a multitude of machine learning approaches have been applied to bioprocesses, all of which are potentially applicable to hybrid modeling provided the limitations of the machine learning component are not exceeded by the problem at hand. For instance, when dealing with systems with high-dimensional inputs, incorporating a dimensionality reduction-based ML model, such as PCA, could be greatly beneficial. On the other hand, if quantified model uncertainty is a critical requirement for later use, then using GP could be the preferred option due to its direct computation of output mean and variance. However, if the process is known to have time-dependent or sequential mechanisms, such as the changing composition of CHO cells [108], then using an RNN would be a better option, as the hidden state is able to simulate internal memory. As such, when deciding the ML component to be used, one needs to first consider not only the amount of the data available but also the ultimate goal of the hybrid model construction.

14.2.6 Dynamic Optimization Using Hybrid Modeling

As previously highlighted in Table 14.2, bioprocess models are frequently implemented for optimization purposes. Dynamic optimization in the context of bioprocesses refers to the identification of optimal operating conditions for a process over time, such as batch, fed-batch, and continuous systems. Since a number of bioprocess models incorporate Ordinary Differential Equations (ODEs), the dynamic optimization of bioprocess must incorporate a form of numerical integration within the optimization algorithm, resulting in non-convex optimization problems that are not trivial to solve. Some optimization problems also deal with a process model with high stiffness, in which case, causing numerical issues with model discretization and resorting to gradient-free optimization algorithms, such as PSO and evolutionary algorithms, despite no guarantee of convergence to even a local optimum. Generally, it is recommended to either conduct gradient-based optimization from multiple initial points or utilize gradient-free optimization to identify a good initial guess from which gradient-based optimization can be employed, thus ensuring sufficient exploration of the solution space.

14.2.6.1 Optimal Feeding in Fed-Batch Bioprocesses

In this section, we outline an optimization framework for a fed-batch bioprocess as an example. When optimizing a fed-batch process, we can consider the degrees of freedom to be the dynamic substrate feeding rate. In this example, we propose maximizing final biomass cell density, for which the optimization problem, incorporating the hybrid model described in Eqs. 14.8a–14.8c, takes the form given in Framework 14.12:

$$\min_{\boldsymbol{F}, C_S^{\text{feed}}} - C_X(t_{\text{f}}), \tag{14.12a}$$

$$\text{s.t.}: \qquad F^{\min} \leq F(t) \leq F^{\max}, \tag{14.12b}$$

$$\frac{dC_X(t)}{dt} = \mu_{\max} \cdot r_S(t) - \theta_{\mu_{\text{d}}}(\boldsymbol{C}(t), F(t)) \cdot C_X(t) - \frac{F(t)}{V(t)} \cdot C_X(t), \tag{14.12c}$$

$$\frac{dC_S(t)}{dt} = \frac{F(t)}{V(t)} \cdot \left(C_S^{\text{feed}} - C_S(t)\right) - v_{\max_S} \cdot r_S(t), \tag{14.12d}$$

$$\frac{dC_P(t)}{dt} = \theta_{q_P}(\boldsymbol{C}(t), F(t)) \cdot C_X(t) - \frac{F(t)}{V(t)} \cdot C_P(t), \tag{14.12e}$$

$$\frac{dV(t)}{dt} = F(t), \tag{14.12f}$$

$$r_S(t) = \frac{C_S(t)}{C_S(t) + \theta_{K_S}(\boldsymbol{C}(t), F(t))} \cdot C_X(t), \tag{14.12g}$$

where C is the vector of state variables $[C_X, C_S, C_P]$, V is the volume of fluid in the reactor, $F(t)$ is the volumetric feeding rate at time t, $C_S^{\text{feed}}$is the concentration of substrate in the feed, and $\boldsymbol{F}$ is the vector of discretized time-varying feeding rate. It should also be noted that optimization packages are often minimization-oriented, and minimizing the negative of a function is equivalent to finding its maximum. Eqs. 14.8a–14.8c have been modified to take the forms shown in Eqs. 14.12c–14.12 g to incorporate the mass balance of fed-batch operation. It should also be noted that introducing feeding can incur stiffness in the optimization framework, particularly with pulse feeding, which can pose challenges for gradient-descent optimization algorithms in some cases.

14.2.6.2 Dynamic Optimization Under Uncertainty

Uncertainty can be incorporated within a dynamic optimization framework through the objective function or constraints, depending on the application. For example, robust optimization requires an optimal solution to meet inequality constraints subject to *all* possible combinations of uncertain parameters within their uncertainty bounds, providing applicability to biosystems where product quality needs to be guaranteed or where certain conditions endanger cultured organisms. Therefore, robust optimization algorithms tend to identify a set of "worst-case" parameters (parameters most likely to violate constraints); the optimum is decided using the "worst-case" parameters for constraints, instead of their nominal values [109]. A study looking at robust optimization algorithms for a penicillin production case study, where cell density and substrate concentrations had maximum allowable values throughout fed-batch operation, was presented in a previous study on robust dynamic optimization [110]. Although robust optima can be found, the solutions can be overly conservative; a similar approach that overcomes this issue is chance-constrained optimization, where a predetermined proportion of solutions are allowed to violate the uncertain inequality constraints [111]. However, this does require assumptions to be made regarding the manner in which parameter uncertainties are distributed. Both robust and chance-constrained optimization methodologies are computationally expensive for dynamic systems, and there is no guarantee of finding a global optimum in most cases [110, 112].

A less computationally expensive approach involves sampling parameters to generate several parallel models; optimization of the nominal objective function is then subject to the constraints for all parallel models, thus approximating robustness or the meeting of chance constraints. This multi-modal approach (i.e. scenario tree-based approach) was proven to violate constraints in fewer instances than solely using the nominal parameters in a previous study [113]; however, it did not achieve total robustness, unlike some more computationally expensive approaches that aim to solve a bilevel optimization problem to identify the "worst-case" parameters, as demonstrated in a subsequent study [110].

When considering the uncertainty arising from the objective function instead of constraints, uncertainty can be accounted for by introducing a penalty term that penalizes the variance of the objective function [114]. Decreasing the penalization of variance allows the user to explore a region lacking in knowledge to possibly design new experiments, while increasing this penalty targets an optimum that is less likely to vary from model predictions. Again, generating a distribution of objective function values can be achieved by generating several models in parallel using different parameter samples, from which a mean and variance of the objective function can be calculated. Finally, given the complexity of dynamic optimization under uncertainty, this is not included in the current book chapter (neither is a detailed uncertainty analysis of hybrid model construction). However, interested readers can refer to [110] for a study on robust optimization of penicillin production, [115] for chance-constrained RL for batch process control, and [116] for optimization under uncertainty for beer fermentation.

14.3 Hybrid Modeling of Microalgal Lutein Production

14.3.1 Introduction to Microalgal Lutein Production

Now, in order to illustrate the potential of hybrid modeling, we will use microalgal lutein production as a case study to apply the hybrid modeling methodology explained in Section 14.2. Lutein is a carotenoid that has several health benefits for humans, including anticancer effects and prevention of cataracts [117–119]. However, humans do not synthesize lutein internally and instead rely on the ingestion of lutein from food and supplements. Although some foods naturally contain lutein, particularly fruit and vegetables [120], consuming lutein supplements can be more convenient and effective. The consumption of lutein supplements has helped drive the global lutein market to an estimated $369 million in 2024, with an expectation to grow to $461 million by 2029 with a Compound Annual Growth Rate (CAGR) of 6.1% [121]. Currently, commercial lutein production is entirely sourced from marigold flower petals [122]. However, microalgae, microscopic photosynthetic organisms that produce lutein for light harvesting [123], have the potential to become a commercially viable option for lutein production due to less labor-intensive harvesting, higher growth rate, and capacity for all-year harvesting (rather than seasonal) compared to marigold flowers [117]. Despite the potential commercial viability, several limitations exist within the microalgal lutein production process, namely the low lutein content within microalgal cells, resulting in low process yields, and high costs of microalgal biomass production compared to the marigold flower, which does not require as much energy or nutrient feeding to grow [117].

Various options are being explored to drive microalgal lutein production toward commercial viability, one of which is heterotrophic growth (the focus of this case study), where the lutein-producing microalgae grow in the absence of light, allowing ultra-high cell densities to be achieved, thus improving the potential yield and production rate of lutein [124]. In order to further optimize the production of lutein under heterotrophic growth conditions, experiments can be conducted to identify some optimal operating conditions, such as temperature and choice of substrate, as demonstrated in a previous study [124]. Additionally, another consideration is to carry out a fed-batch operation and optimize the feeding strategy, which can utilize different flow rates and feed compositions. However, the potential combinations of feed rates and compositions are numerous, making the optimal feeding strategy challenging to identify experimentally due to the multitude of degrees of freedom. Model-based optimization addresses the complex nature of optimizing a fed-batch process by efficiently exploring a wide range of potential feeding conditions, therefore reducing the need for experimental trials. In this section, we demonstrate the application of hybrid modeling to an HCD culture of lutein-producing microalgae for fed-batch optimization.

14.3.2 Experimental Setup and Data Availability

The case study in this section focuses on the growth of *Chlorella sorokiniana* under fed-batch conditions utilizing glucose and urea as substrates. Urea was favored over sodium nitrate as a nitrogen source for achieving higher biomass productivity in a previous study [124]. Glucose and urea are fed in a single combined feed continuously throughout the fed-batch process. The fed-batch reactor has an initial working volume of 2.5 L with a single feed rate capable of reaching up to 0.05 L/h. The feed was designed to permit feed concentrations of glucose and urea as high as 750 g/L and 60 g/L, respectively.

Measurements of cell density were collected every 12 hours, with measurements of glucose and urea concentrations taken every 12 hours for the first 24 hours, followed by more frequent sampling of 3–4 times per subsequent 12-hour interval, and lutein content was measured at 24-hour intervals. Two initial fed-batch experiments were conducted, both starting with the same initial cell density, glucose concentration, and urea concentration. The first experiment commenced with an initial 24-hour batch phase, followed by the maintenance of glucose concentration between 4 g/L and 8 g/L through frequent manual adjustment of feed flow rate until the 168-hour mark passed, when the experiment was terminated. Glucose and urea feed concentrations were fixed at 750 g/L and 46.5 g/L, respectively, throughout the experiment. For the second experiment, which also started with a 24-hour batch phase, the feed rate was only changed every 12 hours, with urea feed concentration being adjusted to a value between 0 g/L and 60 g/L every 12 hours until the experiment was terminated after 144 hours. The experimental data can be seen in Figure 14.4, where lutein content has been multiplied by the cell density to show the overall change in concentration within the reactor.

14.3.3 Constructing a Hybrid Model for Microalgal Lutein Production

Ultimately, the case study aims to identify an optimal fed-batch feeding strategy that maximizes the lutein produced by the end of the process. In order to facilitate fed-batch feed optimization, a model of the fed-batch process will be constructed and subsequently integrated within an optimization framework. The model should be capable of generating accurate predictions despite limited experimental data, enabling an efficient optimization of the fed-batch process. The hybrid model built for this specific case study is constructed in accordance with the methodology discussed throughout Section 14.2. Primarily, kinetic modeling is conducted, which is additionally used as a benchmark for the hybrid model, before introducing an ANN to enhance the model accuracy.

14.3.3.1 Preliminary Kinetic Modeling of Microalgal Lutein Production

A fundamental kinetic model of the case study system takes the form shown in Eqs. 14.13a–14.13f:

$$\frac{dC_X(t)}{dt} = \mu_{\max} \cdot \frac{C_G(t)}{C_G(t)+K_G} \cdot C_X(t) - \mu_{\mathrm{d}} \cdot C_X(t) - \frac{F(t)}{V(t)} \cdot C_X(t), \tag{14.13a}$$

$$\frac{dC_G(t)}{dt} = \frac{F(t)}{V(t)} \cdot \left(C_G^{\mathrm{feed}}(t) - C_G(t)\right) - v_{\max\,G} \cdot \frac{C_G(t)}{C_G(t)+K_G} \cdot C_X(t), \tag{14.13b}$$

$$\frac{dC_N(t)}{dt} = \frac{F(t)}{V(t)} \cdot \left(C_N^{\mathrm{feed}}(t) - C_N(t)\right) - v_{\max\,N} \cdot \frac{C_N(t)}{C_N(t)+K_N} \cdot C_X(t), \tag{14.13c}$$

$$\frac{dL_C(t)}{dt} = Y_{LX} - \mu_{\max} \cdot \frac{C_G(t)}{C_G(t)+K_G} \cdot L_C(t) - d_L \cdot L_C(t), \tag{14.13d}$$

$$C_L(t) = L_C(t) \cdot C_X(t), \tag{14.13e}$$

$$\frac{dV(t)}{dt} = F(t), \tag{14.13f}$$

where C_X represents cell density in g/L, C_G represents glucose concentration in g/L, C_N represents urea concentration in g/L, L_C represents lutein content in mg/(g biomass), and C_L represents the overall lutein concentration within the reactor. The mechanistic model relies on the assumption of Monod kinetics for substrate uptake and growth, as well as first-order cell death. The use of Monod kinetics for the backbone of the model is justified by its successful incorporation within several models of lutein-producing microalgal systems seen in literature [125–127]. Urea was decoupled from cell growth in Eq. 14.13c because the experimental data in Figure 14.4 illustrates the ability of cells to continue growing in the presence of little to no urea; the undisclosed relationship between urea and biomass growth rate will be quantified by the machine learning component when constructing the hybrid model. It should be noted that lutein content is an intracellular concentration. Hence, dilution is accounted for in Eq. 14.13d. The parameters in Eqs. 14.13a–14.13f are estimated using the parameter estimation framework described in Framework 14.3, subject to Eqs. 14.13a–14.13f instead of Eqs. 14.5b–14.5d.

The kinetic model fitting can be seen in Figure 14.4 with the parameter values provided in column two of Table 14.3. Despite fitting cell density well, with an R^2 value of 0.932 across the two fed-batch experiments, the kinetic model particularly struggled to capture the substrate concentration profiles, which is evidenced in the R^2 values of −1.073 and 0.123 for glucose and urea, respectively, across the two fed-batch experiments. The kinetic model fitting of both fed-batch experiments 1 and 2, shown in Figure 14.4, generally shows the kinetic model is underpredicting the substrate concentrations, particularly beyond the first 24 hours, which means that the kinetic model is overpredicting their respective consumption rates. By the 24-hour mark, the cell density has already increased by over 600%, meaning the overprediction of consumption rates could be due to mass transfer limitations or intracellular competition as a result of increased cell density, which the mechanistic model does not explain. Since the mechanistic understanding of the process is evidently incomplete, a data-driven component can be introduced to capture the remaining relationships within the system while not requiring the additional assumptions of a more intricate kinetic model, therefore avoiding the introduction

Table 14.3 Kinetic model parameters and their respective time-varying mean and standard deviation.

Parameter	Nominal value	Time-varying value ($\mu \pm \sigma$)
$\mu_{max}(h^{-1})$	7.85×10^{-2}	$6.20 \times 10^{-2} \pm 3.50 \times 10^{-2}$
$\mu_d(h^{-1})$	2.67×10^{-3}	N/A
$K_G(g\ L^{-1})$	7.92×10^{-2}	N/A
$v_{max_G}(g(g\ biomass)^{-1}h^{-1})$	1.59×10^{-1}	$3.48 \times 10^{-2} \pm 2.41 \times 10^{-2}$
$K_N(g\ L^{-1})$	6.43	N/A
$v_{max_N}(g(g\ biomass)^{-1}h^{-1})$	1.40×10^{-1}	$1.71 \times 10^{-1} \pm 1.45 \times 10^{-1}$
$d_L(h^{-1})$	1.64×10^{-1}	N/A
$Y_{LX}(mg(g\ biomass)^{-1}h^{-1})$	2.88×10^{-1}	N/A

of further inductive bias. The kinetic model constructed in this section provides a foundation for the hybrid model and will henceforth be used as a benchmark for further comparison.

14.3.3.2 Artificial Neural Network Implementation for Time-Varying Parameters

Due to the inadequate fitting of the kinetic model, as discussed in the previous section, a data-driven component is introduced to improve the model accuracy. For this case study, a Feed Forward Neural Network (FFNN) is chosen as the data-driven model due to its ability to capture any nonlinear relationships that we may find in the fed-batch fermentation process. The first step is to identify the time-varying parameters, which in this case were found to be $\mu_{\max}$, $\nu_{\max_G}$, and $\nu_{\max_N}$ (whose time-varying parameters are $\theta_{\mu_{\max}}$, $\theta_{\nu_{\max G}}$, and $\theta_{\nu_{\max_N}}$, respectively) as these parameters presented the highest uncertainties to which the fit of the model also demonstrated sufficient sensitivity to their values. The parameter estimation framework in Framework 14.5, subject to Eqs. 14.13a–14.13f, is then used to identify time-varying parameter profiles.

Following the identifications of time-varying parameter profiles, whose mean (μ) and standard deviation (σ) are shown in Table 14.3, the FFNN is trained to model the relationship between the time-varying parameters as outputs and the state variables and control actions as inputs. In this case, the state variables and control actions used as inputs to the FFNN are C_X, C_G, C_N, F_G (the feed rate of glucose in g/h) and F_N(the feed rate of urea in g/h). Control actions are included in the FFNN input because the state variable values alone (at the time the parameters are updated) do not provide the full information of the system before the next parameter update occurs. The FFNN is trained using the L1 Loss function (Mean Average Error, or MAE), shown in Eq. 14.6, with L1 regularization penalizing the absolute values of the ANN weights and biases, as illustrated in Eq. 14.9. The L1 regularization term converges redundant parameters to a value of zero to reduce overparameterization and overfitting of the FFNN by encouraging the training to focus only on the most influential weights and biases.

The FFNN, consisting of 2 hidden layers with 8 nodes in each layer, was trained using augmented data; each augmented dataset is a copy of the training dataset with the inputs being subject to Gaussian noise. Augmenting data improves the performance of neural networks by expanding the training dataset and making the model more robust to noise, thus reducing overfitting [128, 129]. In this study, 100 augmented datasets were generated using a Gaussian noise distribution and a standard deviation of 0.01 (1% of the standard deviation of the standardized data). It is important to include enough noise to improve the FFNN robustness, while not introducing so much that it inhibits the FFNN fitting capabilities. The combination of parameter regularization and data augmentation helps combat potential overfitting using the very limited dataset of just two fed-batch experiments. Training was conducted over 40,000 epochs, as in a previous hybrid modeling study [130], to ensure convergence over the large augmented dataset. The precision plots for the augmented dataset training are shown in Figure 14.3, which illustrates

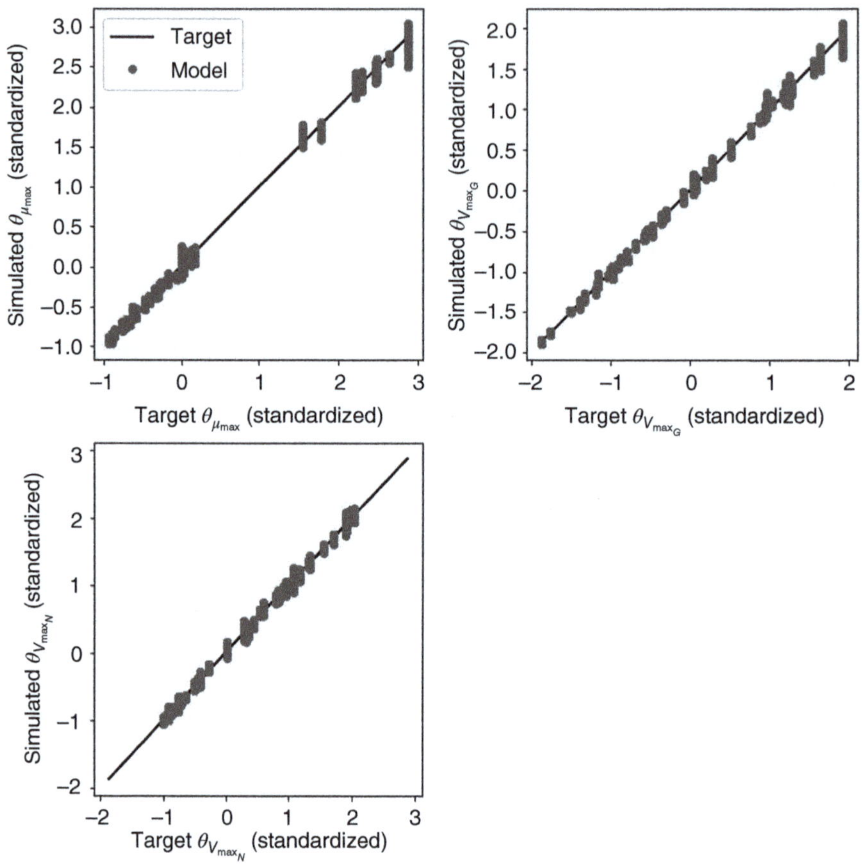

Figure 14.3 Precision plots of the standardized time-varying parameters $\theta_{\mu_{max}}$, $\theta_{v_{max_G}}$, and $\theta_{v_{max_N}}$ from the training of the FFNN using 100 augmented datasets.

the strong fitting capabilities of the FFNN, which converged to an MAE (or L1 loss) of 0.014 and R^2 values of 0.999, 0.998, and 0.999 for the standardized time-varying parameters $\theta_{\mu_{max}}$, $\theta_{v_{max_G}}$, and $\theta_{v_{max_N}}$, respectively.

To illustrate the improvement made by introducing a data-driven component, we must compare the fitting of the benchmark kinetic model to the hybrid model. Figure 14.4 shows the fitting of both the initial kinetic model and the overall hybrid model, which improved the mean R^2 value from 0.219 to 0.974 across the two fed-batch experiments. The first fed-batch experiment (shown in Figure 14.4a) saw a large improvement in the mean R^2 value, going from 0.263 to 0.962. The second fed-batch experiment (shown in Figure 14.4b) saw a slightly larger improvement in the mean R^2 value, going from 0.174 to 0.986. In terms of the simulated state variables, the smallest improvement in R^2 value was from 0.932 to 0.996 for cell density, meaning the original kinetic model provided a good estimation of the yield of cells from the glucose fed to the reactor. Since the overall lutein concentration is found to be strongly linked to cell density, this saw a similar change in R^2 value from 0.892 to 0.965, which further supports the decision to not implement a time-varying

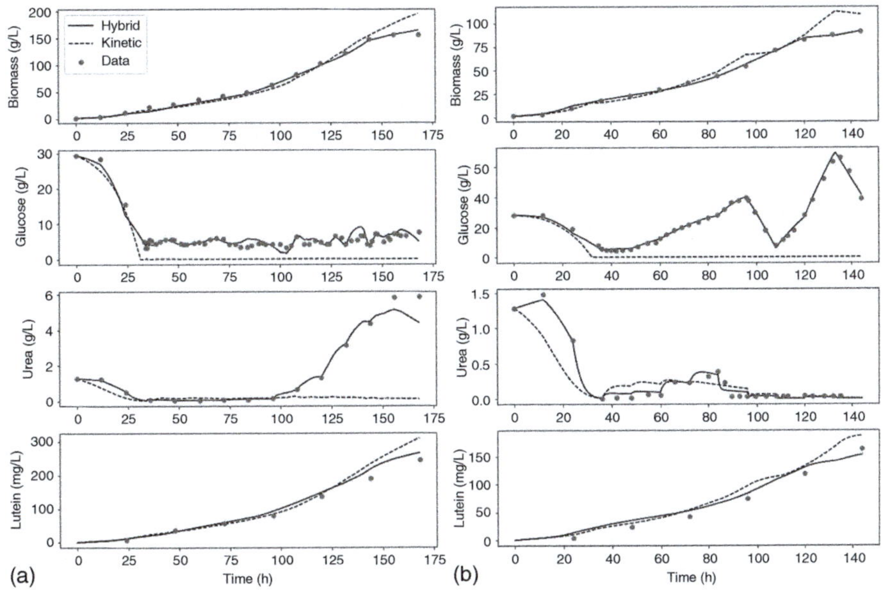

Figure 14.4 Hybrid model fitting compared to the benchmark kinetic model of fed-batch experiment 1 (a) and fed-batch experiment 2 (b).

parameter for lutein content in Eq. 14.13d. However, urea and glucose saw much larger improvements in fitting with their R^2 values increasing from 0.123 to 0.972 and −1.073 to 0.963, respectively, which accounts for the vast majority of the overall fitting improvement. The hybrid model is able to improve the fitting of the substrate concentration profiles because the FFNN can account for changes in parameter values that may be caused by various phenomena, such as metabolic activity or mass transfer limitations, without the need for a detailed physical model. On the other hand, the benchmark kinetic model is limited to the assumed structure and therefore lacks generalizability in comparison to the hybrid model.

Since the implementation of an FFNN introduces additional parameters, the corrected Akaike Information Criterion (AICc) can also be accounted for to compare models of differing sizes, as demonstrated in a previous hybrid modeling study [131]. For fed-batch experiment 1, the hybrid model improved the AICc from a value of 521.0 for the kinetic model to 419.3 for the hybrid model, with fed-batch experiment 2 seeing a similar improvement in the AICc from 460.1 to 381.3, therefore demonstrating that the introduction of an FFNN is justified and has not inappropriately increased the model complexity.

The drastic improvement in the fitting of the dynamic model highlights the potential of incorporating machine learning to capture dynamics that preliminary mechanistic modeling may overlook. Since the model is sufficiently accurate, it is ready to be integrated within an optimization framework to design experiments that maximize the production of lutein, the results of which are discussed in the next section.

14.3.4 Dynamic Optimization of Microalgal Lutein Production

The dynamic optimization framework aims to maximize the final lutein mass in the reactor at the end of the fed-batch process spanning 5 days (120 hours) with feeding updated every 12 hours to allow for automated operation. The overall optimization framework is illustrated in Framework 14.14:

$$\min_{F, C_S^{\text{feed}}, C_N^{\text{feed}}} -m_L(t_f), \tag{14.14a}$$

$$\text{s.t.}: \quad 0 \leq F(t) \leq 0.05 \text{ L h}^{-1}, \tag{14.14b}$$

$$0 \leq C_S^{\text{feed}}(t) \leq 750 \text{ g L}^{-1}, \tag{14.14c}$$

$$0 \leq C_N^{\text{feed}}(t) \leq 60 \text{ g L}^{-1}, \tag{14.14d}$$

$$m_L(t) = C_L(t) \cdot V(t), \tag{14.14e}$$

where the parameters $\mu_{\max}$, $v_{\max_G}$, and $v_{\max_N}$ in Eqs. 14.13a–14.13f are replaced by their corresponding ANN outputs, the overall lutein production m_L is taken as the product of lutein concentration and working volume, as in Eq. 14.14e, and the control vector at time t is defined by $u(t) = [F(t)\ C_S^{\text{feed}}(t), F(t)\ C_N^{\text{feed}}(t)]$.

For the purpose of this case study, a computational experiment has been conducted to examine the optimization results of the current hybrid model using a more comprehensive hybrid model validated over a wide range of operating conditions as the "ground-truth." Figure 14.5 shows the results of optimization Framework 14.14, which predicted a final lutein production of 290.3 mg, with an actual final concentration of 366.0 mg. For comparison, the second fed-batch experiment used for hybrid model construction was also designed with 12-hour feeding stages and achieved a total lutein production of 466.8 mg. Although this is a reduction in overall lutein production due to the shorter batch time, the optimized experiment takes place over a shorter time horizon (120 hours instead of 144 hours) and achieved the highest lutein yield of 0.702 mg/(g glucose) in comparison to the yields of 0.611 mg/(g glucose)$^-$ and 0.675 mg/(g glucose) achieved in fed-batch experiments 1 and 2, respectively. In terms of the predicted feed rate, the steadily increasing feed rate shown in Figure 14.5 does make sense from a physical point of view because the increase in cell density will have a higher demand for substrate to maintain the larger cell density. Both glucose and urea feed concentrations maintained their maximum permitted values of 750 g/L and 60 g/L, respectively, thus ensuring sufficient substrates are provided for growth.

The model prediction mismatch in the latter stages of the simulation in Figure 14.5 highlights where further information is required to better train the model. Dynamic models can be particularly susceptible to error propagation (small errors early in the simulation that snowball into a larger offset further along the simulated time horizon). For example, the largest errors in the glucose and urea concentration profiles shown in Figure 14.5 exist beyond a time of 100 hours, emphasizing the need for improved training in the latter stages of the process. One cause of this

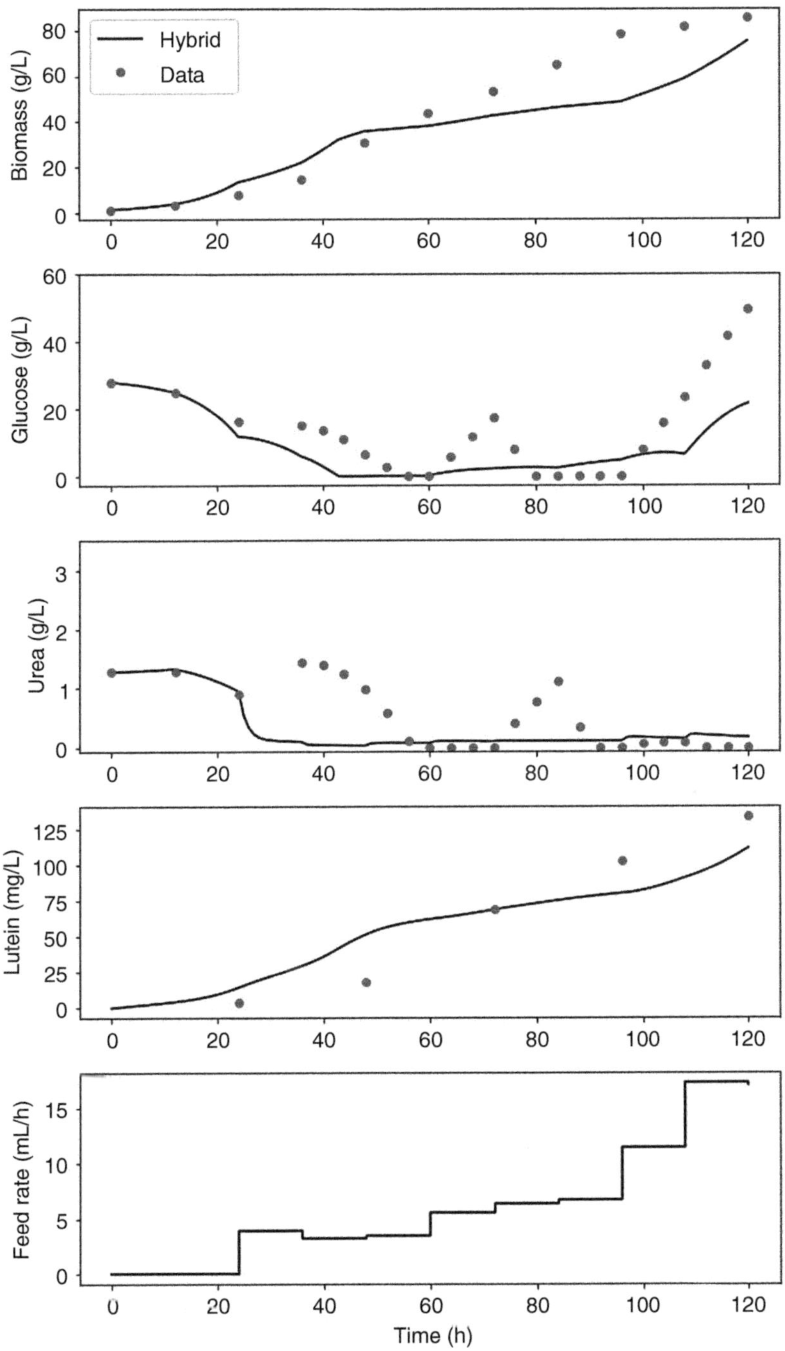

Figure 14.5 Hybrid model predictions against the ground-truth data using the optimal feeding strategy deduced from the optimization framework described in Framework 14.14.

mismatch could be the prediction of cell density, the fit of which has an R^2 value of 0.799, where the hybrid model initially over-predicts cell growth, leading to the over-predictions of substrate consumption rates. Subsequently, the over-prediction of substrate consumption rates means that the substrate concentrations are generally under-predicted from a time of 25–30 hours onward, resulting in an underprediction of cell growth. The suboptimal prediction of cell density is likely to be due to the training dataset; despite augmenting the data with 100 new datasets, Figure 14.3 shows that the time-varying parameter $\theta_{\mu_{max}}$ has a noticeable gap in the augmented training dataset where no training data exists for time-varying parameters between standardized values of 0.5 and 1.5, despite the use of a penalty term that penalizes the deviation of time-varying parameters in the objective function shown in Eq. 14.5a. Moreover, the increasing cell density throughout the process increases the sensitivity of the system to propagated errors, for example the glucose consumption rate in fed-batch experiment 1 increased from 1.91 g/L/h to 6.97 g/L/h from when the feeding commenced at the 24-hour mark to the final time of 168 hours, meaning simulation errors have a larger effect on simulated trajectories later on in the time horizon. A similar trend is shown in the feeding profile in Figure 14.5, which reflects the increased glucose consumption rates in the optimization result over time, thus increasing the sensitivity of the model to errors in the prediction of substrate uptake. The worst performing component for prediction was urea, with an R^2 value of −0.115. However, this is due to two main limitations of the dataset; (1) there exist gaps in the training for $\theta_{\mu_{max}}$, as aforementioned, which directly affects urea (as the consumption rate is shown to be directly proportional to cell density in Eq. 14.13c), and (2) both fed-batch experiments used for training saw the urea sit at low concentrations for prolonged periods, meaning the FFNN anticipates a high consumption rate of urea (and therefore v_{max_N}) regardless of the FFNN inputs, which will not always hold true. The low urea concentrations in the training dataset will also inhibit the ability of the FFNN to recognize the effects of urea on growth – the only information the data provides is that growth can still occur when subject to low urea concentrations. Similarly, the glucose concentration in fed-batch experiment 1 remained in a narrow region between 4 and 8 g/L beyond a time of 36 hours, which again limits the ability of the FFNN to adequately capture the effects that glucose has on the time-varying parameters. The poor predictive capabilities for urea and glucose emphasizes the need for data covering a broad range of operating conditions.

Although the predictive performance may seem suboptimal during certain short phases of the process, it must be emphasized that only two fed-batch experiments were available for training, making the construction of a high-fidelity model extremely challenging. So, in this context, the hybrid model predictions are satisfactory. In cases where a limited number of experiments can be designed, careful DoEs should be conducted to maximize the information obtained from each experiment. In order to improve the hybrid model predictions in the future, constraints could be added to reduce the excess remaining substrates at the end of the experiment to improve the process yield. In addition, feed strategies that differ more from existing experiments could be tried, through the use of a penalty function in the optimization framework objective that encourages the exploration of a wider range of operating conditions for when the model is updated, thus highlighting the importance of Model Based DoEs (MBDoE) in model training and the subsequent process optimization. In its current state, the most effective way to utilize this hybrid model would be with online updating of the data-driven component from real-time data, as retraining the FFNN can be done quickly and accurately.

14.4 Summary

The chapter began by highlighting some key literature that illustrates a variety of ML applications to bioprocess development and optimization. An observation was made that existing models generally fall within the spectrum between mechanistic ("white box") and purely data-driven ("black box") modeling, with their respective strengths and weaknesses explained in depth. In order to simultaneously address the limitations of both mechanistic and data-driven modeling and utilize their respective strengths, the authors present the concept of hybrid modeling (or "grey box") for the integration of ML within a mechanistic model. Several applications of hybrid models in bioprocessing from literature are explored, demonstrating extensive capabilities for the enhancement of bioprocesses.

The chapter then outlines a comprehensive methodology for constructing hybrid models, commencing with a mechanistic model based on mass balances and simple kinetic laws. For the cases where a mechanistic model performs in a dissatisfactory manner, sensitivity and uncertainty analysis are used to identify the most uncertain parameters, which are then reformulated as time-varying functions. Machine learning models can then be trained to estimate the time-varying parameters using experimental state variables and control actions as inputs. Detailed procedures are provided for training ANNs, including regularization, data augmentation, and partitioning into training, validation, and test sets. Bootstrapping is also introduced as a strategy for quantifying uncertainty in hybrid model predictions and improving hybrid model robustness.

The methodology described in this chapter is then applied to a detailed case study on lutein production by *Chlorella. sorokiniana* microalgae under heterotrophic fed-batch conditions. Initial mechanistic modeling based on Monod kinetics could not accurately predict glucose and urea concentrations, despite performing well for cell density, and so an FFNN was introduced to model three key time-varying parameters: maximum specific growth rate, maximum specific glucose uptake rate, and maximum specific urea uptake rate. The hybrid model significantly outperformed the purely mechanistic model, improving mean R^2 values across key state variables from 0.219 to 0.974.

Finally, the hybrid model was incorporated within a dynamic optimization framework to maximize lutein production over a 120-hour fed-batch operation. The optimization degrees of freedom were provided as the dynamic feed flow rate and substrate feed concentrations, demonstrating how model-based optimization can assist in efficiently identifying optimal operating conditions. Overall, the modeling and optimization of the lutein-producing microalgal fed-batch process demonstrate the potential of hybrid modeling for dynamic bioprocesses, particularly under conditions of limited data availability. Although not explicitly explored in this chapter, it is important to note that online model updating can be employed to incorporate new data and enhance predictive accuracy. Equally critical is the availability of informative and diverse datasets to ensure robust hybrid model training.

References

1 Lasi, H., Fettke, P., Kemper, H.G. et al. (2014). Industry 4.0. *Business & information systems engineering* 6: 239–242.

2 Ghobakhloo, M. (2020). Industry 4.0, digitization, and opportunities for sustainability. *J. Clean. Prod.* 252: 119869.

3 Sharma, D., Patel, P., and Shah, M. (2023). A comprehensive study on industry 4.0 in the pharmaceutical industry for sustainable development. *Environ. Sci. Pollut. Res.* 30 (39): 90088–90098.

4 Chen, Y., Yang, O., Sampat, C. et al. (2020). Digital twins in pharmaceutical and biopharmaceutical manufacturing: a literature review. *PRO* 8 (9): 1088.

5 Helgers, H., Hengelbrock, A., Schmidt, A., and Strube, J. (2021). Digital twins for continuous mRNA production. *PRO* 9 (11): 1967.

6 Udugama, I.A., Lopez, P.C., Gargalo, C.L. et al. (2021). Digital twin in biomanufacturing: challenges and opportunities towards its implementation. *Systems Microbiology and Biomanufacturing* 1: 257–274.

7 Juarez, M.G., Botti, V.J., and Giret, A.S. (2021). Digital twins: review and challenges. *J. Comput. Inf. Sci. Eng.* 21 (3): 030802.

8 Noll, P. and Henkel, M. (2020). History and evolution of modeling in biotechnology: modeling & simulation, application and hardware performance. *Comput. Struct. Biotechnol. J.* 18: 3309–3323.

9 Slezov, V.V. (2009). *Kinetics of First Order Phase Transitions*. John Wiley & Sons.

10 Mu'azzam K, da Silva FVS, Murtagh J, Gallagher MS. (2024). A roadmap for model-based bioprocess development. *Biotechnol. Adv.* 108378.

11 Von Stosch, M., Oliveira, R., Peres, J., and de Azevedo, S.F. (2014). Hybrid semi-parametric modeling in process systems engineering: past, present and future. *Comput. Chem. Eng.* 60: 86–101.

12 Han, K. and Levenspiel, O. (1988). Extended Monod kinetics for substrate, product, and cell inhibition. *Biotechnol. Bioeng.* 32 (4): 430–447.

13 Heijnen, J.J. and Verheijen, P.J. (2013). Parameter identification of in vivo kinetic models: Limitations and challenges. *Biotechnol. J.* 8 (7): 768–775.

14 Zou, J., Han, Y., and So, S.S. (2009). Overview of artificial neural networks. *Artificial neural networks: methods and applications* 458: 14–22.

15 Gargalo, C.L., de Las Heras, S.C., Jones, M.N. et al. (2021). Towards the development of digital twins for the bio-manufacturing industry. *Digital twins: Tools and concepts for smart biomanufacturing* 176: 1–34.

16 Greenacre, M., Groenen, P.J., Hastie, T. et al. (2022). Principal component analysis. *Nature Reviews Methods Primers* 2 (1): 100.

17 Nair, V.V., Dhar, H., Kumar, S. et al. (2016). Artificial neural network based modeling to evaluate methane yield from biogas in a laboratory-scale anaerobic bioreactor. *Bioresour. Technol.* 217: 90–99.

18 Marklund, S. and Wikström E, Löfvenius G, Fängmark I, Rappe C. (1994). Emissions of polychlorinated compounds in combustion of biofuel. *Chemosphere* 28 (10): 1895–1904.

19 Notingher, I., Jell, G., Lohbauer, U. et al. (2004). In situ non-invasive spectral discrimination between bone cell phenotypes used in tissue engineering. *J. Cell. Biochem.* 92 (6): 1180–1192.

20 Gunther, J.C., Seborg, D.E., and Conner, J.S. (2006). Fault detection and diagnosis in industrial fed-batch cell culture. *IFAC Proceedings Volumes* 39 (2): 203–208.

21 Perez, B.A. and Toraman, H.E. (2024). Investigating primary decomposition of polypropylene through detailed compositional analysis using two-dimensional gas chromatography and principal component analysis. *J. Anal. Appl. Pyrolysis* 177: 106376.

22 CJF, D., Oliveira, R.S., Nakajima, N.Y. et al. (2020). Integration of principal component analysis and artificial neural network to modeling productive capacity of eucalypt stands from biophysical attributes. *For. Ecol. Manag.* 460: 117862.

23 Mowbray, M., Savage, T., Wu, C. et al. (2021). Machine learning for biochemical engineering: a review. *Biochem. Eng. J.* 172: 108054.

24 Chandra, R. and Kundu, M. (2024). Classification and authentication of operating conditions in different processes using partial least squares. *Chemical Product and Process Modeling* 19 (1): 135–145.

25 Freire, R.S., Ferreira, M.M., and Durán N, Kubota LT. (2003). Dual amperometric biosensor device for analysis of binary mixtures of phenols by multivariate calibration using partial least squares. *Anal. Chim. Acta* 485 (2): 263–269.

26 Duran-Villalobos, C.A., Goldrick, S., and Lennox, B. (2020). Multivariate statistical process control of an industrial-scale fed-batch simulator. *Comput. Chem. Eng.* 132: 106620.

27 Stubbs, S., Zhang, J., and Morris, J. (2018). BioProcess performance monitoring using multiway interval partial least squares. In: *Computer Aided Chemical Engineering*, vol. 41 (ed. R. Singh and Z. Yuan), 243–259. Elsevier.

28 Yang, K.K., Wu, Z., and Arnold, F.H. (2019). Machine learning-guided directed evolution for protein engineering. *Nat. Methods* 16 (8): 687–694.

29 Carinhas, N., Bernal, V., Teixeira, A.P. et al. (2011). Hybrid metabolic flux analysis: combining stoichiometric and statistical constraints to model the formation of complex recombinant products. *BMC Syst. Biol.* 5: 1–13.

30 Zoqi, M. (2024). Generalization of artificial neural network for predicting methane production in laboratory-scale anaerobic bioreactor landfills. *Global Journal of Environmental Science and Management* 10 (1): 225–244.

31 Thibault, J., Van Breusegem, V., and Chéruy A. (1990). On-line prediction of fermentation variables using neural networks. *Biotechnol. Bioeng.* 36 (10): 1041–1048.

32 Quaglio, M., Roberts, L., Jaapar, M.S.B. et al. (2020). An artificial neural network approach to recognise kinetic models from experimental data. *Comput. Chem. Eng.* 135: 106759.

33 del Rio-Chanona, E.A., Wagner, J.L., Ali, H. et al. (2019). Deep learning-based surrogate modeling and optimization for microalgal biofuel production and photobioreactor design. *AICHE J.* 65 (3): 915–923.

34 Senior, A.W., Evans, R., Jumper, J. et al. (2020). Improved protein structure prediction using potentials from deep learning. *Nature* 577 (7792): 706–710.

35 Amidi, A., Amidi, S., Vlachakis, D. et al. (2018). EnzyNet: enzyme classification using 3D convolutional neural networks on spatial representation. *PeerJ.* 6: e4750.

36 Shah, P., Sheriff, M.Z., Bangi, M.S.F. et al. (2022). Deep neural network-based hybrid modeling and experimental validation for an industry-scale fermentation process: identification of time-varying dependencies among parameters. *Chem. Eng. J.* 441: 135643.

37 Kimutai, G., Ngenzi, A., Said, R.N. et al. (2020). An optimum tea fermentation detection model based on deep convolutional neural networks. *Data* 5 (2): 44.

38 Salehinejad H, Sankar S, Barfett J, Colak E, Valaee S. 2017. Recent advances in recurrent neural networks. arXiv preprint arXiv:180101078.

39 Alley, E.C., Khimulya, G., Biswas, S. et al. (2019). Unified rational protein engineering with sequence-based deep representation learning. *Nat. Methods* 16 (12): 1315–1322.

40 Oubounyt, M., Louadi, Z., Tayara, H., and Chong, K.T. (2019). DeePromoter: robust promoter predictor using deep learning. *Front. Genet.* 10: 286.

41 Jiang, H., Deng, J., and Chen, Q. (2024). Monitoring of simultaneous saccharification and fermenta- tion of ethanol by multi-source data deep fusion strategy based on near-infrared spec- tra and electronic nose signals. *Eng. Appl. Artif. Intell.* 127: 107299.

42 MacKay, D.J. et al. (1998). Introduction to Gaussian processes. *NATO ASI series F computer and systems sciences* 168: 133–166.

43 Bradford, E., Schweidtmann, A.M., Zhang, D. et al. (2018). Dynamic modeling and optimization of sustainable algal production with uncertainty using multivariate Gaussian processes. *Comput. Chem. Eng.* 118: 143–158.

44 Bradford, E., Imsland, L., Zhang, D., and del Rio Chanona, E.A. (2020). Stochastic data-driven model predictive control using gaussian processes. *Comput. Chem. Eng.* 139: 106844.

45 di Sciascio, F. and Amicarelli, A.N. (2008). Biomass estimation in batch biotechnological processes by Bayesian Gaussian process regression. *Comput. Chem. Eng.* 32 (12): 3264–3273.

46 Schulz, E., Speekenbrink, M., and Krause, A. (2018). A tutorial on Gaussian process regression: modelling, exploring, and exploiting functions. *J. Math. Psychol.* 85: 1–16.

47 Paterson A, Pestana J, Maini VD. 2025. An Iterative Block Matrix Inversion (IBMI) algorithm for symmetric positive definite matrices with applications to covariance matrices. https://arxiv.org/abs/2502.06377.

48 Sarmah, U., Borah, P., and Bhattacharyya, D.K. (2024). Ensemble learning methods: an empirical study. *SN Computer Science* 5 (7): 924.

49 Bapat PM, Wangikar PP. Optimization of rifamycin B fermentation in shake flasks via a machine-learning-based approach. *Biotechnol. Bioeng.* 2004; 86 (2): 201–8.

50 Acosta-Pavas, J.C., Robles-Rodriguez, C.E., Griol, D. et al. (2024). Soft sensors based on interpretable learners for industrial-scale fed-batch fermentation: learning from simulations. *Comput. Chem. Eng.* 187: 108736.

51 Miraboutalebi, S.M., Kazemi, P., and Bahrami, P. (2016). Fatty acid methyl ester (FAME) composition used for estimation of biodiesel cetane number employing random forest and artificial neural networks: a new approach. *Fuel* 166: 143–151.

52 Chu, X., Sun, T., Li, Q. et al. (2022). Prediction of liquid–liquid phase separating proteins using machine learning. *BMC bioinformatics* 23 (1): 72.

53 Bandpey, A.F., Abdi, J., and Firozjaee, T.T. (2024). Improved estimation of dark fermentation biohydrogen production utilizing a robust categorical boosting machine-learning algorithm. *Int. J. Hydrog. Energy* 52: 190–199.

54 Wise, T., Emery, K., and Radulescu, A. (2024). Naturalistic reinforcement learning. *Trends Cogn. Sci.* 28 (2): 144–158.

55 Manzo, A. and Ciancarini, P. (2023). Enhancing stockfish: a chess engine tailored for training human players. In: *International Conference on entertainment Computing* (ed. P. Ciancarini, D.l. A, H. Hlavacs, and F. Poggi), 275–289. Springer.

56 Li, H., Qiu, T., and You, F. (2024). AI-based optimal control of fed-batch biopharmaceutical process leveraging deep reinforcement learning. *Chem. Eng. Sci.* 292: 119990.

57 Fu, P.C. and Barford, J. (1996). A hybrid neural network—first principles approach for modelling of cell metabolism. *Comput. Chem. Eng.* 20 (6–7): 951–958.

58 Mowbray, M.R., Wu, C., Rogers, A.W. et al. (2023). A reinforcement learning-based hybrid modeling framework for bioprocess kinetics identification. *Biotechnol. Bioeng.* 120 (1): 154–168.

59 Petsagkourakis, P., Sandoval, I.O., Bradford, E. et al. (2022). Chance constrained policy optimization for process control and optimization. *J. Process Control* 111: 35–45.

60 MÁ, d.C.S., Sandoval, I.O., KKM, H. et al. (2024). The automated discovery of kinetic rate models–methodological frameworks. *Digital Discovery* 3 (5): 954–968.

61 Narayanan, H., Bournazou, M.N.C., and Gosálbez GG, Butté A. (2022). Functional-hybrid modeling through automated adaptive symbolic regression for interpretable mathematical expressions. *Chem. Eng. J.* 430: 133032.

62 Rogers AW, Lane A, Mendoza C, Watson S, Kowalski A, Martin P, et al. 2024. Integrating knowledge-guided symbolic regression and model-based design of experiments to automate process flow diagram development. arXiv preprint arXiv:240504592.

63 Agharafeie, R., Ramos, J.R.C., Mendes, J.M., and Oliveira, R. (2023). From shallow to deep bioprocess hybrid modeling: advances and future perspectives. *Fermentation* 9 (10): 922.

64 Herrera-Ruiz, J.F., Fontalvo, J., and Prado-Rubio, O.A. (2024). Advances on hybrid modelling for bioprocesses engineering: insights into research trends and future directions from a bibliometric approach. *Results in Engineering* 103548.

65 Zhang, D., Savage, T.R., and Cho, B.A. (2020). Combining model structure identification and hybrid modelling for photo-production process predictive simulation and optimisation. *Biotechnol. Bioeng.* 117 (11): 3356–3367.

66 Cabaneros Lopez, P., Udugama, I.A., Thomsen, S.T. et al. (2021). Transforming data to information: a parallel hybrid model for real-time state estimation in lignocellulosic ethanol fermentation. *Biotechnol. Bioeng.* 118 (2): 579–591.

67 Zhao, M., Zhao, S., and Liu, F. (2023). Semi-supervised hybrid modeling of the yeast fermentation process. *Machines.* 11 (1): 63.

68 Dong, Y., Fan, Q., Yan, X. et al. (2014). Development of a hybrid model for sodium gluconate fermentation by aspergillus Niger. *J. Chem. Technol. Biotechnol.* 89 (12): 1875–1882.

69 Rogers, A.W., Song, Z., Ramon, F.V. et al. (2023). Investigating 'greyness' of hybrid model for bioprocess predictive modelling. *Biochem. Eng. J.* 190: 108761.

70 Pinto, J., de Azevedo, C.R., Oliveira, R., and von Stosch, M. (2019). A bootstrap-aggregated hybrid semi-parametric modeling framework for bioprocess development. *Bioprocess Biosyst. Eng.* 42: 1853–1865.

71 Bayer, B., Duerkop, M., and Pörtner R, Möller J. (2023). Comparison of mechanistic and hybrid modeling approaches for characterization of a CHO cultivation process: requirements, pitfalls and solution paths. *Biotechnol. J.* 18 (1): 2200381.

72 Botton, A., Barberi, G., and Facco, P. (2022). Data augmentation to support biopharmaceutical process development through digital models—A proof of concept. *PRO* 10 (9): 1796.

73 Vega-Ramon, F., Zhu, X., Savage, T.R. et al. (2021). Kinetic and hybrid modeling for yeast astaxanthin production under uncertainty. *Biotechnol. Bioeng.* 118 (12): 4854–4866.

74 Rydal, T., Frandsen, J., Nadal-Rey, G. et al. (2024). Bringing a scalable adaptive hybrid modeling framework closer to industrial use: application on a multiscale fungal fermentation. *Biotechnol. Bioeng.* 121 (5): 1609–1625.

75 Ge, Y., Tao, J., Wang, Z. et al. (2024). A hybrid approach of anaerobic digestion model no. 1 and machine learning to model and optimize continuous anaerobic digestion processes. *Biomass Bioenergy* 184: 107176.

76 Yang, S., Fahey, W., Truccollo, B. et al. (2024). Hybrid Modeling of fed-batch cell culture using physics-informed neural network. *Ind. Eng. Chem. Res.* 63 (39): 16833–16846.

77 da Silva, P.A. and Pinheiro Á DT, Rocha MVP, Gonçalves LRB, Cartaxo SJM. (2021). Hybrid neural network modeling and particle swarm optimization for improved ethanol production from cashew apple juice. *Bioprocess and Biosyst. Eng.* 44: 329–342.

78 Ferreira, A., Dias, J., Von Stosch, M. et al. (2014). Fast development of Pichia pastoris GS115 Mut+ cultures employing batch-to-batch control and hybrid semi-parametric modeling. *Bioprocess Biosyst. Eng.* 37: 629–639.

79 Teixeira, A.P., Clemente, J.J., Cunha, A.E. et al. (2006). Bioprocess iterative batch-to-batch optimization based on hybrid parametric/nonparametric models. *Biotechnol. Prog.* 22 (1): 247–258.

80 Teixeira, A., Cunha, A., Clemente, J. et al. (2005). Modelling and optimization of a recombinant BHK-21 cultivation process using hybrid grey-box systems. *J. Biotechnol.* 118 (3): 290–303.

81 Reyes, S.J., Durocher, Y., Pham, P.L., and Henry, O. (2022). Modern sensor tools and techniques for monitoring, controlling, and improving cell culture processes. *PRO* 10 (2): 189.

82 Liu, Y. (2007). Overview of some theoretical approaches for derivation of the Monod equation. *Appl. Microbiol. Biotechnol.* 73: 1241–1250.

83 Lee, L.F. and Umberger, B.R. (2016). Generating optimal control simulations of musculoskeletal movement using OpenSim and MATLAB. *PeerJ.* 4: e1638.

84 Hart, W.E., Laird, C.D., Watson, J.P. et al. (2017). *Pyomo-Optimization Modeling in Python*, vol. 67. Springer.

85 Kohl VM, Marcato ALM, de Barros Correia P. (2023). Using nonlinear optimization solvers to improve PAR (p) coefficients estimation in synthetic inflow scenarios generation. In: *2023 58th International Universities Power Engineering Conference (UPEC)*. p. 1–6. IEEE.

86 Thacker, W.C. (1989). The role of the hessian matrix in fitting models to measurements. *J. Geophys. Res. Oceans* 94 (C5): 6177–6196.

87 Montrul, S. (2006). Convergent outcomes in L2 acquisition and L1 loss. In: *First Language Attrition: Interdisciplinary Perspectives on Methodological Issues* (ed. R.W. Eppley, L.P. Atkinson, and W.J. Jenkins), 259–279. John Benjamins Publishing Company.

88 Janocha K, Czarnecki WM. 2017. On loss functions for deep neural networks in classification. arXiv preprint arXiv:170205659.

89 SL, Ö.z., Tan, C.O., and U, Ö.z. (2006). Methodological issues in building, training, and testing artificial neural networks in ecological applications. *Ecol. Model.* 195 (1–2): 83–93.

90 Mi, X., Zou, Y., Wei, W., and Ma, K. (2005). Testing the generalization of artificial neural networks with cross-validation and independent-validation in modelling rice tillering dynamics. *Ecol. Model.* 181 (4): 493–508.

91 Huang, L., Qin, J., Zhou, Y. et al. (2023). Normalization techniques in training dnns: methodology, analysis and application. *IEEE Trans. Pattern Anal. Mach. Intell.* 45 (8): 10173–10196.

92 Collins J, Ballé J, Shlens J. 2019. Accelerating training of deep neural networks with a standardization loss. arXiv preprint arXiv:190300925.

93 Shanker, M., Hu, M.Y., and Hung, M.S. (1996). Effect of data standardization on neural network training. *Omega* 24 (4): 385–397.

94 Santos, C.F.G.D. and Papa, J.P. (2022). Avoiding overfitting: a survey on regularization methods for convolutional neural networks. *ACM Computing Surveys (CSUR)* 54 (10s): 1–25.

95 Shi, G., Zhang, J., Li, H., and Wang, C. (2019). Enhance the performance of deep neural networks via L2 regularization on the input of activations. *Neural. Process. Lett.* 50: 57–75.

96 Brownlee, J. (2018). What is the difference between a batch and an epoch in a neural network. *Machine learning mastery* 20: 1–5.

97 Chen, W., Dong, X., Chen, X. et al. (2024). pommDNN: performance optimal GPU memory management for deep neural network training. *Futur. Gener. Comput. Syst.* 152: 160–169.

98 Prechelt, L. (2002). Early stopping-but when? In: *Neural Networks: Tricks of the Trade*, 55–69. Springer.

99 Wythoff, B.J. (1993). Backpropagation neural networks: a tutorial. *Chemom. Intell. Lab. Syst.* 18 (2): 115–155.

100 Imambi, S., Prakash, K.B., and Kanagachidambaresan, G. (2021). *PyTorch. Programming with TensorFlow: Solution for Edge Computing Applications*, 87–104. Springer. https://doi.org/10.1007/978-3-030-57077-4_10.

101 Chen, W., Baghdasaryan, L., Buranathiti, T., and Cao, J. (2004). Model validation via uncertainty propagation and data transformations. *AIAA J.* 42 (7): 1406–1415.

102 Steyerberg, E.W. (2019). *Overfitting and Optimism in Prediction Models. Clinical Prediction Models: A Practical Approach to Development, Validation, and Updating*, 95–112. Springer.

103 Huang W, Zhang J, Huang K. 2019. Bootstrap estimated uncertainty of the environment model for model-based reinforcement learning. In: *Proceedings of the AAAI conference on artificial intelligence*. vol. 33; p. 3870–7.

104 Bertsimas, D. and Van Parys, B. (2022). Bootstrap robust prescriptive analytics. *Math. Program.* 195 (1): 39–78.

105 Pennington, O., SE, Ŕ., Sebastian, M.T. et al. (2024). A multiscale hybrid modelling methodology for cell cultures enabled by enzyme-constrained dynamic metabolic flux analysis under uncertainty. *Metab. Eng.* 86: 274–287.

106 Kay, H., Vega-Ramon, F., and Zhang, D. (2024). Comparison of machine learning based hybrid modelling methodologies for dynamic simulation of chemical reaction networks. In: *Computer Aided Chemical Engineering*, vol. 53, 133–138. Elsevier.

107 Rogers, A.W., Song, Z., Vega Ramon, F. et al. (2023). Hybrid modelling under uncertainty: effects of model greyness, data quality and data quantity.

In: *Machine Learning and Hybrid Modelling for Reaction Engineering: Theory and Applications*, 208–228. https://doi.org/10.1039/BK9781837670178-00208.

108 D, S., Ruckerbauer, D.E., Galleguillos, S.N. et al. (2020). What CHO is made of: variations in the biomass composition of Chinese hamster ovary cell lines. *Metab. Eng.* 61: 288–300.

109 Nimmegeers P, Telen D, Beetens M, Logist F, Van Impe J. 2016. Parametric uncertainty propagation for robust dynamic optimization of biological networks. In: *2016 Amer- ican Control Conference (ACC)*. IEEE. p. 6929–34.

110 Puschke, J., Djelassi, H., Kleinekorte, J., and Hannemann-Tamás R, Mitsos A. (2018). Robust dynamic optimization of batch processes under parametric uncertainty: utilizing approaches from semi-infinite programs. *Comput. Chem. Eng.* 116: 253–267.

111 Geng, X. and Xie, L. (2019). Data-driven decision making in power systems with probabilistic guarantees: theory and applications of chance-constrained optimization. *Annu. Rev. Control.* 47: 341–363.

112 Bonnans, J.F. and Bonnans, J.F. (2019). Semidefinite and semi-infinite programming. Convex and stochastic. *Optimization* 75–116.

113 Puschke, J., Zubov, A., Kosek, J., and Mitsos, A. (2017). Multi-model approach based on parametric sensitivities–a heuristic approximation for dynamic optimization of semi batch processes with parametric uncertainties. *Comput. Chem. Eng.* 98: 161–179.

114 Geletu, A., M, K., Zhang, H., and Li, P. (2013). Advances and applications of chance constrained approaches to systems optimisation under uncertainty. *Int. J. Syst. Sci.* 44 (7): 1209–1232.

115 Mowbray, M., Petsagkourakis, P., del Rio-Chanona, E.A., and Zhang, D. (2022). Safe chance constrained reinforcement learning for batch process control. *Comput. Chem. Eng.* 157: 107630.

116 Bhonsale, S., Mores, W., and Van Impe, J. (2021). Dynamic optimisation of beer fermentation under parametric uncertainty. *Fermentation* 7 (4): 285.

117 Fu, Y., Wang, Y., Yi, L. et al. (2023). Lutein production from microalgae: a review. *Bioresour. Technol.* 376: 128875.

118 Hu, J., Nagarajan, D., Zhang, Q. et al. (2018). Heterotrophic cultivation of microalgae for pigment production: a review. *Biotechnol. Adv.* 36 (1): 54–67.

119 Manayi, A., Abdollahi, M., Raman, T. et al. (2016). Lutein and cataract: from bench to bedside. *Crit. Rev. Biotechnol.* 36 (5): 829–839.

120 Becerra, M.O., Contreras, L.M., Lo, M.H., and Díaz JM, Herrera GC. (2020). Lutein as a functional food ingredient: stability and bioavailability. *J. Funct. Foods* 66: 103771.

121 Iwamoto, H., Soccol, C.R., Molina-Aulestia, D.T. et al. (2024). Lutein from microalgae: an industrial perspective of its production, downstream processing, and market. *Fermentation* 10 (2): 106.

122 Lin, J.H., Lee, D.J., and Chang, J.S. (2015). Lutein production from biomass: marigold flowers versus microalgae. *Bioresour. Technol.* 184: 421–428.

123 Ren, Y., Deng, J., Lin, Y. et al. (2022). Developing a Chromochloris zofingiensis mutant for enhanced production of lutein under CO_2 aeration. *Mar. Drugs* 20 (3): 194.

124 Xie, Y., Zhang, Z., Ma, R. et al. (2022). High-cell-density heterotrophic cultivation of microalga Chlorella sorokiniana FZU60 for achieving ultra-high lutein production efficiency. *Bioresour. Technol.* 365: 128130.

125 Zhang, X., Shi, X., and Chen, F. (1999). A kinetic model for lutein production by the green microalga Chlorella protothecoides in heterotrophic culture. *J. Ind. Microbiol. Biotechnol.* 23 (6): 503–507.

126 del Rio-Chanona, E.A., Nr, A., Zhang, D. et al. (2017). Kinetic modeling and process analysis for Desmodesmus sp. lutein photo-production. *AICHE J.* 63 (7): 2546–2554.

127 Bentahar, J. and Deschenes, J.S. (2024). A robust multinutrient kinetic model for enhanced lutein and biomass yields in mixotrophic microalgae cultivation: a step towards successful large-scale productions. *Biotechnol. Bioeng.* 121 (5): 1596–1608.

128 Hernández-Garćıa A, König P. 2018. Further advantages of data augmentation on convolutional neural networks. In: *Artificial Neural Networks and Machine Learning–ICANN 2018: 27th International Conference on Artificial Neural Networks*, Rhodes, Greece, (4–7 October 2018) Part I 27. Springer. p. 95–103.

129 Shijie J, Ping W, Peiyi J, Siping H. 2017. Research on data augmentation for image classification based on convolution neural networks. In: *2017 Chinese automation congress (CAC)*. IEEE. p. 4165–70.

130 Ramos, J.R., Pinto, J., Poiares-Oliveira, G. et al. (2024). Deep hybrid modeling of a HEK293 process: combining long short-term memory networks with first principles equations. *Biotechnol. Bioeng.* 121 (5): 1554–1568.

Index

Machine Learning and Big Data-enabled Biotechnology, First Edition. Edited by Hal S. Alper.

b

i

k

l

m

n

r

s

t